MODERN CONTROL SYSTEM THEORY AND APPLICATION

SECOND EDITION

Stanley M. Shinners

The Sperry Division
Division of Sperry Rand Corporation
and
New York Institute of Technology

MODERN CONTROL SYSTEM THEORY AND APPLICATION

SECOND EDITION

ADDISON-WESLEY PUBLISHING COMPANY
Reading, Massachusetts | Menlo Park, California
London | Amsterdam | Don Mills, Ontario | Sydney

This book is in the
ADDISON-WESLEY SERIES IN ELECTRICAL ENGINEERING

Consulting Editors
David K. Cheng, Leonard A. Gould, Fred K. Manasse

To my wife
Doris
and to my children
Sharon, Walter, and Daniel

PREFACE

The goal of *Modern Control System Theory and Application* is to bridge the gap existing between classical and modern control theory. My main objective in writing this volume has been to offer an easily understandable book from the reader's viewpoint. It is written in an integrated manner in order that one can proceed from the first to the last chapter clearly.

This book aims at an audience consisting of the senior undergraduate student, the first-year graduate student, and the practicing control-system engineer. The text, in its first edition, has been used at over one hundred colleges and universities. Its contents correspond to approximately three semester credit hours and are based on courses that I have taught at universities (New York Institute of Technology, The Cooper Union and the Polytechnic Institute of Brooklyn) and in industry (Sperry Rand Corporation).

Modern Control System Theory and Application is distinguished by the following desirable features:

1. A unified blending of the classical and modern approaches. It uses the transfer function and state-space methods in parallel throughout the book, and then introduces modern optimal control theory including such important topics as dynamic programming and the maximum principle.

2. Presentation of working digital-computer programs.

3. The development of theoretical topics is coupled with clear applications of the theory in engineering design. Recognizing that control theory is interdisciplinary and cuts across all specialized engineering fields, I have presented modern illustrations and practical problems from the fields of ecology, sociology, servomechanisms, space-vehicle systems, aircraft, submarines, hydrofoils, economics, management, biomedical engineering, and nuclear reactor control systems. This has been further emphasized in the second edition and should prove to be of interest to electrical, mechanical, aerospace, system, chemical, nuclear, biomedical, and industrial engineers.

4. A coverage of a variety of topics of recent importance. In addition to dynamic programming and the maximum principle, the presentation includes Liapunov's stability criterion, Popov's method, the generalized circle criterion, and linear-state-variable feedback.

5.   An in-depth coverage of linear and nonlinear control systems. This is a flexible book designed for a one-term course that can be easily adapted to the student's training and ability and to various curricula.

6.   A set of problems, which has been greatly expanded for the second edition, with answers to one third of them contained in the book. This is a helpful feature of the book for the student, and also for the practicing control-system engineer who desires to use the book for self-study.

7.   An accompanying Instructor's Guide *and* Solutions Manual containing detailed solutions to the remaining problems. This manual also presents several possible course outlines that are based on the students' level of learning, background, and proficiency.

By means of these features, *Modern Control System Theory and Application* is unique and fills a rather large gap in the existing literature on feedback control-system design.
    Chapter 1 introduces the concept of open-loop and closed-loop control systems, and defines the terminology and nomenclature used throughout the book. Chapter 2 reviews Fourier methods, the Laplace transform, and the transfer-function method. In addition, it introduces the signal-flow diagram and the state-variable concept. The transfer-function and state-variable representations of several common devices found in control systems are derived in Chapter 3. The concepts of conservation and analogy are also introduced in this chapter. Chapter 4 focuses attention on second-order systems since they occur so frequently and since many higher-order systems can be approximated as second-order systems. Chapter 5 considers various performance criteria including the IAE, ISE, ITAE, ITSE, and ISTAE methods. A comprehensive treatment of techniques for determining stability, together with several useful and practical examples, is given in Chapter 6. The techniques presented include the state-space, the Routh-Hurwitz, Nyquist, Bode-diagram, Nichols-chart, and root-locus methods. In addition, digital-computer programs are illustrated which can be used for the routine calculations associated with several of these methods. The concepts of stability, presented in Chapter 6, are applied in great detail to the design of linear control systems in Chapter 7. Chapter 8 discusses the theory of design of nonlinear systems along with practical examples. Included are linearizing approximations, the describing function, piecewise-linear approximations, the state-space analysis technique (the phase plane), Liapunov's stability criterion, Popov's stability criterion, and the generalized circle criterion. The use of analog and digital computers as an aid in several of these methods is illustrated together with working programs. Chapter 9 introduces optimal control theory and concludes the presentation. The following material is covered at an introductory level in this final chapter: controllability, observability, calculus of variations, Bellman's dynamic programming, and Pontryagin's maximum principle.
    The flexibility of *Modern Control System Theory and Application* for adapting to the course level, the various curricula and the students' training and ability is quite

evident from this content summary. For example, if the book is being used in a course where the students have only a limited familiarity with feedback, have not had the Laplace transform previously, and if computers are not available to the students for solving problems, it is recommended that the instructor cover Chapters 1 through 8 in detail deleting the material on computer-aided design. In addition, Sections 9.1 through 9.4 should be covered. However, if the class has good familiarity with feedback and the Laplace transform and computers are available to the students for solving problems, then it is recommended that the instructor should start with Section 2.12 and cover all of the material through Chapter 9. Variations to these general guidelines are contained in the course outlines listed in the Instructor's Guide and Solutions Manual.

Sampled-data and stochastic-control theories have not been included in this book since they are usually not included in an introductory course. It is hoped, however, that the reader will be motivated to continue his study of these and other topics of current interest in the control literature.

I wish to express my sincere appreciation to Dr. Leonard A. Gould of the Massachusetts Institute of Technology for his detailed review of the manuscript and for his valuable comments. A debt of gratitude is also owed to Dr. John G. Truxal of the SUNY at Stony Brook for his thorough review of the manuscript and for his useful suggestions. Thanks are due also to Mr. Gerald Grushow of Sperry and Dr. Walt White of Texas Tech University for their helpful comments.

I am most grateful to my wife Doris and daughter Sharon Rose for their encouragement, understanding, patience, and typing assistance throughout. In addition, I wish to express thanks and appreciation to my parents for their efforts, encouragement, and inspiration.

Jericho, New York   S. M. S.
January 1978

CONTENTS

1 THE GENERAL CONCEPT OF CONTROL-SYSTEM DESIGN

1.1 INTRODUCTION

The desire of man to control nature's forces successfully has been the catalyst for progress throughout history. His goal has been to control these forces in order to help him perform physical tasks which were beyond his own capabilities. During the dynamic and highly motivated twentieth century, the control-system engineer has transformed many of man's hopes and dreams into reality.

The control of systems is an interdisciplinary subject and cuts across all specialized engineering fields. This book recognizes this fact and presents illustrations of control systems from the fields of electrical, mechanical, aeronautical, chemical, nuclear, economics, management, bioengineering, and other related fields. The versatile subject of automatic control ranks today as one of the most promising fields, and its growth potential appears unlimited.

Control systems can be defined as devices which regulate the flow of energy, matter or other resources. Their arrangement, complexity, and appearance vary with their purpose and function. In general, control systems can be categorized as being either open loop or closed loop. The distinguishing feature between these two types of control system is the use of feedback comparison for closed-loop operation.

Properties characteristic of open-loop and closed-loop control systems are discussed in this chapter. We shall give several examples of each type so that the reader may gain a thorough understanding and a good foundation for further studies in this book. Included will be a qualitative, philosophical comparison between the behavior of closed-loop control systems and that of living creatures. The feedback concept can also be applied to model our economic system. A discussion of modern control-system applications, and definitions of nomenclature and symbols used, will conclude the presentation of this chapter.

1.2 OPEN-LOOP CONTROL SYSTEMS

Open-loop control systems represent the simplest and least complex form of controlling devices. Their concept and functioning are illustrated by several simple examples in this section.

Figure 1.1 illustrates a simple tank level control system. We wish to hold the tank level, h, within reasonable acceptable limits even though the outlet flow through valve

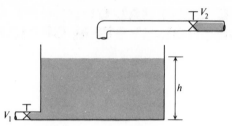

Fig. 1.1   Tank level control system.

V_1 is varied. This can be achieved by irregular manual adjustment of the inlet flow rate by valve V_2. This system is not a precision system since it does not have the capability of accurately measuring the output flow rate through valve V_1, the input flow rate through valve V_2, or the tank level. Figure 1.2 shows the simple relationship which exists in this system between the input (the desired tank level) and the output (the actual tank level). This signal-flow representation of the physical system is called a *block diagram*. Arrows are used to show the input entering and the output leaving the control system. This control system does not have any feedback comparison, and the term *open loop* is used to describe its absence.

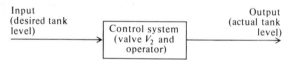

Fig. 1.2   Tank level control system block diagram.

The angular position of a missile launcher being controlled from a remote source is illustrated in Fig. 1.3. Commands from a potentiometer located at a remote location activate the positioning of the missile launcher. The control signal is amplified and drives a motor which is geared to the launcher. The block diagram in Fig. 1.2 is also applicable to this system. The input would be the desired angular position, the output would be the actual angular position, and the control system would consist of the amplifier and motor. For accurate positioning the missile launcher should be

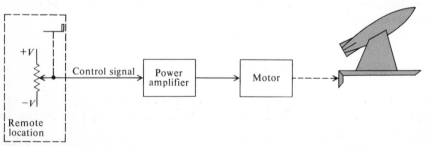

Fig. 1.3   Controlling the position of a missile launcher from a remote location.

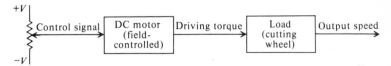

Fig. 1.4 Field-controlled dc motor.

precisely calibrated with reference to the angular position of the potentiometer, and the characteristics of the potentiometer, amplifier, and motor should remain constant. Except for the potentiometer, the components that comprise this open-loop control system are not precision devices. Their characteristics can easily change and result in false calibration and poor accuracies. In practice, simple open-loop control systems are never used for the accurate positioning of fire control systems due to the inherent possibility of inaccuracies and the stakes involved.

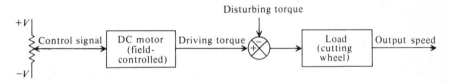

Fig. 1.5 Field-controlled dc motor having a disturbance torque.

Figure 1.4 illustrates a field-controlled dc motor turning a cutting wheel at a constant speed. When a piece of wood is applied to the surface of the cutting wheel, it acts as a disturbing torque to the driving torque of the motor and results in a reduction of the speed of the cutting wheel, assuming that the control signal remains constant. This situation can be represented as shown in Fig. 1.5. The symbol appearing between the motor and the load represents a subtractor.

The effect of disturbance torques, or other secondary inputs, is detrimental to the accurate functioning of an open-loop control system. It has no way of automatically correcting its output since there is no feedback comparison. We must resort to changing the input manually in order to compensate for secondary inputs.

1.3 CLOSED-LOOP CONTROL SYSTEMS

Closed-loop control systems derive their valuable accurate reproduction of the input from feedback comparison. An error detector derives a signal proportional to the differences between the input and output. The closed-loop control system drives the output until it equals the input and the error is zero. Any differences between the actual and desired output will be automatically corrected in a closed-loop control system. Through proper design, the system can be made relatively independent of secondary inputs and changes in component characteristics. This section illustrates

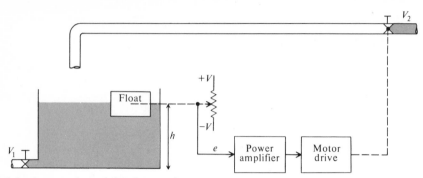

Fig. 1.6 Automatic tank level control system.

the closed-loop control system versions of the open-loop control systems considered in Section 1.2.

Figure 1.6 illustrates an automatic tank level control version of the system shown in Fig. 1.1. It can maintain the desired tank level h within quite accurate tolerances even though the output flow rate through valve V_1 is varied. If the tank level is not correct, an error voltage, e, is developed. This is amplified and applied to a motor drive which adjusts valve V_2 in order to restore the desired tank level by adjusting the inlet flow rate. A block diagram analogous to this system is shown in Fig. 1.7. Because feedback comparison is present, the term *closed loop* is used to describe the system's operation.

Figure 1.8 illustrates an automatic missile launcher position control version of the system shown in Fig. 1.3. This feedback system can position the missile launcher quite accurately on commands from potentiometer R_1. Potentiometer R_2 feeds a signal back to the difference amplifier, which functions as an error detector. Should an error exist, it is amplified and applied to a motor drive which adjusts the output-shaft position until it agrees with the input shaft position, and the error is zero. The block diagram shown in Fig. 1.7 is also applicable to this system. The input would be the desired angular position, the output would be the actual angular position, and the control system would consist of the amplifier and the motor.

An automatic speed control version of the field-controlled dc motor, which was shown in Fig. 1.4, is illustrated in Fig. 1.9. This feedback system has the capability of maintaining the output speed relatively constant even though disturbing torques may

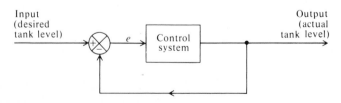

Fig. 1.7 Block diagram of a closed-loop system.

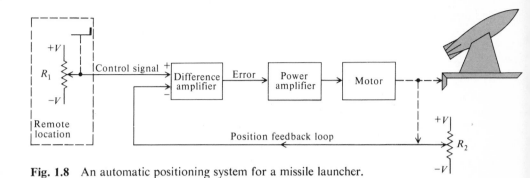

Fig. 1.8 An automatic positioning system for a missile launcher.

occur. A tachometer, which functions as a transducer that transforms speed to voltage, is the feedback element for this control system. Should the output speed differ from the desired speed, the difference amplifier develops an error signal which adjusts the field current of the motor in order to restore the desired output speed.

Feedback control systems used to control position, velocity, and acceleration are very common in industrial and military applications. They have been given the special

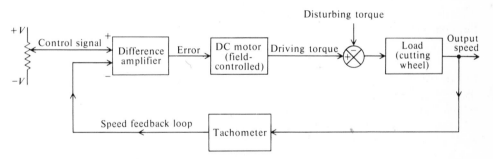

Fig. 1.9 Automatic speed control for a field-controlled dc motor.

name of *servomechanisms*. With all their many advantages, feedback systems have a very serious disadvantage since the closed-loop system may inadvertently act as an oscillator. Through proper design, however, all the advantages of feedback can be utilized without having an unstable system. A major task of this book is to determine how this may be accomplished for several kinds of systems.

1.4 HUMAN CONTROL SYSTEMS

The relation between the behavior of living creatures and the functioning of feedback control systems has recently gained wide attention. Norbert Wiener in his book

Cybernetics [8] implied that all systems, living and mechanical, are both information and feedback control systems. He suggested that the most promising techniques for studying both systems are information theory and feedback control theory.

Several characteristics of feedback control systems can be linked to human behavior. Feedback control systems can "think" in the sense that they can replace, to some extent, human operations. These devices do not have the privilege of freedom in their thinking process and are constrained by the designer to some predetermined function. Adaptive feedback control systems, which are capable of modifying their

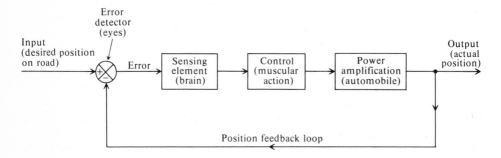

Fig. 1.10   Steering of an automobile—a feedback control system involving a human transfer function.

functioning in order to achieve optimum performance in a varying environment, have recently gained wide attention. These systems are a step closer to the adaptive capability of human behavior [14].

The human body is, indeed, a very complex and highly perfected adaptive feedback control system. Consider, for example, the human actions required to steer an automobile. The driver's object is to keep the automobile traveling in the center of a chosen lane on the road. Changes in the direction of the road are compensated for by the driver turning the steering wheel. His object is to keep the difference between the output (the actual path of the car) and the input (the desired path of the car) as close to zero as is possible.

Figure 1.10 illustrates the block diagram of the feedback control system involved in steering an automobile. The error detector in this case is the driver's eyes. This in turn activates the brain's sensing elements. Signals are then transmitted from the brain to the driver's muscles which control the steering wheel. Power amplification is provided by the automobile's steering mechanism, which controls the position of the front wheels. Of course, this description is very crude—any attempt to construct a mathematical model of the process should somehow account for the adaptability of the human being and the effects of learning, fatigue, motivation, and familiarity with the road.

1.5 MODERN CONTROL-SYSTEM APPLICATIONS

Feedback control systems are to be found in almost every aspect of our daily environment. In the home, the refrigerator utilizes a temperature-control system. The desired temperature is set and a thermostat measures the actual temperature and the error. A compressor motor is utilized for power amplification. Other applications of control in the home are the hot-water heater (see Problem 1.3), the central heating system, and the oven, which all work on a similar principle.

In industry, the term *automation* is very common. Modern industrial plants utilize temperature controls, pressure controls, speed controls, position controls, etc. The chemical process control field is an area where automation has played an important role [1]. Here, the control engineer is interested in controlling temperature, pressure, humidity, thickness, volume, quality, and many other variables. Areas of additional interest include automatic warehousing [2] and inventory control [3] and automation of farming [4].

Modern control concepts are being utilized in an ever increasing degree to help solve various problems. In the transportation field, for example, automatic control systems have been devised to regulate automobile traffic and control high-speed train systems. A very widely acclaimed high-speed rail transportation system is in operation in Japan [5, 6]. This latest innovation to the system of the Japanese National Railways is the Tokyo-to-Osaka super-express train, which is illustrated in Fig. 1.11. The high-speed railroad link between Tokyo and Osaka is commonly called the Tokaido line. It travels over a 320-mile route in 3 hours and 10 minutes. The train can travel at 130 miles per hour over most of the route. The system utilizes a control computer to control the trains in an optimum manner. Figure 1.12 illustrates the general position control concepts of such a high-speed automated train system. Observe from the diagram that it contains a position-measuring loop and a velocity-measuring loop. Position can be measured from the rotation of the train's wheels. Speed can be measured by using velocity-sensing devices such as tachometers. The control computer system monitors the positions and speeds of all trains in the system and issues control signals via a high-speed communication system.

Automatic control systems have been applied to a large degree by the aerospace industry. Modern high-speed aircraft, such as the F-4 long-range all-weather interceptor and attack bomber, illustrated in Fig. 1.13(a), are controlled almost entirely automatically during their missions. Figure 1.13(b) illustrates the cockpit instrumentation needed to control this modern aircraft. Figure 1.14 illustrates one axis of the autopilot for a typical aircraft. The feedback-loop concept is quite evident. The desired heading is compared with the actual heading. An error signal is produced which drives the aircraft's control elements in order to reduce the heading error to zero. Automatic control concepts have also been applied to the problem of stabilizing the attitude of space vehicles. Figure 1.15 illustrates the Apollo space vehicle. A typical block diagram for one axis of a space attitude control system is illustrated in Fig. 1.16. A computer is utilized to close the loop and compare the desired and actual

Fig. 1.11   Tokyo-to-Osaka super-express train.   (Courtesy of Japanese National Railways)

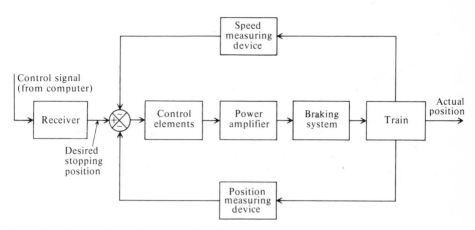

Fig. 1.12   Automatic position control system.

attitude. Observe from this figure that the overall action can be viewed in terms of a simple feedback loop.

The Polaris/Poseidon Fleet Ballistic Missile (FBM) Weapon System is a good example of the application of automatic control to a complex naval weapon system [9]. Figure 1.17 illustrates the underwater launching of the Polaris A-3 strategic ballistic missile. Let us go through the process of launching and guiding this missile in order to illustrate the multitude of automatic control systems involved in the operation. Polaris missiles are launched from the submarine by means of an air or a gas/steam generator ejection system. The missile is propelled from the launch tube, through the water and to the surface. At this point, a control system automatically ignites the missile's first-stage rocket motor and sends the missile on its mission. Two positions must be known very accurately for successfully controlling a missile: that of the target and launcher. In addition, the initial velocity of the launcher must be known very accurately. Since the position and velocity of the ship (launcher) is continuously changing in the FBM system, great emphasis must be placed on the design of the navigation system. The FBM navigation system, managed by the Sperry Systems Management Division, Sperry Rand Corporation, Great Neck, New York, utilizes several methods that complement each other in order to provide a very high order of accuracy in determining the ship's position and velocity. The heart of the system is the Ship's Inertial Navigation System (SINS), a complex system of gyroscopes, accelerometers, and computers, which relate movement and speed of the ship in all directions to true north in order to give a continuous report of ship position and velocity. The missile's guidance package consists of an inertial platform and a digital computer. The inertial platform is a gyro-stabilized set of three accelerometers. Once launch has occurred, the missile computer is in complete control of the missile. During flight, missile accelerations are measured by the inertial platform and integrated into velocities which are continuously fed to the computer. The computer

(a)

(b)

Fig. 1.13 (a) The F-4 aircraft. (b) Cockpit of the F-4. (Official USAF photos)

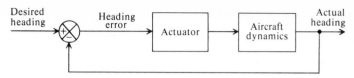

Fig. 1.14 A typical autopilot system.

Fig. 1.15 Apollo 7 undergoing preflight checkout. (Official NASA photo)

11

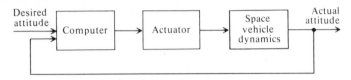

Fig. 1.16   The attitude control system of a typical space vehicle.

Fig. 1.17   Underwater firing of the Polaris A-3 missile.   (Official U.S. Navy photo)

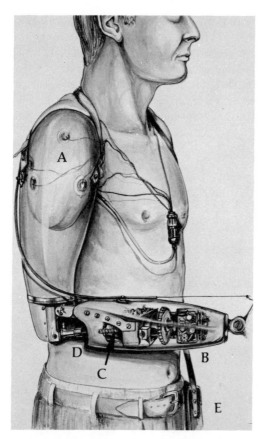

Fig. 1.18 The Boston Arm. The EMG signals (A) are processed by a small computer (B) which compares muscle effort with load on the terminal device and controls motor (C) to achieve lift or lowering of forearm at desired speed or force. Motion of forearm is performed by changing length of ball screw (D) via gear train from motor. Power for motor is stored in small battery pack on belt (E). (Courtesy of the Liberty Mutual Insurance Company)

continuously compares the attained velocity information with that reference velocity which will permit the payload to continue on to the target on a ballistic trajectory. When this desired velocity is attained, the computer automatically issues a signal which commands separation from the second stage motor and the payload continues on a ballistic trajectory to the target.

There have been many applications of feedback control system concepts in the field of bioengineering for prosthetic purposes [10, 11]. As an example of the application of feedback control system concepts to aid amputees, let us consider a recent advance in motor prosthesis. Figure 1.18 illustrates an artificial elbow, known as the Boston Arm, that was developed by the Massachusetts Institute of Technology,

Liberty Mutual Insurance Companies, Massachusetts General Hospital, and Harvard Medical School [12, 13]. This device depends on an electrical signal of about one millivolt, known as the electromyographic (EMG) signal, which is generated by a biochemical reaction whenever a muscle tissue contracts. Even in an above-elbow amputation, some muscle tissue usually remains in the stump which the amputee can contract and thus generate an EMG signal. EMG signals, controlled by the efferent nervous system that transmits information from the brain to the limb, are

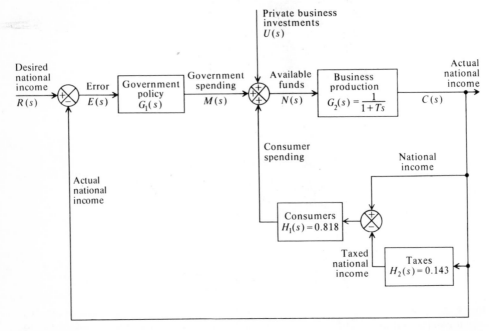

Fig. 1.19 Economic feedback relationship model concerning national income, government policy on spending, private business investment, business production, taxes, and consumer spending.

picked up by electrodes taped to the skin over the muscle. They are then amplified and used to control a battery-powered motor inside the forearm of the prosthesis. The signals from the biceps muscles control the device in flexion, while the triceps muscles control extension. Afferent reflexive feedback to the human nervous system of force-sensing information is achieved by means of a strain-gauge element whose electrical output opposes the EMG signal in this negative-feedback control system. Thus, an amputee can judge the force exerted by the elbow. The angular position of the elbow is also fed back to the control nervous system by means of a tactile display. In one application, this was found to aid an amputee in determining the position of by supplementing the usual visual and aural cues.

Automatic control theory has also been applied for modeling the feedback processes of our economic system in order to understand it better. Our economic system contains many feedback systems and regulatory agencies. Figure 1.19 illustrates a crude model of the economics concerned with national income, government policy on spending, private business investment, business production, taxes, and consumer spending. This type of feedback model assists the analyst in understanding the overall effects of government policy and private business investment on the national income.

Feedback control systems have a very bright and unlimited potential. This important field is one in which the engineer should become knowledgeable and proficient for solving problems found in this modern technological age. Before we begin our discussion of analyzing and synthesizing feedback control systems, let us define the nomenclature which is standard in the field and which we will use.

1.6 DEFINITION OF NOMENCLATURE AND SYMBOLS

The nomenclature and symbols described in this section are based on the standards issued by the American National Standards Institute Committee C85, Terminology for Automatic Control, under the sponsorship of the American Society of Mechanical Engineers [7, 15]. Figure 1.20 represents a general feedback control system, using standard symbols and notations. It would be well worth the reader's time to memorize the nomenclature and symbols and the definitions of these terms, since they are employed universally by the practicing control system engineer.

The *command*, v, is an input which is developed externally and is independent of the feedback control system.

The *reference input elements*, a, produce a signal proportional to the command.

The *reference input*, r, is the signal input which is proportional to the command.

The *primary feedback*, b, is a signal that is a function of the controlled variable and is compared with the reference input in order to obtain the actuating signal.

The *actuating signal*, e, equals the difference between the reference input and the primary feedback.

The *control elements*, G_1, develop the manipulated variable from the actuating signal.

The *manipulated variable*, m, is the quantity obtained from the control elements and applied to the controlled system.

The *disturbance*, u, represents undesired signals that attempt to affect the value of the controlled variable c.

The *controlled system*, G_2, is the device to be controlled.

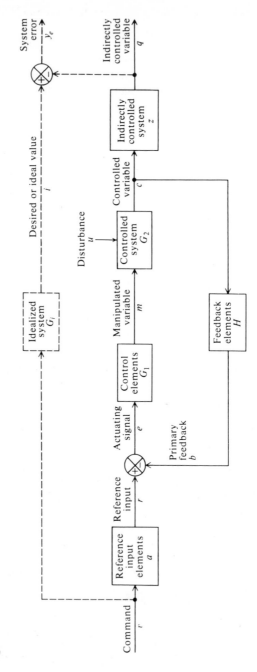

Fig. 1.20 Block diagram of a general feedback control system illustrating notations and terminology.

The *controlled variable*, c, is the quantity of the feedback system that is controlled.

The *indirectly controlled system*, z, is outside the feedback loop and relates the indirectly controlled variables to the controlled variable.

The *indirectly controlled variable*, q, is related to the controlled variable through the indirectly controlled system. It is outside the feedback system and is not directly measured for control.

The *feedback elements*, H, produce the primary feedback from the controlled variable.

The *idealized system*, G_i, is a system whose performance results in a response with an ideal value from the command.

The *ideal value*, i, is the value of the indirectly controlled variable which would result from an idealized system operating from the same command as the actual system.

The *system error*, y_e, is the difference between the ideal value and the indirectly controlled variable.

The dashed-line portion of the block diagram represents an idealized system. It is used to compare the indirectly controlled variable of the system with the ideal, or desired, value. The system error is the difference between these two values. For systems that have unity feedback the ideal system would also equal unity.

The reference input elements and feedback elements usually consist of devices known as *transducers*. These devices convert various types of input signals into other forms of signals. Many examples of transducers are used in industry. The potentiometer is probably the simplest form. Its prime function is to convert a shaft position into a proportional electrical signal. Transducers that convert velocity and acceleration into proportional electrical signals are tachometers and accelerometers, respectively. Pressure and temperature transducers convert pressure and temperature changes, respectively, into proportional electrical signals.

The primary functions of the control elements are to provide amplification to the actuating signals and to modify the frequency characteristics of the signal in order to insure stability. Electronic, electromechanical, magnetic, hydraulic, fluidic [16], and pneumatic devices are used for this purpose.

The controlled system is the portion of the system that responds to the manipulated variable and develops the controlled variable. The controlled system may represent a missile launcher, an aircraft's frame, or a chemical process, for example.

PROBLEMS

1.1 Figure P1.1 illustrates the control system for controlling the angular position of a ship's heading. The desired heading, which is determined by the gyroscope setting, is the reference. An electrical signal proportional to the reference is obtained from a resistor that is fixed to the

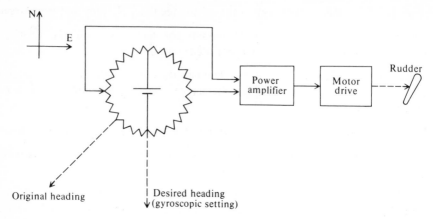

Figure P1.1

ship's frame. Qualitatively explain how this control system operates in following a command for a southbound direction.

1.2 Explain how the rudder positioning system discussed in Problem 1.1 can be modified to become a closed-loop control system.

1.3 An electric hot-water heater is illustrated in Fig. P1.3. The heating element is turned on and off by a thermostatic switch in order to maintain a desired temperature. Any demand for hot water results in hot water leaving the tank and cold water entering. Draw the simple functional block diagram for this closed-loop control system and qualitatively explain how it operates if the reference temperature of the thermostat is changed.

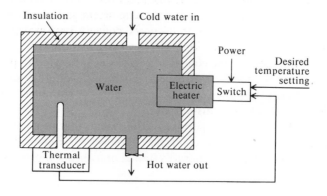

Figure P1.3

1.4 Explain how the system discussed in Problem 1.3 operates if the ambient temperature surrounding the tank suddenly changes. Refer to the block diagram.

1.5 Figure P1.5 illustrates a control system using a human operator as part of the closed-loop control system. Draw the block diagram of this liquid volume rate control system.

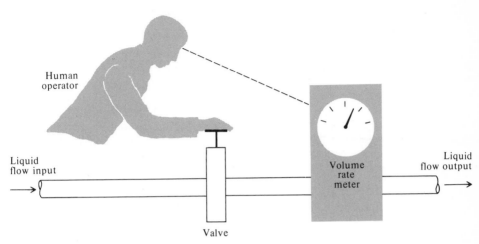

Figure P1.5

1.6 Devise a system that can control the speed of an internal-combustion engine in accordance with a command in the form of a voltage. Explain the operation of your system.

1.7 Devise a system that can control the position, rate, and acceleration of an elevator used in an apartment house. What specifications or limits would you place on the position, velocity, and acceleration capabilities of the system?

1.8 For the control system devised in Problem 1.7, describe what happens when a man weighing 200 lb enters the elevator that has stopped at one of the floors of the apartment building. Utilize a functional block diagram.

1.9 The economic model illustrated in Fig. P1.9 illustrates the relationship between wages, prices, and cost of living. Note that an automatic cost of living increase results in a positive feedback loop. Indicate how additional feedback loops in the form of legislative control can stabilize the economic system.

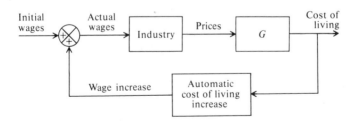

Figure P1.9

1.10 Explain what happens to the automatic position control system illustrated in Fig. 1.12 if the speed-measuring device fails. Can the system still operate?

1.11 Determine what happens in the autopilot system illustrated in Fig. 1.14 if the aircraft suddenly enters a turbulent atmosphere.

1.12 Modify the block diagram of the altitude control system of the space vehicle illustrated in Fig. 1.16, to allow for sudden failure of the computer and manual control of the vehicle.

1.13 The pH factor of a liquid is a very important factor to be controlled in chemical process control systems. Figure P1.13 illustrates an example of a system which controls the pH factor of a liquid flowing through a tank. The pH probe measures the actual pH level and compares it with the desired pH level. If the liquid in the tank does not have the required acidic level, then the motor adjusts the valve which permits additional acid to enter the tank. The limitation of the system shown is that it does not have any provisions for adding a base solution in the event that the liquid in the tank is too acidic. Modify this control system so that it could add an acid or a base to the liquid in the tank, and achieve the desired pH factor for all kinds of liquid pH factors.

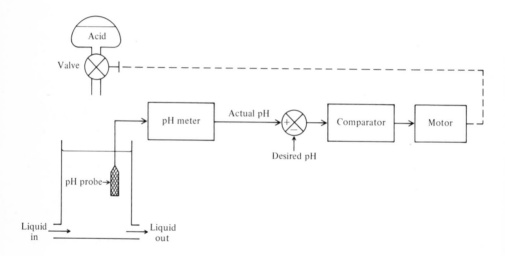

Figure P1.13

1.14 The actual sales price of a commodity in a free economic market is governed by the law of supply and demand that states that the market demand for the item increases as its price decreases. In addition, this concept states that an equilibrium sales price is achieved when the supply is equal to the demand. The control engineer can represent this concept as a feedback system with the actual sales price being represented as the output, and the desired market sales price change as the command input. Draw the block diagram for this economic system which includes the supplier, demander, prices, and market. Under which conditions is an equilibrium operation of supply and demand achieved?

1.15 The level of a liquid in a chemical process control system must be maintained constant within a narrow range. In order to achieve this, a float-operated device is used to sense the level of the liquid and control the valve opening which determines the flow of liquid into the tank to replace that which is consumed in the process. Show how this float-valve control system can be operated in a closed-loop control system.

1.16   Repeat Problem 1.15 if the level of the liquid is controlled by a float switch which opens the valve whenever the level of the liquid reaches a certain predetermined level. The valve is automatically closed by a timer switch four minutes after it is opened by the float switch, even though the liquid may be more or less than the desired level. Is this an open-loop or closed-loop control system?

REFERENCES

1. S. C. Lyman, "Computer directs process control at textile plant," *Control Eng.* **13**, 20–21 (October 1966).
2. J. C. Keebler, "Warehouse automation," *Automation* **13**, 64 (December 1966).
3. S. Hearne, "In-plant data collection tracks material," *Control Eng.* **14**, 100 (May 1967).
4. S. W. R. Cox, "Automation in agriculture," *Control* **9**, 247–52 (May 1965).
5. I. Nakamura and S. Yamazaki, "On the centralized system for train operation and traffic control—Including signaling and routing information," *Railway Technical Research Institute*, **5**, No. 1 (1964).
6. *Technical Aspects on the New Tokaido Line*, Japanese National Railways, Tokyo (October 1966).
7. ASA-C85.1, *Terminology for Automatic Control*, American National Standards Institute, New York (1963).
8. N. Wiener, *Cybernetics or Control and Communication in the Animal and the Machine* (2nd Edn.), The MIT Press and Wiley, New York (1961).
9. Polaris/Poseidon Fleet Ballistic Missile Weapon System, Fact Sheet, U.S. Navy Special Projects Office, Washington, D.C. (September 1967).
10. N. A. Coulter, Jr., and O. L. Updike, Jr., "Biomedical control developments," in *Proceedings of the 1965 Joint Automatic Control Conference*, pp. 258–72.
11. R. W. Mann, "Efferent and afferent control of an electromyographic, proportional-rate, force sensing artificial elbow with cutaneous display of joint angle," *Proc. Inst. Mech. Engrs.* **183**, 86–91 (1968–69).
12. *The Boston Arm*, Liberty Mutual Insurance Company, Research Center, 175 Berkeley Street, Boston.
13. "Designers still grope for efficient 'limbs'," *Electronic Design* **18**, U92–U93 (March 1970).
14. S. M. Shinners, *Techniques of System Engineering*, McGraw-Hill, New York (1967).
15. USAS-C85.1a, *Supplement to Terminology for Automatic Control*, American National Standards Institute, New York (1966).
16. S. M. Shinners, "Fluidics," *Electro-Technology* **79**, 81–94 (March 1967).

2   MATHEMATICAL TECHNIQUES FOR THE CONTROL ENGINEER

2.1 INTRODUCTION

The design of linear, continuous, feedback control systems is dependent on mathematical techniques such as the Laplace transformation, the signal-flow diagram, and the state-space concept. In addition to these techniques, the design of linear, sampled-data, feedback control systems requires a knowledge of the z-transform and some aspects of information theory. The design of nonlinear, continuous, feedback control systems is dependent on mathematical techniques such as the Fourier transform, the signal-flow diagram, and the state-space concept. The design of systems having random inputs requires a knowledge of probability theory and statistical theory. The scope of this book does not permit a detailed discussion of all of these mathematical devices. The philosophy followed here is to review the theory of those techniques necessary for understanding the design of linear and nonlinear continuous systems, and to focus attention on the specific application of these mathematical tools to these classes of control systems.

This chapter reviews the theory and application of complex-variable theory, the Fourier transform, the Laplace transformation, the signal-flow diagram, and matrix techniques. In addition, it introduces the state-space concept.

2.2 COMPLEX VARIABLES AND THE s-PLANE

The design of control systems depends greatly on the application of complex-variable theory. In what follows, the complex variable s is composed of a real part σ and an imaginary part ω. In the complex s-plane, σ is usually plotted horizontally and ω vertically. An arbitrary function $F(s)$ is considered to be a function of the complex variable s if there is at least one value of $F(s)$ for every value of s. If there is only one value of $F(s)$ for every value of s, the function $F(s)$ is called a *single-valued function*. However, if there is more than one point in the $F(s)$-plane for every value of s, then $F(s)$ is a *multivalued function*. Figure 2.1 illustrates the mapping of a single-valued function from the s-plane to the $F(s)$-plane. Figure 2.2 illustrates the corresponding mapping for a multivalued function.

A. Key Notions Four notions of complex-variable theory which are important to the control systems engineer are those of analytic functions, singularities, poles, and

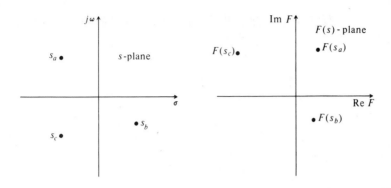

Fig. 2.1 Mapping of a single-valued function.

zeros of a function. A complex-variable function $F(s)$ is *analytic* in a region if the function and all of its derivatives exist at every point in that region. As an example, the function

$$F(s) = \frac{1}{s + 4} \tag{2.1}$$

is analytic at every point in the s-plane except at the point $s = -4$. *Singularities* are defined as points in the s-plane where the function, or its derivatives, do not exist. An important example of a singularity is a *pole*. If a function $F(s)$ is analytic in the region of s_j, except at the poles of s_j, then $F(s)$ has a pole of order q (where q is finite) at $s = s_j$ if

$$\lim_{s \to s_j} [F(s)(s - s_j)^q] \tag{2.2}$$

is finite. Therefore, the denominator of $F(s)$ must contain the factor $(s - s_j)^q$, and the function becomes infinite when $s = s_j$. In Eq. (2.1), the function has a simple

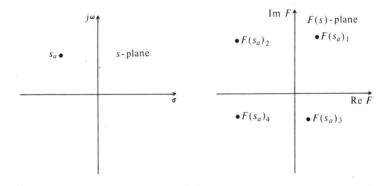

Fig. 2.2 Mapping of a multivalued function.

pole (i.e., a pole of order one) at $s = -4$. As a more complex example, consider the function

$$F(s) = \frac{100(s + 1)(s + 8)^2}{s(s + 4)(s + 10)(s + 20)^3} \,. \tag{2.3}$$

This function has simple poles at $s = 0$, -4, and -10. In addition, it has a pole of order three at $s = -20$.

Finally, we consider the concept of *zeros*. If a function $F(s)$ is analytic at $s = s_j$, then $F(s)$ has a zero of order q (where q is finite) at $s = s_j$, if

$$\lim_{s \to s_j} [F(s)(s - s_j)^{-q}] \tag{2.4}$$

is finite. Therefore, the numerator of $F(s)$ must contain the factor $(s - s_j)^q$, and the function becomes zero where $s = s_j$. For example, in Eq. (2.3), there is a simple zero (i.e., a zero of order one) at $s = -1$ and a zero of order two at $s = -8$.

B. The Residue Theorem   An important aspect of complex-variable theory is the residue theorem. This will be used later to prove Nyquist's theorem—a very important relationship for determining the stability of linear systems.

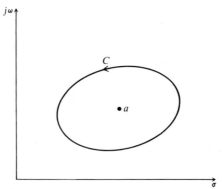

Fig. 2.3   A contour C around an analytic function containing a pole of order n at a.

In order to develop this theorem, consider a function $f(s)$ which is analytic on and within a closed contour C, where a pole exists of order n at a [16]. This is illustrated in Fig. 2.3. For this situation the expression for $f(s)$ is modified to the following form:

$$f(s) = \frac{f(s)(s - a)^n}{(s - a)^n} = \frac{F(s)}{(s - a)^n} \,. \tag{2.5}$$

The new function $F(s)$ is analytic on and within the closed contour including the point a. Since the function $F(s)$ is analytic at point a (and thus all of its derivatives at point *a* exist), it can be expanded about point *a* in a Taylor series as follows:

$$F(s) = F(a) + F'(a)(s - a) + \frac{F''(a)}{2!}(s - a)^2 + \cdots$$

$$+ \frac{F^{(n)}(a)}{n!}(s - a)^n + \cdots. \tag{2.6}$$

Substituting Eq. (2.6) into Eq. (2.5), we obtain the following series:

$$f(s) = \frac{F(a)}{(s - a)^n} + \frac{F'(a)}{(s - a)^{n-1}} + \frac{F''(a)}{2!\,(s - a)^{n-2}} + \cdots$$

$$+ \frac{F^{(n-1)}(a)}{(n - 1)!\,(s - a)} + \frac{F^{(n)}(a)}{n!} + \frac{F^{n+1}(a)}{(n + 1)!}(s - a) + \cdots. \tag{2.7}$$

Equation (2.7) is the Laurent expansion of $f(s)$ at a. Let us now consider the integration of $f(s)$, in its Laurent-expansion form, around a closed contour which encloses the point a and on and within which there are no other singularities.

$$\frac{1}{2\pi j} \oint f(s)\,ds = \frac{F(a)}{2\pi j} \oint \frac{ds}{(s - a)^n} + \frac{F'(a)}{2\pi j} \oint \frac{ds}{(s - a)^{n-1}} + \cdots$$

$$+ \frac{F^{(n-1)}(a)}{(n - 1)!\,2\pi j} \oint \frac{ds}{(s - a)} + \frac{F^{(n)}(a)}{n!\,2\pi j} \oint ds. \tag{2.8}$$

The notation $\oint$ is used to denote an integration around a closed curve; the arrow indicates that the integration is positive in the counterclockwise direction. In Eq. (2.8), it can be shown [16] that all integrals on the right-hand side of the equation are zero except for

$$\frac{1}{2\pi j} \oint \frac{ds}{(s - a)} = 1. \tag{2.9}$$

Therefore, Eq. (2.8) reduces to

$$\frac{1}{2\pi j} \oint f(s)\,ds = \frac{F^{(n-1)}(a)}{(n - 1)!}. \tag{2.10}$$

The right-hand side of Eq. (2.10) represents the only coefficient of the Laurent expansion which affects the value of the integral of $f(s)$ around the closed contour, and is denoted as the *residue* $R^{(n)}$ of $f(s)$ in the pole at $s = a$. Hence we have

$$\frac{1}{2\pi j} \oint f(s)\,ds = \frac{F^{(n-1)}(a)}{(n - 1)!} = R^{(n)}. \tag{2.11}$$

For cases where there are more than one pole on or within the contour, the path of integration can be deformed as illustrated in Fig. 2.4 and the results just obtained can

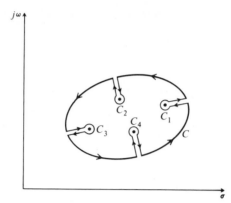

Fig. 2.4 A contour C around an analytic function containing four poles.

be applied. Since the contour excludes the poles, with circles of radii approaching zero, Cauchy's theorem can be applied. This theorem states that the line integral around a closed contour is zero if the function $f(s)$ is analytic within and upon the contour [16]. Therefore, the following expression can be written:

$$\frac{1}{2\pi j}\oint f(s)\,ds = 0 = \frac{1}{2\pi j}\oint_C f(s)\,ds - \frac{1}{2\pi j}\oint_{C_1} f(s)\,ds$$

$$-\frac{1}{2\pi j}\oint_{C_2} f(s)\,ds - \frac{1}{2\pi j}\oint_{C_3} f(s)\,ds - \frac{1}{2\pi j}\oint_{C_4} f(s)\,ds. \quad (2.12)$$

Therefore, the integral around the outer contour C is given by

$$\frac{1}{2\pi j}\oint_C f(s)\,ds = R_1 + R_2 + R_3 + R_4. \quad (2.13)$$

It can thus be concluded that if $f(s)$ is analytic except at the poles, the integral around the closed path equals the sum of the residues in the poles inside the contour:

$$\frac{1}{2\pi j}\oint_C f(s)\,ds = \sum_{i=1}^{4} R_i. \quad (2.14)$$

From Eqs. (2.5) and (2.11), the residue of a pole of order n is given by

$$R^{(n)} = \frac{F^{(n-1)}(a)}{(n-1)!} = \lim_{s\to a} \frac{d^{n-1}}{ds^{n-1}}\left[\frac{f(s)(s-a)^n}{(n-1)!}\right]. \quad (2.15)$$

As an example of applying the residue theorem, consider the evaluation of the following integral around a closed contour C which encloses the point $s = -1$:

$$\frac{1}{2\pi j}\oint_C \frac{s^2}{s+1}\,ds. \quad (2.16)$$

In this example, the function $f(s)$ being considered is given by

$$f(s) = \frac{s^2}{s + 1}, \tag{2.17}$$

which has a single pole at $s = -1$. From Eq. (2.15), the residue at $s = -1$ is given by

$$R^{(1)} = \lim_{s \to -1} \frac{s^2}{s + 1} \frac{s + 1}{0!} = 1. \tag{2.18}$$

Therefore, from Eq. (2.16), the integral around any closed contour enclosing the pole at $s = -1$ is given by

$$\oint_C \frac{s^2}{s + 1} \, ds = 2\pi j. \tag{2.19}$$

2.3 FOURIER SERIES AND THE FOURIER TRANSFORM

In the analysis of the behavior of systems evolving in time, it is often convenient to introduce mathematical transformations that take us from the time domain to a new domain called the frequency domain. Such transformations are called transforms. Here we will focus on the Fourier series, which is used to analyze periodic functions of time, and the Fourier transform, which is used to examine aperiodic time functions of a restricted class. Section 2.4 introduces the Laplace transform, which is of great value in the analysis of time functions which vanish for negative time.

A. The Fourier Series Given a periodic function, $f(t)$, whose period is T, then

$$f(t) = f(t + T). \tag{2.20}$$

Functions which satisfy Eq. (2.20) can be represented by a Fourier series provided the function is bounded and contains only a finite number of discontinuities in a finite interval. The classical Fourier series is given by

$$f(t) = \frac{A_0}{2} + \sum_{K=1}^{K=\infty} A_K \cos K\omega t + \sum_{K=1}^{K=\infty} B_K \sin K\omega t, \tag{2.21}$$

where

$$A_K = \frac{2}{T} \int_{-T/2}^{T/2} f(t) \cos K\omega t \, dt, \qquad K = 0, 1, 2, 3, \ldots, \tag{2.22}$$

$$B_K = \frac{2}{T} \int_{-T/2}^{T/2} f(t) \sin K\omega t \, dt, \qquad K = 1, 2, 3, \ldots. \tag{2.23}$$

If $f(t) = -f(-t)$, then the function is odd and $A_K = 0$; Fig. 2.5(a) illustrates an odd function. If $f(t) = f(-t)$, then the function is even and $B_K = 0$; Fig. 2.5(b) illustrates an even function. The two terms in the series whose frequency is $K\omega$ when added comprise the Kth *harmonic* of $f(t)$. The amplitude of the harmonic is given by

$$\sqrt{A_K^2 + B_K^2}, \tag{2.24}$$

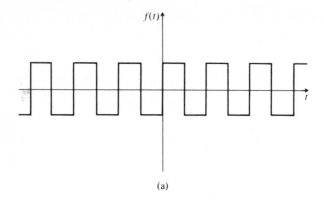

(a)

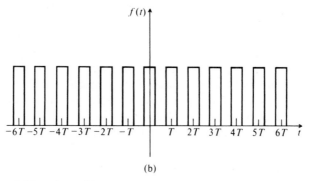

(b)

Fig. 2.5 (a) An odd function. (b) An even function.

and the phase is given by

$$\tan^{-1}\left(-B_K/A_K\right). \tag{2.25}$$

As an example of the application of the Fourier series, consider the periodic waveshape illustrated in Fig. 2.6(a). This function represents a periodic switching function whose period is $2\pi/\omega_c$. In addition, it is an even function since $f(t) = f(-t)$, so that the B_K terms are zero. The evaluation of A_0 and A_K follows from Eq. (2.22):

$$A_0 = \frac{1}{\pi/\omega_c}\left[\int_{-\pi/\omega_c}^{-\pi/2\omega_c}(-1)\,dt + \int_{-\pi/2\omega_c}^{\pi/2\omega_c}(1)\,dt + \int_{\pi/2\omega_c}^{\pi/\omega_c}(-1)\,dt\right] = 0 \tag{2.26}$$

$$A_K = \frac{1}{\pi/\omega_c}\left[\int_{-\pi/\omega_c}^{-\pi/2\omega_c}(-1)\cos K\omega_c t\,dt\right.$$

$$\left. + \int_{-\pi/2\omega_c}^{\pi/2\omega_c}(1)\cos K\omega_c t\,dt + \int_{\pi/2\omega_c}^{\pi/\omega_c}(-1)\cos K\omega_c t\,dt\right] \tag{2.27}$$

$$A_K = \frac{2\sin\left(K\pi/2\right)}{K\pi/2}. \tag{2.28}$$

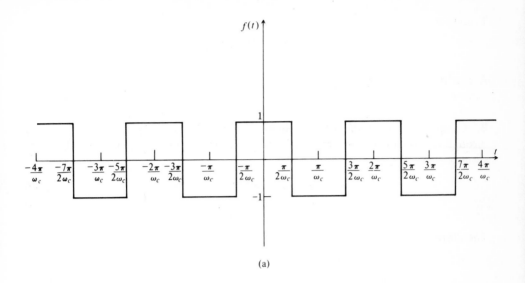

(a)

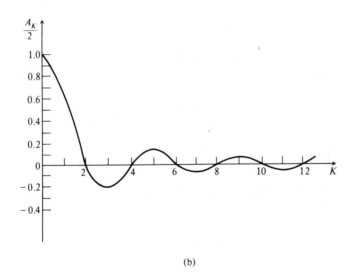

(b)

Fig. 2.6   (a) A periodic function.   (b) Envelope of the function

$$\frac{A_K}{2} = \frac{\sin K\pi/2}{K\pi/2}$$

showing the $(\sin x)/x$ pattern.

Therefore,

$$f(t) = 2 \sum_{K=1}^{K=\infty} \frac{\sin (K\pi/2)}{K\pi/2} \cos K\omega t. \qquad (2.29)$$

A plot of the envelope of the function A_K, which follows the familiar $(\sin x/x)$ wave pattern, is illustrated in Fig. 2.6(b).

B. Complex Form of the Fourier Series The Fourier series given by Eqs. (2.21) through (2.23) can be converted to a complex form by means of the following substitutions:

$$\sin K\omega t = \frac{1}{2j} (e^{jK\omega t} - e^{-jK\omega t}), \qquad (2.30)$$

$$\cos K\omega t = \tfrac{1}{2}(e^{jK\omega t} + e^{-jK\omega t}). \qquad (2.31)$$

Therefore,

$$f(t) = \frac{A_0}{2} + \sum_{K=1}^{\infty} \frac{A_K}{2} (e^{jK\omega t} + e^{-jK\omega t}) + \sum_{K=1}^{\infty} \frac{B_K}{2j} (e^{jK\omega t} - e^{-jK\omega t}).$$

This can also be written as

$$f(t) = \frac{A_0}{2} + \frac{1}{2} \sum_{K=1}^{\infty} (A_K - jB_K)e^{jK\omega t} + \frac{1}{2} \sum_{K=1}^{\infty} (A_K + jB_K)e^{-jK\omega t}. \qquad (2.32)$$

Note that from Eqs. (2.22) and (2.23),

$$(A_K - jB_K) = \frac{2}{T} \int_{-T/2}^{T/2} f(t)e^{-jK\omega t}\, dt$$

and

$$(A_K + jB_K) = \frac{2}{T} \int_{-T/2}^{T/2} f(t)e^{jK\omega t}\, dt.$$

Therefore, Eq. (2.32) becomes

$$f(t) = \frac{A_0}{2} + \sum_{K=1}^{\infty} \frac{e^{jK\omega t}}{T} \int_{-T/2}^{T/2} f(t)e^{-jK\omega t}\, dt + \sum_{K=1}^{\infty} \frac{e^{-jK\omega t}}{T} \int_{-T/2}^{T/2} f(t)e^{jK\omega t}\, dt. \qquad (2.33)$$

Since the second and third terms on the right side of Eq. (2.33) differ by a sign, the complex form of the Fourier series can be written as

$$f(t) = \sum_{K=-\infty}^{K=\infty} \frac{e^{jK\omega t}}{T} \int_{-T/2}^{T/2} f(t)e^{-jK\omega t}\, dt. \qquad (2.34)$$

This equation is usually written in the following form:

$$f(t) = \frac{1}{T} \sum_{K=-\infty}^{K=\infty} C_K e^{jK\omega t}, \qquad (2.35)$$

where

$$C_K = \int_{-T/2}^{T/2} f(t)e^{-jK\omega t}\,dt. \tag{2.36}$$

Equation (2.35) gives the complex Fourier series of $f(t)$, and Eq. (2.36) defines the complex Fourier coefficient, C_K.

As an example of the application of the complex form of the Fourier series, consider the periodic pulse train illustrated in Fig. 2.7(a). The complex Fourier

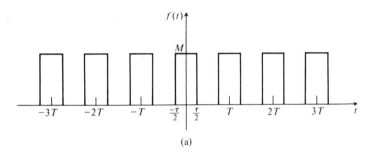

(a)

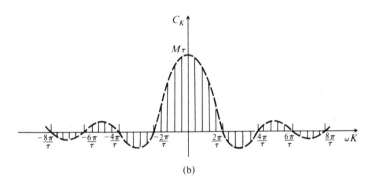

(b)

Fig. 2.7 (a) Periodic pulse train. (b) Complex Fourier coefficient C_K for the pulse train of Fig. 2.7(a).

coefficient of this periodic pulse train is given by

$$C_K = \int_{-\tau/2}^{\tau/2} Me^{-jK\omega t}\,dt = \left[-\frac{Me^{-jK\omega t}}{jK\omega}\right]_{-\tau/2}^{\tau/2} = \frac{2M}{K\omega}\sin \omega K\tau/2. \tag{2.37}$$

Equation (2.37) can also be written as

$$C_K = M\tau \frac{\sin \frac{1}{2}\omega K\tau}{\frac{1}{2}\omega K\tau}.$$

A plot of C_K for all values of frequency is illustrated in Fig. 2.7(b). The envelope of C_K follows a $(\sin x/x)$ function and the spacing between lines is given by $\Delta\omega K = 2\pi/T$.

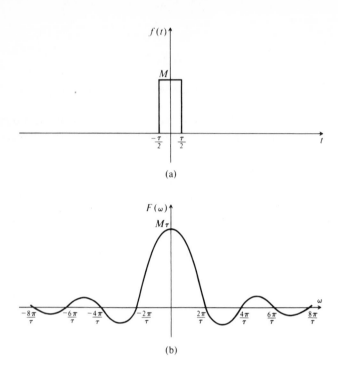

Fig. 2.8 (a) A nonperiodic pulse train. (b) Fourier transform for the pulse illustrated in Fig. 2.8(a).

C. The Fourier Integral The control-system engineer finds many functions of interest which are not periodic. For these functions, the Fourier series cannot be applied. However, the Fourier integral can be used to analyze a wide class of aperiodic functions.

Let us assume that instead of the periodic pulse train in Fig. 2.7(a), we have only one nonperiodic pulse as illustrated in Fig. 2.8(a). By considering this single pulse as the fundamental function during a period and assuming that its period is infinity, this function can be represented by the Fourier integral. Basically, the approach is to obtain the Fourier series of the function of Fig. 2.8a assuming that it is periodic with a period of infinity. Let us write Eq. (2.35) as

$$f(t) = \frac{1}{2\pi} \sum_{K=-\infty}^{K=\infty} C_K e^{jK\omega t} \Delta K\omega$$

by letting $T = 2\pi/\Delta K\omega$. As T approaches infinity, $\Delta K\omega$ approaches zero and we have the following relationships:

$$\lim_{\substack{T \to \infty \\ \Delta K\omega \to 0}} f(t) = \lim_{\substack{T \to \infty \\ \Delta K\omega \to 0}} \frac{1}{2\pi} \sum_{K=-\infty}^{K=\infty} C_K e^{jK\omega t} \Delta K\omega, \qquad (2.38)$$

$$\lim_{\substack{T \to \infty \\ \Delta K\omega \to 0}} f(t) = \frac{1}{2\pi} \int_{-\infty}^{\infty} C_K e^{j\omega t} d\omega. \qquad (2.39)$$

Therefore, the discrete lines of C_K in Fig. 2.7(b) merge into a continuous frequency spectrum as illustrated in Fig. 2.8(b). As the period T approaches infinity, Eq. (2.36) can be written as:

$$C_K = F(\omega) = \int_{-\infty}^{\infty} f(t)e^{-j\omega t}\, dt. \qquad (2.40)$$

This is the definition of the Fourier transform of a nonperiodic function $f(t)$, and is commonly denoted as $F(\omega)$. It is important to note here that the function must satisfy the following condition of absolute convergence before the Fourier transform of a function is taken:

$$\int_{-\infty}^{\infty} |f(t)|\, dt < \infty. \qquad (2.41)$$

As an example of the calculation of a Fourier integral, consider the following exponential time function, where $\theta < 0$.

$$f(t) = e^{\theta t} \qquad \text{for} \quad t > 0 \qquad (2.42)$$
$$f(t) = 0 \qquad \text{for} \quad t < 0. \qquad (2.43)$$

Let us first check on the convergence of the function. From Eq. (2.41),

$$\int_{-\infty}^{\infty} |e^{\theta t}|\, dt < \infty, \qquad \text{since} \qquad \theta < 0. \qquad (2.44)$$

Therefore, this exponential function is absolutely integrable and its Fourier transform can be obtained from Eq. (2.40), with the lower limit replaced by zero, as follows:

$$F(\omega) = \int_{0}^{\infty} e^{\theta t} e^{-j\omega t}\, dt. \qquad (2.45)$$

Integrating, we obtain

$$F(\omega) = \left[\frac{e^{t(\theta - j\omega)}}{\theta - j\omega}\right]_{0}^{\infty}, \qquad (2.46)$$

$$F(\omega) = \frac{1}{j\omega - \theta}. \qquad (2.47)$$

The Fourier transform is a very powerful mathematical tool that is used to a great extent in engineering. However, the limitation defined by Eq. (2.41) restricts its use in important situations. For example, the control engineer is usually interested in the response of a control system to unit step, ramp, and parabolic time functions denoted by $U(t)$, $tU(t)$, and $t^2U(t)$, respectively. Unfortunately the Fourier transforms of these functions do not exist. For these types of function, the engineer modifies the Fourier transform by adding a convergence factor $e^{-\sigma t}$ where σ is a real number that is large enough to maintain absolute convergence. Therefore, the new transform is given by

$$F(\sigma, \omega) = \int_{0}^{\infty} f(t)e^{-\sigma t}e^{-j\omega t}\, dt. \qquad (2.48)$$

Notice that the lower limit is defined as zero rather than minus infinity, so that the new transform only applies to time functions that vanish for negative time. However, this is not a serious limitation in control problems since the time reference is usually chosen to be $t = 0$. By defining

$$s = \sigma + j\omega, \tag{2.49}$$

Eq. (2.48) can be written as

$$F(s) = \int_0^\infty f(t)e^{-st}\, dt, \tag{2.50}$$

where $F(s)$ is defined as the Laplace transform of the function $f(t)$.

2.4 THE LAPLACE TRANSFORM

The Laplace transform [1, 2] is helpful in the solution of ordinary differential equations describing the behavior of systems. When the transform operates on a differential equation, a "transformed" equation results. It is expressed in terms of an arbitrary complex variable, s. The resulting transformed equation is in purely algebraic terms which can be easily manipulated to obtain a solution for the desired quantity as an explicit function of the complex variable. In order to obtain a solution in terms of the original variable, it is necessary to carry out an inversion process to determine the desired time function. The inverse Laplace transform is given by

$$f(t) = \frac{1}{2\pi j} \int_{\sigma - j\infty}^{\sigma + j\infty} F(s)e^{st}\, ds, \tag{2.51}$$

where σ is a real constant greater than the real part of any singularity of $F(s)$. The evaluation of this integral is usually difficult, and the inverse transformations are usually obtained by utilizing a table of Laplace transforms. The Laplace transform $F(s)$ of a certain function of time $f(t)$ is conventionally written as

$$F(s) \triangleq \mathscr{L}[f(t)]. \tag{2.52}$$

From the definition of the Laplace transform given by Eq. (2.50), the integral exists if

$$\int_0^\infty |f(t)e^{-\sigma t}|\, dt < \infty. \tag{2.53}$$

2.5 USEFUL LAPLACE TRANSFORMS

The Laplace transforms for various time functions will now be considered. These are readily obtainable through a direct application of Eq. (2.50).

A. The Laplace Transform of a Unit Step For the unit step function defined by

$$f(t) = U(t), \qquad f(t) = \begin{cases} 0 & \text{for} \quad t < 0, \\ 1 & \text{for} \quad t > 0, \end{cases}$$

the Laplace transform is

$$\mathcal{L}[f(t)] = \int_0^\infty (1)e^{-st}\, dt = \left[-\frac{1}{s} e^{-st} \right]_0^\infty.$$

Therefore,

$$\mathcal{L}[U(t)] = 1/s, \qquad \text{Re}\,[s] > 0. \tag{2.54}$$

From here on we assume that $f(t) = 0$ for $t < 0$.

B. The Laplace Transform of an Exponential Decay For the function

$$f(t) = e^{-\alpha t},$$

we have the Laplace transform

$$\mathcal{L}[f(t)] = \int_0^\infty e^{-(s+\alpha)t}\, dt = \left[-\frac{1}{s+\alpha} e^{-(s+\alpha)t} \right]_0^\infty.$$

Therefore,

$$\mathcal{L}(e^{-\alpha t}) = \frac{1}{s+\alpha} \qquad \text{Re}\,[s] > -\alpha. \tag{2.55}$$

C. The Laplace Transform of a Unit Ramp For the function

$$f(t) = t,$$

the Laplace transform is

$$\mathcal{L}[f(t)] = \int_0^\infty te^{-st}\, dt. \tag{2.56}$$

Integrating by parts,

$$\int u\, dv = uv - \int v\, du, \tag{2.57}$$

with $u = t$, $dv = e^{-st}\, dt$, the following is obtained:

$$\int_0^\infty te^{-st}\, dt = \left[t\frac{e^{-st}}{-s} \right]_0^\infty - \int_0^\infty \frac{e^{-st}}{-s}\, dt.$$

Therefore,

$$\mathcal{L}(t) = \frac{1}{s^2}, \qquad \text{Re}\,[s] > 0. \tag{2.58}$$

D. The Laplace Transform of a Sinusoidal Function For the function

$$f(t) = \sin \omega t,$$

the Laplace transform is

$$\mathcal{L}[f(t)] = \mathcal{L}(\sin \omega t) = \int_0^\infty \sin \omega t \, e^{-st}\, dt. \tag{2.59}$$

The solution to Eq. (2.59) is simplified by using the exponential form of sin ωt,

$$\sin \omega t = \frac{e^{j\omega t} - e^{-j\omega t}}{2j}.$$

Therefore,

$$\int_0^\infty \sin \omega t \, e^{-st} \, dt = \int_0^\infty \frac{e^{j\omega t} - e^{-j\omega t}}{2j} e^{-st} \, dt$$

$$= \frac{1}{2j} \int_0^\infty (e^{-(s-j\omega)t} - e^{-(s+j\omega)t}) \, dt$$

$$= \frac{1}{2j} \left(\frac{1}{s - j\omega} - \frac{1}{s + j\omega} \right). \qquad (2.60)$$

Therefore,

$$\mathscr{L}(\sin \omega t) = \frac{\omega}{s^2 + \omega^2}, \qquad \text{Re } [s] > 0. \qquad (2.61)$$

Once the Laplace transform for any function $f(t)$ is obtained and tabulated, it need not be derived again. The foregoing results and other important transform pairs useful to the control engineer appear in Table 2.1 on page 40. An extended table is shown in Appendix A. In addition, the location of the poles of the transformed function in the complex plane is listed.

2.6 IMPORTANT PROPERTIES OF THE LAPLACE TRANSFORM

The Laplace transform has been introduced in order to simplify several mathematical operations. These operations center upon the solution of linear differential equations. Several basic properties of the Laplace transform are given here.

A. Addition and Subtraction If the Laplace transforms of $f_1(t)$ and $f_2(t)$ are $F_1(s)$ and $F_2(s)$, respectively, then

$$\mathscr{L}[f_1(t) \pm f_2(t)] = F_1(s) \pm F_2(s).$$

B. Multiplication by a Constant If the Laplace transform of $f(t)$ is $F(s)$, then multiplication of the function $f(t)$ by a constant K results in a Laplace transform $KF(s)$.

C. Direct Transforms of Derivatives If the Laplace transform of $f(t)$ is $F(s)$, the transform of the time derivative $\dot{f}(t)$ of $f(t)$ is given by

$$\mathscr{L}[\dot{f}(t)] = sF(s) - f(0^+), \qquad (2.62)$$

where $f(0^+)$ is the initial value of $f(t)$, evaluated as $t \to 0$ from the positive region. The Laplace transform of the nth derivative of a function is given by

$$\mathscr{L}\left\{ \frac{d^n f}{dt^n} \right\} = s^n F(s) - s^{n-1} f(0^+) - s^{n-2} \dot{f}(0^+) - \cdots - f^{(n-1)}(0^+), \qquad (2.63)$$

where $\dot{f}(0^+)$ is the first derivative of $f(t)$ evaluated at $t = 0^+$. The notation $f^{(n-1)}(0^+)$ represents the $(n-1)$th derivative of f with respect to time evaluated at $t = 0^+$.

D. Direct Transforms of Integrals If the Laplace transform of $f(t)$ is $F(s)$, the transform of the time integral of $f(t)$ is given by

$$\mathscr{L}\left[\int f(t)\, dt\right] = \frac{F(s)}{s} + \frac{1}{s}\left[\int f(t)\, dt\right]_{t=0^+}, \tag{2.64}$$

where $[\int f(t)\, dt]_{t=0^+}$ signifies that the integral is evaluated as $t \to 0$ from the positive region. In general, for nth-order integration,

$$\mathscr{L}\left[\int\int \cdots \int f(t)\, dt^n\right] = \frac{F(s)}{s^n} + \frac{1}{s^n}\left[\int f(t)\, dt\right]_{t=0^+}$$

$$+ \frac{1}{s^{n-1}}\left[\int\int f(t)\, dt^2\right]_{t=0^+} + \cdots + \frac{1}{s}\left[\int\int \cdots \int f(t)\, dt^n\right]_{t=0^+}. \tag{2.65}$$

E. Shifting Theorem The Laplace transform of a time function $f(t)$ delayed in time by T equals the Laplace transform of $f(t)$ multiplied by e^{-sT}:

$$\mathscr{L}[f(t-T)U(t-T)] = e^{-sT}F(s), \qquad T \geqslant 0. \tag{2.66}$$

F. The Initial-Value Theorem If the Laplace transform of $f(t)$ is $F(s)$, and if $\lim_{s \to \infty} sF(s)$ exists, then the initial value of the time function is given by

$$\lim_{t \to 0} f(t) = \lim_{s \to \infty} sF(s). \tag{2.67}$$

G. The Final-Value Theorem If the Laplace transform of $f(t)$ is $F(s)$, and if $sF(s)$ is analytic on the imaginary axis and in the right half-plane, then the final value of the time function is given by

$$\lim_{t \to \infty} f(t) = \lim_{s \to 0} sF(s). \tag{2.68}$$

2.7 INVERSION BY PARTIAL FRACTION EXPANSION

The time response is the quantity of ultimate interest to the control-system designer. The process of inversion of a function $F(s)$ to find the corresponding time function $f(t)$ is denoted symbolically by

$$\mathscr{L}^{-1}F(s) = f(t). \tag{2.69}$$

In applications, $F(s)$ is usually a rational function of the form

$$F(s) = \frac{A_X s^X + A_{X-1} s^{X-1} + \cdots + A_1 s + A_0}{s^Y + B_{Y-1} s^{Y-1} + \cdots + B_1 s + B_0}. \tag{2.70}$$

In practical systems, the order of the polynomial in the denominator is equal to, or greater than, that of the numerator. For the cases where $Y > X$, partial fraction expansion is directly applicable. When $Y = X$, it is necessary to reduce $F(s)$ to a proper fraction by long division.

The simplest method for obtaining inverse transformations is to use a table of transforms. Unfortunately, many forms of $F(s)$ are not found in the usual table of Laplace transform pairs. When the form of the solution cannot be readily reduced to a form available in a table, we must use the technique known as partial fraction expansion. This method permits the expansion of the algebraic equation into a series of simpler terms whose transforms are available from a table. It is then possible to obtain the inverse transformation of the original algebraic expression by adding together the inverse transformations of the terms in the expansion. Equation (2.71) expresses this operation symbolically. The function $F(s)$ represents the original algebraic expression and $F_1(s)$, $F_2(s)$, $F_3(s)$, ..., $F_n(s)$ are terms of the partial fraction expansion:

$$\mathcal{L}^{-1}[F(s)] = \mathcal{L}^{-1}[F_1(s)] + \mathcal{L}^{-1}[F_2(s)] + \cdots + \mathcal{L}^{-1}[F_n(s)]. \tag{2.71}$$

As an example of the method, consider the transform

$$F(s) = \frac{As + B}{(s + C)(s + D)}. \tag{2.72}$$

In this equation A, B, C, and D are constants. This function can be expanded into partial fractions:

$$\frac{As + B}{(s + C)(s + D)} = \frac{K_1}{s + C} + \frac{K_2}{s + D}. \tag{2.73}$$

To determine K_1, both sides of Eq. (2.73) are multiplied by $s + C$, yielding

$$\frac{As + B}{s + D} = K_1 + \frac{K_2(s + C)}{s + D}. \tag{2.74}$$

By substituting $s = -C$, the last term vanishes and a numerical value for K_1 can be obtained. An analogous procedure leads to the value for K_2. One finds

$$K_1 = \frac{B - AC}{D - C}, \qquad K_2 = \frac{B - AD}{C - D}. \tag{2.75}$$

An expression for the expanded form of $F(s)$ can now be obtained by substituting this result into Eq. (2.73):

$$F(s) = \frac{B - AC}{D - C} \frac{1}{s + C} + \frac{B - AD}{C - D} \frac{1}{s + D}. \tag{2.76}$$

It is now a simple process to obtain the inverse Laplace transform from this last equation. The corresponding time function can be obtained by merely inspecting the terms and comparing them with transform pairs listed in Table 2.1:

$$f(t) = \frac{B - AC}{D - C} e^{-Ct} + \frac{B - AD}{C - D} e^{-Dt}. \tag{2.77}$$

Transforms with multiple poles are occasionally encountered. For example, consider the transform

$$F(s) = \frac{As + B}{(s + C)^2(s + D)}.$$ (2.78)

In this equation A, B, C, and D are constants. The partial fraction expansion is written as

$$\frac{As + B}{(s + C)^2(s + D)} = \frac{K_1}{(s + C)^2} + \frac{K_2}{s + C} + \frac{K_3}{s + D}.$$ (2.79)

To find K_1, both sides of Eq. (2.79) are multiplied by $(s + C)^2$:

$$\frac{As + B}{s + D} = K_1 + K_2(s + C) + \frac{K_3(s + C)^2}{s + D}.$$ (2.80)

The constant K_1 can now be evaluated by simply substituting $s = -C$.

$$K_1 = \frac{B - AC}{D - C}.$$ (2.81)

In order to determine the constant K_2, both sides of Eq. (2.80) must be differentiated with respect to s, and s is then set equal to $-C$:

$$\left[\frac{d}{ds}\frac{As + B}{s + D}\right]_{s=-C} = K_2 + K_3\left[\frac{d}{ds}\frac{(s + C)^2}{s + D}\right]_{s=-C}.$$ (2.82)

The resulting numerical value for K_2 is given by

$$K_2 = \left[\frac{d}{ds}\frac{As + B}{s + D}\right]_{s=-C} = \frac{AD - B}{(D - C)^2}.$$ (2.83)

The constant K_3 can be obtained by the same procedure used in evaluating the expression given by Eq. (2.72). Its value is

$$K_3 = \frac{B - AD}{(C - D)^2}.$$ (2.84)

2.8 SOLUTION OF DIFFERENTIAL EQUATIONS USING THE LAPLACE TRANSFORM

Now that the physical relationship of a linear system has been described by means of its integrodifferential equation, the analysis of the system's dynamic behavior can be carried out by solving the equations and incorporating the initial conditions into the solution. Two examples are given in this section to illustrate the application of the Laplace transform to solve a linear differential equation. In general, we take the Laplace transform of each term in the differential equation. This step eliminates time and all of the time derivatives from the original equation and results in an algebraic equation in s. The resulting equation is then solved for the transform of the desired time function. The final step involves obtaining the inverse Laplace transform which yields the solution directly.

Example 1   Consider the following linear differential equation:

$$\frac{d^2y}{dt^2} + 5\frac{dy}{dt} + 6y = 6. \tag{2.85}$$

Assume the initial conditions are

$$\dot{y}(0^+) = 2, \qquad y(0^+) = 2.$$

By taking the Laplace transform of both sides of Eq. (2.85), the following equation is obtained (using Eqs. (2.62) and (2.63)):

$$s^2 Y(s) - sy(0^+) - \dot{y}(0^+) + 5s Y(s) - 5y(0^+) + 6 Y(s) = 6/s. \tag{2.86}$$

Substituting the values of the initial conditions, and solving for $Y(s)$, yields the following equation:

$$Y(s) = \frac{2s^2 + 12s + 6}{s(s^2 + 5s + 6)} = \frac{2s^2 + 12s + 6}{s(s + 3)(s + 2)}. \tag{2.87}$$

If Eq. (2.87) is expanded by means of partial fractions as discussed previously, the following expansion is obtained:

$$Y(s) = \frac{1}{s} - \frac{4}{s + 3} + \frac{5}{s + 2}. \tag{2.88}$$

The inverse Laplace transform of Eq. (2.88) is given, using Table 2.1, by

$$y(t) = 1 - 4e^{-3t} + 5e^{-2t}, \qquad t > 0. \tag{2.89}$$

This solution is composed of two portions: the steady-state solution given by 1, and the transient solution given by $-4e^{-3t} + 5e^{-2t}$. As a check of the steady-state solution, we can apply the final-value theorem given by Eq. (2.68):

$$\lim_{t\to\infty} y(t) = \lim_{s\to 0} s Y(s) = \lim_{s\to 0} \frac{2s^2 + 12s + 6}{(s + 3)(s + 2)} = 1. \tag{2.90}$$

Example 2   As a second example, the following differential equation is considered and the inapplicability of the final-value theorem when the function is not analytic in the right half-plane is illustrated:

$$\frac{d^2y}{dt^2} + \frac{dy}{dt} = e^{4t}. \tag{2.91}$$

The initial conditions are assumed to be

$$y(0^+) = 2, \qquad \dot{y}(0^+) = 0. \tag{2.92}$$

Table 2.1 Important Laplace transform pairs

Name of function	Time function, $f(t)$	Laplace transform, $F(s)$	Location of poles in s-plane
1. Unit impulse at $t = 0$	$\delta(t)$	1	None
2. Unit step	$U(t)$	$\dfrac{1}{s}$	One pole at the origin
3. Unit ramp	t	$\dfrac{1}{s^2}$	Double pole at the origin
4. Parabolic	t^2	$\dfrac{2}{s^3}$	Triple pole at the origin
5. nth order ramp	t^n	$\dfrac{n!}{s^{n+1}}$	Pole of multiplicity $(n + 1)$ at the origin
6. Exponential decay	$e^{-\alpha t}$	$\dfrac{1}{s + \alpha}$	One pole on the real axis at $-\alpha$
7. Sine wave	$\sin \omega t$	$\dfrac{\omega}{s^2 + \omega^2}$	Two poles on the imaginary axis at $\pm j\omega$
8. Cosine wave	$\cos \omega t$	$\dfrac{s}{s^2 + \omega^2}$	Two poles on the imaginary axis at $\pm j\omega$
9. Exponentially decaying sine wave	$e^{-\alpha t} \sin \omega t$	$\dfrac{\omega}{(s + \alpha)^2 + \omega^2}$	Two complex poles located at $-\alpha \pm j\omega$

Taking the Laplace transform of both sides of Eq. (2.91), the following equation is obtained:

$$[s^2 Y(s) - sy(0^+) - \dot{y}(0^+)] + sY(s) - y(0^+) = \frac{1}{s - 4}. \qquad (2.93)$$

By substituting the values of the initial conditions, and solving for $Y(s)$, the following equation is obtained:

$$Y(s) = \frac{2s^2 - 6s - 7}{s(s + 1)(s - 4)}. \qquad (2.94)$$

Expansion of Eq. (2.94) by means of partial fractions, as discussed previously, gives

$$Y(s) = \frac{\frac{7}{4}}{s} + \frac{\frac{1}{5}}{s + 1} + \frac{\frac{1}{20}}{(s - 4)}. \qquad (2.95)$$

From Table 2.1, the inverse Laplace transform of Eq. (2.95) is given by

$$y(t) = 1.75 + \tfrac{1}{5}e^{-t} + \tfrac{1}{20}e^{4t}. \qquad (2.96)$$

It is obvious from Eq. (2.96) that the final value of this function is infinite. However, if one were to apply the final-value theorem to Eq. (2.94), the incorrect final value of

1.75 would be obtained. This example, therefore, illustrates very clearly that Eq. (2.68) cannot be applied when the function $F(s)$ is not analytic in the right half-plane.

2.9 THE TRANSFER-FUNCTION CONCEPT

For analysis and design, control systems are usually described by a set of differential equations. A block diagram is a device for displaying the interrelationships of the equations pictorially. Each component is described by its *transfer function*. This is defined as the ratio of the transform of the output of the component to the transform

Fig. 2.9   Block diagram of a simple linear system.

of the input. The component is assumed to be at rest prior to excitation, and all initial values are assumed to be zero when determining the transfer function.

Consider the block diagram of the simple system shown in Fig. 2.9. The only assumption made concerning this system is that the input and output are related by a linear differential equation whose coefficients are constant and can be written in the form

$$A_n \frac{d^n c}{dt^n} + \cdots + A_1 \frac{dc}{dt} + A_0 c = B_m \frac{d^m r}{dt^m} + \cdots + B_1 \frac{dr}{dt} + B_0 r. \quad (2.97)$$

The Laplace transform of Eq. (2.97), assuming zero initial conditions, can be written as

$$(A_n s^n + \cdots + A_1 s + A_0)C(s) = (B_m s^m + \cdots + B_1 s + B_0)R(s). \quad (2.98)$$

The ratio $C(s)/R(s)$ is called the transfer function of the element and completely characterizes its performance. Designating the transfer function of the element as $G(s)$, we obtain

$$G(s) = \frac{C(s)}{R(s)} = \frac{B_m s^m + \cdots + B_1 s + B_0}{A_n s^n + \cdots + A_1 s + A_0}. \quad (2.99)$$

Therefore, assuming that the initial conditions are zero, the Laplace transform of the output is

$$C(s) = G(s)R(s). \quad (2.100)$$

In general, the function $G(s)$ is the ratio of two polynomials in s:

$$G(s) = P(s)/Q(s). \quad (2.101)$$

The transfer function $G(s)$ is a property of the system elements only, and is not dependent on the excitation and initial conditions. In addition, transfer functions can be used to represent closed-loop as well as open-loop systems.

2.10 TRANSFER FUNCTIONS OF COMMON NETWORKS

The control engineer depends heavily on simple passive networks to modify the transfer function of the feedback control system in order to promote stability, improve closed-loop performance, and minimize the effects of noise and other undesirable signals.

Figure 2.10(a) represents an electrical network which is used for integration or to provide a phase lag. This circuit obtains its integrating property from the fact that

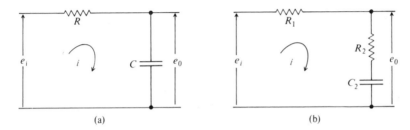

(a)   (b)

Fig. 2.10   (a) An integrating or phase-lag network. (b) An integrating network with fixed high-frequency attenuation.

the voltage across the capacitor is proportional to the integral of the current through it. To determine the behavior of this network we must determine the transfer function relating input and output signals. The integrodifferential equations describing the behavior of this network are given by

$$e_i = Ri + \frac{1}{C} \int i \, dt, \tag{2.102}$$

$$e_o = \frac{1}{C} \int i \, dt. \tag{2.103}$$

These differential equations can be solved for the relation between e_o and e_i by means of the Laplace transform; $e_o(0^+)$ is assumed equal to zero. Thus one finds

$$E_i(s) = RI(s) + I(s)/Cs, \tag{2.104}$$

$$E_o(s) = I(s)/Cs. \tag{2.105}$$

Table 2.2 Transfer functions of common networks

	Network	Transfer function
1		$\dfrac{RCs}{RCs+1}$
2		$\dfrac{1}{RCs+1}$
3		$\dfrac{R_2}{R_1+R_2}\dfrac{1+R_1C_1s}{1+\dfrac{R_2}{R_1+R_2}R_1C_1s}$
4		$\dfrac{R_2C_2s+1}{(R_1+R_2)C_2s+1}$
5		$\dfrac{(1+R_1C_1s)(1+R_2C_2s)}{R_1R_2C_1C_2s^2+(R_1C_1+R_2C_2+R_1C_2)s+1}$

6		$$\dfrac{R_2(R_1+R_3)C_1C_2s^2 + (R_1C_1+R_2C_2+R_3C_1)s+1}{(R_1R_2+R_2R_3+R_1R_3)C_1C_2s^2 + (R_1C_1+R_2C_2+R_1C_2+R_3C_1)s+1}$$
7		$$\dfrac{\dfrac{R_1^2R_3}{R_2+R_3}C_1C_2s^2 + \left[R_2C_1+\dfrac{R_2^2R_3C_2}{R_1(R_2+R_3)}\right]s+\dfrac{R_2+R_3}{R_1}}{R_2R_3C_1C_2s^2 + \left(R_3C_1+R_2C_1+\dfrac{R_2R_3}{R_1}C_2\right)s+\dfrac{R_2+R_3}{R_1}+1}$$
8		$$\dfrac{s(L_1C_1s+R_1C_1)+1}{L_1C_1s^2+(R_1+R_2)C_1s+1}$$
9		$$\dfrac{(L_1/R_1)s+1}{\dfrac{L_1}{R_1}R_2C_1s^2 + \left(\dfrac{L_1}{R_1}+R_2C_1\right)s+\dfrac{R_1+R_2}{R_1}}$$

Eliminating $I(s)$ results in the transfer function

$$\frac{E_o(s)}{E_i(s)} = \frac{1}{RCs + 1}. \tag{2.106}$$

Notice from Eq. (2.106) that integration is possibly only for those frequencies where $|RCs| \gg 1$. It is important to note that this same transfer function can be obtained quite simply by determining the impedance of each circuit element and using the voltage divider rule. Thus

$$\frac{E_o(s)}{E_i(s)} = \frac{1/(Cs)}{1/(Cs) + R} = \frac{1}{RCs + 1}. \tag{2.107}$$

The impedance of the capacitance approaches a short circuit at very high frequencies, and ultimately it will have no output. This circuit is basically a low-pass filter whose high-frequency response is attenuated to a very large degree. Very often it is undesirable to have such a large attenuation at high frequencies. The circuit of Fig. 2.10(b) limits this attenuation to a value of $R_2/(R_1 + R_2)$. The transfer function of the network is obtained directly by using the voltage divider rule. Thus,

$$\frac{E_o(s)}{E_i(s)} = \frac{R_2 + (1/C_2 s)}{R_1 + R_2 + (1/C_2 s)} = \frac{R_2 C_2 s + 1}{(R_1 + R_2)C_2 s + 1}. \tag{2.108}$$

A tabulation of the foregoing results, together with other useful transfer functions which the control engineer will most likely encounter, is shown in Table 2.2. Network 1 is known as a differentiating or phase-lead network. Notice that it is basically a high-pass filter possessing very large attenuation at low frequencies. This attenuation can be limited to a finite value of $R_2/(R_1 + R_2)$ by network 3. A lag–lead network, which provides a phase lag at low frequencies and a phase lead at high frequencies, is shown as network 5. Networks 6 and 7 are slight modifications of network 5. Network 8 is used for eliminating unwanted frequency bands. Network 9 is used for passing signals in a narrow band of frequencies.

2.11 TRANSFER FUNCTIONS OF SYSTEMS

In order to determine the transfer function of complex systems, it is necessary to eliminate intermediate variables of the elements that comprise the system. This will enable the designer to obtain a relation between the input and output of the overall system. This section will consider the transfer functions of cascaded elements, single-loop feedback systems, and multiple-loop feedback systems.

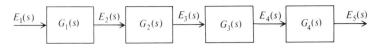

Fig. 2.11   A cascaded system.

A cascaded system is shown in Fig. 2.11. The transfer function of the overall system can be obtained by solving the following set of equations:

$$E_2(s) = G_1(s)E_1(s), \tag{2.109}$$

$$E_3(s) = G_2(s)E_2(s), \tag{2.110}$$

$$E_4(s) = G_3(s)E_3(s), \tag{2.111}$$

$$E_5(s) = G_4(s)E_4(s). \tag{2.112}$$

By inspection, it can be seen that the transfer function of the cascaded system is the product of the transfer functions of the individual elements:

$$E_5(s) = G_1(s)G_2(s)G_3(s)G_4(s)E_1(s). \tag{2.113}$$

Consider the elementary linear feedback system in Fig. 2.12. $G(s)$ and $H(s)$ represent the transfer functions of the direct-transmission and feedback portions of the loop, respectively. They may be individually composed of cascaded elements and minor feedback loops.

The following three equations are required in order to compute the overall system transfer function:

$$B(s) = H(s)C(s), \tag{2.114}$$

$$E(s) = R(s) - B(s), \tag{2.115}$$

$$C(s) = G(s)E(s). \tag{2.116}$$

Solution of these three equations results in the following transfer function relating $R(s)$ and $C(s)$:

$$\frac{C(s)}{R(s)} = \frac{G(s)}{1 + G(s)H(s)}. \tag{2.117}$$

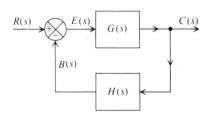

Fig. 2.12   General block diagram of a single-loop feedback system.

For cases where $|G(s)H(s)| \gg 1$, the closed-loop transfer function can be approximated by

$$\frac{C(s)}{R(s)} \simeq \frac{1}{H(s)}. \tag{2.118}$$

This implies that the closed-loop transfer function is independent of the direct transmission transfer function $G(s)$, and only depends on the feedback transfer function $H(s)$. Use is made of this characteristic in feedback amplifier design in order that the overall amplifier gain may be insensitive to tube, transistor, and/or integrated circuit parameter variations. Note also that the approximate transfer function is the inverse of the feedback transfer function. This property is used for producing behavior which may be difficult to achieve directly.

The characteristic equation for the system can be obtained by setting the denominator of the system transfer function equal to zero:

$$1 + G(s)H(s) = 0. \tag{2.119}$$

This equation determines system stability, and it will receive much attention in later chapters.

Another useful relationship is the transfer function $E(s)/R(s)$ when $H(s) = 1$. Then $E(s)$ represents the error $R(s) - C(s)$. One finds

$$\left.\frac{E(s)}{R(s)}\right|_{H(s)=1} = \frac{1}{1 + G(s)}. \tag{2.120}$$

When $|G(s)| \gg 1$, this can be approximated by

$$\left.\frac{E(s)}{R(s)}\right|_{H(s)=1} \simeq \frac{1}{G(s)}. \tag{2.121}$$

Thus the error is small when the magnitude of the open-loop transfer function is large.

Practical feedback systems usually contain multiple feedback loops and several inputs. All multiple-loop systems can be reduced to the basic form shown in Fig. 2.12 by means of step-by-step feedback loop reduction or by means of *signal-flow diagrams* which are considered in later sections of this chapter. Multiple inputs, which are present in all control systems because unwanted inputs (such as noise and drift) are present, can occur anywhere in the feedback system. Successive block diagram reduction techniques permit the designer to determine their effect on overall performance. Table 2.3 illustrates several transformations that can be used to simplify the reduction of a multiple feedback system. The technique of multiple feedback loop reduction can best be understood by means of an example.

Figure 2.13 illustrates a multiple-loop feedback system containing three feedback loops. The original feedback system is illustrated in Fig. 2.13(a). Figure 2.13(b) through (e) shows successive steps in reducing this system using the transformations

Table 2.3 Block diagram transformations

Transformation	Original block diagram	Equivalent block diagram
1. Moving a pickoff point behind a block		
2. Moving a pickoff point ahead of a block		
3. Moving a summing point behind a block		
4. Moving a summing point ahead of a block		
5. Eliminating a feedback loop		

illustrated in Table 2.3. Figure 2.13(f) shows the closed-loop transfer function of the system.

Block diagram reduction techniques get very tedious and time consuming as the number of feedback paths increases, as is illustrated in Fig. 2.13. In order to solve complex problems, it is much simpler to make use of the theorems and properties of signal-flow graphs, which permit a solution almost by inspection.

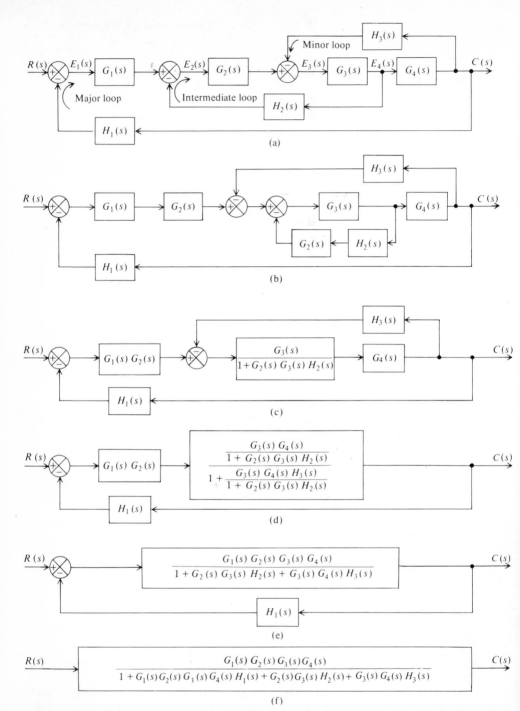

Fig. 2.13 Reducing a multiple-loop system containing complex paths. (a) The original system. (b) Rearrangement of the summing points of the intermediate and minor loops. (c) Reduction of the equivalent intermediate loop. (d) Reduction of the equivalent minor loop. (e) The equivalent feedback system. (f) The system transfer function.

2.12 THE SIGNAL-FLOW DIAGRAM AND MASON'S THEOREMS

Signal-flow graphs and Mason's theorems [3, 4] enable the control engineer to determine the response of a complicated linear, multiloop system to any input much more rapidly than do block diagram reduction techniques.

A signal-flow graph is a topological representation of a set of linear equations having the form

$$y_i = \sum_{j=1}^{n} a_{ij} y_j, \qquad i = 1, 2, \ldots, n. \tag{2.122}$$

This equation expresses each of the n variables in terms of the others and themselves. A signal-flow graph represents a set of equations of this type by means of branches and nodes. A node is assigned to each variable of interest in the system. For

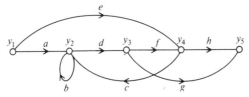

Fig. 2.14 Signal-flow diagram.

example, node i represents variable y_i. Branches are used to relate the different variables. For example, branch ij relates variable y_i to y_j, where the branch originates at node i and terminates at node j. Consider the following set of linear equations

$$y_2 = ay_1 + by_2 \qquad + cy_4 \qquad , \tag{2.123}$$
$$y_3 = \qquad dy_2 \tag{2.124}$$
$$y_4 = ey_1 \qquad + fy_3 \tag{2.125}$$
$$y_5 = \qquad gy_3 + hy_4. \tag{2.126}$$

The signal-flow graph which represents this set of equations is shown in Fig. 2.14. Here y_1 can be interpreted as the input to the system and y_5 as its output. Usually we would be interested in obtaining the ratio of y_5/y_1.

Before proceeding further, several terms used in signal-flow diagrams must be defined.

a) A *source* is a node having only outgoing branches, such as y_1 in the preceding illustration.

b) A *sink* is a node having only incoming branches, such as y_5.

c) A *path* is a group of connected branches having the same sense of direction. In Fig. 2.14, *eh*, *adfh*, and *b* are paths.

d) *Forward paths* are paths which originate from a source and terminate at a sink and along which no node is encountered more than once, as *eh*, *ecdg*, *adg*, and *adfh*.

e) *Path gain* is the product of the coefficients associated with the branches along the path.

f) *Feedback loop* is a path originating from a node and terminating at the same node. In addition, a node cannot be encountered more than once. In the preceding example b and dfc are feedback loops.

g) *Loop gain* is the product of the coefficients associated with the branches forming a feedback loop.

2.13 REDUCTION OF THE SIGNAL-FLOW DIAGRAM

Several preliminary simplifications can be made to the complex signal-flow graphs of a system by means of the following signal-flow graph algebra.

a) *Addition*

   1. The signal-flow diagram in Fig. 2.15(a) represents the linear equation

$$y_3 = ay_1 + by_2. \tag{2.127}$$

   2. The signal-flow diagram in Fig. 2.15(b) represents the linear equation

$$y_2 = (a + b)y_1. \tag{2.128}$$

b) *Multiplication.* The signal-flow diagram in Fig. 2.15(c) represents the linear equation

$$y_4 = abcy_1. \tag{2.129}$$

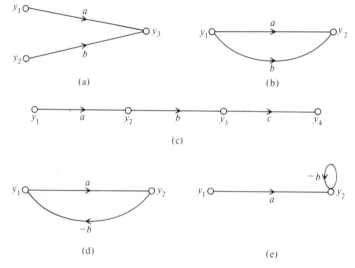

Fig. 2.15   Signal-flow graph algebra.

c) *Feedback loops*
    1. The signal-flow diagram in Fig. 2.15(d) represents the linear equation

$$y_2 = \frac{a}{1 + ab}\, y_1. \tag{2.130}$$

    2. The signal-flow diagram in Fig. 2.15(e) represents the linear equation

$$y_2 = \frac{a}{1 + b}\, y_1. \tag{2.131}$$

It is possible to apply the preceding signal-flow diagram algebra to a complicated graph and reduce it to one containing only a source and a sink. This process requires repeated applications until the final desired form is obtainable. An interesting property of network and system topology, based on Mason's theorems [3, 4] permits the writing of the desired answer almost by inspection. The general expression for signal-flow graph gain is given by

$$G = \frac{\sum_K G_K \Delta_K}{\Delta}, \tag{2.132}$$

where $\Delta = 1 - \sum L_1 + \sum L_2 - \sum L_3 + \cdots + (-1)^m \sum L_m$

$\quad L_1 =$ gain of each closed loop in the graph
$\quad L_2 =$ product of the loop gains of any two nontouching closed loops (loops are considered nontouching if they have no node in common)

$\qquad \cdot \quad \cdot \quad \cdot \quad \cdot \quad \cdot \quad \cdot \quad \cdot \quad \cdot \quad \cdot \quad \cdot \quad \cdot \quad \cdot \quad \cdot \quad \cdot \quad \cdot$

$\quad L_m =$ product of the loop gains of any m nontouching loops
$\quad G_K =$ gain of the Kth forward path
$\quad \Delta_K =$ the value of Δ for that part of the graph not touching the Kth forward path (value of Δ remaining when the path producing G_K is removed).

Δ is known as the determinant of the graph and Δ_K is the cofactor of the forward path K. Basically, Δ consists of the sum of the products of loop gains taken none at a time (1), one at a time (with a minus sign), two at a time (with a plus sign), etc.; Δ_K contains the products of the nontouching loops. The proof of this general gain expression is contained in Reference 4. A few examples follow in order to show how this expression may be used.

Example 1 For Fig. 2.16(a),

$$\Delta = 1 - bd,$$
$$G_1 = abc,$$
$$\Delta_1 = 1.$$

Therefore,

$$G = \frac{abc}{1 - bd}.$$

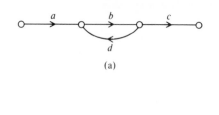

(a)

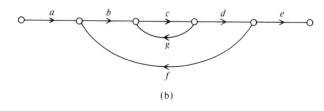

(b)

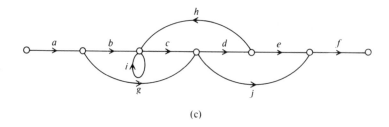

(c)

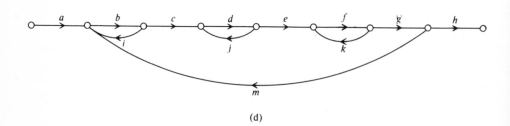

(d)

Fig. 2.16   Signal-flow diagram examples: (a) Example 1, (b) Example 2, (c) Example 3, (d) Example 4.

Example 2 For Fig. 2.16(b),

$$\Delta = 1 - cg - bcdf,$$
$$G_1 = abcde,$$

Therefore,

$$\Delta_1 = 1,$$
$$G = \frac{abcde}{1 - cg - bcdf}.$$

Example 3 For Fig. 2.16c,

$$\Delta = 1 - (i + cdh),$$
$$G_1 = abcdef,$$
$$G_2 = agdef,$$
$$G_3 = agjf,$$
$$G_4 = abcjf,$$
$$\Delta_1 = 1, \qquad \Delta_3 = 1 - i,$$
$$\Delta_2 = 1 - i, \qquad \Delta_4 = 1.$$

Therefore,

$$G = \frac{abcdef + agdef(1 - i) + agjf(1 - i) + abcjf}{1 - (i + cdh)}.$$

Example 4 For Fig. 2.16d,

$$\Delta = 1 - (bi + dj + fk + bcdefgm) + (bidj + bifk + djfk) - bidjfk,$$
$$G_1 = abcdefgh,$$
$$\Delta_1 = 1.$$

Therefore,

$$G = \frac{abcdefgh}{1 - (bi + dj + fk + bcdefgm) + (bidj + bifk + djfk) - bidjfk}.$$

2.14 APPLICATION OF MASON'S THEOREMS AND THE SIGNAL-FLOW DIAGRAM TO MULTIPLE FEEDBACK SYSTEMS

It is important at this point to differentiate between signal-flow diagrams and block diagrams. Basically, the signal-flow diagram represents a detailed picture of a system's topological structure, whereas the block diagram focuses on the transfer functions that comprise the various elements of the system. The signal-flow diagram is useful in analyzing multiple-loop feedback systems and in determining the effect of a particular element or parameter in an overall feedback system, whereas the block diagram is useful in the design and analysis of sections of a feedback system.

Essentially both present the same information in different ways, and Mason's theorems can be applied to both. However, Mason's theorems are conventionally used with the signal-flow diagram since the topology is more clearly depicted by the signal-flow diagram.

The signal-flow diagram for Fig. 2.13(a) is shown in Fig. 2.17. By inspection, the overall system transfer function is

$$\Delta = 1 + G_1(s)G_2(s)G_3(s)G_4(s)H_1(s) + G_2(s)G_3(s)H_2(s) + G_3(s)G_4(s)H_3(s),$$
$$G_A = G_1(s)G_2(s)G_3(s)G_4(s),$$
$$\Delta_A = 1.$$

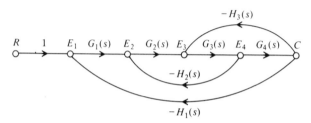

Figure 2.17

Therefore,

$$\frac{C(s)}{R(s)} = G = \frac{G_1(s)G_2(s)G_3(s)G_4(s)}{1 + G_1(s)G_2(s)G_3(s)G_4(s)H_1(s) + G_2(s)G_3(s)H_2(s) + G_3(s)G_4(s)H_3(s)}.$$

This result agrees with the transfer function shown in Fig. 2.13(f).

The foregoing examples illustrate the simplifications made possible by use of Mason's theorems in conjunction with the signal-flow diagrams. It will be extended to the state-space concept in this chapter and, in the remaining chapters, the signal-flow diagram approach will be used to simplify the solutions of problems. In addition, further properties and applications of this powerful tool will be demonstrated.

2.15 REVIEW OF MATRIX ALGEBRA

The classical methods of describing a linear system by means of transfer functions, block diagrams, and signal-flow diagrams have thus far been presented in this chapter. An inherent characteristic of this type of representation is that the system dynamics are described by definable input–output relationships. Disadvantages of these techniques, however, are that the initial conditions have been neglected and intermediate variables lost. The methods cannot be used for nonlinear, or time-varying systems. Furthermore, working in the frequency domain is not convenient for applying modern optimal control theory, discussed in Chapter 9, which is based on the time domain. The use of digital computers also serves to focus on time-domain methods.

Therefore, a different set of tools for describing the system in the time domain is needed and is provided by state-space methods. As a necessary preliminary, matrix algebra is reviewed in this section [15].

A matrix **A** is a rectangular array of elements defined by

$$\mathbf{A} = \begin{bmatrix} a_{11} & a_{12} & \cdots & a_{1n} \\ a_{21} & a_{22} & \cdots & a_{2n} \\ \cdot & \cdot & & \cdot \\ \cdot & \cdot & & \cdot \\ \cdot & \cdot & & \cdot \\ a_{m1} & a_{m2} & \cdots & a_{mn} \end{bmatrix}. \tag{2.133}$$

A matrix having m rows and n columns is referred to as an $m \times n$ matrix. A square matrix is one for which $m = n$. A column matrix, or vector, is one for which $n = 1$, and is represented in the following manner:

$$\mathbf{a} = \begin{bmatrix} a_1 \\ a_2 \\ \cdot \\ \cdot \\ \cdot \\ a_m \end{bmatrix}. \tag{2.134}$$

Capital letters are used to denote matrices and lower-case letters to denote vectors. Matrix **A** equals matrix **B** if element a_{ij} equals element b_{ij} for each i and each j, where the subscripts refer to elements in row i and column j of the respective matrices.

In order to present the basic types and properties of matrices, matrix **A** will be considered in the following discussion where the representative element is a_{ij}.

A. Identity Matrix The identity matrix, or unit matrix, is a square matrix whose principal diagonal elements are unity, all other elements being zero. This matrix, which is denoted by **I**, is given by

$$\mathbf{I} = \begin{bmatrix} 1 & 0 & \cdots & 0 \\ 0 & 1 & \cdots & 0 \\ \cdot & \cdot & & \cdot \\ \cdot & \cdot & & \cdot \\ \cdot & \cdot & & \cdot \\ 0 & 0 & \cdots & 1 \end{bmatrix}. \tag{2.135}$$

An interesting property of the identity matrix is that multiplication of any matrix **A** by an identity matrix **I** results in the original matrix **A**:

$$\mathbf{AI} = \mathbf{A}, \tag{2.136}$$

where matrix multiplication is defined later on.

B. Diagonal Matrix A diagonal matrix, which is denoted by diag x_i, is given by

$$\text{diag } \mathbf{x}_i = \begin{bmatrix} x_1 & 0 & \cdots & 0 \\ 0 & x_2 & \cdots & 0 \\ \cdot & \cdot & & \cdot \\ \cdot & \cdot & & \cdot \\ \cdot & \cdot & & \cdot \\ 0 & 0 & \cdots & x_n \end{bmatrix}. \tag{2.137}$$

C. Transpose of a Matrix A matrix is transposed by interchanging its rows and columns. To form the transpose of a matrix, element a_{ij}, which is the element of row i and column j, is interchanged with element a_{ji}, which is the element of row j and column i, for all i and j. For example, the transpose of $\mathbf{A}$, where

$$\mathbf{A} = \begin{bmatrix} a_{11} & a_{12} & a_{13} \\ a_{21} & a_{22} & a_{23} \\ a_{31} & a_{32} & a_{33} \end{bmatrix} \tag{2.138}$$

is written as $\mathbf{A}^T$ where

$$\mathbf{A}^T = \begin{bmatrix} a_{11} & a_{21} & a_{31} \\ a_{12} & a_{22} & a_{32} \\ a_{13} & a_{23} & a_{33} \end{bmatrix}. \tag{2.139}$$

Note that the transpose of $\mathbf{A}^T$ is $(\mathbf{A}^T)^T = \mathbf{A}$.

D. Symmetric Matrix A symmetric matrix is defined by the condition

$$a_{ij} = a_{ji}. \tag{2.140}$$

This matrix, which is denoted by $\mathbf{A}_s$, is represented by

$$\mathbf{A}_s = \begin{bmatrix} a_{11} & a_{12} & \cdots & a_{1n} \\ a_{12} & a_{22} & \cdots & a_{2n} \\ \cdot & \cdot & & \cdot \\ \cdot & \cdot & & \cdot \\ \cdot & \cdot & & \cdot \\ a_{1n} & a_{2n} & \cdots & a_{nn} \end{bmatrix}. \tag{2.141}$$

Notice that it is symmetrical about the principal diagonal, and that a symmetric matrix and its transpose are identical.

E. Skew-Symmetric Matrix A skew-symmetric matrix is defined by the condition

$$a_{ij} = -a_{ji}, \qquad a_{ii} = 0. \tag{2.142}$$

This matrix, which is denoted by $\mathbf{A}_a$, is represented by

$$\mathbf{A}_a = \begin{bmatrix} 0 & a_{12} & \cdots & a_{1n} \\ -a_{12} & 0 & \cdots & a_{2n} \\ \cdot & \cdot & & \cdot \\ \cdot & \cdot & & \cdot \\ \cdot & \cdot & & \cdot \\ -a_{1n} & -a_{2n} & \cdots & 0 \end{bmatrix}. \tag{2.143}$$

Note that a skew-symmetric matrix is equal to the negative of its transpose.

F. Zero Matrix The zero matrix or null matrix, denoted by $\mathbf{0}$, is defined as the matrix whose elements are all zero. It has the property that

$$\mathbf{A} + \mathbf{0} = \mathbf{A}, \tag{2.144}$$

where addition of matrices is defined later on.

G. Adjoint Matrix The adjoint of a square matrix is formed by replacing each element of the matrix by its corresponding cofactor, and then transposing the result. For example, if a matrix $\mathbf{A}$ is given by

$$\mathbf{A} = \begin{bmatrix} a_{11} & a_{12} & a_{13} \\ a_{21} & a_{22} & a_{23} \\ a_{31} & a_{32} & a_{33} \end{bmatrix}, \tag{2.145}$$

then the cofactors of the elements a_{11}, a_{21} are given respectively by the determinants

$$A_{11} = \begin{vmatrix} a_{22} & a_{23} \\ a_{32} & a_{33} \end{vmatrix} = a_{22}a_{33} - a_{23}a_{32}, \tag{2.146}$$

$$A_{21} = -\begin{vmatrix} a_{12} & a_{13} \\ a_{32} & a_{33} \end{vmatrix} = a_{32}a_{13} - a_{12}a_{33}. \tag{2.147}$$

The element a_{ij} has for its cofactor A_{ij} with the proper algebraic sign, $(-1)^{i+j}$, prefixed. The new matrix formed by replacing the original elements in the matrix with their corresponding cofactors is given by

$$\begin{bmatrix} A_{11} & A_{12} & A_{13} \\ A_{21} & A_{22} & A_{23} \\ A_{31} & A_{32} & A_{33} \end{bmatrix}. \tag{2.148}$$

The transpose of this matrix results in the expression for the adjoint (adj) of matrix $\mathbf{A}$:

$$\text{adj } \mathbf{A} = \begin{bmatrix} A_{11} & A_{21} & A_{31} \\ A_{12} & A_{22} & A_{32} \\ A_{13} & A_{23} & A_{33} \end{bmatrix}. \tag{2.149}$$

In order to present the basic operations of matrix analysis, three matrices **A**, **B**, and **C** will be considered in the following discussion where the representative elements are denoted by a_{ij}, b_{ij}, c_{ij}.

A. Addition or Subtraction The sum (or difference) of two matrices **A** and **B** with the same numbers of rows and columns is obtained by adding (or subtracting) corresponding elements. The result is a new matrix **C**, where

$$\mathbf{C} = \mathbf{A} + \mathbf{B}, \tag{2.150}$$

and

$$c_{ij} = a_{ij} + b_{ij}. \tag{2.151}$$

For example, if matrices **A** and **B** are given by

$$\mathbf{A} = \begin{bmatrix} a_{11} & a_{12} & \cdots & a_{1n} \\ a_{21} & a_{22} & \cdots & a_{2n} \\ \cdot & \cdot & & \cdot \\ \cdot & \cdot & & \cdot \\ \cdot & \cdot & & \cdot \\ a_{n1} & a_{n2} & \cdots & a_{nn} \end{bmatrix}, \quad \mathbf{B} = \begin{bmatrix} b_{11} & b_{12} & \cdots & b_{1n} \\ b_{21} & b_{22} & \cdots & b_{2n} \\ \cdot & \cdot & & \cdot \\ \cdot & \cdot & & \cdot \\ \cdot & \cdot & & \cdot \\ b_{n1} & b_{n2} & \cdots & b_{nn} \end{bmatrix}, \tag{2.152}$$

then the sum of matrix **A** and matrix **B** is given by

$$\mathbf{C} = \mathbf{A} + \mathbf{B} = \begin{bmatrix} (a_{11} + b_{11}) & (a_{12} + b_{12}) & \cdots & (a_{1n} + b_{1n}) \\ (a_{21} + b_{21}) & (a_{22} + b_{22}) & \cdots & (a_{2n} + b_{2n}) \\ \cdot & \cdot & & \cdot \\ \cdot & \cdot & & \cdot \\ \cdot & \cdot & & \cdot \\ (a_{n1} + b_{n1}) & (a_{n2} + b_{n2}) & \cdots & (a_{nn} + b_{nn}) \end{bmatrix}. \tag{2.153}$$

B. Multiplication of a Matrix by a Scalar Multiplication of a matrix **A** by a scalar d is equivalent to multiplying each element of the matrix by d. The result is a new matrix **C**, where

$$\mathbf{C} = d\mathbf{A} \tag{2.154}$$

and

$$c_{ij} = da_{ij}. \tag{2.155}$$

For example, if matrix **A** is given by

$$\mathbf{A} = \begin{bmatrix} a_{11} & a_{12} & \cdots & a_{1n} \\ a_{21} & a_{22} & \cdots & a_{2n} \\ \cdot & \cdot & & \cdot \\ \cdot & \cdot & & \cdot \\ \cdot & \cdot & & \cdot \\ a_{n1} & a_{n2} & \cdots & a_{nn} \end{bmatrix} \tag{2.156}$$

then the multiplication of matrix $\mathbf{A}$ by scalar d is given by

$$\mathbf{C} = d\mathbf{A} = \begin{bmatrix} da_{11} & da_{12} & \cdots & da_{1n} \\ da_{21} & da_{22} & \cdots & da_{2n} \\ \cdot & \cdot & & \cdot \\ \cdot & \cdot & & \cdot \\ \cdot & \cdot & & \cdot \\ da_{n1} & da_{n2} & \cdots & da_{nn} \end{bmatrix}, \tag{2.157}$$

C. Multiplication of Two Matrices Postmultiplication* of matrix $\mathbf{A}$ by matrix $\mathbf{B}$ results in a new matrix $\mathbf{C}$, where

$$\mathbf{C} = \mathbf{AB} \tag{2.158}$$

and

$$c_{ij} = \sum_{k=1}^{n} a_{ik}b_{kj}. \tag{2.159}$$

These equations state that the result of postmultiplication of matrix $\mathbf{A}$ by matrix $\mathbf{B}$ is a matrix $\mathbf{C}$ whose element located in row i and column j is obtained by multiplying each element in row i of matrix $\mathbf{A}$ by the corresponding element in column j of matrix $\mathbf{B}$ and then adding the results. It is important to note that the number of columns in matrix $\mathbf{A}$ must equal the number of rows in matrix $\mathbf{B}$ so that matrices $\mathbf{A}$ and $\mathbf{B}$ may be multiplied together. For example, if matrices $\mathbf{A}$ and $\mathbf{B}$ are given by

$$\mathbf{A} = \begin{bmatrix} a_{11} & a_{12} & a_{13} \\ a_{21} & a_{22} & a_{23} \\ a_{31} & a_{32} & a_{33} \end{bmatrix}, \quad \mathbf{B} = \begin{bmatrix} b_{11} & b_{12} \\ b_{21} & b_{22} \\ b_{31} & b_{32} \end{bmatrix} \tag{2.160}$$

then the product of matrices $\mathbf{A}$ and $\mathbf{B}$ is given by

$$\mathbf{C} = \mathbf{AB} = \begin{bmatrix} (a_{11}b_{11} + a_{12}b_{21} + a_{13}b_{31}) & (a_{11}b_{12} + a_{12}b_{22} + a_{13}b_{32}) \\ (a_{21}b_{11} + a_{22}b_{21} + a_{23}b_{31}) & (a_{21}b_{12} + a_{22}b_{22} + a_{23}b_{32}) \\ (a_{31}b_{11} + a_{32}b_{21} + a_{33}b_{31}) & (a_{31}b_{12} + a_{32}b_{22} + a_{33}b_{32}) \end{bmatrix}. \tag{2.161}$$

Matrix multiplication is associative and distributive with respect to addition, but in general not commutative:

$$\mathbf{A(BC)} = \mathbf{(AB)C} \qquad \text{(associative)}, \tag{2.162}$$
$$\mathbf{A(B + C)} = \mathbf{AB} + \mathbf{AC} \qquad \text{(distributive)}, \tag{2.163}$$
$$\mathbf{AB} \neq \mathbf{BA} \qquad \text{(commutative)}. \tag{2.164}$$

D. Inverse of a Square Matrix The inverse of a square matrix $\mathbf{B}$ is denoted by $\mathbf{B}^{-1}$ and has the property

$$\mathbf{BB}^{-1} = \mathbf{B}^{-1}\mathbf{B} = \mathbf{I}. \tag{2.165}$$

---

* The terms post- or premultiplication are used to indicate whether the matrix is multiplied from the right or the left, respectively.

It can be shown that the product of an adjoint matrix with the matrix itself has the property that it is equal to the product of an identity matrix and the determinant of the matrix:

$$\mathbf{B} \text{ adj } \mathbf{B} = \mathbf{I} \, |\mathbf{B}|. \tag{2.166}$$

Using these two relationships, we can derive the expression for the inverse matrix $\mathbf{B}^{-1}$. By solving for $\mathbf{I}$ from Eq. (2.166), the following relationship is obtained:

$$\mathbf{I} = \frac{\mathbf{B} \text{ adj } \mathbf{B}}{|\mathbf{B}|}, \qquad |\mathbf{B}| \neq 0. \tag{2.167}$$

Using Eqs. (2.165) and (2.167), we obtain

$$\mathbf{B}^{-1}\mathbf{B} = \mathbf{B}\mathbf{B}^{-1} = \frac{\mathbf{B} \text{ adj } \mathbf{B}}{|\mathbf{B}|}, \qquad |\mathbf{B}| \neq 0. \tag{2.168}$$

Solving for the inverse matrix $\mathbf{B}^{-1}$ in terms of the adjoint matrix and the determinant of the matrix, we obtain the following expression:

$$\mathbf{B}^{-1} = \frac{\text{adj } \mathbf{B}}{|\mathbf{B}|}, \qquad |\mathbf{B}| \neq 0, \tag{2.169}$$

or

$$\mathbf{B}^{-1} = \frac{1}{|\mathbf{B}|} \begin{bmatrix} B_{11} & B_{21} & \cdots & B_{n1} \\ B_{12} & B_{22} & \cdots & B_{n2} \\ \cdot & \cdot & & \cdot \\ \cdot & \cdot & & \cdot \\ \cdot & \cdot & & \cdot \\ B_{1n} & B_{2n} & \cdots & B_{nn} \end{bmatrix}, \qquad |\mathbf{B}| \neq 0 \tag{2.170}$$

Note that $\mathbf{B}^{-1}$ does not exist if $|\mathbf{B}| = 0$, although Eq. (2.166) still holds.

E. Differentiation of a Matrix The usual concepts regarding the differentiation of scalar variables carry over to the differentiation of a matrix. Let $\mathbf{A}$ be an $m \times n$ matrix whose elements $a_{ij}(t)$ are differentiable functions of the scalar variable t. The derivative of $\mathbf{A}$ with respect to the variable t is given by

$$\frac{d}{dt}[\mathbf{A}] = \dot{\mathbf{A}} = \begin{bmatrix} \dfrac{da_{11}(t)}{dt} & \dfrac{da_{12}(t)}{dt} & \cdots & \dfrac{da_{1n}(t)}{dt} \\[2mm] \dfrac{da_{21}(t)}{dt} & \dfrac{da_{22}(t)}{dt\cdot} & \cdots & \dfrac{da_{2n}(t)}{dt} \\[1mm] \cdot & \cdot & & \cdot \\ \cdot & \cdot & & \cdot \\ \cdot & \cdot & & \cdot \\[1mm] \dfrac{da_{m1}(t)}{dt} & \dfrac{da_{m2}(t)}{dt} & \cdots & \dfrac{da_{mn}(t)}{dt} \end{bmatrix}. \tag{2.171}$$

In addition, the derivative of the sum of two matrices is the sum of the derivatives of the matrices:

$$\frac{d}{dt}[\mathbf{A} + \mathbf{B}] = \dot{\mathbf{A}} + \dot{\mathbf{B}}. \tag{2.172}$$

F. Integration of a Matrix The usual concepts regarding the integration of scalar variables also carry over to the integration of a matrix. Again, let $\mathbf{A}$ be an $m \times n$ matrix whose elements $a_{ij}(t)$ are integrable functions of the scalar variable t. The integration of $\mathbf{A}$ with respect to the variable t is given by

$$\int \mathbf{A}\, dt = \begin{bmatrix} \int a_{11}(t)\, dt & \int a_{12}(t)\, dt & \cdots & \int a_{1n}(t)\, dt \\ \int a_{21}(t)\, dt & \int a_{22}(t)\, dt & \cdots & \int a_{2n}(t)\, dt \\ \cdot & & & \cdot \\ \cdot & & & \cdot \\ \cdot & & & \cdot \\ \int a_{m1}(t)\, dt & \int a_{m2}(t)\, dt & \cdots & \int a_{mn}(t)\, dt \end{bmatrix}. \tag{2.173}$$

2.16 STATE-SPACE CONCEPTS

In the analysis of a system via the state-space approach, the system is characterized by a set of first-order differential or difference [2] equations that describe its "state" variables. System analysis and design can be accomplished by solving a set of first-order equations rather than a single, higher-order equation. This approach simplifies the problem and has several advantages when utilizing a digital computer for solution.

What is meant by the state of a system? It is defined as the minimum set of variables, denoted by $x_1, \ldots, x_n$ and specified at time $t = t_0$, which together with the given inputs $u_1, \ldots, u_m$, determine the state at any future time $t \geqslant t_0$ [6–9, 10, 22].

Physically, this means that a set of state variables $x_1(t_0), x_2(t_0), \ldots, x_n(t_0)$ define the initial state of the system based on past history. In addition, the set of state variables $x_1, x_2, x_3, \ldots, x_n$ characterize the future behavior of the system once the inputs for $t \geqslant t_0$ are specified, together with the knowledge of the initial states. We can view the state of a system, therefore, as describing the past, present, and future behavior of the system. It is important to emphasize that state variables are not necessarily system outputs and may not always be accessible, measurable, observable, or controllable.

The state-space approach has the following advantages.

1. The solution to a set of first-order differential or difference equations is much easier to determine on a digital computer than the solution of the equivalent higher-order differential or difference equation.

2. The state-space concept greatly simplifies the mathematical notation by utilizing vector matrix notation for the set of first-order equations.

3. The inclusion of the initial conditions of a system in the analysis of control systems, which is difficult using conventional techniques, can be accounted for readily in the state-space approach.

4. The state-space approach can be applied to the solution of most nonlinear, time-varying, stochastic, and sampled-data system configurations.

5. The state-space representation lends itself to system synthesis using modern control techniques that are discussed in later chapters.

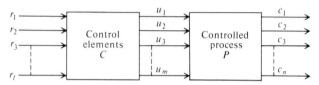

Fig. 2.18   The optimal control problem.

Let us consider the block diagram of Fig. 2.18 in order to define the nomenclature utilized in the state-space approach [10, 22]. This open-loop system is used to introduce the characteristics of the basic optimal control problem in Chapter 9. Briefly, the optimal control problem is concerned with the determination of the control elements C so that the controlled process P performs optimally with respect to a selected performance S for the anticipated inputs $r_1, r_2, r_3, \ldots, r_l$, subject to limitations on the controlled processes inputs $u_1, u_2, u_3, \ldots, u_m$. Initially, attention will be focused on the controlled process.

It is assumed that u and c are time functions, and that $u_i(t)$ and $c_j(t)$ are used to represent the values of u_i and c_j at time t. It is convenient to represent the inputs $u_1, u_2, u_3, \ldots, u_m$ by an *input vector* $\mathbf{u}$, where

$$\mathbf{u} = \begin{bmatrix} u_1 \\ u_2 \\ \cdot \\ \cdot \\ \cdot \\ u_m \end{bmatrix}, \tag{2.174}$$

and the outputs $c_1, c_2, c_3, \ldots, c_n$ by an *output vector* $\mathbf{c}$, where

$$\mathbf{c} = \begin{bmatrix} c_1 \\ c_2 \\ \cdot \\ \cdot \\ \cdot \\ c_n \end{bmatrix}. \tag{2.175}$$

The *input function space* of P refers to the set of all possible input functions **u** that can be applied to P. The *input space* of P refers to the set of all possible values that the vector **u** can have at some time T. The *initial state* of P refers to its initial conditions at the starting time t_0.

The output of P, which is denoted by **c**, is a function of the input **u** and the initial conditions of P. For example, if it is assumed that the input **u** is applied to P from t_0 to T (where $T \geqslant t_0$), then the segments of the time functions **u** and **c** over the observation interval (t_0, T) are denoted by $\mathbf{u}(t_0, T)$ and $\mathbf{c}(t_0, T)$. In addition, the initial state of P is denoted by $\mathbf{p}(t_0)$. Therefore, the value of $\mathbf{c}(t_0, T)$ depends on $\mathbf{u}(t_0, T)$ and $\mathbf{p}(t_0)$.

How can a state vector be associated with a controlled process, and then the state-equations found for P? Firstly, by associating with P a vector $\mathbf{x}(t_0)$, a set of equations which are satisfied by **x**, **u**, and **c** can then be determined. If the resulting set of equations have the property that **x** and **c** are uniquely determined by $\mathbf{x}(t_0)$ and $\mathbf{u}(t_0, T)$, then **x** qualifies as a state vector of P.

Consider the state equations for the system illustrated in Fig. 2.18. It is assumed in this derivation that P is a linear, time-invariant system characterized by the following differential equation:

$$A_n \frac{d^n c}{dt^n} + A_{n-1} \frac{d^{n-1} c}{dt^{n-1}} + \cdots + A_0 c = u, \tag{2.176}$$

where all the coefficients are constant and $A_n \neq 0$. Here $c^{(n)}$ refers to the nth derivative of c with respect to time, $c^{(n-1)}$ is the $(n-1)$th derivative, etc. It is further assumed that the initial time, t_0, is zero. Taking the Laplace transform of Eq. (2.176) results in

$$(A_n s^n + A_{n-1} s^{n-1} + \cdots + A_0) C(s) = U(s) + A_n c^{(n-1)}(0)$$
$$+ (A_n s + A_{n-1}) c^{(n-2)}(0) + \cdots + (A_n s^{n-1} + \cdots + A_1) c(0), \tag{2.177}$$

where $c^{(m)}(0)$ represents the initial value of $c^{(m)}$ at $t_0 = 0$. Rearranging Eq. (2.177), the following relationship is obtained:

$$C(s) = \frac{U(s)}{E(s)}$$

$$+ \frac{A_n c^{(n-1)}(0) + (A_n s + A_{n-1}) c^{(n-2)}(0) + \cdots + (A_n s^{n-1} + \cdots + A_1) c(0)}{E(s)},$$
$$\tag{2.178}$$

where

$$E(s) = A_n s^n + A_{n-1} s^{n-1} + \cdots + A_0. \tag{2.179}$$

This equation states that the Laplace transform of the output $C(s)$, and thus the output in the time domain $c(t)$ [for $t > 0$], is determined by the input **u** (for $t > 0$) and the values of the output **c** and all its derivatives up to and including order $n - 1$ at

$t = 0$. The initial state $\mathbf{x}(0)$ can be represented by the following n vector:

$$\mathbf{x}(0) = \big(c(0), \ldots, c^{(n-1)}(0)\big). \tag{2.180}$$

In addition, the vector $\mathbf{x}(t)$ can be represented by

$$\mathbf{x}(t) = \big(c(t), \ldots, c^{(n-1)}(t)\big). \tag{2.181}$$

It will be demonstrated that Eq. (2.181) determines a state vector for P.

A mathematical description of the dynamic relationships between the inputs and outputs of the controlled process can be derived from Eq. (2.181). This vector equation can be rewritten as the following set of linear equations:

$$x_1 = c, \; x_2 = \dot{c}, \ldots, x_n = c^{(n-1)}. \tag{2.182}$$

In addition, Eq. (2.182) can be written in the following form:

$$\dot{x}_1 = x_2, \ldots, \dot{x}_n = c^{(n)} = \frac{1}{A_n}(u - A_0 x_1 - \cdots - A_{n-1} x_n). \tag{2.183}$$

This set of first-order differential equations can be represented in vector form by

$$\dot{\mathbf{x}} = \mathbf{Px} + \mathbf{B}u \tag{2.184}$$

$$\mathbf{c} = \mathbf{Lx} \tag{2.185}$$

where

$$\mathbf{x} = \begin{bmatrix} x_1 \\ x_2 \\ \cdot \\ \cdot \\ \cdot \\ x_{n-1} \\ x_n \end{bmatrix}, \quad \mathbf{P} = \begin{bmatrix} 0 & 1 & 0 & \cdots & 0 \\ 0 & 0 & 1 & \cdots & 0 \\ \cdot & \cdot & \cdot & & \cdot \\ \cdot & \cdot & \cdot & & \cdot \\ 0 & 0 & 0 & \cdots & 1 \\ -\dfrac{A_0}{A_n} & -\dfrac{A_1}{A_n} & -\dfrac{A_2}{A_n} & \cdots - & \dfrac{A_{n-1}}{A_n} \end{bmatrix}, \tag{2.186}$$

$$\dot{\mathbf{x}} = \begin{bmatrix} \dot{x}_1 \\ \dot{x}_2 \\ \cdot \\ \cdot \\ \cdot \\ \dot{x}_n \end{bmatrix}, \quad \mathbf{B} = \begin{bmatrix} 0 \\ 0 \\ \cdot \\ \cdot \\ \cdot \\ 1 \\ \dfrac{1}{A_n} \end{bmatrix}, \quad \mathbf{L} = [1, 0, \ldots, 0].$$

The representation given in Eq. (2.186) is known as the phase-variable canonical form and the matrix $\mathbf{P}$ is called the companion matrix. The solution to Eq. (2.184), $\mathbf{x}(t)$, is uniquely determined by $\mathbf{x}(0)$ and $\mathbf{u}(0, T)$. Reference 7 shows that this is a necessary and sufficient condition to qualify $\mathbf{x}(t)$ as a state vector for P. Note that P can have infinitely many different state-vector representations.

Equations (2.184) and (2.185) are the state and output equations for linear time-invariant systems only. The state and output equations of general nonlinear and/or time-varying systems are given by

$$\dot{\mathbf{x}} = \mathbf{f}(\mathbf{x}, \mathbf{u}, t), \tag{2.187}$$

$$\mathbf{c} = \mathbf{g}(\mathbf{x}, \mathbf{u}, t). \tag{2.188}$$

In these equations, $\mathbf{u}$ is an m-vector, $\mathbf{x}$ is an n-vector, and $\mathbf{c}$ is a p-vector.

A time description of the controlled process can be obtained by solving the differential equations (2.184) or (2.187). The solution is represented as

$$\mathbf{x}(t) = \boldsymbol{\phi}_u(t, \mathbf{x}_0, t_0). \tag{2.189}$$

The above equation is interpreted as the value of $\mathbf{x}$ at time t after starting at time t_0, in state $\mathbf{x}_0$, and governed by the control input $\mathbf{u}$, defined for the interval $t_0 \leqslant t \leqslant T$.

In the remainder of this section, several examples are given for converting the dynamics of a system (given in any of several forms) into the state-space forms given by [6]

$$\dot{x}_1 = f_1(x_1, \ldots, x_n; u_1, \ldots, u_m),$$
$$\cdots \tag{2.190}$$
$$\dot{x}_n = f_n(x_1, \ldots, x_n; u_1, \ldots, u_m),$$

or in the vector state-space form given by

$$\dot{\mathbf{x}} = \mathbf{f}(\mathbf{x}, \mathbf{u}, t). \tag{2.191}$$

The linear, time-invariant forms of Eqs. (2.190) and (2.191) are represented as follows:

$$\dot{x}_1 = A_{11}x_1 + \cdots + A_{1n}x_n + B_{11}u_1 + \cdots + B_{1m}u_m,$$
$$\cdots \tag{2.192}$$
$$\dot{x}_n = A_{n1}x_1 + \cdots + A_{nn}x_n + B_{n1}u_1 + \cdots + B_{nm}u_m,$$

and

$$\dot{\mathbf{x}} = \mathbf{P}\mathbf{x} + \mathbf{B}\mathbf{u}. \tag{2.193}$$

In Eq. (2.193), $\mathbf{x}$ is a column matrix (vector) whose components are $x_1, \ldots, x_n$; $\mathbf{u}$ is a column matrix whose components are $u_1, \ldots, u_m$; $\mathbf{P}$ is an $n \times n$ matrix with entries A_{ij}, and $\mathbf{B}$ is an $n \times m$ matrix with entries B_{ij}.

In the first example used to illustrate the representation of the dynamics of a system in state-space form, consider the problem of rocket flight in two dimensions. Representing the vertical and horizontal axes by v and r, respectively, the describing equations are given by

$$\ddot{r} = F \cos \theta, \tag{2.194}$$

$$\ddot{v} = F \sin \theta - g, \tag{2.195}$$

where F is thrust force per unit mass, θ is thrust direction, and g is the gravitational force. The control inputs are considered to be F and θ. Defining

$$x_1 = r, \qquad x_2 = \dot{r},$$
$$x_3 = v, \qquad x_4 = \dot{v},$$
$$u_1 = F, \qquad u_2 = \theta,$$

we find that the dynamics are described by

$$\dot{x}_1 = x_2,$$
$$\dot{x}_2 = u_1 \cos u_2,$$
$$\dot{x}_3 = x_4,$$
$$\dot{x}_4 = u_1 \sin u_2 - g.$$

This system can also be described in vector form by

$$\dot{\mathbf{x}} = \mathbf{Px} + \mathbf{Bu}, \tag{2.196}$$

$$\mathbf{c} = \mathbf{Lx}, \tag{2.197}$$

where

$$\mathbf{x} = \begin{bmatrix} x_1 \\ x_2 \\ x_3 \\ x_4 \end{bmatrix}, \qquad \dot{\mathbf{x}} = \begin{bmatrix} \dot{x}_1 \\ \dot{x}_2 \\ \dot{x}_3 \\ \dot{x}_4 \end{bmatrix}, \qquad \mathbf{P} = \begin{bmatrix} 0 & 1 & 0 & 0 \\ 0 & 0 & 0 & 0 \\ 0 & 0 & 0 & 1 \\ 0 & 0 & 0 & 0 \end{bmatrix}, \qquad \mathbf{c} = \begin{bmatrix} r \\ v \end{bmatrix}$$

$$\mathbf{B} = \begin{bmatrix} 0 & 0 & 0 & 0 \\ 0 & 1 & 0 & 0 \\ 0 & 0 & 0 & 0 \\ 0 & 0 & 0 & 1 \end{bmatrix}, \qquad \mathbf{u} = \begin{bmatrix} 0 \\ u_1 \cos u_2 \\ 0 \\ u_1 \sin u_2 - g \end{bmatrix}, \qquad \mathbf{L} = \begin{bmatrix} 1 & 0 & 0 & 0 \\ 0 & 0 & 1 & 0 \end{bmatrix}.$$

For the second example, consider a system whose transfer function is given by

$$P(s) = \frac{C(s)}{U(s)} = \frac{5}{s^3 + 8s^2 + 9s + 2}. \tag{2.198}$$

The differential equation corresponding to this system is given by

$$\frac{d^3c}{dt^3} + 8\frac{d^2c}{dt^2} + 9\frac{dc}{dt} + 2c = 5u. \tag{2.199}$$

Defining the state variables as

$$x_1 = c, \qquad x_2 = \dot{c}, \qquad x_3 = \ddot{c}, \tag{2.200}$$

the system can now be described by the following three first-order differential equations:

$$\dot{x}_1 = x_2 = \dot{c}, \tag{2.201}$$

$$\dot{x}_2 = x_3 = \ddot{c}, \tag{2.202}$$

$$\dot{x}_3 = -2x_1 - 9x_2 - 8x_3 + 5u. \tag{2.203}$$

Therefore, the system can be described in vector form by

$$\dot{\mathbf{x}} = \mathbf{P}\mathbf{x} + \mathbf{B}\mathbf{u}, \tag{2.204}$$

$$\mathbf{c} = \mathbf{L}\mathbf{x}, \tag{2.205}$$

where

$$\mathbf{P} = \begin{bmatrix} 0 & 1 & 0 \\ 0 & 0 & 1 \\ -2 & -9 & -8 \end{bmatrix}, \quad \dot{\mathbf{x}} = \begin{bmatrix} \dot{x}_1 \\ \dot{x}_2 \\ \dot{x}_3 \end{bmatrix},$$

$$\mathbf{B} = \begin{bmatrix} 0 & 0 & 0 \\ 0 & 0 & 0 \\ 0 & 0 & 5 \end{bmatrix}, \quad \mathbf{u} = \begin{bmatrix} 0 \\ 0 \\ u \end{bmatrix},$$

$$\mathbf{x} = \begin{bmatrix} x_1 \\ x_2 \\ x_3 \end{bmatrix}, \quad \mathbf{L} = \begin{bmatrix} 1 & 0 & 0 \end{bmatrix}.$$

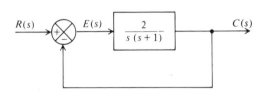

Fig. 2.19   A feedback control system.

As a third example concerned with obtaining the state-space equation of a system, consider the closed-loop system shown in Fig. 2.19. The closed-loop transfer function of this system is given by

$$\frac{C(s)}{R(s)} = \frac{2}{s^2 + s + 2}. \tag{2.206}$$

The corresponding differential equation is given by

$$\frac{d^2c}{dt^2} + \frac{dc}{dt} + 2c = 2r. \tag{2.207}$$

By defining the state variables as

$$x_1 = c, \quad x_2 = \dot{c}, \tag{2.208}$$

the system can be described by the following two first-order differential equations:

$$\begin{aligned} \dot{x}_1 &= x_2 = \dot{c} \\ \dot{x}_2 &= -2x_1 - x_2 + 2r. \end{aligned} \tag{2.209}$$

Therefore, the entire system can be described in vector form by

$$\dot{\mathbf{x}} = \mathbf{P}\mathbf{x} + \mathbf{B}r, \tag{2.210}$$

$$\mathbf{c} = \mathbf{L}\mathbf{x}, \tag{2.211}$$

where

$$\mathbf{P} = \begin{bmatrix} 0 & 1 \\ -2 & -1 \end{bmatrix}, \quad \mathbf{B} = \begin{bmatrix} 0 \\ 2 \end{bmatrix}, \quad \mathbf{x} = \begin{bmatrix} x_1 \\ x_2 \end{bmatrix}, \quad \dot{\mathbf{x}} = \begin{bmatrix} \dot{x}_1 \\ \dot{x}_2 \end{bmatrix}, \quad \mathbf{L} = \begin{bmatrix} 1 & 0 \end{bmatrix}.$$

At this point, it is important to re-emphasize that the state variables are not necessarily the outputs of a system. The system output can be measured or observed, but
a state variable may not always be measurable or observable. Sometimes, a state
variable can be viewed as an output if it is measurable or observable. The concept
of observability is discussed in Chapter 9 when optimal control theory is presented.

2.17 THE STATE-VARIABLE DIAGRAM

The state-variable diagram provides a physical picture that is useful in understanding
the state-space concept. In addition, the differential equations relating the state
variables are easily obtained by inspection of the diagram. A state-variable diagram
consists of integrators, summing devices, and amplifiers. Outputs from the integrators
denote the state variables. It should be noted that the state-variable diagram is the
same as an analog computer simulation diagram [23].
    As an example of determining the state-variable diagram, consider a system whose
transfer function is given by

$$P(s) = \frac{C(s)}{U(s)} = \frac{s^2 + 4s + 1}{s^3 + 9s^2 + 8s}. \tag{2.212}$$

Dividing top and bottom by s^3, we obtain the following

$$P(s) = \frac{C(s)}{U(s)} = \frac{s^{-1} + 4s^{-2} + s^{-3}}{1 + 9s^{-1} + 8s^{-2}}. \tag{2.213}$$

Defining

$$E(s) = \frac{U(s)}{1 + 9s^{-1} + 8s^{-2}}, \tag{2.214}$$

Eq. (2.213) may be rewritten as follows:

$$C(s) = (s^{-1} + 4s^{-2} + s^{-3})E(s). \tag{2.215}$$

From Eq. (2.215) and the relation

$$E(s) = U(s) - 9s^{-1}E(s) - 8s^{-2}E(s), \tag{2.216}$$

the state-variable diagram can easily be obtained as indicated in Fig. 2.20.* The state-variables are indicated in the diagrams as x_1, x_2, and x_3. Also, the differential equations relating the state variables are easily obtained from Fig. 2.20 by inspection. From the state-variable diagram, the differential equations relating the state variables are as follows:

$$\dot{x}_1 = x_2,$$
$$\dot{x}_2 = x_3, \qquad\qquad (2.217)$$
$$\dot{x}_3 = u - 8x_2 - 9x_3.$$

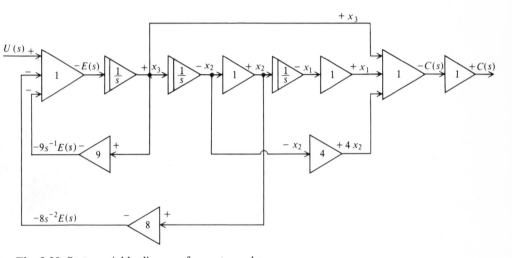

Fig. 2.20 State-variable diagram for system where

$$P(s) = \frac{s^2 + 4s + 1}{s^3 + 9s^2 + 8s}.$$

Therefore, the entire system can be described in vector form by

$$\dot{\mathbf{x}} = \mathbf{P}\mathbf{x} + \mathbf{B}u, \qquad\qquad (2.218)$$

$$\mathbf{c} = \mathbf{L}\mathbf{x}, \qquad\qquad (2.219)$$

---

* This analog computer simulation diagram contains more inverters than are actually required to implement Eqs. (2.215) and (2.216). It has been presented in this manner to indicate the proper phase relationships existing among the various states, as indicated in Eq. (2.217). From now on, all state-variable diagrams will also contain a sufficient number of inverters to indicate the proper phase relationships among the various states. It is left as an exercise to show how an analog computer simulation of Eqs. (2.215) and (2.216) can be performed using the same number of integrators and summers as shown in Fig. 2.20, but with only one inverter.

where

$$\mathbf{P} = \begin{bmatrix} 0 & 1 & 0 \\ 0 & 0 & 1 \\ 0 & -8 & -9 \end{bmatrix}, \quad \mathbf{B} = \begin{bmatrix} 0 \\ 0 \\ 1 \end{bmatrix},$$

$$\mathbf{x} = \begin{bmatrix} x_1 \\ x_2 \\ x_3 \end{bmatrix}, \quad \dot{\mathbf{x}} = \begin{bmatrix} \dot{x}_1 \\ \dot{x}_2 \\ \dot{x}_3 \end{bmatrix}, \quad \mathbf{L} = \begin{bmatrix} 1 & 4 & 1 \end{bmatrix}.$$

The output $c(t)$ can be obtained by a linear combination of the three state variables as follows:

$$c(t) = x_1(t) + 4x_2(t) + x_3(t). \tag{2.220}$$

As a second example for determining the state-variable diagram, consider a system whose transfer function is given by

$$P(s) = \frac{C(s)}{U(s)} = \frac{2}{s^2(s^2 + s + 1)}. \tag{2.221}$$

Dividing through by s^4 we obtain

$$P(s) = \frac{C(s)}{U(s)} = \frac{2s^{-4}}{1 + s^{-1} + s^{-2}}. \tag{2.222}$$

Defining

$$E(s) = \frac{2U(s)}{1 + s^{-1} + s^{-2}},$$

Eq. (2.222) may be rewritten as

$$C(s) = s^{-4}E(s). \tag{2.223}$$

From Eq. (2.223) and the relation

$$E(s) = 2U(s) - s^{-1}E(s) - s^{-2}E(s), \tag{2.224}$$

the state variable diagram for this system can easily be obtained (see Fig. 2.21). The state-variables are referred to as x_1, x_2, x_3, and x_4. They are defined as:

$$x_1 = c, \quad x_2 = \dot{c}, \quad x_3 = \ddot{c}, \quad x_4 = \dddot{c}. \tag{2.225}$$

The differential equations relating the state variables are as follows:

$$\begin{aligned} \dot{x}_1 &= x_2 = \dot{c}, \\ \dot{x}_2 &= x_3 = \ddot{c}, \\ \dot{x}_3 &= x_4 = \dddot{c}, \\ \dot{x}_4 &= 2u - x_3 - x_4. \end{aligned} \tag{2.226}$$

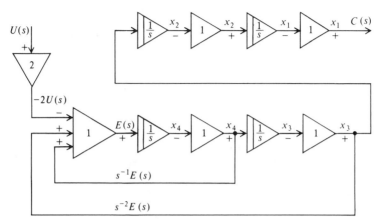

Fig. 2.21 State-variable diagram for system where

$$P(s) = \frac{2}{s^2(s^2 + s + 1)}$$

The corresponding vector form is given by

$$\dot{x} = Px + Bu, \tag{2.227}$$
$$c = Lx, \tag{2.228}$$

where

$$P = \begin{bmatrix} 0 & 1 & 0 & 0 \\ 0 & 0 & 1 & 0 \\ 0 & 0 & 0 & 1 \\ 0 & 0 & -1 & -1 \end{bmatrix}, \qquad B = \begin{bmatrix} 0 \\ 0 \\ 0 \\ 2 \end{bmatrix},$$

$$x = \begin{bmatrix} x_1 \\ x_2 \\ x_3 \\ x_4 \end{bmatrix}, \qquad \dot{x} = \begin{bmatrix} \dot{x}_1 \\ \dot{x}_2 \\ \dot{x}_3 \\ \dot{x}_4 \end{bmatrix}, \qquad L = [1 \quad 0 \quad 0 \quad 0].$$

From this discussion it can be seen that it is possible to apply the signal-flow diagram technique to state-space analysis. Mason's theorems can be applied to the signal-flow diagram which is obtained directly by inspection of the state-variable

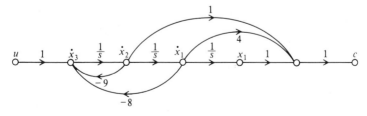

Fig. 2.22 Signal-flow diagram corresponding to the state-variable diagram of Fig. 2.20.

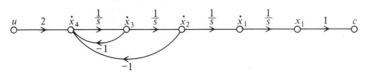

Fig. 2.23 Signal-flow diagram corresponding to the state-variable diagram of Fig. 2.21.

diagram. The signal-flow diagram also provides a physical interpretation of the state-space concept since its nodes actually represent the different states of the system.

For example, the signal-flow diagrams corresponding to the state-variable diagrams of Figs. 2.20 and 2.21 are given in Figs. 2.22 and 2.23, respectively. The physical meaning of system state is quite clear from these diagrams.

2.18 DIGITAL-COMPUTER EVALUATION OF THE TIME RESPONSE

The state-space representation of a system's dynamics easily lends itself to analysis by means of a digital computer. The technique involves the division of the time axis into sufficiently small increments $t = 0, T, 2T, 3T, 4T, \ldots$, where T is the incremental time of evaluation $\Delta\tau$. This time increment must be made small enough for accurate results. Round-off errors in the computer, however, limit how small the time increment can be.

To illustrate the procedure, let us consider the equation

$$\dot{\mathbf{x}}(t) = \mathbf{P}\mathbf{x}(t) + \mathbf{B}\mathbf{u}(t). \tag{2.229}$$

By definition of a derivative,

$$\dot{\mathbf{x}}(t) = \lim_{\Delta\tau \to 0} \frac{\mathbf{x}(t + \Delta\tau) - \mathbf{x}(t)}{\Delta\tau}. \tag{2.230}$$

Utilizing this definition, the value of $\mathbf{x}(t)$ when t is subdivided into the increments $\Delta\tau$ can be determined. Since $\Delta\tau = T$, we can say (approximately) that

$$\dot{\mathbf{x}} = \frac{\mathbf{x}(t + T) - \mathbf{x}(t)}{T}. \tag{2.231}$$

Substituting Eq. (2.231) into Eq. (2.229), we obtain

$$\frac{\mathbf{x}(t + T) - \mathbf{x}(t)}{T} = \mathbf{P}\mathbf{x}(t) + \mathbf{B}\mathbf{u}(t). \tag{2.232}$$

Equation (2.232) may be solved for $\mathbf{x}(t + T)$ as follows:

$$\mathbf{x}(t + T) = T\mathbf{P}\mathbf{x}(t) + \mathbf{x}(t) + T\mathbf{B}\mathbf{u}(t). \tag{2.233}$$

This equation can be rewritten as

$$\mathbf{x}(t + T) = (T\mathbf{P} + \mathbf{I})\mathbf{x}(t) + T\mathbf{Bu}(t). \tag{2.234}$$

To generalize this expression for the intervals mT, let

$$t = mT, \tag{2.235}$$

where $m = 0, 1, 2, 3, 4, \ldots$. Therefore, Eq. (2.234) can be rewritten as the recurrence relation

$$\mathbf{x}[(m + 1)T] = (T\mathbf{P} + \mathbf{I})\mathbf{x}(mT) + T\mathbf{Bu}(mT). \tag{2.236}$$

Equation (2.236) states that the value of the state vector at time $(m + 1)T$ is based on the values of $\mathbf{x}$ and $\mathbf{u}$ at time mT. This resulting recurrence relation is a sequential series of calculations that is very suitable for digital-computer operation. Note that this is a very crude scheme—more sophisticated schemes involve more refined approximations to $\dot{\mathbf{x}}$ [24, 25, 26].

2.19 THE TRANSITION MATRIX

The transition matrix relates the state of a system at $t = t_0$ to its state at some subsequent time t, when the input $\mathbf{u} = 0$. In order to define the transition matrix of a system, let us consider the general form of the state-space equation (see Eq. 2.193):

$$\dot{\mathbf{x}} = \mathbf{Px} + \mathbf{Bu}. \tag{2.237}$$

The Laplace transform of Eq. (2.237) is given by

$$s\mathbf{X}(s) - \mathbf{x}(0^+) = \mathbf{PX}(s) + \mathbf{BU}(s), \tag{2.238}$$

where $\mathbf{X}(s)$ is the Laplace transform of $\mathbf{x}(t)$ and $\mathbf{U}(s)$ is the Laplace transform of $\mathbf{u}(t)$. Solving for $\mathbf{X}(s)$, we obtain

$$\mathbf{X}(s) = [s\mathbf{I} - \mathbf{P}]^{-1}\mathbf{x}(0^+) + [s\mathbf{I} - \mathbf{P}]^{-1}\mathbf{BU}(s). \tag{2.239}$$

The inverse Laplace transform of Eq. (2.239) is given by

$$\mathbf{x}(t) = \mathbf{\Phi}(t)\mathbf{x}(0^+) + \int_0^t \mathbf{\Phi}(t - \tau)\mathbf{Bu}(\tau)\, d\tau, \tag{2.240}$$

where

$$\mathbf{\Phi}(t) = \mathscr{L}^{-1}\{[s\mathbf{I} - \mathbf{P}]^{-1}\}, \tag{2.241}$$

and the second term on the right-hand side of Eq. (2.240) is the convolution integral. Equation (2.240) is known as the state-transition equation of the system. When the input $\mathbf{u} = 0$, Eq. (2.240) reduces to

$$\mathbf{x}(t) = \mathbf{\Phi}(t)\mathbf{x}(0^+). \tag{2.242}$$

The matrix $\mathbf{\Phi}(t)$ is defined as the transition matrix since it relates the transition of the system state at time $t_0 = 0$, to the state at some subsequent time t. It has the following properties:

$$\mathbf{\Phi}(0) = \mathbf{I}, \tag{2.243}$$

$$\mathbf{\Phi}(t_2 - t_0) = \mathbf{\Phi}(t_2 - t_1)\mathbf{\Phi}(t_1 - t_0), \tag{2.244}$$

$$\mathbf{\Phi}(t + \tau) = \mathbf{\Phi}(t)\mathbf{\Phi}(\tau), \tag{2.245}$$

$$\mathbf{\Phi}^{-1}(t) = \mathbf{\Phi}(-t). \tag{2.246}$$

Very often it is desired to use a more general initial time, t_0. Equation (2.240) can be modified by letting $t = t_0$. Solving for $\mathbf{x}(0^+)$, we obtain the following expression:

$$\mathbf{x}(0^+) = \mathbf{\Phi}^{-1}(t_0)\mathbf{x}(t_0) - \mathbf{\Phi}^{-1}(t_0)\int_0^{t_0} \mathbf{\Phi}(t_0 - \tau)\mathbf{Bu}(\tau)\, d\tau.$$

Using Eq. (2.246), this equation can be rewritten as

$$\mathbf{x}(0^+) = \mathbf{\Phi}(-t_0)\mathbf{x}(t_0) - \mathbf{\Phi}(-t_0)\int_0^{t_0} \mathbf{\Phi}(t_0 - \tau)\mathbf{Bu}(\tau)\, d\tau. \tag{2.247}$$

Substituting Eq. (2.247) into Eq. (2.240), the following expression is obtained:

$$\mathbf{x}(t) = \mathbf{\Phi}(t)\mathbf{\Phi}(-t_0)\mathbf{x}(t_0) - \mathbf{\Phi}(t)\mathbf{\Phi}(-t_0)\int_0^{t_0} \mathbf{\Phi}(t_0 - \tau)\mathbf{Bu}(\tau)\, d\tau$$

$$+ \int_0^{t} \mathbf{\Phi}(t - \tau)\mathbf{Bu}(\tau)\, d\tau. \tag{2.248}$$

Using Eq. (2.244), Eq. (2.248) can be reduced to

$$\mathbf{x}(t) = \mathbf{\Phi}(t - t_0)\mathbf{x}(t_0) + \int_{t_0}^{t} \mathbf{\Phi}(t - \tau)\mathbf{Bu}(\tau)\, d\tau. \tag{2.249}$$

Equation (2.249) is the state-transition equation of the system for $t \geqslant t_0$.

As an example of determining the transition matrix, consider the system where the transfer function of the controlled process is given by

$$P(s) = \frac{C(s)}{U(s)} = \frac{1}{s^2}. \tag{2.250}$$

Its corresponding differential equation is given by $\ddot{c} = u$. Defining the state variables as

$$x_1 = c, \qquad x_2 = \dot{c}, \tag{2.251}$$

the system can be described by the following two first-order differential equations:

$$\dot{x}_1 = x_2 = \dot{c}, \qquad \dot{x}_2 = u. \tag{2.252}$$

Therefore, the entire system can be described by the equation

$$\dot{\mathbf{x}} = \mathbf{Px} + \mathbf{B}u, \tag{2.253}$$

where

$$\mathbf{P} = \begin{bmatrix} 0 & 1 \\ 0 & 0 \end{bmatrix}, \quad \mathbf{B} = \begin{bmatrix} 0 \\ 1 \end{bmatrix}, \quad \mathbf{x} = \begin{bmatrix} x_1 \\ x_2 \end{bmatrix}, \quad \dot{\mathbf{x}} = \begin{bmatrix} \dot{x}_1 \\ \dot{x}_2 \end{bmatrix}. \tag{2.254}$$

The transition matrix, which is defined by

$$\mathbf{\Phi}(t) = \mathcal{L}^{-1}\{[s\mathbf{I} - \mathbf{P}]^{-1}\} \tag{2.255}$$

can be obtained from Eq. (2.254). We find

$$[s\mathbf{I} - \mathbf{P}] = \begin{bmatrix} s & 0 \\ 0 & s \end{bmatrix} - \begin{bmatrix} 0 & 1 \\ 0 & 0 \end{bmatrix} = \begin{bmatrix} s & -1 \\ 0 & s \end{bmatrix}. \tag{2.256}$$

From Eq. (2.169), we know that

$$\mathbf{B}^{-1} = \frac{\text{adj } \mathbf{B}}{|\mathbf{B}|}. \tag{2.257}$$

Therefore

$$[s\mathbf{I} - \mathbf{P}]^{-1} = \frac{\text{adj }[s\mathbf{I} - \mathbf{P}]}{|s\mathbf{I} - \mathbf{P}|} = \frac{\begin{bmatrix} s & 1 \\ 0 & s \end{bmatrix}}{\begin{vmatrix} s & -1 \\ 0 & s \end{vmatrix}} = \frac{\begin{bmatrix} s & 1 \\ 0 & s \end{bmatrix}}{s^2} = \begin{bmatrix} \dfrac{1}{s} & \dfrac{1}{s^2} \\ 0 & \dfrac{1}{s} \end{bmatrix}. \tag{2.258}$$

The transition matrix defined by Eq. (2.241) is the inverse transform of this matrix. It is given by

$$\mathbf{\Phi}(t) = \mathcal{L}^{-1}\{[s\mathbf{I} - \mathbf{P}]^{-1}\} = \begin{bmatrix} U(t) & t \\ 0 & U(t) \end{bmatrix}. \tag{2.259}$$

The transition matrix may also be derived directly from the state-variable diagram. As an example of the technique, consider a system described by the differential equation

$$\ddot{c} + 4\dot{c} + 3c = r. \tag{2.260}$$

The Laplace transform of Eq. (2.260) yields

$$\frac{C(s)}{R(s)} = \frac{1}{s^2 + 4s + 3}, \tag{2.261}$$

and dividing top and bottom by s^2, we obtain

$$\frac{C(s)}{R(s)} = \frac{s^{-2}}{1 + 4s^{-1} + 3s^{-2}}. \tag{2.262}$$

Defining

$$E(s) = \frac{R(s)}{1 + 4s^{-1} + 3s^{-2}}, \tag{2.263}$$

Eq. (2.262) may be rewritten as

$$C(s) = s^{-2}E(s). \tag{2.264}$$

From Eq. (2.264) and the relation

$$E(s) = R(s) - 4s^{-1} E(s) - 3s^{-2} E(s), \tag{2.265}$$

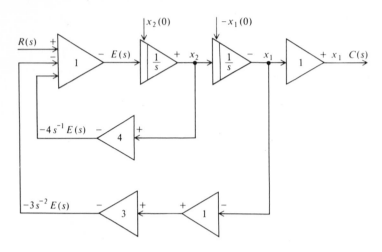

Fig. 2.24 State-variable diagram for system where

$$\frac{C(s)}{R(s)} = \frac{1}{s^2 + 4s + 3}.$$

the state-variable diagram for this system is obtained as illustrated in Fig. 2.24. In addition, for generality, it is assumed that the states of the system, x_1 and x_2, have the initial conditions $x_1(0)$ and $x_2(0)$, respectively. The corresponding signal-flow diagram is given by Fig. 2.25. The transformed state-transition equations of the system are obtained from this state-variable signal-flow diagram using Mason's formula (Eq. 2.132):

$$X_1(s) = \frac{s^{-1}(1 + 4s^{-1})x_1(0)}{\Delta} + \frac{s^{-2}x_2(0)}{\Delta} + \frac{s^{-2}R(s)}{\Delta} \qquad (2.266)$$

$$X_2(s) = \frac{-3s^{-2}x_1(0)}{\Delta} + \frac{s^{-1}x_2(0)}{\Delta} + \frac{s^{-1}R(s)}{\Delta} \qquad (2.267)$$

where

$$\Delta = 1 - (-4s^{-1} - 3s^{-2}) = 1 + 4s^{-1} + 3s^{-2}. \qquad (2.268)$$

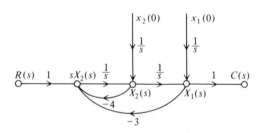

Fig. 2.25 State-variable signal-flow diagram corresponding to the state-variable diagram of Fig. 2.24.

Simplifying Eqs. (2.266), (2.267), and (2.268), we obtain the following pair of equations:

$$X_1(s) = \frac{s + 4}{s^2 + 4s + 3} x_1(0) + \frac{1}{s^2 + 4s + 3} x_2(0) + \frac{R(s)}{s^2 + 4s + 3}, \quad (2.269)$$

$$X_2(s) = \frac{-3}{s^2 + 4s + 3} x_1(0) + \frac{s}{s^2 + 4s + 3} x_2(0) + \frac{sR(s)}{s^2 + 4s + 3}. \quad (2.270)$$

These two equations can be put into the following form:

$$\begin{bmatrix} X_1(s) \\ X_2(s) \end{bmatrix} = \frac{1}{(s + 1)(s + 3)} \begin{bmatrix} s + 4 & 1 \\ -3 & s \end{bmatrix} \begin{bmatrix} x_1(0) \\ x_2(0) \end{bmatrix} + \begin{bmatrix} \dfrac{1}{(s + 1)(s + 3)} \\ \dfrac{s}{(s + 1)(s + 3)} \end{bmatrix} R(s). \quad (2.271)$$

From Eq. (2.271), we can easily obtain the transition matrix by taking the inverse Laplace transform. It is assumed in the following solution that $r(t) = U(t)$ and $R(s) = 1/s$:

$$\begin{bmatrix} x_1(t) \\ x_2(t) \end{bmatrix} = \begin{bmatrix} 1.5e^{-t} - 0.5e^{-3t} & 0.5e^{-t} - 0.5e^{-3t} \\ -1.5e^{-t} + 1.5e^{-3t} & -0.5e^{-t} + 1.5e^{-3t} \end{bmatrix} \begin{bmatrix} x_1(0) \\ x_2(0) \end{bmatrix}$$
$$+ \begin{bmatrix} 0.33U(t) - 0.5e^{-t} + 0.167e^{-3t} \\ 0.5e^{-t} - 0.5e^{-3t} \end{bmatrix}. \quad (2.272)$$

Therefore, the transition matrix is given by

$$\Phi(t) = \begin{bmatrix} 1.5e^{-t} - 0.5e^{-3t} & 0.5e^{-t} - 0.5e^{-3t} \\ -1.5e^{-t} + 1.5e^{-3t} & -0.5e^{-t} + 1.5e^{-3t} \end{bmatrix}. \quad (2.273)$$

Once the transition matrix is obtained, the evaluation of the time response is easily obtained.

This technique unfortunately shows that the use of the state-variable diagram and signal-flow graph method for determining the transition matrix is extremely inefficient compared with calculating it directly from Eq. (2.241). However, the method utilizing the state-variable diagram in conjunction with the signal-flow graph does have advantages in certain situations since it offers other choices of state variables and permits application of Mason's theorems for determining the relationships among the various states required in the transition matrix.

2.20 APPLICATION OF THE STATE-SPACE METHOD

The purpose of this section is to illustrate how one may obtain the complete solution for the output in the time domain of a control system utilizing the state-space method. In this example, we will want to determine the complete solution by evaluating Eq. (2.240).

Consider a system described by the following differential equation:

$$\ddot{c} + 2\dot{c} + c = \dot{r} + r. \tag{2.274}$$

It is desired to determine the output $c(t)$, given that the input $r(t)$ is given by

$$r(t) = \sin t \tag{2.275}$$

and the initial conditions are $x_1(0) = 1$ and $x_2(0) = 0$. The technique employed is to determine the transition matrix from Eq. (2.241) and then evaluate Eq. (2.240) for $\mathbf{x}(t)$. The output $\mathbf{c}(t)$ is then evaluated from

$$\mathbf{c}(t) = \mathbf{L}\mathbf{x}(t). \tag{2.276}$$

If the state variables are defined by

$$x_1 = c, \qquad x_2 = \dot{c}, \tag{2.277}$$

and u by

$$u = r + \dot{r},$$

then the system can be described by the following two first-order differential equations:

$$\begin{aligned} \dot{x}_1 &= x_2, \\ \dot{x}_2 &= -2x_2 - x_1 + u. \end{aligned} \tag{2.278}$$

Therefore, the system can be described by

$$\dot{\mathbf{x}} = \mathbf{P}\mathbf{x} + \mathbf{B}u, \tag{2.279}$$

where

$$\mathbf{P} = \begin{bmatrix} 0 & 1 \\ -1 & -2 \end{bmatrix}, \quad \mathbf{B} = \begin{bmatrix} 0 \\ 1 \end{bmatrix}, \quad \mathbf{x} = \begin{bmatrix} x_1 \\ x_2 \end{bmatrix}, \quad \dot{\mathbf{x}} = \begin{bmatrix} \dot{x}_1 \\ \dot{x}_2 \end{bmatrix}. \tag{2.280}$$

The transition matrix, which is defined by Eq. (2.241), can be obtained from Eq. (2.280). We find

$$[s\mathbf{I} - \mathbf{P}] = \begin{bmatrix} s & 0 \\ 0 & s \end{bmatrix} - \begin{bmatrix} 0 & 1 \\ -1 & -2 \end{bmatrix} = \begin{bmatrix} s & -1 \\ 1 & s+2 \end{bmatrix}. \tag{2.281}$$

From Eq. (2.169), we know that

$$\mathbf{B}^{-1} = \frac{\text{adj } \mathbf{B}}{|\mathbf{B}|}. \tag{2.282}$$

Therefore,

$$[s\mathbf{I} - \mathbf{P}]^{-1} = \frac{\text{adj } [s\mathbf{I} - \mathbf{P}]}{|s\mathbf{I} - \mathbf{P}|} = \frac{\begin{bmatrix} s+2 & 1 \\ -1 & s \end{bmatrix}}{(s+1)^2} = \begin{bmatrix} \dfrac{s+2}{(s+1)^2} & \dfrac{1}{(s+1)^2} \\ -\dfrac{1}{(s+1)^2} & \dfrac{s}{(s+1)^2} \end{bmatrix}. \tag{2.283}$$

The transition matrix defined by Eq. (2.241) is the inverse transform of this matrix. It is given by

$$\boldsymbol{\Phi}(t) = \mathcal{L}^{-1}\{[s\mathbf{I} - \mathbf{P}]^{-1}\} = \begin{bmatrix} e^{-t}(t + 1) & te^{-t} \\ -te^{-t} & e^{-t}(1 - t) \end{bmatrix}. \tag{2.284}$$

The full solution for the output can be obtained from Eqs. (2.240) and (2.276) as follows:

$$\mathbf{x}(t) = \boldsymbol{\Phi}(t)\mathbf{x}(0^+) + \int_0^t \boldsymbol{\Phi}(t - \tau)\mathbf{B}\mathbf{u}(\tau)\, d\tau, \tag{2.285}$$

$$\mathbf{c}(t) = \mathbf{L}\mathbf{x}(t). \tag{2.286}$$

Substituting Eq. (2.285) into Eq. (2.286), we obtain the following relationship for the output in terms of the transition matrix:

$$\mathbf{c}(t) = \mathbf{L}\boldsymbol{\Phi}(t)\mathbf{x}(0^+) + \int_0^t \mathbf{L}\boldsymbol{\Phi}(t - \tau)\mathbf{B}\mathbf{u}(\tau)\, d\tau. \tag{2.287}$$

We know $\boldsymbol{\Phi}(t)$ from Eq. (2.284). We have looked at many similar systems in this chapter, and should know by inspection now that

$$\mathbf{L} = [1 \quad 0], \quad \mathbf{x}(0^+) = \begin{bmatrix} x_1(0) \\ x_2(0) \end{bmatrix}. \tag{2.288}$$

For this system, the input function $\mathbf{u}(\tau)$ is obtained as follows:

$$u(\tau) = r(\tau) + \dot{r}(\tau) = \sin \tau + \cos \tau. \tag{2.289}$$

Substituting all of these values into Eq. (2.287), we obtain the following expression:

$$c(t) = [1 \quad 0]\begin{bmatrix} e^{-t}(t + 1) & te^{-t} \\ -te^{-t} & e^{-t}(1 - t) \end{bmatrix}\begin{bmatrix} 1 \\ 0 \end{bmatrix}$$
$$+ \int_0^t [1 \quad 0]\begin{bmatrix} e^{-(t-\tau)}(t - \tau + 1) & (t - \tau)e^{-(t-\tau)} \\ -(t - \tau)e^{-(t-\tau)} & e^{-(t-\tau)}(1 - t + \tau) \end{bmatrix}\begin{bmatrix} 0 \\ 1 \end{bmatrix}(\sin \tau + \cos \tau)\, d\tau. \tag{2.290}$$

On simplifying, the result becomes

$$c(t) = e^{-t}(t + 1) + \int_0^t [(t - \tau)e^{-(t-\tau)}][\sin \tau + \cos \tau]\, d\tau. \tag{2.291}$$

Integrating and simplifying, we finally obtain the output as

$$c(t) = \tfrac{3}{2}e^{-t} + te^{-t} + \tfrac{1}{2}\sin t - \tfrac{1}{2}\cos t. \tag{2.292}$$

2.21 DIGITAL-COMPUTER EVALUATION OF THE TRANSITION MATRIX

The transition matrix may be readily evaluated by means of digital-computer techniques. Several methods have been proposed for its numerical evaluation. References 11 and 12 discuss one type of computational algorithm developed by Fadeev for accomplishing this. However, this approach requires the Laplace-transform inversion of $\Phi(s)$. Unfortunately, this approach is very tedious for matrices of any size. This section presents a straightforward method which evaluates the transition matrix based on its infinite matrix series definition [13]. Direct application of the series definition gives a very efficient and fast method that depends only on matrix multiplication.

In order to derive the series definition of the transition matrix [14], let us consider the following equation:

$$\dot{\Phi}(t) = P\Phi(t), \tag{2.293}$$

where

$$\Phi(0) = I.$$

The transition matrix, $\Phi(t)$, represents a solution to this equation, and let us assume that it is given by

$$\Phi(t) = e^{Pt}, \tag{2.294}$$

where

$$e^{Pt} = I + Pt + \frac{P^2 t^2}{2!} + \cdots + \frac{P^k t^k}{k!} + \cdots. \tag{2.295}$$

We shall now work backwards and prove that Eq. (2.294) is indeed the correct solution to Eq. (2.293). Following this procedure, the value of $\dot{\Phi}(t)$ is given by

$$\frac{d}{dt}[e^{Pt}] = P + P^2 t + \frac{P^3 t^2}{2!} + \cdots + \frac{P^{k+1} t^k}{k!} + \cdots. \tag{2.296}$$

A comparison of Eqs. (2.295) and (2.296) indicates that

$$\frac{d}{dt}[e^{Pt}] = Pe^{Pt}. \tag{2.297}$$

Therefore, from the definition of Eq. (2.294), we find that

$$\dot{\Phi}(t) = P\Phi(t), \tag{2.298}$$

so that

$$\Phi(t) = e^{Pt} = \sum_{k=0}^{\infty} \frac{P^k t^k}{k!} \tag{2.299}$$

is indeed a correct solution to Eq. (2.293).

From this derivation, we can now extend our original definition of the transition matrix (see Eq. 2.241) to the following:

$$\Phi(t) = \mathcal{L}^{-1}\{[sI - P]^{-1}\} = e^{Pt} = \sum_{k=0}^{\infty} \frac{P^k t^k}{k!}. \tag{2.300}$$

Since the matrix series is uniformly convergent for any finite interval, the transition matrix can be determined within prescribed accuracy using only a finite number of terms [15].

An iterative procedure for evaluating $e^{\mathbf{P}t}$, based on the definition of Eq. (2.299), is now presented, and is readily adapted for digital computer computation [13]. Let $e^{\mathbf{P}t}$ be represented as

$$e^{\mathbf{P}t} = \mathbf{M} + \mathbf{R}, \tag{2.301}$$

where $\mathbf{M}$ is the approximating matrix for $e^{\mathbf{P}t}$,

$$\mathbf{M} = \sum_{k=0}^{K} \frac{\mathbf{P}^k T^k}{k!}, \quad (T = t), \quad \longleftarrow \tag{2.302}$$

and $\mathbf{R}$ is the remainder matrix

$$\mathbf{R} = \sum_{k=K+1}^{\infty} \frac{\mathbf{P}^k T^k}{k!}. \tag{2.303}$$

Assuming that each element in the matrix $e^{\mathbf{P}t}$ is required to within an accuracy of at least b significant digits, then

$$|r_{ij}| \leqslant 10^{-b} |m_{ij}|, \tag{2.304}$$

where r_{ij} and m_{ij} represent elements of the matrices $\mathbf{R}$ and $\mathbf{M}$.

Let the norm of matrix $\mathbf{P}$ be given by

$$\|\mathbf{P}\| = \sum_{i,j=1}^{m} |k_{ij}|. \tag{2.305}$$

Then, it can be shown that

$$\|\mathbf{P}^k\| \leqslant \|\mathbf{P}\|^k, \qquad k = 1, 2, 3, \ldots. \tag{2.306}$$

Therefore, each element of the matrix $\mathbf{P}^k$ is less than or equal to $\|\mathbf{P}\|^k$. It follows that

$$|r_{ij}| \leqslant \sum_{k=K+1}^{\infty} \frac{\|\mathbf{P}\|^k T^k}{k!}. \tag{2.307}$$

Let us define the ratio of the second term to the first term of the previous series to be ϵ as follows:

$$\epsilon = \frac{\|\mathbf{P}\| T}{K + 2}. \quad \longleftarrow \tag{2.308}$$

Therefore,

$$\frac{\|\mathbf{P}\| T}{k} \leqslant \epsilon, \qquad k \geqslant K + 2. \tag{2.309}$$

Substituting Eq. (2.309) into Eq. (2.307), we obtain the following expression:

$$|r_{ij}| \leqslant \frac{\|\mathbf{P}\|^{K+1} T^{K+1}}{(K + 1)!}(1 + \epsilon + \epsilon^2 + \cdots). \tag{2.310}$$

Equation (2.310) can be rewritten in closed form as

$$|r_{ij}| \leqslant \frac{(\|\mathbf{P}\| T)^{K+1}}{(K + 1)!} \frac{1}{1 - \epsilon}. \tag{2.311}$$

Let us summarize the steps of this iterative procedure for evaluating e^{Pt} before applying it to a problem:

a) An initial value of K is chosen arbitrarily.

b) The value of m_{ij} is evaluated by means of Eq. (2.302).

c) The value of ϵ is determined by means of Eq. (2.308).

d) The upper bound of $|r_{ij}|$ is calculated from Eq. (2.311).

e) Each element of $\mathbf{M}$, obtained from Eq. (2.302), is compared with the upper bound of $|r_{ij}|$ obtained from Eq. (2.311).

f) If the inequality of Eq. (2.304) is not satisfied, the value of K is increased, and steps (a) through (e) are repeated; otherwise, the procedure is ended.

As an example of applying this procedure, let us evaluate $\mathbf{\Phi}(t)$ numerically for the following example [13]:

$$\mathbf{P} = \begin{bmatrix} 0 & 1 & 0 \\ 0 & 0 & 1 \\ -0.75 & -2.75 & -3 \end{bmatrix}$$

$$T = 0.1.$$

Let us assume that each element in the matrix e^{PT} is required to within an accuracy of at least four significant digits and each number carries six significant digits. The transition matrix is obtained approximately, using the procedure indicated:

$$e^{PT} \simeq \mathbf{M} = \begin{bmatrix} 0.999884 & 0.995717 \times 10^{-1} & 0.452513 \times 10^{-3} \\ -0.339385 \times 10^{-2} & 0.987440 & 0.859963 \times 10^{-1} \\ -0.644972 \times 10^{-1} & -0.239884 & 0.729451 \end{bmatrix}.$$

In this example, $b = 4$. When $K = 9$, the upper bound of $|r_{ij}|$ from Eq. (2.311) is 0.587945×10^{-7}. Therefore, $10^{b} |r_{ij}| = 0.587945 \times 10^{-3} < |m_{ij}|$ $(i, j = 1, 2, \ldots)$, where m_{ij} are the elements of $\mathbf{M}$ given in the example. This illustrative example indicates the simplicity and accuracy of the procedure for obtaining the transition matrix utilizing its series definition in conjunction with a digital computer.

2.22 SUMMARY

Many mathematical techniques have been presented in this chapter for use by the control-system engineer. Starting with complex-variable theory, we then developed the Fourier transform and Laplace transform. The transfer function, block diagram, and signal-flow diagrams were then presented. It was pointed out that these concepts were not applicable to the more general, nonlinear, time-varying system. For this class of systems, the state-space concept was then presented. Matrix algebra was reviewed, and the state-variable diagram and the transition matrix were presented. It is reasonable for the reader at this point to ask which methods he should use.

There are no hard and fast guidelines but reasonable rules of thumb can be outlined. In general, if the problem is one of analysis, if the system has one input and one output, and if its differential equation can be described by a linear differential equation having constant coefficients, then the engineer can use the simple transfer-function/block-diagram approach or the state-space method, each technique complementing the other. On the other hand, if the analysis problem involves nonlinearities, time-varying characteristics and multivariable inputs and outputs, then the state-space approach should be used. If the problem is one of synthesis involving optimal control theory, then again the state-space approach should be used.

Since the main purpose of this book is pedagogical, both the transfer function/block diagram and state-space concept will be used wherever possible. For example, in the following chapter, where it is desired to represent mathematically various linear physical components, both approaches are used. The reader should also develop this dual capability and be able to handle a problem from either point of view.

PROBLEMS

2.1 Prove that the Laplace transform of an exponentially decaying sine wave, $e^{-\alpha t} \sin \omega t$, is given by

$$\frac{\omega}{(s + \alpha)^2 + \omega^2}.$$

This transform is shown as item 9 in Table 2.1.

2.2 Obtain the direct Laplace transform for the following differential and integral equations:

a) $L \dfrac{di(t)}{dt} + Ri(t) + \dfrac{1}{C} \displaystyle\int i(t)\, dt = e(t)$,

b) $M \dfrac{d^2x(t)}{dt^2} + B \dfrac{dx(t)}{dt} + Kx(t) = 3t$,

c) $J \dfrac{d^2\theta(t)}{dt^2} + B \dfrac{d\theta(t)}{dt} + K\theta(t) = 10 \sin \omega t$.

2.3 Obtain the inverse Laplace transform for the following expressions:

a) $F_A(s) = \dfrac{20}{(s + 2)^2(s^2 + 12s + 16)}$,

b) $F_B(s) = \dfrac{10(s + 2)}{(s^2 - 16)(s + 1)}$,

c) $F_C(s) = \dfrac{2(s + 1)}{s(s^2 + 8s + 4)}$.

2.4 Prove that the transfer function for the network illustrated in Table 2.2, item 3, is given by the expression shown.

2.5 Prove that the transfer function for the network illustrated in Table 2.2, item 6, is given by the expression shown.

2.6 Prove that the transfer function for the network illustrated in Table 2.2, item 9, is given by the expression shown.

2.7 By means of block-diagram reduction techniques, find the transfer function of the system $C(s)/R(s)$ for the configuration illustrated in Fig. P2.7.

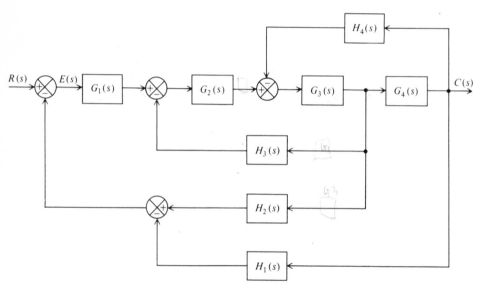

Figure P2.7

2.8 For the system illustrated in Fig. P2.7, what is the effect on the system transfer function of inserting a sign reverser in series with $H_3(s)$?

2.9 Determine the transfer function of the system shown in Fig. P2.7 relating error $E(s)$ and input $R(s)$.

2.10 Repeat Problem 2.9 with a sign reverser inserted in series with $H_3(s)$. What is the significance of this result as compared to that obtained in Problem 2.9?

2.11 Repeat Problem 2.7 using the signal-flow diagram method.

2.12 Repeat Problem 2.8 using the signal-flow diagram method.

2.13 Repeat Problem 2.9 using the signal-flow diagram method.

2.14 Repeat Problem 2.10 using the signal-flow diagram method.

2.15 Determine the transfer function of the overall system shown in Fig. P2.15, $C(s)/R(s)$, using the signal-flow diagram.

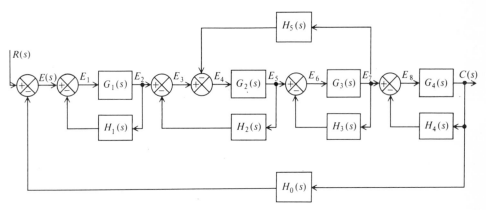

Figure P2.15

2.16 Repeat Problem 2.15 with the feedback path containing element $H_5(s)$ removed.

2.17 Determine the transfer function of the system shown in Fig. P2.15 relating error $E(s)$ and input $R(s)$.

2.18 Repeat Problem 2.17 with the feedback path containing element $H_5(s)$ removed.

2.19 Determine the transfer function of the overall system shown in Fig. P2.19, $C(s)/R(s)$, using the signal-flow diagram.

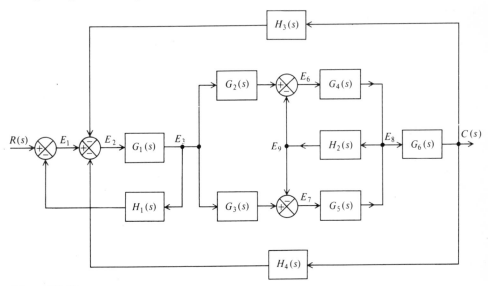

Figure P2.19

2.20 Determine the transfer function of the system shown in Fig. P2.19 relating error $E_1(s)$ and input $R(s)$.

2.21 Repeat Problem 2.19, with the feedback path containing element $H_1(s)$ removed.

2.22 Repeat Problem 2.20, with the feedback path containing element $H_1(s)$ removed.

2.23 Determine the system transmission $T = C/R$ for the signal-flow diagram shown in Fig. P2.23.

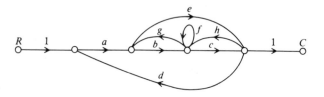

Figure P2.23

2.24 Synthesize a signal-flow diagram for the following source-to-sink transmission T, such that each branch transmission in the graph is a different letter.

$$T = \frac{C}{R} = \frac{ah(1 - cf - dg)}{(1 - be)(1 - dg) - cf}.$$

2.25 Synthesize a flow graph for the following source-to-sink transmission expression, such that each branch transmission in the graph is a different letter.

$$T = \frac{C}{R} = \frac{(ag + adi + e)(f + bh + bcj) + ij(1 - abcd)}{1 - abcd}.$$

2.26 The signal-flow diagram of a control system is illustrated in Fig. P2.26.
a) Determine the overall transmission $T = C/R$.
b) If the K' branch were made zero, the same overall transmission could be obtained by appropriately modifying the G branch. Determine the required modification.

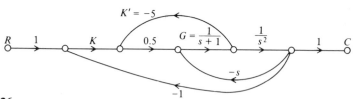

Figure P2.26

2.27 Determine the vector equations for the systems characterized by the following differential equations:

a) $\dfrac{d^2c}{dt^2} + 2\dfrac{dc}{dt} + c = 0,$

b) $\dfrac{d^2c}{dt^2} + 2\dfrac{dc}{dt} + c = A,$

c) $\dfrac{d^3c}{dt^3} + 3\dfrac{d^2c}{dt^2} + 2\dfrac{dc}{dt} + 2c = 0,$

d) $\dfrac{d^3c}{dt^3} + 3\dfrac{d^2c}{dt^2} + 2\dfrac{dc}{dt} + 2c = A.$

2.28 The approximate linear equations for a spherical satellite are given by

$$I\ddot{\theta}_1 + \omega_0 I \dot{\theta}_3 = L_1,$$

$$I\ddot{\theta}_2 \qquad\quad = L_2,$$

$$I\ddot{\theta}_3 - \omega_0 I \dot{\theta}_1 = L_3,$$

where θ_1, θ_2, θ_3 represent angular deviations of the satellite from a set of axes with fixed orientation, L_1, L_2, L_3 represent applied torques, I represents the moment of inertia, and ω_0 represents the angular frequency of the oriented axis. Determine the vector form of the system's dynamics.

2.29 The signal-flow diagram of Fig. P2.29 illustrates the process of interest accrual in a savings account. The initial deposit is represented as $r(t)$ and the total savings is represented as $c(t)$. The interest is assumed to be a constant of $K\%$ per year. It is interesting to note from this representation that it represents a positive feedback process.

a) Write the state equation of this system and determine the total savings as a function of deposit(s) and the interest rate.
b) Compare the total savings at the end of a ten year period for the following two conditions:
   1. An initial deposit of $10,000 held in a savings account over a ten-year period.
   2. A yearly deposit of $1,000, totaling $10,000 over a ten-year period, in a savings account.
   In each case assume that the interest rate K is 5% per year.
c) How do each of these cases compare with the case of savings without interest.

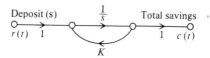

Figure P2.29

2.30 The voltage build-up of a simple vacuum-tube oscillator is given by the Van der Pol equation as follows:

$$\ddot{v} - u(1 - v^2)\dot{v} + v = 0.$$

Determine the plant dynamics.

2.31 The APOLLO 11 mission, in which Astronauts Neil Armstrong and Edwin Aldrin successfully soft landed the Lunar Excursion Module (LEM) on the lunar surface, was a historic event. Figure P2.31(a) is a photograph of the LEM vehicle taken by Astronaut Michael Collins from the Apollo Command Module window after they separated. In order to obtain the state-space vector equations of the LEM during the terminal soft-landing phase, let us consider the basic physics involved [17]. Figure P2.31(b) illustrates the forces acting on the LEM, assuming that the vehicle is vertical and subject to the following conditions:

a) The only forces acting on the vehicle are its own weight and the thrust which acts as a braking force.
b) The moon is flat in the vicinity of the desired landing point.
c) The propulsion system is capable of producing a mass flow rate, $\dot{m}$.

(a)

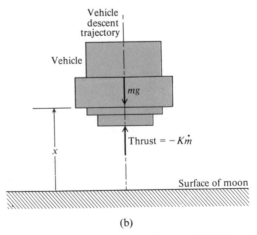

(b)

Fig. P2.31 (a) Apollo 11 Astronauts Neil Armstrong and Edwin Aldrin are inside the lunar module separated from the Apollo command module. (Official NASA photo) (b) Forces acting on LEM.

Based on these assumptions, the motion of the vehicle is governed by the following relation:

$$\ddot{x} = -\frac{K\dot{m}}{m} + g,$$

where

x = altitude
m = total mass
$\dot{m}$ = mass flow rate $\leqslant 0$
g = acceleration of gravity at the surface
K = velocity of exhaust gases = constant > 0.

Defining the state variables as

$$x = x_1, \qquad x_2 = \dot{x}_1,$$

$$x_3 = m, \qquad u_1 = \dot{m},$$

determine the vector equation of the system during the terminal soft-landing phase of the mission.

2.32 Determine the vector equation for the system shown in Fig. P2.32.

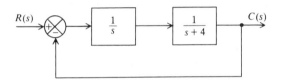

Figure P2.32

2.33 The landing of an aircraft consists of several phases [18]. First, the aircraft is guided toward the airport with approximately the correct heading by radio direction-finding equipment. Within a few miles of the airport, radio contact is made with the radio beam of the instrument landing system (ILS). In following this beam, the pilot guides the aircraft along a glide path angle of approximately $3°$ toward the runway. Finally, at an altitude of approximately 100 ft, the flare-out phase of the landing begins. During this final phase of the landing, the ILS radio beam is no longer effective, nor is the $-3°$ glide path angle desirable from the viewpoint of safety and comfort. Therefore, the pilot must guide the aircraft along the desired flare path by making visual contact with the ground. Let us consider, in this problem, the states of the aircraft during the final phase of the landing (the last 100 ft of the aircraft's descent). Figure P2.33(a) defines the aircraft's coordinates and angles. Figure P2.33(b) illustrates the aircraft's block diagram in terms of measurable state signals from the elevator deflection angle δ_e to the altitude $h(t)$. The state signals being fed back are all measurable. For example, the altitude $h(t)$ can be measured with a radar altimeter, and the rate of ascent $\dot{h}(t)$ can be measured with a barometric ratemeter. In addition, the pitch angle, $\theta(t)$, and the pitch rate, $\dot{\theta}(t)$, can be easily measured using gyros. Defining the states of the aircraft-landing system as

$$x_1(t) = \dot{\theta}(t), \qquad x_2(t) = \theta(t),$$

$$x_3(t) = \dot{h}(t), \qquad x_4(t) = h(t), \qquad u_1(t) = \delta_e(t),$$

determine the vector equation of this system.

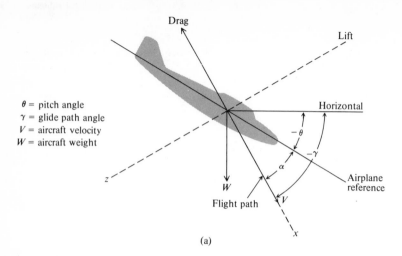

θ = pitch angle
γ = glide path angle
V = aircraft velocity
W = aircraft weight

(a)

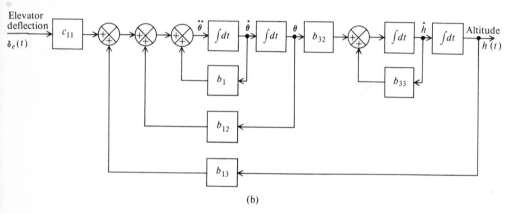

(b)

$$b_{11} = \frac{1}{T_s} - 2\zeta\omega_s, \quad b_{13} = \frac{1}{VT_s^2} - \frac{2\zeta\omega_s}{VT_s} + \frac{\omega_s^2}{V}, \quad b_{33} = -\frac{1}{T_s}$$

$$b_{12} = \frac{2\zeta\omega_s}{T_s} - \omega_s^2 - \frac{1}{T_s^2}, \quad b_{32} = \frac{V}{T_s} \quad c_{11} = \omega_s^2 K_s T_s$$

where

K_s = short period gain of the aircraft,
T_s = path time constant,
ω_s = short period resonant frequency of the aircraft,
ζ = short period damping factor of the aircraft.

Figure P2.33

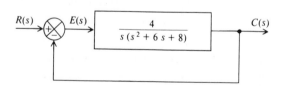

Figure P2.34

2.34 Determine the vector equation for the system shown in Fig. P2.34.

2.35 A finless torpedo is directionally unstable without control. However, its controlled performance is usually made better than that of a torpedo with fins by utilizing optimal control theory. It is possible to synthesize an intercept trajectory that overcomes the problems associated with random disturbances, measurement noise, and drift [19]. Figure P2.35 illustrates the motion of such a torpedo in the yaw axis. If it is assumed that the optimal

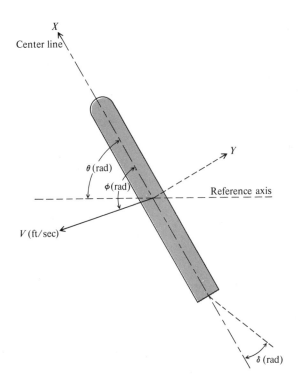

Figure P2.35

intercept trajectory is a straight line, then the usual small-angle assumptions are reasonably accurate and the problem can be treated by linear techniques. The following equations describe yaw acceleration, $\ddot{\theta}$, side-slip effects, ϕ, and the command input, δ_c, for the finless torpedo:

$$M_2\dot{\phi} + Y_\phi\phi + (Y_r - M_1)\dot{\theta} + Y_\delta\delta = 0,$$

$$I_2\ddot{\theta} - N_\phi\phi - N_r\dot{\theta} - N_\delta\delta = 0,$$

$$t'\dot{\delta} = -\delta + \delta_c.$$

The value t' is a normalized time parameter. Since the problem has been presented on the assumption that only small perturbations are being considered, no bounds need to be placed on δ_c. Determine the vector equation for this system with the coefficients, normalized with respect to ship length, defined as follows:

M_1 = longitudinal virtual mass coefficient = 1.56
M_2 = lateral virtual mass coefficient = 2.92
I_2 = virtual moment of inertia coefficient = 0.138
Y_ϕ = static side force rate coefficient = 0.60
N_ϕ = static yaw moment rate coefficient = 0.99
Y_r = damping force rate coefficient = 0.20
N_r = damping moment rate coefficient = −0.08
Y_δ = rudder force rate coefficient = 0.10
N_δ = rudder moment rate coefficient = $Y_\delta e_R$ = −0.50
e_R = dimensionless coordinate of rudder side force = −0.50.

2.36 Given

$$\mathbf{A} = \begin{bmatrix} 1 & 2 & 4 \\ 2 & 1 & 1 \\ 2 & 2 & 2 \end{bmatrix} \quad \text{and} \quad \mathbf{B} = \begin{bmatrix} 2 & 4 & 2 \\ 2 & 1 & 1 \\ 4 & 2 & 4 \end{bmatrix},$$

determine $\mathbf{A}\,\mathbf{B}^{-1}$.

2.37 How can the characteristic equation of the control system illustrated in Fig. P2.37 be determined utilizing matrix techniques?

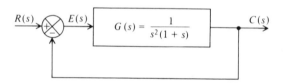

Figure P2.37

2.38 Determine the vector equation for an open-loop system where

$$\frac{C(s)}{R(s)} = \frac{1}{s^2(s + 10)}.$$

2.39 Repeat Problem 2.38 with

$$\frac{C(s)}{R(s)} = \frac{1}{s(s + 1)(s + 8)}.$$

2.40 Repeat Problem 2.38 with

$$\frac{C(s)}{R(s)} = \frac{5}{s(s^2 + 4s + 2)}.$$

2.41 Determine the state-variable diagram and the vector equation for the following open-loop system:

$$P(s) = \frac{C(s)}{U(s)} = \frac{s + 1}{s(s^2 + 7s + 1)}.$$

2.42 The automatic depth control of a submarine is an interesting control problem. Figure P2.42(a) illustrates a representative problem, where the actual depth of the submarine is denoted as C. In practice, the actual depth of a submarine is measured by a pressure trans-ducer. This measurement is then compared with the desired depth, R. Any differences are amplified in a control system which appropriately adjusts the stern plane actuator angle θ. An equivalent block diagram of such a system is illustrated in Fig. P2.42(b). Determine the state-variable signal-flow diagram and the vector equation for this system.

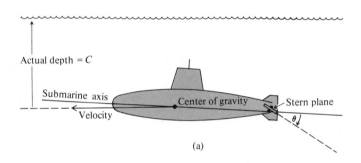

(a)

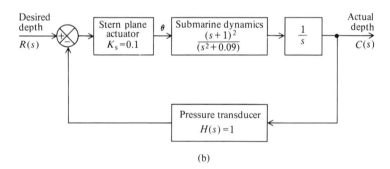

(b)

Figure P2.42

2.43 Determine analytically the transition matrix of the system considered in Problem 2.41.

2.44 Determine the transition matrix of the following open-loop control system analytically:

$$P(s) = \frac{C(s)}{U(s)} = \frac{1}{s(s + 2)(s + 5)}.$$

2.45 Repeat Problem 2.44 with

$$P(s) = \frac{C(s)}{U(s)} = \frac{1}{s^2(s + 6)(s + 10)}.$$

2.46 A nuclear reactor has been operating in equilibrium for a long period of time at a high thermal-neutron flux level, and is suddenly shut down. At shut-down, the density of iodine 135 (I) is 5×10^{16} atoms per unit volume and that of xenon 135 (X) is 2×10^{15} atoms per unit volume. The equations of decay are given by the following state equations:

$$\dot{I} = -0.1I, \qquad \dot{X} = -I - 0.05X.$$

a) Determine the transition matrix.
b) Determine the system response equations, $I(t)$ and $X(t)$.
c) Determine the half-life time of I 135. The unit of time for $\dot{I}$ and $\dot{X}$ is hours.

2.47 Control-system concepts have recently been applied in attempts to solve some problems of sociology and of economics. Let us consider one such example concerned with the problem of underdeveloped countries [20, 21]. Representing the number of underdeveloped countries by C_u, and developed countries by C_d, the state variable x_1 is defined as

$$x_1 = C_u/C_d.$$

The state variable x_1 is an indication of the development of the countries in the world. A second state variable x_2 is used to represent the tendency towards underdevelopment. In writing a set of dynamical equations for these state variables, it is important to recognize that the tendency towards underdevelopment can be reduced by means of technical-assistance programs, education, etc. However, it is interesting to note that studies indicate that the gap between the developed and underdeveloped nations is growing, since the underdeveloped nations tend to remain in their present state relative to the developed area [21]. It has been proposed that the following set of state equations may be used to represent this process [20]:

$$\dot{x}_1 = -Ax_1 - Bx_2,$$
$$\dot{x}_2 = Cx_1 - Dx_2.$$

a) Determine the transition matrix of this system when $A = 4$, $B = C = 1$, and $D = 2$.
b) Determine the response of this system when $x_1(0) = 2$ and $x_2(0) = 1$.

2.48 Determine the transition matrix of the following open-loop control system analytically:

$$P(s) = \frac{C(s)}{U(s)} = \frac{1}{s^2(s+1)^2}.$$

2.49 The effectiveness of a mass-marketing campaign to sell a product can be predicted by analyzing its state-equation representation using control techniques. Consider the promotion of a new automobile in a city by the mass media including television, radio, newspapers, and periodicals. Consider the population under study to be made up of the following three groups:

x_1 represents the group who might be receptive to the automobile;
x_2 represents the group who buys the automobile;
x_3 represents the group removed from groups x_1 and x_2 due to death or other factors that cause them to be isolated or removed from the groups x_1 and x_2.

The rate at which new receptives are added to the population is equal to u_1, and the rate at which new buyers are added to the population is equal to u_2. The model of this system can be represented by the following set of equations:

$$\dot{x}_1 = -mx_1 - nx_2 + u_1,$$

$$\dot{x}_2 = nx_1 - px_2 + u_2,$$

$$\dot{x}_3 = mx_1 + px_2.$$

Assume a closed population model ($u_1 = u_2 = 0$).

a) Determine the transition matrix of this system when $m = 4$, $n = 2$, and $p = 1$.

b) Forecast how well the automobile will sell and whether the marketing campaign will be successful by determining the response of $x_2(t)$ of this system for the following initial conditions: $x_1(0) = 100{,}000$; $x_2(0) = 20{,}000$; $x_3(0) = 2{,}000$. What conclusions can you reach from your results?

2.50 Control techniques are being utilized to solve problems in the aquatic systems field. These ecological problems require a mathematical model to provide an understanding of the characteristics within the system. An aquatic ecosystem model [27] for Lago Pond, a Georgia farm pond, has been developed. This pond is a man-made pond created by an earth-fill dam and is used for sport fishing. The pond is fertilized in the spring and periodically throughout the summer in order to cause algal blooms; this increase in primary productivity travels up the food chain and causes an increase in the fish population. A constant coefficient model of Lago Pond, which assumes a constant temperature environment, is given by the following set of equations:

$$\dot{x}_1 = -30.8x_1 + 616 + 200 \sin 0.524t$$

$$\dot{x}_2 = 2.13x_1 - 17.4x_2 + 0.0458x_9$$

$$\dot{x}_3 = 1.67x_1 - 8.66x_3 + 0.0553x_9$$

$$\dot{x}_4 = 0.457x_2 + 0.553x_3 - 0.941x_4 + 0.148x_6$$

$$\dot{x}_5 = 0.0925x_3 + 0.0814x_4 - 0.349x_5 + 0.00224x_6$$

$$\dot{x}_6 = 5.97x_2 - 1.94x_6$$

$$\dot{x}_7 = 0.346x_2 + 2.73x_3 + 0.0193x_6 - 0.314x_7 + 0.23$$

$$\dot{x}_8 = 0.0898x_4 + 0.0166x_5 + 0.0166x_7 - 0.104x_8$$

$$\dot{x}_9 = 13.5x_1 + 5.45x_2 + 4.23x_3 + 0.213x_4 + 0.0703x_5 + 0.382x_6 + 0.0628x_7 + 0.0207x_8 - 0.816x_9$$

The states x_4, x_5, x_6, and x_7 represent the levels of redbreast, warmouth, chaoborus, and bluegill sunfish in the lake; the state x_8 represents the level of bass in the lake. States x_1, x_2, and x_3 represent nutrient factors for the fish as a result of pond fertilization, and state x_9 represents detritus (rocks and other small material worn or broken away from a mass due to the action of water).

a) Determine the companion matrix **P**, and **Bu**, for the vector equation of the system:

$$\dot{x} = Px + Bu.$$

b) (optional) Simulate this constant coefficient model of Lago Pond on a digital computer, and determine the time responses of the states for 12 months using the algorithm given in Section 2.18 (see Eq. 2.236). Assume the following initial conditions for the nine states (units are in Kcal/m^2): $x_1 = 20$; $x_2 = 3.5$; $x_3 = 6.4$; $x_4 = 7.16$; $x_5 = 3.44$; $x_6 = 10.8$; $x_7 = 61$; $x_8 = 16.5$; $x_9 = 400$. What conclusions can you reach from your results?

2.51 A very interesting ecological problem is that of rabbits and foxes in a controlled environment. If the number of rabbits were left alone, they would grow indefinitely until the food supply was exhausted. Representing the number of rabbits by x_1, their growth rate is given by

$$\dot{x}_1 = Ax_1.$$

However, rabbit-eating foxes in the environment change this relationship to the following:

$$\dot{x}_1 = Ax_1 - Bx_2$$

where x_2 represents the fox population. In addition, if foxes must have rabbits to exist, then their growth rate is given by

$$\dot{x}_2 = -Cx_2 + Dx_1$$

a) Assume that $A = 1$, $B = 2$, $C = 2$, and $D = 4$. Determine the transition matrix for this ecological model.
b) From the transition matrix, determine the response of this ecological model when $x_1(0) = 100$ and $x_2(0) = 50$. Explain your results.

2.52[28] A very important ecological control problem is the treatment of wastes in water. Control engineers are attempting to solve this problem by adding bacteria, which grow by using the pollutant as a nutrient, to the waste flow. In this manner, wastes in sewage treatment plants can be eliminated by this natural biological process. Microbial growth models currently being used to analyze this kind of problem are based on differential equations. Defining x as the bacteria concentration required for pollutant removal, s as the pollutant serving as the nutrient substance for the bacteria, and D as the rate of flow/volume (dilution rate), the normalized material balance equations can be approximated by the following two state equations:

$$\dot{x} = ux - Dx,$$

$$\dot{s} = -ux - Ds.$$

a) Assuming that the specific growth rate u is a constant, determine the transition matrix of this ecological control system.
b) Analyze the meaning of the transition matrix obtained for this problem. Under what conditions does a balance exist in this ecological control system?

REFERENCES

1. R. V. Churchill, *Modern Operational Mathematics in Engineering*, McGraw-Hill, New York (1944).

2. M. F. Gardner and J. L. Barnes, *Transients in Linear Systems*, Vol. 1, Wiley, New York (1942).

3. S. J. Mason, "Feedback theory: Some properties of signal flow graphs," *Proc. IRE* **41**, 1144 (1953).

4. S. J. Mason, "Feedback theory: Further properties of signal flow graphs," *Proc. IRE* **44**, 920 (1956).

5. F. E. Horn, *Elementary Matrix Algebra* (2nd Edn.) Macmillan, New York (1964).

6. C. A. Desoer, "An introduction to state space techniques in linear systems," in *Proceedings of the 1962 Joint Automatic Control Conference*, New York, pp. 10-2-1 to 10-2-5.

7. L. A. Zadeh, "An introduction to state-space techniques," in *Proceedings of the 1962 Joint Automatic Control Conference*, New York, pp. 10-1-1 to 10-1-5.

8. G. Leitmann (Ed.), *Optimization Techniques*, Academic, New York (1962).

9. P. DeRusso, R. J. Roy, and C. M. Close, *State Variables for Engineers*, Wiley, New York (1967).

10. B. C. Kuo, "State transition flow graphs of continuous and sampled dynamic systems," in *Proceedings of the 1962 Western Electronic Show and Convention*, August 21–24, 1962, Los Angeles.

11. D. K. Faddeev and V. N. Faddeeva, *Computational Methods of Linear Algebra*, Freeman, San Francisco (1963).

12. B. S. Morgan, Jr., "Sensitivity analysis and synthesis of multivariable systems," *IEEE Trans. Automatic Control* **11**, 506–12 (1966).

13. M. I. Liou, "A novel method of evaluating transient response," *Proc. IEEE* **54**, 20–23 (1966).

14. B. O. Watkins, *Introduction to Control Systems*, Macmillan, New York (1969), pp. 258–9.

15. R. Bellman, *Stability Theory of Differential Equations*, McGraw-Hill, New York (1953), pp. 12–13.

16. W. T. Thomson, *Laplace Transformation, Theory and Engineering Applications*, Prentice-Hall, Englewood Cliffs, N.J. (1950).

17. J. S. Meditch, "On the problem of optimal thrust programming for a lunar soft landing," *IEEE Trans. Automatic Control* **AC-9**, 477–84 (1964).

18. F. J. Ellert and C. W. Merriam, III, "Synthesis of feedback controls using optimization theory—An example," in *Proceedings of the 1962 Joint Automatic Control Conference*, p. 19–1.

19. N. W. Rees, "An application of optimal control theory to the guided torpedo problem," in *Proceedings of the 1968 Joint Automatic Control Conference*, pp. 820–25.

20. R. C. Dorf, *Modern Control Systems*, Addison-Wesley Publishing Company, Reading, Mass. (1967), pp. 292–94.

21. D. Rockefeller, "The population problem and economic progress," *Vital Speeches* **32**, 366–70 (1966).

22. L. A. Zadeh and C. A. Desoer, *Linear System Theory—The State Space Approach*, McGraw-Hill, New York (1963).

23. G. A. Korn and T. M. Korn, *Electronic Analog and Hybrid Computers*, McGraw-Hill, New York (1964).

24. R. W. Hamming, *Numerical Methods for Scientists and Engineers*, McGraw-Hill, New York (1962).

25. D. D. McCracken and W. S. Dorn, *Numerical Methods and FORTRAN Programming*, Wiley, New York (1964).

26. H. H. Rosenbrick and C. Storey, *Computational Techniques for Chemical Engineers*, Pergamon, Oxford (1966).
27. W. R. Emanuel and R. J. Mulholland, "Energy Based Dynamic Model for Lago Pond, Georgia," in *Proceedings of the 1974 Joint Automatic Control Conference*, pp. 354–62.
28. G. D'Ans, P. W. Kotovic, and D. Gottlieb, "A Nonlinear Regulator Problem for a Model of Biological Water Treatment," *IEEE Trans. Automatic Control* **AC-16**, 341–47 (1971).

3 STATE EQUATIONS AND TRANSFER-FUNCTION REPRESENTATION OF PHYSICAL LINEAR CONTROL-SYSTEM ELEMENTS

3.1 INTRODUCTION

The purpose of this chapter is to illustrate the procedures used for deriving the state equations and transfer-function representation for several common linear control-system elements. This is a very important step that must be mastered before considering the determination of performance and stability of various kinds of control systems.

In general, the devices that are encountered can be classified as being electrical, mechanical, electromechanical, hydraulic, thermal, etc. The emphasis here will be to describe the state equations and transfer functions for linear control system elements that are representative of these classes. Nonlinear devices are discussed in Chapter 8 on Nonlinear Feedback Control-System Design.

3.2 STATE EQUATIONS OF ELECTRICAL NETWORKS

Before introducing the techniques for formulating the state equations and transfer-function representation of some common linear control-system devices, the subject of electrical networks is considered [1]. Specifically, it is the purpose of this section to illustrate how the engineer can represent the state equations of electrical networks [2]. The very same method is applicable to the analysis of control-system devices. Network equations of electrical circuits can be formulated from the basic laws of Kirchhoff. It is assumed that the reader is familiar with the corresponding loop and nodal techniques for analyzing electrical networks [1].

The loop equation of the basic RLC circuit illustrated in Fig. 3.1 is given by

$$e(t) = Ri(t) + L\frac{di(t)}{dt} + \frac{1}{C}\int i(t)\,dt. \tag{3.1}$$

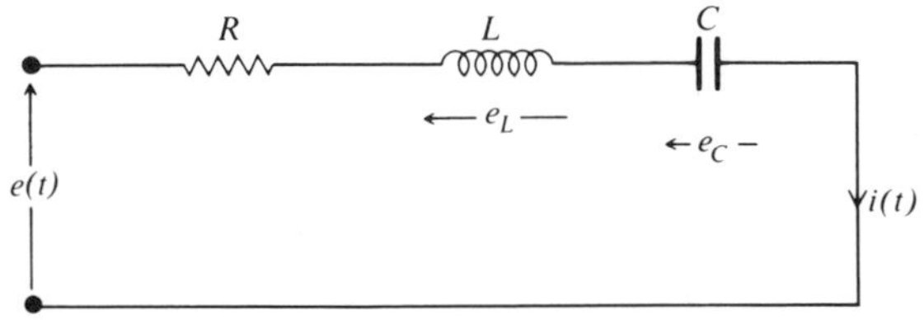

Fig. 3.1 An RLC circuit.

The current in the inductor and the voltage across the capacitor are usually considered to be the state variables of a network. Then the resulting state equations for this circuit can be formulated by relating the voltage across the inductor L and the current in the capacitor C as follows:

$$L\frac{di(t)}{dt} = e(t) - e_c(t) - Ri(t), \tag{3.2}$$

$$i(t) = C\frac{de_c(t)}{dt}. \tag{3.3}$$

Defining the state variables as

$$x_1 = e_c(t), \tag{3.4}$$

$$x_2 = i(t), \tag{3.5}$$

the following state equations of the circuit are obtained:

$$\dot{x}_1 = \frac{1}{C}x_2, \tag{3.6}$$

$$\dot{x}_2 = -\frac{1}{L}x_1 - \frac{R}{L}x_2 + \frac{e(t)}{L}. \tag{3.7}$$

These equations can be written in vector form as

$$\dot{x} = Px + Br, \tag{3.8}$$

where

$$x = \begin{bmatrix} x_1 \\ x_2 \end{bmatrix}, \quad \dot{x} = \begin{bmatrix} \dot{x}_1 \\ \dot{x}_2 \end{bmatrix}, \quad P = \begin{bmatrix} 0 & \dfrac{1}{C} \\ -\dfrac{1}{L} & -\dfrac{R}{L} \end{bmatrix}, \quad B = \begin{bmatrix} 0 \\ \dfrac{1}{L} \end{bmatrix}, \quad r = e(t).$$

As a second example, consider the electrical network of Fig. 3.2. The state equations are obtained by writing the equation for the currents in the capacitors and

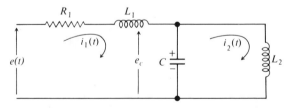

Fig. 3.2   An electrical network.

the voltage across the inductors as follows:

$$i_1(t) - i_2(t) = C \frac{de_c(t)}{dt}, \tag{3.9}$$

$$L_1 \frac{di_1(t)}{dt} = e(t) - R_1 i_1(t) - e_c(t), \tag{3.10}$$

$$L_2 \frac{di_2(t)}{dt} = e_c(t). \tag{3.11}$$

If the state variables are defined as

$$x_1 = e_c(t), \tag{3.12}$$

$$x_2 = i_1(t), \tag{3.13}$$

$$x_3 = i_2(t), \tag{3.14}$$

the following state equations for this electrical network are obtained:

$$\dot{x}_1 = \frac{1}{C} x_2 - \frac{1}{C} x_3, \tag{3.15}$$

$$\dot{x}_2 = -\frac{1}{L_1} x_1 - \frac{R_1}{L_1} x_2 + \frac{1}{L_1} e(t), \tag{3.16}$$

$$\dot{x}_3 = \frac{1}{L_2} x_1. \tag{3.17}$$

These equations can be written in vector form as

$$\dot{\mathbf{x}} = \mathbf{P}\mathbf{x} + \mathbf{B}r, \tag{3.18}$$

where

$$\mathbf{x} = \begin{bmatrix} x_1 \\ x_2 \\ x_3 \end{bmatrix}, \quad \dot{\mathbf{x}} = \begin{bmatrix} \dot{x}_1 \\ \dot{x}_2 \\ \dot{x}_3 \end{bmatrix}, \quad \mathbf{P} = \begin{bmatrix} 0 & \dfrac{1}{C} & -\dfrac{1}{C} \\ -\dfrac{1}{L_1} & -\dfrac{R_1}{L_1} & 0 \\ \dfrac{1}{L_2} & 0 & 0 \end{bmatrix}, \quad \mathbf{B} = \begin{bmatrix} 0 \\ \dfrac{1}{L_1} \\ 0 \end{bmatrix}, \quad r = e(t).$$

These simple examples illustrate the general method for analyzing electrical networks utilizing the state-space technique. The reader is referred to Reference 2 for a more detailed discussion of this particular subject. The same general method is illustrated next for the analysis of commonly used linear control-system devices.

3.3 TRANSFER-FUNCTION AND STATE-SPACE
REPRESENTATION OF TYPICAL MECHANICAL CONTROL-SYSTEM DEVICES

Mechanical control-system devices can generally be classified as being either translational or rotational. The major difference between the two is that we talk of forces and translational units in the former, and torque and angular units in the latter. Newton's three laws of motion [3] govern the action of both types of mechanical system. Basically, these laws state that the sum of the applied forces, or torques, must equal the sum of the reactive forces, or torques, for a body whose acceleration is zero. Another way of stating this is that the sum of the forces must equal zero for a body at rest or moving at a constant velocity. We shall consider some representative translational systems and then some rotational systems. The basic concepts illustrated and developed here should be sufficient to enable the reader to handle more complex systems.

1. Mechanical Translation Systems The three basic characteristics of a mechanical translational system are mass, stiffness, and damping. Mass represents an element having inertia. Stiffness represents the restoring force action such as that of a spring. Damping, or viscous friction, represents a characteristic of an element that absorbs energy. We use the English system of units as our standard. The symbols for the various quantities together with their respective units are shown in Table 3.1.

Table 3.1 Mechanical translation symbols and units

Quantity	Symbol	English units
Distance	y	ft
Velocity	v	ft/sec
Acceleration	a	ft/sec^2
Force	f	lb
Mass	M	slugs
Damping factor	B	lb/(ft/sec)
Stiffness factor	K	lb/ft

a) *Force Applied to a Mechanical System Containing a Mass, Spring, and Damper.* The case of a force $f(t)$ applied to a mass, spring, and damper is shown in Fig. 3.3. The system produces a displacement of the mass, $y(t)$, measured from a reference terminal, y_0, which is assumed to be stationary. The application of Newton's law to this system yields

$$M \frac{d^2y}{dt^2} + B \frac{dy}{dt} + Ky = f(t). \tag{3.19}$$

This second-order differential equation can be written as two first-order differential equations by defining

$$x_1 = y \tag{3.20}$$

and

$$x_2 = \dot{y} \tag{3.21}$$

The resulting state equations are given by

$$\dot{x}_1 = x_2, \tag{3.22}$$

$$\dot{x}_2 = -\frac{K}{M} x_1 - \frac{B}{M} x_2 + \frac{1}{M} f(t). \tag{3.23}$$

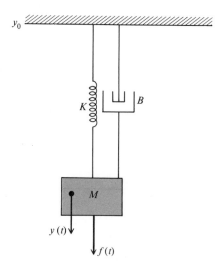

Fig. 3.3   Force applied to a system containing a spring, a mass, and a damper.

In vector form, these equations become

$$\dot{\mathbf{x}} = \mathbf{P}\mathbf{x} + \mathbf{B}r, \tag{3.24}$$

$$y = \mathbf{L}\mathbf{x}, \tag{3.25}$$

where

$$\mathbf{x} = \begin{bmatrix} x_1 \\ x_2 \end{bmatrix}, \quad \dot{\mathbf{x}} = \begin{bmatrix} \dot{x}_1 \\ \dot{x}_2 \end{bmatrix}, \quad \mathbf{P} = \begin{bmatrix} 0 & 1 \\ -\dfrac{K}{M} & -\dfrac{B}{M} \end{bmatrix}, \quad \mathbf{B} = \begin{bmatrix} 0 \\ \dfrac{1}{M} \end{bmatrix},$$

$$r = f(t), \quad \mathbf{L} = [1 \quad 0].$$

Assuming the system is initially at rest, the transfer function of this mechanical translation system, defined as the ratio of the output $Y(s)$ divided by the input $F(s)$, is readily found to be

$$\frac{Y(s)}{F(s)} = \frac{1/M}{s^2 + (B/M)s + K/M},$$ (3.26)

and its simple block diagram appears in Fig. 3.4.

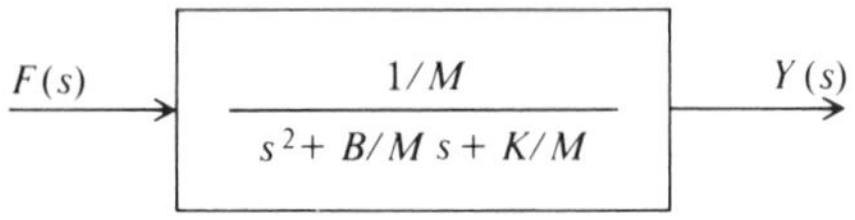

F(s) → [1/M / (s² + B/M s + K/M)] → Y(s)

Fig. 3.4   Block diagram for the spring–mass–damper system.

b) *Force Applied to a Complex Mechanical System.* Consider next the complex mechanical system illustrated in Fig. 3.5. A force $f(t)$ is applied to a mass, spring, and damper which in turn applies a force to another mass, spring, and damper. Mass M_2 is displaced a distance $y_2(t)$, and mass M_1 is displaced a distance $y_1(t)$. In order to apply Newton's laws to this system, it is advantageous to first separate the mechanical system into a set of free-body diagrams as illustrated in Fig. 3.6. Newton's equations for this resulting system are given by

$$f(t) = M_2 \frac{d^2 y_2(t)}{dt^2} + B_2 \left[\frac{dy_2(t)}{dt} - \frac{dy_1(t)}{dt} \right] + K_2[y_2(t) - y_1(t)],$$ (3.27)

$$0 = B_2 \left[\frac{dy_2(t)}{dt} - \frac{dy_1(t)}{dt} \right] + K_2[y_2(t) - y_1(t)]$$

$$- M_1 \frac{d^2 y_1(t)}{dt^2} - B_1 \frac{dy_1(t)}{dt} - K_1 y_1(t).$$ (3.28)

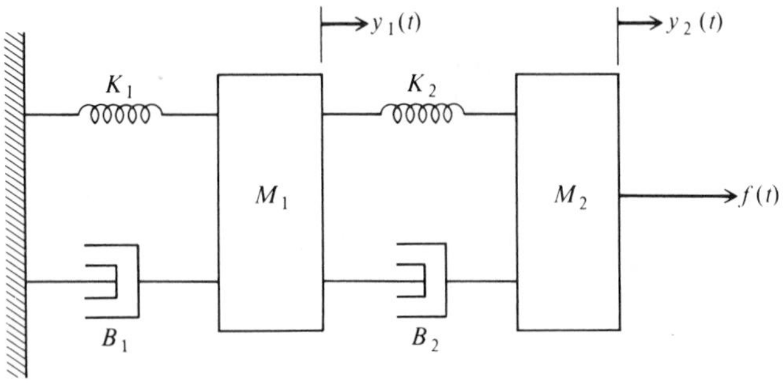

Fig. 3.5   A mechanical system.

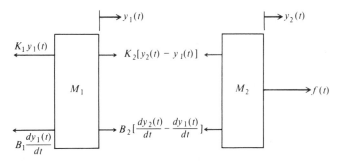

Fig. 3.6 Free-body diagram for the system illustrated in Fig. 3.5.

These two second-order differential equations can be transformed into four first-order differential equations by defining

$$x_1 = y_1(t), \tag{3.29}$$

$$x_2 = y_2(t), \tag{3.30}$$

$$x_3 = \frac{dy_1(t)}{dt}, \tag{3.31}$$

$$x_4 = \frac{dy_2(t)}{dt}. \tag{3.32}$$

The resulting state equations are given by

$$\dot{x}_1 = x_3, \tag{3.33}$$

$$\dot{x}_2 = x_4, \tag{3.34}$$

$$\dot{x}_3 = -\frac{1}{M_1}(K_1 + K_2)x_1 + \frac{K_2}{M_1}x_2$$
$$\quad - \frac{1}{M_1}(B_1 + B_2)x_3 + \frac{B_2}{M_1}x_4, \tag{3.35}$$

$$\dot{x}_4 = \frac{K_2}{M_2}x_1 - \frac{K_2}{M_2}x_2 + \frac{B_2}{M_2}x_3 - \frac{B_2}{M_2}x_4 + \frac{f(t)}{M_2}. \tag{3.36}$$

In vector form, these equations become

$$\dot{\mathbf{x}} = \mathbf{Px} + \mathbf{Br}, \tag{3.37}$$

$$\mathbf{y} = \mathbf{Lx}, \tag{3.38}$$

where

$$\mathbf{x} = \begin{bmatrix} x_1 \\ x_2 \\ x_3 \\ x_4 \end{bmatrix}, \quad \dot{\mathbf{x}} = \begin{bmatrix} \dot{x}_1 \\ \dot{x}_2 \\ \dot{x}_3 \\ \dot{x}_4 \end{bmatrix},$$

$$\mathbf{P} = \begin{bmatrix} 0 & 0 & 1 & 0 \\ 0 & 0 & 0 & 1 \\ -\dfrac{1}{M_1}(K_1 + K_2) & \dfrac{K_2}{M_1} & -\dfrac{1}{M_1}(B_1 + B_2) & \dfrac{B_2}{M_1} \\ \dfrac{K_2}{M_2} & -\dfrac{K_2}{M_2} & \dfrac{B_2}{M_2} & -\dfrac{B_2}{M_2} \end{bmatrix}$$

$$\mathbf{B} = \begin{bmatrix} 0 \\ 0 \\ 0 \\ \dfrac{1}{M_2} \end{bmatrix} \quad \mathbf{L} = \begin{bmatrix} 1 & 0 & 0 & 0 \\ 0 & 1 & 0 & 0 \end{bmatrix}, \quad r = f(t), \quad \mathbf{y} = \begin{bmatrix} y_1 \\ y_2 \end{bmatrix}.$$

2. Mechanical Rotational Systems   The three basic characteristics of a mechanical rotational system are moment of inertia, stiffness, and damping. Rotational systems are quite similar to translational systems except that torque equations are used to describe system equilibrium instead of force equations, and we use angular displacement, velocity, and acceleration quantities. The symbols for the various quantities and their respective units are shown in Table 3.2.

Table 3.2   Mechanical rotation symbols and units

Quantity	Symbol	English units
Angle	θ	radians (rad)
Angular velocity	ω	rad/sec
Angular acceleration	α	rad/sec^2
Torque	T	lb ft
Moment of inertia	J	lb ft sec^2
Damping factor	B	lb ft/(rad/sec)
Stiffness factor	K	lb ft/rad

a) Torque Applied to a Body Having a Moment of Inertia, a Twisting Shaft, and a Damping Device. The configuration of a torque $T(t)$ applied to a body having a moment of inertia J, a twisting shaft having a stiffness factor K, and a damper having a damping factor B, is shown in Fig. 3.7. The system produces a displacement, $\theta(t)$, measured from an equilibrium position θ_0, which is assumed to be zero. The reference end of the damping device is assumed to be stationary. Applying Newton's law to this system yields

$$J\frac{d^2\theta(t)}{dt^2} + B\frac{d\theta(t)}{dt} + K\theta(t) = T(t). \tag{3.39}$$

This second-order differential equation can be written as two first-order differential equations by defining

$$x_1 = \theta(t), \tag{3.40}$$

$$x_2 = \dot{\theta}(t). \tag{3.41}$$

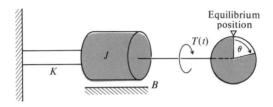

Fig. 3.7 Torque applied to a moment of inertia, a twisting shaft, and a damping device.

The resulting state equations are given by

$$\dot{x}_1 = x_2, \tag{3.42}$$

$$\dot{x}_2 = -\frac{K}{J}x_1 - \frac{B}{J}x_2 + \frac{1}{J}T(t). \tag{3.43}$$

In vector form, these equations become

$$\dot{\mathbf{x}} = \mathbf{P}\mathbf{x} + \mathbf{B}r, \tag{3.44}$$

$$\boldsymbol{\theta} = \mathbf{L}\mathbf{x}, \tag{3.45}$$

where

$$\mathbf{x} = \begin{bmatrix} x_1 \\ x_2 \end{bmatrix}, \quad \dot{\mathbf{x}} = \begin{bmatrix} \dot{x}_1 \\ \dot{x}_2 \end{bmatrix}, \quad \mathbf{P} = \begin{bmatrix} 0 & 1 \\ -\dfrac{K}{J} & -\dfrac{B}{J} \end{bmatrix}, \quad \mathbf{B} = \begin{bmatrix} 0 \\ \dfrac{1}{J} \end{bmatrix},$$

$$\mathbf{L} = [1 \quad 0], \quad r = T(t).$$

3.4 TRANSFER-FUNCTION AND STATE-SPACE REPRESENTATION OF TYPICAL ELECTRICAL CONTROL-SYSTEM DEVICES

This section illustrates the procedure used for deriving the transfer function and state-space representation of some commonly used electrical devices from their basic differential equations. We specifically analyze a dc generator, armature-controlled dc servomotor, Ward-Leonard system, field-controlled dc servomotor, and a two-phase ac induction servomotor. The analysis is limited to the linear operating range of these devices.

1. DC Generator A dc generator [4, 5] is commonly used in control systems for power amplification. The armature, which is driven at a constant speed n, is capable of producing a relatively large controllable current, i_a, as the field current i_f is varied.

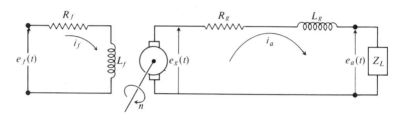

Fig. 3.8 DC generator schematic diagram.

The exact value of i_a is dependent on the load circuit impedance, Z_L. A schematic diagram of the configuration is shown in Fig. 3.8. The symbols R_f, L_f and R_g and L_g represent the resistive and inductive components of the field and armature circuits, respectively.

The voltage induced in the armature, $e_g(t)$, is a function of the speed of rotation n and the flux developed by the field, ϕ. It can be expressed as

$$e_g(t) = K_1 n \phi. \tag{3.46}$$

The flux depends on the field current and the characteristics of the iron used in the field. This is a linear relationship up to a certain saturation point and can be expressed as

$$\phi = K_2 i_f. \tag{3.47}$$

By substituting Eq. (3.47) into Eq. (3.46), and assuming that the armature rotation speed is constant, the relation between the induced armature voltage $e_g(t)$ and the field current i_f can be expressed as

$$e_g(t) = K_g i_f, \tag{3.48}$$

where $K_g = K_1 K_2 n$ = generator constant having units of V/A. The equation relating the field voltage $e_f(t)$, and field current $i_f(t)$, is

$$e_f(t) = R_f i_f + L_f \frac{di_f}{dt}. \tag{3.49}$$

The field current can be eliminated between Eqs. (3.48) and (3.49) and an expression relating $e_f(t)$ and the induced armature voltage $e_g(t)$ can be obtained.

$$e_f(t) = \frac{R_f}{K_g} e_g(t) + \frac{L_f}{K_g} \frac{de_g(t)}{dt}. \tag{3.50}$$

By transforming, there results

$$E_f(s) = \frac{1}{K_g} (R_f + L_f s) E_g(s). \tag{3.51}$$

The transfer function of this device, defined as the ratio of the output $E_g(s)$ to the input $E_f(s)$, is given by

$$\frac{E_g(s)}{E_f(s)} = \frac{K_g/L_f}{s + R_f/L_f} \tag{3.52}$$

and its simple block diagram is shown in Fig. 3.9.

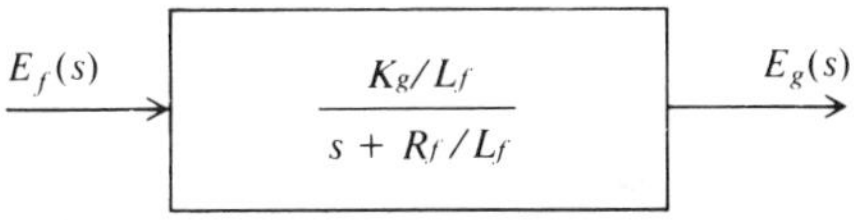

$$E_f(s) \longrightarrow \boxed{\frac{K_g/L_f}{s + R_f/L_f}} \longrightarrow E_g(s)$$

Fig. 3.9   Block diagram of a DC generator.

If it is desired to obtain the transfer function $E_a(s)/E_f(s)$, we must first determine the nature of the actual load connected to the armature. For example, assume that the load impedance is $Z_L(s)$. Then the transfer function $E_a(s)/E_g(s)$ becomes

$$\frac{E_a(s)}{E_g(s)} = \frac{Z_L(s)}{R_g + L_g s + Z_L(s)} \tag{3.53}$$

and the overall transfer function of the dc generator, $E_a(s)/E_f(s)$, is

$$\frac{E_a(s)}{E_f(s)} = \frac{E_g(s)}{E_f(s)} \times \frac{E_a(s)}{E_g(s)} = \frac{K_g/L_f}{s + R_f/L_f} \times \frac{Z_L(s)}{R_g + L_g s + Z_L(s)}. \tag{3.54}$$

2. Armature-Controlled DC Servomotor   Armature-controlled dc servomotors [6] are quite commonly used in control systems. As a matter of fact, a dc generator driving an armature-controlled dc servomotor is known as the Ward-Leonard system. We will study this configuration in the next section drawing on the relations derived for the dc generator and armature-controlled dc servomotor.

     A schematic diagram of the armature-controlled dc servomotor is shown in Fig. 3.10(a). The symbols R_m and L_m represent the resistive and inductive components of the armature circuit. The field excitation is constant, being supplied from a dc source. The motor is shown driving a load having an inertia J and damping B.

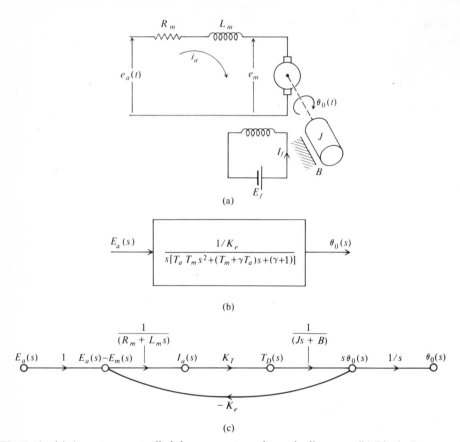

Fig. 3.10 (a) Armature-controlled dc servomotor schematic diagram. (b) Block diagram of an armature-controlled dc motor. (c) Signal-flow diagram representation for the armature-controlled dc servomotor.

As the armature rotates, it develops an induced voltage e_m which is in a direction opposite to $e_a(t)$. The induced voltage is proportional to the speed of rotation n and the flux created by the field current. Since we are assuming that the field current is held constant, the flux must be constant. Therefore, the induced armature voltage is only dependent on the speed of rotation and can be expressed as

$$e_m = K_e n = K_e \frac{d\theta_0}{dt}, \tag{3.55}$$

where K_e = voltage constant of the motor having units of V/(rad/sec). The voltage equation of the armature circuit is

$$e_a(t) = R_m i_a + L_m \frac{di_a}{dt} + e_m. \tag{3.56}$$

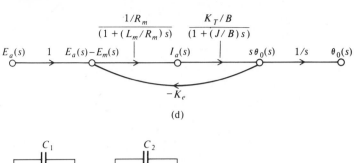

(d)

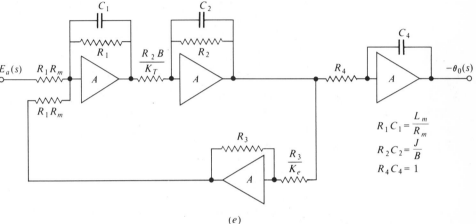

(e)

Fig. 3.10 (d) Modification of the signal-flow diagram to a more compact form. (e) Analog computer circuit simulation for an armature-controlled dc servomotor.

Substituting Eq. (3.55) into Eq. (3.56) and taking the Laplace transform, we obtain

$$E_a(s) = (R_m + L_m s)I_a(s) + K_e s\theta_0(s). \qquad (3.57)$$

The developed torque of the motor, T_D, is a function of the flux developed by the field current, the armature current, and the length and number of the conductors. Since we are assuming that the field current is held constant, the developed torque T_D can be expressed as

$$T_D = K_T i_a \qquad (3.58)$$

where K_T = torque constant of the motor having units of lb ft/A.* The developed torque drives the mechanical load and the torque equation is

$$T_D = J \frac{d^2\theta_0}{dt^2} + B \frac{d\theta_0}{dt}. \qquad (3.59)$$

Substituting Eq. (3.58) into Eq. (3.59) and taking the Laplace transform, we obtain

$$K_T I_a(s) = (Js^2 + Bs)\theta_0(s). \qquad (3.60)$$

---

* The torque constant is a very important characteristic of dc motors.

The overall system transfer function $\theta_0(s)/E_a(s)$ obtained by eliminating $I_a(s)$ between Eqs. (3.57) and (3.60) is

$$\frac{\theta_0(s)}{E_a(s)} = \frac{K_T}{JL_m s^3 + (R_m J + L_m B)s^2 + (R_m B + K_e K_T)s}. \tag{3.61}$$

Equation (3.61) can be defined in terms of an armature time constant T_a, a motor time constant T_m, and a damping factor γ, as follows:

$$\frac{\theta_0(s)}{E_a(s)} = \frac{1/K_e}{s[T_a T_m s^2 + (T_m + \gamma T_a)s + (\gamma + 1)]} \tag{3.62}$$

where

$$T_a = \frac{L_m}{R_m}, \qquad T_m = \frac{JR_m}{K_e K_T}, \qquad \gamma = \frac{R_m B}{K_e K_T}. \tag{3.63}$$

The simple block diagram of this system is illustrated in Fig. 3.10(b).

The state-space model can be derived directly for the armature-controlled dc servomotor from the original set of defining scalar equations, (3.55), (3.56), (3.58), and (3.59). Defining the state variables as

$$x_1 = \theta_0, \tag{3.64}$$

$$x_2 = \dot{\theta}_0, \tag{3.65}$$

$$x_3 = i_a, \tag{3.66}$$

the resulting state equations are given by

$$\dot{x}_1 = x_2, \tag{3.67}$$

$$\dot{x}_2 = -\frac{B}{J}x_2 + \frac{K_T}{J}x_3, \tag{3.68}$$

$$\dot{x}_3 = -\frac{K_e}{L_m}x_2 - \frac{R_m}{L_m}x_3 + \frac{1}{L_m}e_a. \tag{3.69}$$

In vector form, these equations become

$$\dot{\mathbf{x}} = \mathbf{P}\mathbf{x} + \mathbf{B}r, \tag{3.70}$$

$$\mathbf{c} = \mathbf{L}\mathbf{x}, \tag{3.71}$$

where

$$\mathbf{x} = \begin{bmatrix} x_1 \\ x_2 \\ x_3 \end{bmatrix}, \qquad \dot{\mathbf{x}} = \begin{bmatrix} \dot{x}_1 \\ \dot{x}_2 \\ \dot{x}_3 \end{bmatrix}, \qquad \mathbf{B} = \begin{bmatrix} 0 \\ 0 \\ \dfrac{1}{L_m} \end{bmatrix}, \qquad \mathbf{P} = \begin{bmatrix} 0 & 1 & 0 \\ 0 & -\dfrac{B}{J} & \dfrac{K_T}{J} \\ 0 & -\dfrac{K_e}{L_m} & -\dfrac{R_m}{L_m} \end{bmatrix},$$

$$\mathbf{c} = \begin{bmatrix} \theta_0 \\ i_a \end{bmatrix}, \qquad \mathbf{L} = \begin{bmatrix} 1 & 0 & 0 \\ 0 & 0 & 1 \end{bmatrix}, \qquad r = e_a.$$

It is usually quite interesting and revealing to study the signal-flow diagram for such a device, especially for purposes of analog-computer simulation. The governing equations from which the signal-flow diagram can be drawn for the armature-controlled dc servomotor are given by

$$E_m(s) = K_e s \theta_0(s) \qquad \text{[from Eq. (3.55)]} \qquad (3.72)$$

$$I_a(s) = \frac{E_a(s) - E_m(s)}{R_m + L_m s} \qquad \text{[from Eq. (3.56)]} \qquad (3.73)$$

$$T_D(s) = K_T I_a(s) \qquad \text{[from Eq. (3.58)]} \qquad (3.74)$$

$$\theta_0(s) = \frac{1}{s} \frac{T_D(s)}{(Js + B)} \qquad \text{[from Eq. (3.59)]} \qquad (3.75)$$

The signal-flow diagram is illustrated in Fig. 3.10(c). It clearly illustrates the inherent feedback (back electromotive force) of this device. This property is sometimes used to stabilize a feedback control system (see Problem 3.8). It is left as an exercise to the reader to prove that the transfer function of this system, as derived from the signal-flow diagram, agrees with Eqs. (3.62) and (3.63), and that the state equations derived from the state-variable diagram agree with Eqs. (3.70) and (3.71) (see Problem 3.21).

Since the signal-flow graphs represent the mathematical structure of a system, they represent a type of basic analog. Likewise, an analog computer may be thought of as the physical realization of a signal-flow graph. However, any arbitrary signal-flow graph is not necessarily a suitable representation as the program for an analog computer. This is usually due to the fact that the control engineer avoids simulating differentiations which have wide bandwidths and associated noise problems. For the case of the armature-controlled dc servomotor, the signal-flow diagram of Fig. 3.10(c) is modified to a more compact form in Fig. 3.10(d) which is more appropriate for analog simulation. The analog-computer simulation circuitry is shown in Fig. 3.10(e) [16, 17].

3. The Ward-Leonard System A configuration having a dc generator driving an armature-controlled dc motor is known as a Ward-Leonard system [7, 8]. The dc generator acts as a rotating power amplifier that supplies the power which, in turn, drives the servomotor. Variations of the conventional Ward-Leonard system are known as the Amplidyne, the Metadyne, the Rotorol, and the Regulex. The reader is referred to more authoritative books on dc machinery [7] for a description of these devices.

A schematic diagram of the basic Ward-Leonard system is shown in Fig. 3.11. The notations used are the same as those of Figs. 3.8 and 3.10. Many sophisticated variations of this basic configuration, using compensating windings, exist [8].

To enable us to combine the transfer-function relationships derived previously for the dc generator and armature-controlled dc motor, we assume that the generator voltage $e_g(t)$ is applied directly to the armature of the motor. Therefore, we are

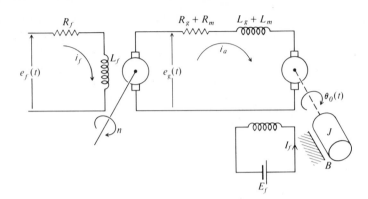

Fig. 3.11   Ward-Leonard system schematic diagram.

interested in applying the generator Eq. (3.52) rather than (3.54). In order to apply the motor transfer function (3.62), we must first combine the resistive and inductive components of the generator's and motor's armatures. This will result in a set of new, modified motor constants as follows:

$$\gamma' = \frac{(R_g + R_m)B}{K_e K_T}, \qquad T'_a = \frac{L_g + L_m}{R_g + R_m}, \qquad T'_m = \frac{J(R_g + R_m)}{K_e K_T}. \tag{3.76}$$

Therefore, Eq. (3.62) may be written as follows in this case:

$$\frac{\theta_0(s)}{E_g(s)} = \frac{1/K_e}{s[T'_a T'_m s^2 + (T'_m + \gamma' T'_a)s + (\gamma' + 1)]}. \tag{3.77}$$

It is now relatively simple to obtain the transfer-function representation of the configuration shown in Fig. 3.11. Defining $e_f(t)$ as the input and $\theta_0(t)$ as the output of the system, we need merely to combine the transfer functions given by Eqs. (3.52) and (3.77). Therefore, the system transfer function is given by

$$\frac{\theta_0(s)}{E_f(s)} = \frac{K_g/L_f}{s + R_f/L_f} \frac{1/K_e}{s[T'_a T'_m s^2 + (T'_m + \gamma' T'_a)s + (\gamma' + 1)]} \tag{3.78}$$

and its simple block diagram is illustrated in Fig. 3.12.

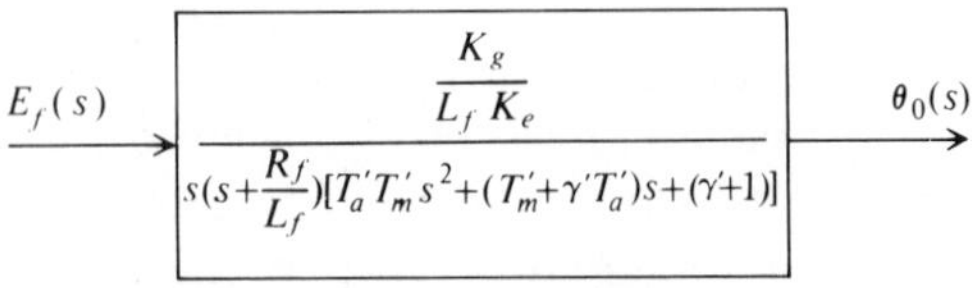

$$E_f(s) \longrightarrow \boxed{\dfrac{\dfrac{K_g}{L_f K_e}}{s\left(s + \dfrac{R_f}{L_f}\right)[T'_a T'_m s^2 + (T'_m + \gamma' T'_a)s + (\gamma' + 1)]}} \longrightarrow \theta_0(s)$$

Fig. 3.12   Block diagram of the Ward-Leonard system.

4. Field-Controlled DC Servomotor A dc servomotor [9] can also be controlled by varying the field current and maintaining a constant armature current. The schematic diagram of such a configuration, known as the field-controlled dc servomotor, is shown in Figure 3.13. The notations are the same as those of Fig. 3.10.

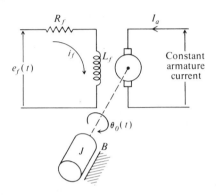

Fig. 3.13   Field-controlled dc servomotor schematic diagram.

The developed torque of the motor, T_D, is a function of the flux developed by the armature current, the field current, and the length of the conductors. Since we are assuming that the armature current is held constant, the developed torque T_D can be expressed as

$$T_D = K_T i_f \tag{3.79}$$

where K_T = torque constant of the motor having units of lb ft/A. The developed torque is used to drive the mechanical load with inertia J and damping B. The torque equation is

$$T_D = J \frac{d^2\theta_0}{dt^2} + B \frac{d\theta_0}{dt}. \tag{3.80}$$

Substituting Eq. (3.79) into Eq. (3.80), we obtain a differential equation relating the field current and the output shaft position:

$$i_f = \frac{J}{K_T} \frac{d^2\theta_0}{dt^2} + \frac{B}{K_T} \frac{d\theta_0}{dt}. \tag{3.81}$$

An expression for the value of the field current i_f can be obtained from the voltage equation of the field circuit.

$$e_f(t) = R_f i_f + L_f \frac{di_f}{dt}. \tag{3.82}$$

Substituting Eq. (3.81) into Eq. (3.82) and taking the Laplace transform, we obtain

$$E_f(s) = (R_f + L_f s)\left(\frac{J}{K_T} s^2 + \frac{B}{K_T} s\right)\theta_0(s). \tag{3.83}$$

The transfer function of the device, defined as the output $\theta_0(s)$ divided by the input $E_f(s)$, is given by

$$\frac{\theta_0(s)}{E_f(s)} = \frac{K_T/R_f B}{s(1 + T_f s)(1 + T_m s)}, \tag{3.84}$$

where

$$T_f = \frac{L_f}{R_f} = \text{field time constant,}$$

$$T_m = \frac{J}{B} = \text{motor time constant,}$$

and its simple block diagram is illustrated in Fig. 3.14(a).

The governing equations from which the signal-flow diagram can be drawn for the field-controlled dc servomotor are given by

$$I_f(s) = \frac{E_f(s)}{R_f + L_f s} \qquad \text{[from Eq. (3.82)]} \tag{3.85}$$

$$T_D(s) = K_T I_f(s) \qquad \text{[from Eq. (3.79)]} \tag{3.86}$$

$$\theta_0(s) = \frac{1}{s} \frac{T_D(s)}{Js + B} \qquad \text{[from Eq. (3.80)]} \tag{3.87}$$

The simple signal-flow diagram of this device, useful for analog computer simulations, is illustrated in Fig. 3.14(b).

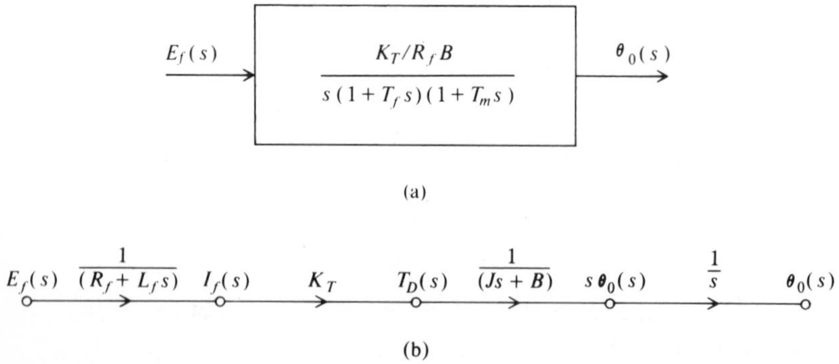

(a)

(b)

Fig. 3.14 (a) Block diagram of a field-controlled dc servomotor. (b) Signal-flow diagram representation for the field-controlled dc servomotor.

The state-space representation of the field-controlled dc servomotor can be obtained either by starting from the defining scalar equations, (3.81) and (3.82), or by utilizing the resultant transfer function derived in Eq. (3.84). The first method is usually preferable. However, since the first method was utilized in the case of the armature-controlled dc servomotor, the second method is utilized here for the field-controlled dc servomotor to permit comparison of the two approaches.

Equation (3.84) can be written in differential-equation form as follows:

$$T_f T_m \frac{d^3\theta_0(t)}{dt^3} + (T_f + T_m) \frac{d^2\theta_0(t)}{dt^2} + \frac{d\theta_0(t)}{dt} = \frac{K_T}{R_f B} e_f(t). \qquad (3.88)$$

This third-order differential equation can be written as three first-order differential equations by defining

$$x_1 = \theta_0(t), \qquad (3.89)$$
$$x_2 = \dot{\theta}_0(t), \qquad (3.90)$$
$$x_3 = \ddot{\theta}_0(t). \qquad (3.91)$$

The resulting state equations are given by

$$\dot{x}_1 = x_2, \qquad (3.92)$$
$$\dot{x}_2 = x_3, \qquad (3.93)$$
$$\dot{x}_3 = -\frac{1}{T_f T_m} x_2 - \frac{(T_f + T_m)}{T_f T_m} x_3 + \frac{K_T}{R_f B T_f T_m} e_f(t). \qquad (3.94)$$

In vector form, these equations become

$$\dot{\mathbf{x}} = \mathbf{P}\mathbf{x} + \mathbf{B}r, \qquad (3.95)$$
$$\boldsymbol{\theta}_0 = \mathbf{L}\mathbf{x}, \qquad (3.96)$$

where

$$\mathbf{x} = \begin{bmatrix} x_1 \\ x_2 \\ x_3 \end{bmatrix}, \quad \dot{\mathbf{x}} = \begin{bmatrix} \dot{x}_1 \\ \dot{x}_2 \\ \dot{x}_3 \end{bmatrix}, \quad \mathbf{P} = \begin{bmatrix} 0 & 1 & 0 \\ 0 & 0 & 1 \\ 0 & -\dfrac{1}{T_f T_m} & -\dfrac{(T_f + T_m)}{T_f T_m} \end{bmatrix}, \quad \mathbf{B} = \begin{bmatrix} 0 \\ 0 \\ \dfrac{K_T}{R_f B T_f T_m} \end{bmatrix},$$

$$\mathbf{L} = [1 \quad 0 \quad 0], \qquad r = e_f(t).$$

Notice that the resultant vector equation has eliminated the field current. This is a measurable state and it is preferable to retain it in the state-space model. If the state-space representation had proceeded from the defining scalar equations, (3.81) and (3.82), then the field current would have been retained—just as the armature current was retained as a state in the case of the armature-controlled dc servomotor (see Eq. 3.66).

5. Two-Phase AC Servomotor The two-phase ac servomotor [10–12] is probably the most commonly used type of servomotor. Its popularity stems from the fact that many error-sensing devices are carrier-frequency (ac) devices. By using two-phase ac servomotors, demodulation need not be performed and ac amplification can be used throughout the electrical portion of the system.

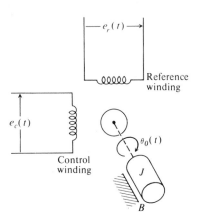

Fig. 3.15   Two-phase ac servomotor schematic diagram.

The ac servomotor is a two-phase induction motor having its two stator coils separated by 90 electrical degrees with a high resistance rotor. A control signal is applied to one phase (the control winding) while the other phase (the reference winding) is supplied with a fixed signal that is phase-shifted by $90°$ relative to the control signal. The motor is used primarily for relatively low-power applications. A schematic diagram of an ac servomotor driving a load of inertia J and damping B is shown in Fig. 3.15. The reference field voltage is denoted as $e_r(t)$ and the control field voltage is denoted as $e_c(t)$. The developed torque of this motor is proportional to $e_r(t)$, $e_c(t)$, and the sine of the angle between $e_r(t)$ and $e_c(t)$.

As the control voltage is varied, the torque T_D and speed n vary. A set of torque–speed curves for various values of control voltage are shown in Fig. 3.16. Notice that these curves show a very large torque for zero speed which is desirable in developing a very rapid acceleration.

Unfortunately, the torque–speed curves are not straight lines. Therefore we cannot write a linear differential equation to represent them. However, by linearizing these characteristics, reasonable accuracy can be achieved.* Since the developed torque T_D is a function of the speed n and the control voltage $e_c(t)$, we use a Taylor

---

* The subject of linearizing approximations is discussed from a more general viewpoint in Section 8.6.

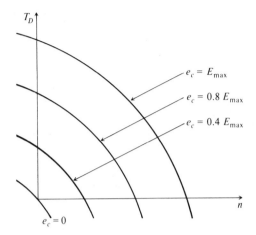

Fig. 3.16   Two-phase ac servomotor torque-speed characteristics.

series expansion of the nonlinear function $T_D = T_D(n, e_c)$ to get an approximate linear function:

$$\frac{\partial T_D}{\partial n} n + \frac{\partial T_D}{\partial e_c} e_c(t) = T_D(n, e_c). \tag{3.97}$$

By defining

$$\frac{\partial T_D}{\partial e_c} \equiv K_e, \qquad \frac{\partial T_D}{\partial n} \equiv K_n \qquad \text{(where } K_n \text{ is a negative number)}$$

and substituting $n = d\theta_0/dt$, we can rewrite Eq. (3.97) as

$$K_n \frac{d\theta_0}{dt} + K_e e_c(t) = T_D. \tag{3.98}$$

The developed torque is used to drive the mechanical load and the torque equation is

$$T_D = J \frac{d^2\theta_0}{dt^2} + B \frac{d\theta_0}{dt}. \tag{3.99}$$

Substituting Eq. (3.99) into Eq. (3.98) and taking the Laplace transform, we obtain

$$K_n s\theta_0(s) + K_e E_c(s) = Js^2\theta_0(s) + Bs\theta_0(s). \tag{3.100}$$

The transfer function of the device, defined as the output $\theta_0(s)$ divided by the input $E_c(s)$, is given by

$$\frac{\theta_0(s)}{E_c(s)} = \frac{K_m}{s(T_m s + 1)}, \tag{3.101}$$

where

$$K_m = \frac{K_e}{B - K_n} = \text{motor constant,}$$

$$T_m = \frac{J}{B - K_n} = \text{motor time constant,}$$

and its simple block diagram is illustrated in Fig. 3.17. Although the torque–speed characteristics are nonlinear, as illustrated in Fig. 3.16, the value of K_n is usually obtained graphically by drawing a straight line through the two end points. The

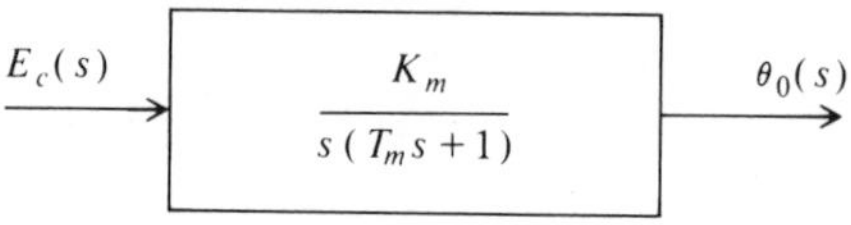

$$E_c(s) \longrightarrow \boxed{\dfrac{K_m}{s(T_m s + 1)}} \longrightarrow \theta_0(s)$$

Fig. 3.17 Block diagram of a two-phase ac servomotor.

accuracy of this approach, compared with drawing a line tangent to the curve at various other points, is analyzed further in Section 8.6 (see Problem 8.1).

Equation (3.101) can be written in differential-equation form as follows:

$$T_m \frac{d^2\theta_0(t)}{dt^2} + \frac{d\theta_0(t)}{dt} = K_m e_c(t). \tag{3.102}$$

This second-order differential equation can be written as two first-order differential equations by defining

$$x_1 = \theta_0(t), \tag{3.103}$$
$$x_2 = \dot{\theta}_0(t). \tag{3.104}$$

The resulting state equations are given by

$$\dot{x}_1 = x_2, \tag{3.105}$$

$$\dot{x}_2 = -\frac{x_2}{T_m} + \frac{K_m}{T_m} e_c(t). \tag{3.106}$$

In vector form, these equations become

$$\dot{\mathbf{x}} = \mathbf{Px} + \mathbf{B}r, \tag{3.107}$$
$$\theta_0 = \mathbf{Lx}, \tag{3.108}$$

where

$$\mathbf{x} = \begin{bmatrix} x_1 \\ x_2 \end{bmatrix}, \quad \dot{\mathbf{x}} = \begin{bmatrix} \dot{x}_1 \\ \dot{x}_2 \end{bmatrix}, \quad \mathbf{P} = \begin{bmatrix} 0 & 1 \\ 0 & -\dfrac{1}{T_m} \end{bmatrix}, \quad \mathbf{B} = \begin{bmatrix} 0 \\ \dfrac{K_m}{T_m} \end{bmatrix},$$

$$\mathbf{L} = [1 \quad 0], \quad r = e_c(t).$$

It is left as an exercise to the reader to prove that the state equations derived from the state-variable diagram agree with Eqs. (3.107) and (3.108) (see Problem 3.23).

6. Servomotors with Gear Reducers The two-phase ac servomotor is inherently a high-speed, low-torque device. In control-system practice, however, what is usually needed is a low-speed, high-torque device. Therefore, a gear reduction is usually required between the high-speed, low-torque servomotor and the load to obtain speed reduction and torque magnification.

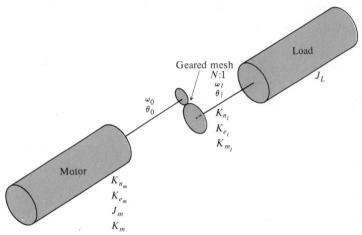

Fig. 3.18 Servomotor geared to load.

In order to analyze the modifications necessary to the transfer function derived in Eq. (3.101), let us consider the configuration illustrated in Fig. 3.18. The sub-subscripts m and l are used to denote motor and load shafts, respectively. It is assumed initially that the inertias of the gears are negligible, the efficiency of trans-mission is perfect, and there is only one gear mesh between the load and the motor. It is also assumed that a servomotor and gear train have been selected which will provide the required torque to obtain the desired acceleration, and to drive the load at the maximum required speed. Basically what is desired is a set of relationships to include the effect of the gear reduction and the load inertia.

There are basically two approaches to the problem. In one method, the parameters K_{n_m}, K_{e_m}, and J_m are reflected to the output shaft, θ_l. Another manner of solving the problem is to reflect the load inertia to the motor shaft. Both methods will be illustrated here, and it will be shown that they are equivalent.

Let us first consider the reflection of the parameters K_{n_m}, K_{e_m}, and J_m to the output shaft, θ_l. K_{n_m} is the ratio of motor torque to motor speed and the value of K_{n_m} is given by

$$K_{n_m} = \frac{\partial T_D}{\partial n}. \qquad (3.109)$$

Since the driving torque at the load is multiplied by the gear reduction factor N, and the motor speed is reduced by this factor at the load shaft, the following relationship is obtained:

$$K_{n_l} = \frac{N}{1/N} \frac{\partial T_D}{\partial n} = N^2 K_{n_m},\qquad(3.110)$$

where K_{n_l} denotes K_n referred to the load shaft.

The parameter K_{e_m} is the ratio of motor torque to control voltage. Since the driving torque of the load is multiplied by the gear reduction factor N, and the control voltage e_c is independent of N,

$$K_{e_l} = N K_{e_m},\qquad(3.111)$$

where K_{e_m} indicates the value of K_e at the motor shaft and K_{e_l} denotes its value referred to the load shaft. The effect of the motor inertia at the load shaft can be obtained by considering the conservation of energy equation:

$$\tfrac{1}{2}J_m \omega_0^2 = \tfrac{1}{2}J_{m_r}\omega_l^2,\qquad(3.112)$$

where J_{m_r} denotes the value of the motor inertia referred to the output shaft. On substitution, the following is obtained:

$$\tfrac{1}{2}J_m \omega_0^2 = \tfrac{1}{2}J_{m_r}\left(\frac{\omega_0}{N}\right)^2.\qquad(3.113)$$

Therefore,

$$J_{m_r} = J_m N^2.\qquad(3.114)$$

This relationship indicates that the motor inertia referred to the output shaft is amplified by a factor of N^2. This factor must be added to the load inertia, J_L, in order to obtain the total output inertia:

$$J_{\text{total}_L} = J_m N^2 + J_L.$$

Based on these relationships, the values of K_m and T_m for the overall combination of servomotor, gear train and load inertia can now be obtained. To do this, it is assumed that the damping factor B in Eq. (3.101) is zero (it is usually negligible compared with the motor damping K_n):

$$K_{m_l} = \frac{N K_{e_m}}{-N^2 K_{n_m}} = \frac{1}{N} K_m,\qquad(3.115)$$

where K_{m_l} denotes the value of K_m referred to the output shaft. This indicates that the system gain is reduced by a factor of N. Furthermore,

$$T_{m_{\text{total}}} = \frac{J_m N^2 + J_L}{-N^2 K_{n_m}} = T_m - \frac{J_L}{N^2 K_{n_m}},\qquad K_{n_m} < 0.\qquad(3.116)$$

This indicates an increase in the servomotor time constant due to the added load inertia. Therefore, the net effect of the gear reduction and the load inertia has been to reduce K_m by a factor of N, and to increase the time constant by the added term $J_L/N^2 K_{n_m}$.

Now let us attack the problem by referring all quantities to the motor shaft and check the result. For this case, the only two factors that have to be considered are the effect of the gear reduction N and the load inertia J_L. As we have just observed, the effect of the gear reduction N is basically a loss in system gain N. Therefore, it must be treated as a box having a transfer function $1/N$. What about the load inertia J_L? Let us reconsider Eq. (3.112) and reflect the load inertia to the servomotor:

$$\tfrac{1}{2} J_{L_r}(N\omega_l)^2 = \tfrac{1}{2} J_L \omega_l^2,$$

where J_{L_r} is the load inertia referred to the motor shaft. Therefore,

$$J_{L_r} = \frac{J_L}{N^2}.$$

This indicates that the load inertia at the motor shaft is reduced by a factor of N^2. We are now ready to determine the values of K_m and T_m of the servomotor and gear reduction combination. As before, B is assumed to be zero. Therefore,

$$K_{m_l} = \frac{1}{N} K_m, \tag{3.117}$$

and

$$T_{m_{total}} = \frac{J_m}{-K_{n_m}} + \frac{J_L}{-N^2 K_{n_m}} = T_m - \frac{J_L}{N^2 K_{n_m}}, \qquad K_{n_m} < 0. \tag{3.118}$$

Observe the agreement between Eqs. (3.115) and (3.117) and between Eqs. (3.116) and (3.118).

The results derived here can easily be extended to multimesh gear trains as illustrated in Fig. 3.19. In this figure, a two-mesh system is illustrated, and the inertia of the respective shafts and gearing combinations are indicated as J_1, J_2, and J_3. It is left as an exercise to the reader to show that the equivalent inertia reflected to the input (motor) shaft is

$$J_{eq_i} = J_1 + \frac{J_2}{N_1^2} + \frac{J_3}{N_1^2 N_2^2},$$

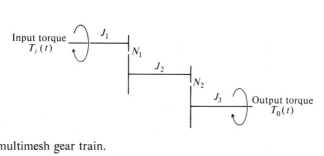

Fig. 3.19 A multimesh gear train.

and the equivalent inertia reflected to the output shaft is given by

$$J_{eq_0} = J_1 N_1^2 N_2^2 + J_2 N_2^2 + J_3.$$

Before leaving the subject of gear ratios, the question is posed as to what is the optimum gear ratio for a given motor and load configuration? This problem can be answered by reconsidering the servomotor, gear train, and load configuration illustrated in Fig. 3.18. The torque developed by the motor in accelerating its own inertia, J_m, and the load inertia, J_L, through the gear ratio N, is given by

$$T_D = J_M(\alpha_l)N + \frac{J_L}{N^2}(\alpha_l)N, \tag{3.119}$$

where α_l represents the maximum required acceleration of the load. Equation (3.119) represents the developed torque at the motor shaft with the load inertia component appropriately reflected to it. In order to find the optimum gear ratio N which will minimize the required motor torque and therefore the size of the motor, Eq. (3.119) is differentiated with respect to N and is set equal to zero:

$$\alpha_l \left[J_m - \frac{J_L}{N^2} \right] = 0. \tag{3.120}$$

Therefore, the gear ratio required to minimize the motor torque is given by

$$N = \sqrt{\frac{J_L}{J_m}}, \tag{3.121}$$

where the gear ratio is the geometric mean of the ratio of the load to motor inertias. Equation (3.121) may also be interpreted from the viewpoint that the value of motor inertia for minimum motor torque required is given by

$$J_m = \frac{J_L}{N^2}. \tag{3.122}$$

Therefore, Eq. (3.122) states that the motor torque is minimized when the motor and load inertias are equal when referred to a common shaft.

Note that this result assumes that the only effect that is important is acceleration of the load. If a maximum load velocity is specified and the motor can only reach a given maximum velocity, the choice of optimum gear ratio is a more complex problem.

3.5 TRANSFER-FUNCTION AND STATE-SPACE REPRESENTATION OF TYPICAL HYDRAULIC DEVICES

Hydraulic components are commonly found in control systems that are either all hydraulic or a combination of electromechanical and hydraulic devices [6, 13]. The procedures used for deriving the transfer function representation of some commonly used hydraulic control system devices from their basic differential equations are illustrated in this section. We specifically consider hydraulic motors, pumps, and valves.

1. Hydraulic Motor and Pump There is no essential difference between a hydraulic pump and motor, just as there is no essential difference between a dc generator and a dc motor. Basically, the hydraulic device is classified as a motor if the input is hydraulic flow or pressure and the output is mechanical position; or a pump if the input is mechanical torque and the output is hydraulic flow or pressure.

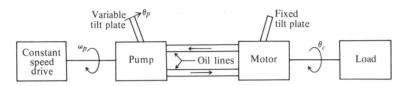

Fig. 3.20   Hydraulic power transmission system.

Figure 3.20 illustrates a commonly used hydraulic power transmission system. This device, which is capable of controlling large torques, consists of a variable displacement pump that is driven at a constant speed. A control stroke, which determines the quantity of oil pumped, also controls the direction of fluid flow. The angular velocity of the hydraulic motor is proportional to the volumetric flow and is in the same direction as the oil flow from the pump. A functional diagram of the hydraulic transmission is illustrated in Fig. 3.21.

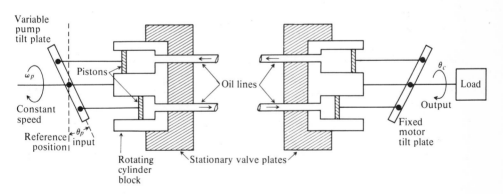

Fig. 3.21   Functional block diagram of a hydraulic power transmission system.

The amount of oil displaced per revolution of the hydraulic pump is a function of the tilt angle θ_p. When $\theta_p = 0°$, there is no flow in the oil lines. As θ_p is increased in the positive direction, more oil flows in the lines with the direction shown. When θ_p is negative, the direction of oil flow reverses.

In order to derive the differential equation relating $\theta_p(t)$ and $\theta_c(t)$, we must define certain hydraulic quantities. The volume of oil flowing from the pump, Q_p, is composed of flow of oil through the motor, Q_m, leakage flow around the motor, Q_l, and compressibility flow, Q_c, that is,

$$Q_p = Q_m + Q_l + Q_c. \tag{3.123}$$

It can be shown that

$$Q_p = K_p\theta_p, \tag{3.124}$$

where

K_p = volumetric pump flow per second per angular displacement of θ_p
θ_p = displacement of the pump stroke;

$$Q_m = V_m\omega_c, \tag{3.125}$$

where

V_m = volumetric motor displacement
ω_c = angular velocity of motor shaft;

$$Q_l = LP_L, \tag{3.126}$$

where

L = leakage coefficient of complete system (ft³/sec)/(lb/ft²),
P_L = load-induced pressure drop across motor (lb/ft²); and

$$Q_c = \frac{dV}{dt} = \frac{V}{K_B}\frac{dP_L}{dt}, \tag{3.127}$$

where

V = total volume of liquid under compression (ft³),
K_B = bulk modulus of oil, (lb/ft²).

Substituting Eqs. (3.124) to (3.127) into Eq. (3.123), we obtain the relationship

$$K_p\theta_p = V_m\omega_c + LP_L + \frac{V}{K_B}\frac{dP_L}{dt}. \tag{3.128}$$

The torque available to drive the motor inertia is

$$\text{torque} = V_mP_L = J\frac{d^2\theta_c}{dt^2}. \tag{3.129}$$

Substituting Eq. (3.129) into (3.128), we obtain

$$K_p\theta_p = V_m\frac{d\theta_c}{dt} + \frac{L}{V_m}J\frac{d^2\theta_c}{dt^2} + \frac{V}{K_B}\frac{J}{V_m}\frac{d^3\theta_c}{dt^3}. \tag{3.130}$$

The Laplace transform of Eq. (3.130) is given by

$$K_p\theta_p(s) = V_ms\theta_c(s) + \frac{LJ}{V_m}s^2\theta_c(s) + \frac{VJ}{K_BV_m}s^3\theta_c(s). \tag{3.131}$$

The transfer function of this device, defined as the ratio of the output $\theta_c(s)$ to the input $\theta_p(s)$, is given by

$$\frac{\theta_c(s)}{\theta_p(s)} = \frac{K_p/V_m}{s[(VJ/K_BV_m^2)s^2 + (LJ/V_m^2)s + 1]},$$
(3.132)

and its simple block diagram is illustrated in Fig. 3.22. It is important to emphasize, however, that even a relatively small amount of air in the oil lines would lower K_B. This would cause the resonant frequency of the system to decrease sharply and reduce its capabilities. In addition, a large volume of oil in the lines between the hydraulic pump and motor has a similar effect. Therefore, these lines should be as short and narrow as possible.

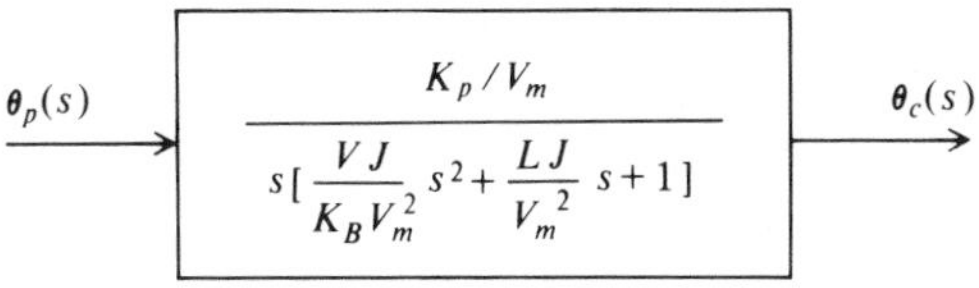

$$\theta_p(s) \longrightarrow \boxed{\frac{K_p/V_m}{s[\dfrac{VJ}{K_BV_m^2}s^2 + \dfrac{LJ}{V_m^2}s + 1]}} \longrightarrow \theta_c(s)$$

Fig. 3.22   Block diagram of a pump-controlled hydraulic transmission system.

Equation (3.132) can be written in differential-equation form as follows:

$$\frac{VJ}{K_BV_m^2}\frac{d^3\theta_c(t)}{dt^3} + \frac{LJ}{V_m^2}\frac{d^2\theta_c(t)}{dt^2} + \frac{d\theta_c(t)}{dt} = \frac{K_p}{V_m}\theta_p(t).$$
(3.133)

This third-order differential equation can be written as three first-order differential equations by defining

$$x_1 = \theta_c(t),$$
(3.134)

$$x_2 = \dot{\theta}_c(t),$$
(3.135)

$$x_3 = \ddot{\theta}_c(t).$$
(3.136)

The resulting state equations are given by

$$\dot{x}_1 = x_2,$$
(3.137)

$$\dot{x}_2 = x_3,$$
(3.138)

$$\dot{x}_3 = -\frac{K_BV_m^2}{VJ}x_2 - \frac{K_BL}{V}x_3 + \frac{K_BK_pV_m}{VJ}\theta_p(t).$$
(3.139)

In vector form, these equations become

$$\dot{x} = Px + Br,$$
(3.140)

$$\theta_c = Lx,$$
(3.141)

where

$$\mathbf{x} = \begin{bmatrix} x_1 \\ x_2 \\ x_3 \end{bmatrix}, \qquad \dot{\mathbf{x}} = \begin{bmatrix} \dot{x}_1 \\ \dot{x}_2 \\ \dot{x}_3 \end{bmatrix}, \qquad \mathbf{P} = \begin{bmatrix} 0 & 1 & 0 \\ 0 & 0 & 1 \\ 0 & -\dfrac{K_B V_m^2}{VJ} & -\dfrac{K_B L}{V} \end{bmatrix},$$

$$\mathbf{b} = \begin{bmatrix} 0 \\ 0 \\ \dfrac{K_B K_p V_m}{VJ} \end{bmatrix}, \qquad \mathbf{L} = [1 \quad 0 \quad 0], \qquad r = \theta_p(t).$$

2. Hydraulic Valve-Controlled Motor Another method of controlling a hydraulic motor is with a constant-pressure source and a valve that controls the flow of oil through it. A valve-controlled hydraulic system is usually smaller than a pump-controlled system and less efficient. Due to the increased losses, the time constants are greatly reduced, and the speed of response is relatively greater. Valve-controlled systems do, however, have the disadvantages associated with devices whose characteristics are nonlinear.

Figure 3.23 illustrates a valve-controlled hydraulic system. A fluid source, at constant pressure, is provided at the center of the control valve. Fluid-return lines are located on each side of this pressure source. When the control valve is moved to the right, hydraulic fluid flows through line A into the hydraulic motor. This results in a pressure differential across the piston of the motor which causes it also to move to the right. This action causes fluid to be pushed back into the valve through line B which returns it to the sump through line E. Similar operation occurs when the control valve is moved to the left. Observe that all fluid flows are blocked when the control valve is in the neutral position, as shown in Fig. 3.23.

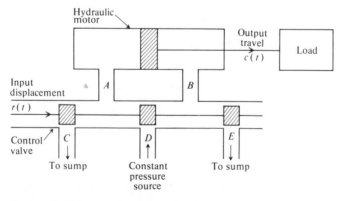

Fig. 3.23 A valve-controlled hydraulic system.

Figure 3.24 represents the characteristics for the valve-controlled hydraulic motor. The pressure between lines going to the motor is denoted by P, the flow through these lines is denoted by Q, and r denotes the displacement of the valve from its neutral position. Although these characteristics are nonlinear, it will be assumed that they are linear for small input displacements. This is basically an application of small-signal theory, used so frequently by circuit designers. For small excursions from a given quiescent operating point,

$$\Delta Q = \frac{\partial Q}{\partial P} \Delta P + \frac{\partial Q}{\partial r} \Delta r. \tag{3.142}$$

At any given quiescent operating point, it will be assumed that $\partial Q/\partial P$ and $\partial Q/\partial r$ are constants.

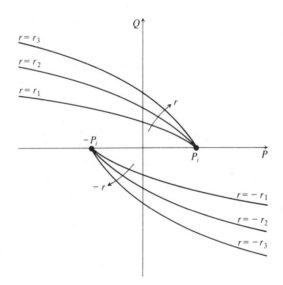

Fig. 3.24   Valve characteristics.

The transfer function relating the input r and output c can be obtained by comparing the valve-controlled hydraulic system with the pump-controlled hydraulic system. Studying these two systems carefully, it is observed that ΔQ is analogous to Q_p, M is analogous to J, ΔP is analogous to P_L, and Δr is analogous to θ_p. Using these analogies, the transfer function can easily be found to be given by

$$\frac{C(s)}{R(s)} = \frac{(1/V_m)(\partial Q/\partial r)}{s[(VM/K_B V_m^2)s^2 + (M/V_m^2)(L - \partial Q/\partial P)s + 1]}. \tag{3.143}$$

The term $L - \partial Q/\partial P$ in the denominator of this equation is always positive since Fig. 3.24 indicates that $\partial Q/\partial P$ is always negative. The simple block diagram for this

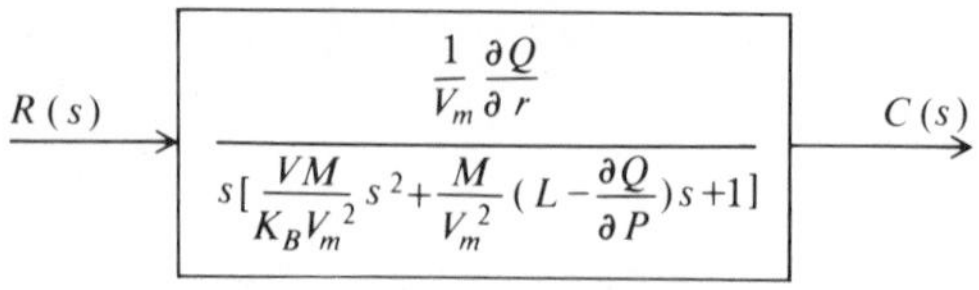

$$s\left[\dfrac{VM}{K_B V_m^{\,2}}s^2+\dfrac{M}{V_m^{\,2}}\left(L-\dfrac{\partial Q}{\partial P}\right)s+1\right]$$

$$\dfrac{\dfrac{1}{V_m}\dfrac{\partial Q}{\partial r}}{}$$

$R(s)$ → [block] → $C(s)$

Fig. 3.25 Block diagram of a valve-controlled hydraulic transmission system.

system is illustrated in Fig. 3.25. It is left as an exercise to the reader to determine what the transfer function of the system is if a spring and damper are attached to the control rod (see Problem 3.11), and to determine the state-space vector form of Eq. (3.143).

3.6 TRANSFER-FUNCTION REPRESENTATION OF THERMAL SYSTEMS

If the assumption is made that the temperature of a body is uniform, then a number of thermal systems can be represented by linear ordinary differential equations [14]. This approximation is reasonably correct for relatively small configurations. This section specifically considers a hot-water heating system as an example of a typical thermal system.

Figure 3.26 illustrates an electric hot-water heating system. The object of this system may be, typically, to supply hot water in a home. Any demand for hot water in the home causes hot water to leave and cold water to enter the tank. In order to reduce heat loss to the surrounding air, the tank is insulated. A thermostatic switch turns an electrical heating element on or off in order to maintain a desired reference temperature.

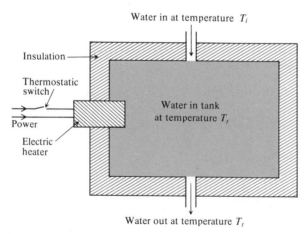

Fig. 3.26 Electric hot-water heating system.

The law of conservation of energy requires that the heat added to the system equals the heat stored plus the heat lost. This can be expressed by the relationship

$$Q_h = Q_c + Q_o - Q_i + Q_l, \tag{3.144}$$

where

Q_h = heat flow supplied by heating element
Q_c = heat flow into storage in the water in the tank
Q_o = heat flow lost by hot water leaving tank
Q_i = heat flow carried in by cold water entering tank
Q_l = heat flow through insulation.

It can be shown that

$$Q_c = C \frac{dT_t}{dt}, \tag{3.145}$$

where

C = thermal capacity of water in tank
T_t = temperature of water in tank;

$$Q_o = VHT_t, \tag{3.146}$$

where

V = water flow from tank
H = specific heat of water;

$$Q_i = VHT_i, \tag{3.147}$$

where

T_i = temperature of water entering tank; and

$$Q_l = \frac{T_t - T_e}{R}, \tag{3.148}$$

where

T_e = temperature of air surrounding tank
R = thermal resistance of insulation, stagnant air film, and stagnant liquid film.

Substitution of Eqs. (3.145) through (3.148) into Eq. (3.144) yields the expression

$$Q_h = C \frac{dT_t}{dt} + VH(T_t - T_i) + \frac{T_t - T_e}{R}. \tag{3.149}$$

The thermal model presented, so far, has considered T_t, T_i, T_e, and V as variables. For the specific condition where V is a constant and $T_e = T_i$, Eq. (3.149) reduces to the expression

$$Q_h = C \frac{dT_r}{dt} + \left(VH + \frac{1}{R}\right) T_r, \tag{3.150}$$

where T_r = temperature of the water in the tank above the reference T_e. The Laplace transform of Eq. (3.150) is given by

$$Q_h(s) = T_r(s)\left(Cs + VH + \frac{1}{R}\right). \tag{3.151}$$

The transfer function of this system, defined as the ratio of the output $T_r(s)$ to the input $Q_h(s)$ is given by

$$\frac{T_r(s)}{Q_h(s)} = \frac{1}{Cs + VH + 1/R} \tag{3.152}$$

and its simple block diagram is illustrated in Fig. 3.27.

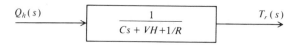

Fig. 3.27   Block diagram for the system shown in Fig. 3.26.

3.7 A GENERALIZED APPROACH FOR MODELING— THE PRINCIPLES OF CONSERVATION AND ANALOGY

Due to the large variety of control system components that occur in practice, a generalized approach is useful for obtaining their mathematical model. Therefore, rather than pursue the presentation of further specific control system components, this section will provide a generalized approach to deriving mathematical models.

There are several general principles that can be useful in serving as guides. The most important are the principle of conservation and the concept of analogous circuits.

1. Principle of Conservation The principle of conservation is a very important guideline for the derivation of a mathematical model. A statement of this concept is that

$$\text{accumulation} = \text{inflow} - \text{outflow}. \tag{3.153}$$

In terms of rates, the principle of conservation is stated as follows:

$$\text{rate of increase (of storage)} = \text{rate of inflow} - \text{rate of outflow}. \tag{3.154}$$

Exactly what is being conserved depends on the application. However, this principle is usually used to establish a balance or inventory of mass, energy, momentum or charge.

The principle of conservation has been used several times throughout this chapter. For example, let us reconsider the hydraulic motor and pump power transmission system illustrated in Fig. 3.20 and the electric hot-water heating system of Fig. 3.26. In the case of the hydraulic motor and pump power transmission system, Eq. 3.123 was derived. It related the volume of oil flowing from the pump Q_p to its distribution in flow through the motor, Q_m, leakage around the motor, Q_l, and compressibility flow (accumulation), Q_c. Therefore, this equation was an application of the principle of conservation where Eq. (3.153) was modified to the form

$$\text{inflow} = \text{outflow} + \text{leakage} + \text{accumulation}. \tag{3.155}$$

Similarly, in the electric hot-water heating system problem, Eq. (3.144) was derived. Let us rearrange it into the following form in order to demonstrate the principle of conservation:

$$Q_h + Q_i = Q_c + Q_o + Q_l. \tag{3.156}$$

This equation states that the heat flow supplied by the heating element, Q_h, plus the heat flow carried in by cold water entering the tank, Q_i, is distributed as heat flow into storage in the water in the tank, Q_c, heat flow lost by hot water leaving the tank, Q_o, and heat flow through the insulation, Q_l. Therefore, the principle of conservation was applied, and the form of the principle applied was the same as Eq. (3.155).

2. The Circuit Concept and Analogy An alternative viewpoint to the principle of conservation is the concept of an analogous circuit. The basis for applying the principle of analogy is that two different physical systems can be described by the same mathematical model. This permits a generalization of ideas specific to a particular field in order that a broader understanding of a variety of apparently unrelated situations can be achieved.

Table 3.3   Analogous quantities

Electrical	Mechanical (linear motion)	Mechanical (rotation)	Thermal	Hydraulic
Current	Force	Torque	Heat flow	Flow
Voltage	Linear velocity	Angular velocity	Temperature	Pressure
Inductance	Spring	Spring	—	Inertia
Capacitance	Mass	Inertia	Capacitance	Compression
Resistance	Dashpot	Dashpot	Resistance	Resistance

The analogy concept can best be understood by focusing attention on some of the devices that have been covered in this chapter. For example, let us look at the electrical circuit of the armature-controlled dc servomotor illustrated in Fig. 3.10(a) and the hydraulic motor and pump illustrated in Fig. 3.20. Equations (3.61) and (3.132) described their respective mathematical models. The direct analogy between Eqs. (3.61) and (3.132) is self-evident. Furthermore, the development of these two models follows a very similar process. It is important to note how their respective relations in both cases enabled one to relate the two coupled sets of physical variables: electrical and mechanical for the armature-controlled dc servomotor, and hydraulic and mechanical for the hydraulic motor and pump transmission system.

This concept of analogy can be extended to relate electrical, mechanical (linear motion), mechanical (rotational), thermal, and hydraulic systems. For purposes of comparison, Table 3.3 illustrates a brief table of analogous quantities in different physical systems. It is important to note that there are two possible analogies between mechanical and electrical systems. If torque (or force) is chosen to be analogous to current, then the mechanical circuit and electrical circuits look alike with inertia being analogous to capacitance and a spring being analogous to inductance. This is the system illustrated in Table 3.3. Another approach could be to choose torque (or force) to be analogous to voltage. Then the mechanical and electrical circuits will be analogous, with inertia being analogous to inductance and a spring being analogous to capacitance: it makes no difference which viewpoint is taken.

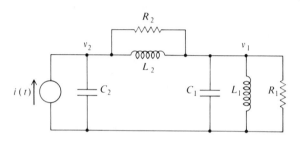

Fig. 3.28   Electrical analog of Fig. 3.5.

By using the method of analogs, complex mechanical (or hydraulic, etc.) systems can be drawn as equivalent circuit diagrams, for which Kirchhoff's voltage and current laws can be utilized to obtain the mathematical model of the system. As an example of this approach, let us reconsider the complex mechanical system of Fig. 3.5, using Kirchhoff's voltage and current laws. By using the analogs of Table 3.3, the mechanical network of Fig. 3.5 is redrawn as an equivalent electrical circuit in Fig. 3.28. The node equations for Fig. 3.28 are written by inspection as

$$\left(C_2 s + \frac{1}{R_2} + \frac{1}{L_2 s}\right)V_2(s) - \left(\frac{1}{R_2} + \frac{1}{L_2 s}\right)V_1(s) = I(s), \qquad (3.157)$$

$$-\left(\frac{1}{R_2} + \frac{1}{L_2 s}\right)V_2(s) + \left(C_1 s + \frac{1}{L_1 s} + \frac{1}{R_1} + \frac{1}{R_2} + \frac{1}{L_2 s}\right)V_1(s) = 0, \qquad (3.158)$$

where

$$i(t) = f(t), \qquad L_1 = 1/K_1, \qquad C_2 = M_2,$$
$$R_1 = 1/B_1, \qquad R_2 = 1/B_2, \qquad v_1 = \dot{y}_1(t),$$
$$L_2 = 1/K_2, \qquad v_2 = \dot{y}_2(t), \qquad C_1 = M_1.$$

Equations (3.157) and (3.158) are analogous to Eqs. (3.27) and (3.28), respectively. Substituting the analogous quantities into Eqs. (3.157) and (3.158), and taking the inverse Laplace transforms we obtain the following set of equations:

$$f(t) = M_2 \frac{d^2 y_2(t)}{dt^2} + B_2 \left[\frac{d y_2(t)}{dt} - \frac{d y_1(t)}{dt} \right] + K_2[y_2(t) - y_1(t)], \qquad (3.159)$$

$$0 = B_2 \left[\frac{d y_2(t)}{dt} - \frac{d y_1(t)}{dt} \right] + K_2[y_2(t) - y_1(t)]$$

$$- M_1 \frac{d^2 y_1(t)}{dt^2} - B_1 \frac{d y_1(t)}{dt} - K_1 y_1(t). \qquad (3.160)$$

Observe that Eqs. (3.159) and (3.160) are identical to Eqs. (3.27) and (3.28), respectively.

Electrical analogies have the advantage that they can be set up very easily in the laboratory. For example, a change in a particular parameter can be accomplished very easily in the electric circuit to determine its overall effects and the electric circuit can be approximately adjusted for the desired response. Afterwards, the parameters in the mechanical (or hydraulic, or thermal, etc.) system can be adjusted by an analogous amount to obtain the same desired response.

PROBLEMS

3.1 a) For the mechanical translational system illustrated in Fig. P3.1, write the differential equation relating the position $y(t)$ and the applied force $f(t)$.

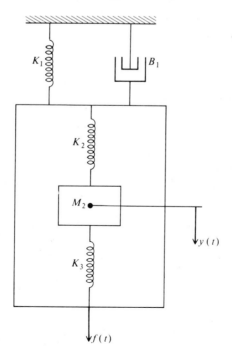

Figure P3.1

b) Determine the transfer function $Y(s)/F(s)$.
c) Determine the state-space vector form of this system.

3.2 a) For the mechanical translational system illustrated in Fig. P3.2, write the differential equation relating the position $y(t)$ and the applied force $f(t)$.
b) Determine the transfer function $Y(s)/F(s)$.
c) Determine the state-space vector form of this system.

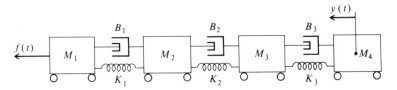

Figure P3.2

3.3 a) For the mechanical rotational system illustrated in Fig. P3.3, write the differential equation relating $T(t)$ and $\theta(t)$.
b) Determine the transfer function $\theta(s)/T(s)$.

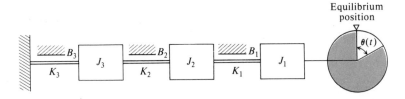

Figure P3.3

3.4 Figure P3.4 represents the diagram of a gyroscope which is used quite frequently in autopilots, stabilized fire control systems, and so on. Assume that the rotor speed is constant, that the total developed torque about the output axis is given by

$$T_o = K' \frac{d\theta_i(t)}{dt}$$

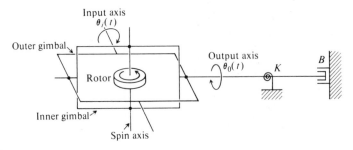

Figure P3.4

where K' is a constant, and that the inner gimbal's moment of inertia about the output axis is J.

a) Write the differential equation relating $\theta_i(t)$ and $\theta_o(t)$.
b) Determine the transfer function $\theta_o(s)/\theta_i(s)$.
c) Determine the vector differential equation of this system.

3.5 The Ward-Leonard system shown in Fig. 3.11 has the constants and characteristics shown in Table P3.5. Derive the transfer function for the system, $\theta_o(s)/E_f(s)$.

Table P3.5

Generator	Motor
$R_f = 1.1\ \Omega$	$R_m = 2\ \Omega$
$L_f = 0.25\ H$	$L_m = 0.05\ H$
$R_g = 0.20\ \Omega$	$K_e = 0.1\ V/(rad/sec)$
$L_g = 0.01\ H$	$K_T = 2\ ft\ lb/A$
$K_g = 1\ V/A$	$J = 0.1\ slug\ ft^2$
	$B = negligible$

3.6 Repeat Problem 3.5 for the constants and characteristics shown in Table P3.6.

Table P3.6

Generator	Motor
$R_f = 1\ \Omega$	$R_m = 1\ \Omega$
$L_f = 0.1\ H$	$L_m = 0.01\ H$
$R_g = 0.1\ \Omega$	$K_e = 10\ V/(rad/sec)$
$L_g = 0.1\ H$	$K_T = 10\ ft\ lb/A$
$K_g = 10\ V/A$	$J = 1\ slug\ ft^2$
	$B = negligible$

3.7 The time constant for highly inductive devices such as the armature circuit of the Ward-Leonard system shown in Fig. 3.11 is usually too long for high-performance control systems. A simple technique for decreasing the time constant of such a system uses feedback, which controls the current of the inductive device. Figure P3.7 illustrates how this can be practically accomplished for a Ward-Leonard system by inserting a resistor R in series with the armature circuit. Other notations used in this diagram are the same as those used in Section 3.4. Assume that the armature current is much greater than the field current.

a) Write the differential equation relating $e_f(t)$ and $\theta_o(t)$.
b) Determine the transfer function $\theta_o(s)/E_f(s)$.
c) Compare the answer to part (b) with Eq. (3.78) and show that the armature time constant has been reduced by means of feedback.

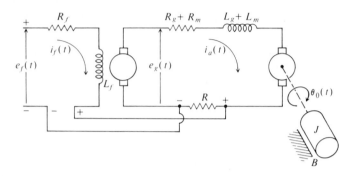

Figure P3.7

3.8 Since it is proportional to velocity, the back emf of a motor is sometimes directly used as a stabilizing voltage. Figure P3.8 illustrates a practical bridge-type circuit which can be used for this purpose. The resistors R_1 and R_2, which have very high resistance relative to the armature circuit resistance, are adjustable in order to obtain the desired value of $e_b(t)$. R_{AC} represents the resistance of the armature and commutator voltage drop; R_{CF} and L_{CF} represent the resistance and inductance of the commutating fields, respectively. K_e, K_T, J, and B have the same significance as in the text of this chapter. For the armature-

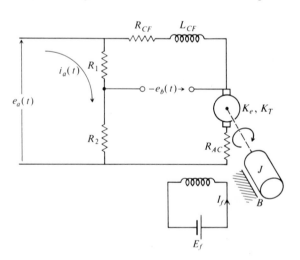

Figure P3.8

controlled dc servomotor illustrated in Fig. P3.8, assume that the field current I_f remains constant.

a) Write the differential equation relating $e_a(t)$ and $e_b(t)$.
b) Determine the transfer function $E_b(s)/E_a(s)$.

3.9 If the armature current of the field-controlled dc servomotor is not held constant, the developed torque is proportional to both the field and the armature currents (see Fig. 3.13).

This essentially results in a nonlinear device because of the multiplication effect of field and armature current.

a) Draw the signal-flow diagram for the system.

b) What approximations can be made to linearize the operation?

3.10 Draw the signal-flow diagram and analog-computer simulation circuitry for the following devices:

a) dc generator

b) the Ward-Leonard system

c) two-phase ac servomotor

3.11 Derive the transfer function $C(s)/R(s)$ for the valve-controlled hydraulic transmission system shown in Fig. 3.23 if a spring in parallel with a viscous damper is attached between the right end of the valve control rod and some stationary reference point. Assume that the spring stiffness factor is K lb/ft and the viscous damping factor is B lb(ft/sec).

3.12 The control of paper color is a very interesting and important problem in the paper processing industry. Figure P3.12(a) illustrates a functional diagram of the problem [15]. This color-control method depends on the availability of a precise, reliable, on-line colorimeter. As indicated in the diagram, dyestuff concentrations are added at various stages of the process. In this diagram, the following nomenclature is used:

$$f = \text{water flow rate, in liters/min}$$
$$\alpha = \text{consistencies (weight of dry fiber/weight of stock)}$$
$$V_t = \text{header tank volume}$$
$$K_v = \text{constant at the dry end}$$
$$V_p = \text{pipe volume}$$
$$V_s = \text{stirred tank volume}$$
$$c = \text{dyestuff concentration, in gm/liter}$$
$$v = \text{machine speed, in m/min}$$
$$m = \text{dyestuff flow rate, in liters/min}$$

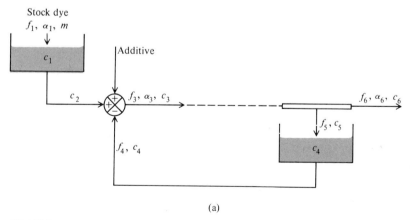

(a)

Figure P3.12(a)

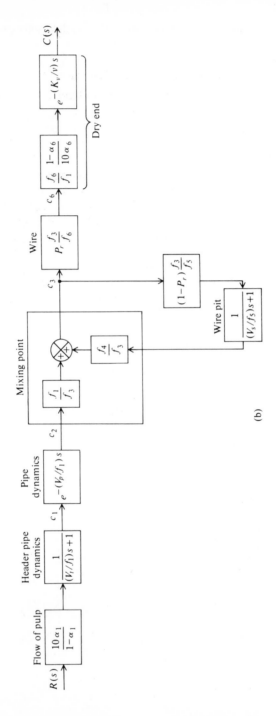

Figure P3.12(b)

Figure P3.12(*b*) illustrates an equivalent block diagram of the color control process [15]. The factor P_r, indicated at several stages of the block diagram, is the retention factor of the dye. It varies with the concentration, with the time allowed for dyeing, and with such factors as pH, alum, resin, temperature and dye interaction. Determine the transfer function of this system, $C(s)/R(s)$, where

$$R(s) = \text{dye concentration ratio} = \text{percent of dye/weight of fiber}$$

$$C(s) = \text{weight concentration of dye in the sheet at the output.}$$

3.13 A two-phase ac servomotor has the following specifications: 115/115 V; 60 Hz; 2 phases; 4 poles; rotor moment of inertia, 0.09 oz in², rated stalled torque, 6 oz in; no-load speed, 2500 rev/min; load inertia and coefficient of viscous friction are negligible. Calculate the transfer function for this servomotor, $\theta_o(s)/E_c(s)$.

3.14 It is common practice to place a gear reduction between a servomotor and the load in order to convert the high-speed, low-torque motor characteristics to a low-speed, high-torque device. Assuming a gear reduction of $N:1$ and a load inertia of J_2, derive the transfer function $\theta_o(s)/E_c(s)$ for this system and compare it with Eq. (3.101). Use the same terminology and motor characteristics as in Section 3.4, part 5. Assume B is finite.

3.15 Based on the derivation of Problem 3.14, repeat Problem 3.13 if the load has an inertia of 0.40 oz in², and a gear ratio of 36:1 is used.

3.16 Repeat Problem 3.15 with the load inertia doubled. What conclusions can you draw from your results?

3.17 A two-phase ac servomotor has the following specifications: 115/115 V; 400 Hz; 2 phases; 4 poles; rotor moment of inertia, 0.04 oz in², rated stall torque, 2 oz in; no-load speed, 5000 rev/min; load inertia, 4 oz in² geared to the motor through a gear reduction of 9:1. Determine the transfer function for the overall servomotor and load. Assume the inertia of the gears is negligible.

3.18 Repeat Problem 3.17 without assuming that the gearing inertia can be neglected. Assume that the gear reduction of 9:1 is achieved in one gear pass and each gear has an inertia of 0.01 oz in². What conclusions can you reach from your results?

3.19 Repeat Problem 3.17 without assuming that the gearing inertia can be neglected. Assume that the gear reduction of 9:1 is achieved in two gear passes of 3:1 each, and that each gear has an inertia of 0.01 oz in². What conclusions can you draw from your results?

3.20 Derive the transfer function of the armature-controlled dc servomotor from its signal-flow diagram.

3.21 Starting with Eq. (3.62), derive the state equations of the armature-controlled dc servomotor from its state-variable diagram.

3.22 Starting with Eq. (3.78), derive the state equations of the Ward-Leonard system from the state-variable diagram.

3.23 Starting with Eq. (3.101), derive the state equations of the two-phase ac servomotor from its state-variable diagram.

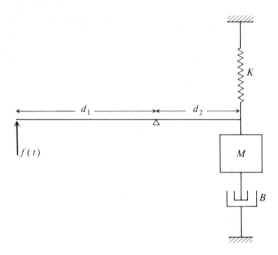

Figure P3.24

3.24 For the mechanical system illustrated in Fig. P3.24, draw an analogous electric circuit and determine the differential equation relating the force $f(t)$ and the position $x(t)$.

3.25 For the mechanical system illustrated in Fig. P3.25, draw an analogous electric circuit and determine the differential equation relating $v_1(t)$, $v_2(t)$, and $f(t)$.

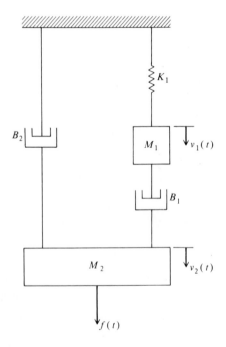

Figure P3.25

3.26 With the advent of modern man-machine control systems utilizing humans to perform various manual tasks, the knowledge of the human transfer function is very important in order to enable the prediction of the system's performance. Many researchers have attempted to determine the human transfer function. For manual control at relatively low frequency, the human characteristics indicate a gain K, an anticipation time constant T_A, an operator's error-smoothing lag time constant T_L, an operator's short neuromuscular delay time constant T_N, and an operator's time-delay factor D. Write the form of the human transfer function. What is the meaning of the time-delay factor D?

REFERENCES

1. M. E. Van Valkenberg, *Network Analysis* (2nd Edn.), Prentice-Hall, Englewood Cliffs, N.J. (1964).
2. R. J. Schwarz and B. Friedland, *Linear Systems*, McGraw-Hill, New York (1965).
3. A. Higdon and W. B. Stiles, *Engineering Mechanics*, Prentice-Hall, Englewood Cliffs, N.J. (1949), Chapter 9.
4. R. M. Saunders, "The dynamo electric amplifier—Class A operation," *Trans. AIEE* **68,** 1368–73 (1949).
5. B. Litman, "An analysis of rotating amplifiers," *Trans. AIEE* **68,** Pt. 2, 111 (1949).
6. H. Chestnut and R. W. Mayer, *Servomechanisms and Regulating System Design* (2nd Edn.), Vol. 1, Wiley, New York (1959).
7. R. C. Kloeffler, R. M. Kerchner, and J. L. Brenneman, *Direct-Current Machinery*, Macmillan, New York (1950).
8. J. C. Gille, N. J. Pelegrin, and P. Decaulne, *Feedback Control Systems*, McGraw-Hill, New York (1959), Chapter 33.
9. H. Chestnut and R. W. Mayer, *Servomechanisms and Regulating System Design*, Vol. 2, Wiley, New York (1955), Chapter 7.
10. R. J. W. Koopman, "Operating characteristics of two-phase servomotors," *Trans. AIEE* **68,** Pt. 1, 319 (1949).
11. A. M. Hopkin, "Transient response of small two-phase servomotors," *Trans. AIEE* **70,** Pt. 1, 881 (1951).
12. M. Liwschitz-Garik and C. C. Whipple, *Electric Machinery*, Vol. 2, *A-C Machines*, Van Nostrand, Princeton, N.J., (1946), Chapter 7.
13. G. J. Brown and D. P. Campbell, *Principles of Servomechanisms*, Wiley, New York (1948).
14. J. D. Trimmer, *Response of Physical Systems*, Wiley, New York (1950).
15. P. R. Bélanger, "Sensitivity design of a paper color control system," in *Proceedings of the 1969 Joint Automatic Control Conference*, pp. 99–106.
16. G. A. Korn and T. M. Korn, *Electronic Analog and Hybrid Computers*, McGraw-Hill, New York (1964).
17. S. M. Shinners, "Which computer—Analog, digital, or hybrid?" *Machine Design* **43,** 104–111 (January 21, 1971).

4   SECOND-ORDER SYSTEMS

4.1 INTRODUCTION

From the frequency domain viewpoint, system order refers to the highest power of s in the denominator of the closed-loop transfer function of a system. In the time domain, system order refers to the highest derivative of the controlled quantity in the equation describing the control system's dynamics. System order is a very significant parameter for characterizing a system.

Second-order systems are very important to the control-system engineer. This type of system characterizes the dynamics of many control-system applications found in the fields of servomechanisms, space-vehicle control, chemical process control, bioengineering, aircraft control systems, ship controls, etc. It is interesting to note that most control system designs are based on second-order system analysis. Even if the system is of higher order, as it usually is, the system may be approximated by a second-order system in order to obtain a first approximation for preliminary design purposes with reasonable accuracy. A more exact solution can then be obtained in terms of departures from the performance of a second-order system.

Due to the importance of second-order systems, this chapter is devoted to presenting its characteristic response in the time domain, and analyzing its state-variable signal-flow diagram. In addition, several important control system definitions are presented. The closed-loop frequency response of second-order systems is presented in Chapter 6 where techniques for obtaining the closed-loop frequency characteristics are derived.

4.2 CHARACTERISTIC RESPONSES
OF TYPICAL FEEDBACK CONTROL SYSTEMS

The purpose of this section is to describe the transient response of a typical feedback control system. We consider a very common configuration in which a two-phase ac servomotor, whose transfer function is given by Eq. (3.101) is enclosed by a simple unity feedback loop. Figure 4.1 illustrates the block diagram of this second-order system. For purposes of simplicity, the reference input elements, control elements, feedback elements, and the indirectly controlled system are assumed to have unity transfer functions.

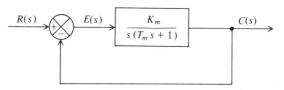

Fig. 4.1 Second-order feedback system containing a two-phase ac induction motor.

The closed-loop transfer function of this system is given by

$$\frac{C(s)}{R(s)} = \frac{K_m/T_m}{s^2 + (1/T_m)s + K_m/T_m} .$$ (4.1)

By defining the undamped natural frequency ω_n and the damping factor ζ as

$$\omega_n^2 = \frac{K_m}{T_m} \quad \text{and} \quad \zeta = \frac{1}{2\omega_n T_m} ,$$ (4.2)

Eq. (4.1) can be rewritten as

$$\frac{C(s)}{R(s)} = \frac{\omega_n^2}{s^2 + 2\zeta\omega_n s + \omega_n^2} .$$ (4.3)

The parameters ω_n and ζ are very important for characterizing a system's response. Note from Eq. (4.3) that ω_n turns out to be the radian frequency of oscillation when $\zeta = 0$. As ζ increases in value from 0, the oscillation decays and becomes more damped. When $\zeta \geqslant 1$, an oscillation does not occur.

We assume that the initial conditions are zero and the input is a unit step. Therefore, $R(s) = 1/s$, and the Laplace transform of the output can be written as

$$C(s) = \frac{\omega_n^2}{s(s^2 + 2\zeta\omega_n s + \omega_n^2)} .$$ (4.4)

Factoring the denominator, we obtain

$$C(s) = \frac{\omega_n^2}{s(s + \zeta\omega_n - \omega_n\sqrt{\zeta^2 - 1})(s + \zeta\omega_n + \omega_n\sqrt{\zeta^2 - 1})} .$$ (4.5)

The exact solution for the output in the time domain is dependent on the value of ζ. When $\zeta \geqslant 1$, the second-order system has poles which lie along the negative real axis of the complex plane. When $\zeta < 1$, however, a pair of complex conjugate poles result. We shall determine the output response to a step input for the three cases: where the damping factor equals unity, is greater than unity, and is less than unity.

Case A. Damping Factor Equals Unity When $\zeta = 1$, Eq. (4.5) reduces to

$$C(s) = \frac{\omega_n^2}{s(s + \omega_n)^2} .$$ (4.6)

The time-domain response can be obtained by utilizing the solution obtained for Eq. (2.78). The partial-fraction expansion of Eq. (4.6) is given by

$$C(s) = \frac{K_1}{(s + \omega_n)^2} + \frac{K_2}{s + \omega_n} + \frac{K_3}{s}. \tag{4.7}$$

Using Eqs. (2.81), (2.83), and (2.84) we find that

$$K_1 = -\omega_n, \tag{4.8}$$

$$K_2 = -1, \tag{4.9}$$

$$K_3 = 1. \tag{4.10}$$

Substituting these constants into Eq. (4.7), we obtain

$$C(s) = \frac{-\omega_n}{(s + \omega_n)^2} - \frac{1}{s + \omega_n} + \frac{1}{s}. \tag{4.11}$$

The time-domain response of the output, $c(t)$, may be obtained by utilizing the table of Laplace transforms given in Appendix A:

$$c(t) = -\omega_n t e^{-\omega_n t} - e^{-\omega_n t} + 1. \tag{4.12}$$

Figure 4.2(a) illustrates the output response together with the unit step input. Notice that the output response exhibits no overshoots when $\zeta = 1$. This response is described as being critically damped.

Case B. Damping Factor Greater than Unity When $\zeta > 1$, the time-domain response can be obtained quite simply from its partial-fraction expansion. This can be expressed as

$$C(s) = \frac{K_1}{s} + \frac{K_2}{s + \zeta\omega_n - \omega_n\sqrt{\zeta^2 - 1}} + \frac{K_3}{s + \zeta\omega_n + \omega_n\sqrt{\zeta^2 - 1}}, \tag{4.13}$$

where $\zeta\omega_n - \omega_n\sqrt{\zeta^2 - 1}$ and $\zeta\omega_n + \omega_n\sqrt{\zeta^2 - 1}$ are positive real numbers.

The constants K_1, K_2, and K_3 can be evaluated quite simply by the methods described in Chapter 2. Their values are

$$K_1 = 1,$$

$$K_2 = [2(\zeta^2 - \zeta\sqrt{\zeta^2 - 1} - 1)]^{-1}, \tag{4.14}$$

$$K_3 = [2(\zeta^2 + \zeta\sqrt{\zeta^2 - 1} - 1)]^{-1}.$$

Therefore Eq. (4.13) can be written as

$$C(s) = s^{-1} + [2(\zeta^2 - \zeta\sqrt{\zeta^2 - 1} - 1)]^{-1}(s + \zeta\omega_n - \omega_n\sqrt{\zeta^2 - 1})^{-1}$$
$$+ [2(\zeta^2 + \zeta\sqrt{\zeta^2 - 1} - 1)]^{-1}(s + \zeta\omega_n + \omega_n\sqrt{\zeta^2 - 1})^{-1}. \tag{4.15}$$

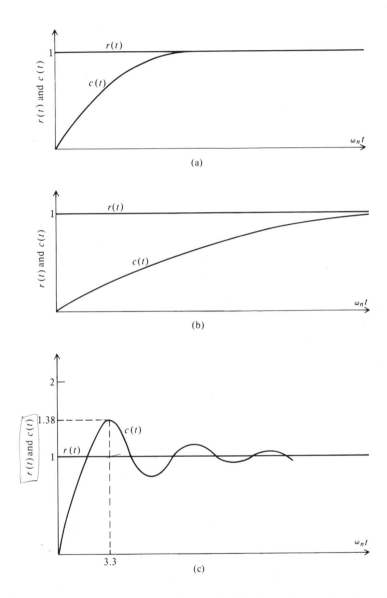

Fig. 4.2 (a) Input and output response for a critically damped second-order system; (b) Input and output response for an overdamped second-order system; (c) Input and output response for an underdamped second-order system ($\zeta = 0.3$).

The time-domain response of the output, $c(t)$, may be obtained by utilizing the table of Laplace transforms given in Appendix A. It can be expressed as

$$c(t) = 1 + [2(\zeta^2 - \zeta\sqrt{\zeta^2 - 1} - 1)]^{-1}e^{-(\zeta - \sqrt{\zeta^2 - 1})\omega_n t}$$
$$+ [2(\zeta^2 + \zeta\sqrt{\zeta^2 - 1} - 1)]^{-1}e^{-(\zeta + \sqrt{\zeta^2 - 1})\omega_n t}. \qquad (4.16)$$

Figure 4.2(b) illustrates the output response together with the unit step input. Notice that when $\zeta > 1$, the output response exhibits no overshoots and takes longer to reach its final value than when $\zeta = 1$. This response is described as being overdamped.

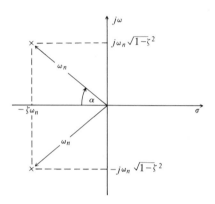

Fig. 4.3   Location of the conjugate complex poles in the complex plane.

Case C.   Damping Factor Less than Unity   When $\zeta < 1$, the time-domain response can be obtained in an analogous manner. The solution is slightly more complex, however, since we now have a pair of complex conjugate poles. The partial-fraction expansion of Eq. (4.5) can be written as

$$C(s) = \frac{K_1}{s} + \frac{K_2}{s + \zeta\omega_n - j\omega_n\sqrt{1 - \zeta^2}} + \frac{K_3}{s + \zeta\omega_n + j\omega_n\sqrt{1 - \zeta^2}}. \qquad (4.17)$$

The constants K_1, K_2, and K_3 can be evaluated in an analogous manner by the method described in Chapter 2, but the algebra becomes a little complicated. In order to simplify this situation somewhat, use is made of the relationship between the location of the conjugate complex poles in the complex plane and the damping factor ζ. The geometry of the configuration is illustrated in Fig. 4.3. Notice that the distance from the origin to either pole equals ω_n. In addition, the angle α has the following trigonometric properties:

$$\cos \alpha = -\zeta, \qquad (4.18)$$

$$\sin \alpha = \sqrt{1 - \zeta^2}. \qquad (4.19)$$

Utilizing the relations given by Eqs. (4.18) and (4.19), the constants K_1, K_2, and K_3 can be expressed as

$$K_1 = 1,$$

$$K_2 = \frac{e^{-j\alpha}}{2j \sin \alpha}, \tag{4.20}$$

$$K_3 = -\frac{e^{j\alpha}}{2j \sin \alpha}.$$

Therefore Eq. (4.17) can be written as

$$C(s) = \frac{1}{s} + \frac{e^{-j\alpha}}{2j \sin \alpha}(s + \zeta\omega_n - j\omega_n\sqrt{1 - \zeta^2})^{-1}$$

$$-\frac{e^{j\alpha}}{2j \sin \alpha}(s + \zeta\omega_n + j\omega_n\sqrt{1 - \zeta^2})^{-1}. \tag{4.21}$$

The time-domain response of the output, $c(t)$, may be obtained by utilizing Appendix A. It can be expressed as

$$c(t) = 1 + \frac{e^{-j\alpha}}{2j \sin \alpha}e^{-(\zeta\omega_n - j\omega_n\sqrt{1-\zeta^2})t}$$

$$-\frac{e^{j\alpha}}{2j \sin \alpha}e^{-(\zeta\omega_n + j\omega_n\sqrt{1-\zeta^2})t}. \tag{4.22}$$

This can be simplified to

$$c(t) = 1 + \frac{e^{-\zeta\omega_n t}}{\sqrt{1 - \zeta^2}} \frac{e^{j(\omega_n t\sqrt{1-\zeta^2} - \alpha)} - e^{-j(\omega_n t\sqrt{1-\zeta^2} - \alpha)}}{2j}, \tag{4.23}$$

or

$$c(t) = 1 + \frac{e^{-\zeta\omega_n t}}{\sqrt{1 - \zeta^2}} \sin(\omega_n\sqrt{1 - \zeta^2}\, t - \alpha). \tag{4.24}$$

Figure 4.2(c) illustrates the output response for a value of ζ approximately equal to 0.3, together with the unit step input. Notice that the output response exhibits several overshoots before finally settling out. This response, which is characteristic of an exponentially damped sinusoid, is described as being underdamped.

The time to the first overshoot and the value of the first overshoot are two interesting identifying characteristics for this type of response. We shall next derive these values in terms of the undamped natural frequency, ω_n, and the damping factor ζ.

Equation (4.24) indicates that the radian frequency of oscillation of the system, ω_m, is

$$\omega_m = \omega_n\sqrt{1 - \zeta^2}. \tag{4.25}$$

The cyclic frequency of oscillation of the system, f_m, is

$$f_m = \frac{\omega_m}{2\pi} = \frac{\omega_n\sqrt{1 - \zeta^2}}{2\pi}. \tag{4.26}$$

The period of oscillation of the system, t_m, is

$$t_m = \frac{1}{f_m} = \frac{2\pi}{\omega_n\sqrt{1 - \zeta^2}}. \tag{4.27}$$

The time at which the peak overshoot occurs, t_p, can be found by differentiating $c(t)$, given by Eq. (4.24), with respect to time and setting the derivative equal to zero:

$$\frac{dc(t)}{dt} = -\zeta\omega_n e^{-\zeta\omega_n t}(\sqrt{1 - \zeta^2})^{-1} \sin(\omega_n\sqrt{1 - \zeta^2}\,t - \alpha)$$
$$+ \omega_n e^{-\zeta\omega_n t} \cos(\omega_n\sqrt{1 - \zeta^2}\,t - \alpha) = 0.$$

This derivative is zero when

$$\omega_n\sqrt{1 - \zeta^2}\,t = 0, \pi, 2\pi, \ldots.$$

The peak overshoot occurs at the first value after zero, provided there are zero initial conditions. Therefore,

$$t_p = \frac{\pi}{\omega_n\sqrt{1 - \zeta^2}}, \tag{4.28}$$

or

$$\omega_n t_p = \frac{\pi}{\sqrt{1 - \zeta^2}}. \tag{4.29}$$

For the case illustrated in Fig. 4.2(c), where $\zeta = 0.3$, the time to the first overshoot is approximately $3.3/\omega_n$.

Substituting Eq. (4.29) into Eq. (4.24) yields the value for the maximum instantaneous value of the output, $c(t)$:

$$c(t) = 1 + \frac{\exp(-\zeta\pi/\sqrt{1 - \zeta^2})}{\sqrt{1 - \zeta^2}} \sin(\pi - \alpha). \tag{4.30}$$

This can be simplified by substituting

$$\sin(\pi - \alpha) = \sin\alpha \quad \text{and} \quad \sin\alpha = \sqrt{1 - \zeta^2}.$$

Therefore,

$$c(t)_{max} = 1 + \exp\left(-\frac{\zeta\pi}{\sqrt{1 - \zeta^2}}\right). \tag{4.31}$$

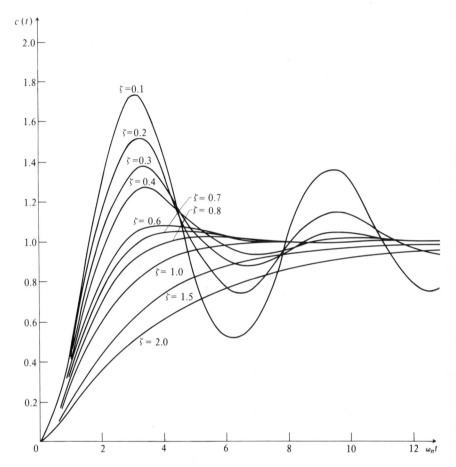

Fig. 4.4 Transient response curves of a second-order system to a step input.

This is usually expressed as a percentage of the input. Therefore, for a unit step input,

$$\text{Maximum percent overshoot} = \exp\left(-\frac{\zeta\pi}{\sqrt{1-\zeta^2}}\right) \times 100. \qquad (4.32)$$

For the case illustrated in Fig. 4.2(c), where $\zeta = 0.3$, the maximum percent overshoot is approximately 38%.

The second-order system is a very common and popular one. In order for the reader to become more familiar with its typical characteristic responses, Figs. 4.4 and 4.5 are shown to illustrate the resulting transient responses, and percent maximum overshoots, respectively, for several values of damping factor.

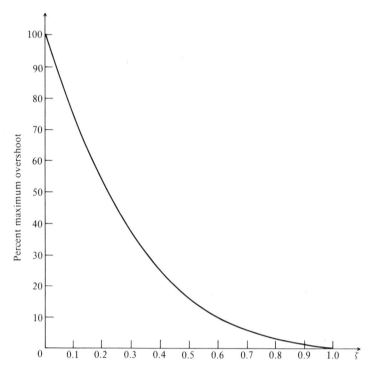

Fig. 4.5 Per cent maximum overshoot versus damping factor for a second-order system.

It is interesting to compare the sketches of Fig. 4.2(a), (b), and (c). The critically damped system appears to be a compromise among the three systems shown. Although it does take a longer time to reach the desired value of unity than the underdamped system, it does not exhibit overshoots. The underdamped system, however, oscillates several times around the desired value before it finally settles to its steady-state value. Depending on the value of the damping factor, the underdamped system may reach its final value faster than the critically damped system. Overdamped systems are hardly ever used in practice. As a matter of interest, practical systems are usually designed to be somewhat underdamped. Chapter 5 will discuss this in greater detail in terms of the various performance criteria.

4.3 STATE-VARIABLE SIGNAL-FLOW DIAGRAM OF A SECOND-ORDER SYSTEM

In the previous section, the derivation of the time response of the second-order system was based on the transfer-function derivation. Therefore, an implicit assumption was made that the initial conditions were zero. This section illustrates how the

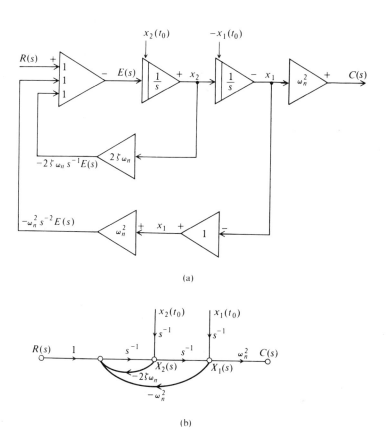

(a)

(b)

Fig. 4.6 (a) State-variable diagram of a second-order system. (b) Signal-flow diagram for a second-order system.

state-variable signal-flow diagram method of the underdamped case yields a more versatile and complete solution in the time domain including the initial condition terms.

The state-variable diagram for this system can be derived from Eq. (4.3) as follows:

$$\frac{C(s)}{R(s)} = \frac{\omega_n^2}{s^2 + 2\zeta\omega_n s + \omega_n^2}. \tag{4.33}$$

Dividing through by s^2, we obtain

$$\frac{C(s)}{R(s)} = \frac{\omega_n^2 s^{-2}}{1 + 2\zeta\omega_n s^{-1} + \omega_n^2 s^{-2}}. \tag{4.34}$$

Defining

$$E(s) = \frac{R(s)}{1 + 2\zeta\omega_n s^{-1} + \omega_n^2 s^{-2}}, \tag{4.35}$$

we may rewrite Eq. (4.34) as

$$C(s) = \omega_n^2 s^{-2} E(s). \tag{4.36}$$

From Eq. (4.36) and

$$E(s) = R(s) - 2\zeta\omega_n s^{-1} E(s) - \omega_n^2 s^{-2} E(s), \tag{4.37}$$

the state-variable diagram of this system can easily be obtained as illustrated in Fig. 4.6(a). The corresponding signal-flow diagram of the system is illustrated in Fig. 4.6(b). The initial conditions are assumed to occur at $t = t_0$. Application of Mason's theorem (Eq. 2.132) to this diagram permits us to write the state transition equation directly as follows:

$$X_1(s) = \frac{s^{-1}(1 + 2\zeta\omega_n s^{-1})}{\Delta} x_1(t_0) + \frac{s^{-2} x_2(t_0)}{\Delta} + \frac{s^{-2} R(s)}{\Delta}, \tag{4.38}$$

$$X_2(s) = \frac{-\omega_n^2 s^{-2}}{\Delta} x_1(t_0) + \frac{s^{-1} x_2(t_0)}{\Delta} + \frac{s^{-1} R(s)}{\Delta}, \tag{4.39}$$

where

$$\Delta = 1 + 2\zeta\omega_n s^{-1} + \omega_n^2 s^{-2}. \tag{4.40}$$

Simplifying Eqs. (4.38), (4.39), and (4.40), we obtain the following set of equations:

$$X_1(s) = \frac{s + 2\zeta\omega_n}{s^2 + 2\zeta\omega_n s + \omega_n^2} x_1(t_0) + \frac{x_2(t_0)}{s^2 + 2\zeta\omega_n s + \omega_n^2}$$
$$+ \frac{1}{s^2 + 2\zeta\omega_n s + \omega_n^2} R(s), \tag{4.41}$$

$$X_2(s) = \frac{-\omega_n^2}{s^2 + 2\zeta\omega_n s + \omega_n^2} x_1(t_0) + \frac{s}{s^2 + 2\zeta\omega_n s + \omega_n^2} x_2(t_0)$$
$$+ \frac{s}{s^2 + 2\zeta\omega_n s + \omega_n^2} R(s). \tag{4.42}$$

These two equations can be put into the following vector form:

$$\begin{bmatrix} X_1(s) \\ X_2(s) \end{bmatrix} = \begin{bmatrix} \dfrac{s + 2\zeta\omega_n}{s^2 + 2\zeta\omega_n s + \omega_n^2} & \dfrac{1}{s^2 + 2\zeta\omega_n s + \omega_n^2} \\ \dfrac{-\omega_n^2}{s^2 + 2\zeta\omega_n s + \omega_n^2} & \dfrac{s}{s^2 + 2\zeta\omega_n s + \omega_n^2} \end{bmatrix} \begin{bmatrix} x_1(t_0) \\ x_2(t_0) \end{bmatrix}$$
$$+ \begin{bmatrix} \dfrac{1}{s^2 + 2\zeta\omega_n s + \omega_n^2} \\ \dfrac{s}{s^2 + 2\zeta\omega_n s + \omega_n^2} \end{bmatrix} R(s). \tag{4.43}$$

The inverse Laplace transform of Eq. (4.43) is given by the following expression (it is assumed that $R(s) = 1/s$ and that $\zeta < 1$):

$$
\begin{bmatrix} x_1(t) \\ x_2(t) \end{bmatrix} =
\begin{bmatrix}
\dfrac{1}{\sqrt{1-\zeta^2}} e^{-\zeta\omega_n(t-t_0)} \sin\left[\omega_n\sqrt{1-\zeta^2}\,(t-t_0) + \phi_1\right] & \dfrac{1}{\omega_n\sqrt{1-\zeta^2}} e^{-\zeta\omega_n(t-t_0)} \sin \omega_n\sqrt{1-\zeta^2}\,(t-t_0) \\[3ex]
\dfrac{-\omega_n}{\sqrt{1-\zeta^2}} e^{-\zeta\omega_n(t-t_0)} \sin \omega_n\sqrt{1-\zeta^2}\,(t-t_0) & \dfrac{1}{\sqrt{1-\zeta^2}} e^{-\zeta\omega_n(t-t_0)} \sin\left[\omega_n\sqrt{1-\zeta^2}\,(t-t_0) + \phi_2\right]
\end{bmatrix}
\begin{bmatrix} x_1(t_0) \\ x_2(t_0) \end{bmatrix}
$$

$$
+ \begin{bmatrix}
\dfrac{1}{\omega_n^2}\left\{1 + \dfrac{1}{\sqrt{1-\zeta^2}} e^{-\zeta\omega_n(t-t_0)} \sin\left[\omega_n\sqrt{1-\zeta^2}\,(t-t_0) - \phi_2\right]\right\} \\[3ex]
\dfrac{1}{\omega_n\sqrt{1-\zeta^2}} e^{-\zeta\omega_n(t-t_0)} \sin \omega_n\sqrt{1-\zeta^2}\,(t-t_0)
\end{bmatrix}, \quad (4.44)
$$

where

$$
\phi_1 = \tan^{-1}\frac{\sqrt{1-\zeta^2}}{\zeta}, \qquad \phi_2 = \tan^{-1}\frac{\sqrt{1-\zeta^2}}{-\zeta}.
$$

The corresponding output response of this second-order system having a unit step input is given by the following expression:

$$
c(t) = \omega_n^2 x_1(t)
$$

$$
= \frac{\omega_n^2}{\sqrt{1-\zeta^2}} e^{-\zeta\omega_n(t-t_0)} \sin\left[\omega_n\sqrt{1-\zeta^2}\,(t-t_0) + \phi_1\right] x_1(t_0)
$$

$$
+ \frac{\omega_n}{\sqrt{1-\zeta^2}} e^{-\zeta\omega_n(t-t_0)} \sin\left[\omega_n\sqrt{1-\zeta^2}\,(t-t_0)\right] x_2(t_0)
$$

$$
+ 1 + \frac{1}{\sqrt{1-\zeta^2}} e^{-\zeta\omega_n(t-t_0)} \sin\left[\omega_n\sqrt{1-\zeta^2}\,(t-t_0) - \phi_2\right]. \quad (4.45)
$$

Equation (4.45) is a much more complete solution to the classical second-order system than that previously derived in Eq. (4.24), since the initial conditions are now accounted for. Observe from Eq. (4.45) that the solution, when the initial conditions are zero, reduces to that of Eq. (4.24) with $\alpha = \phi_2$.

PROBLEMS

4.1 The two-phase ac induction motor of Problem 3.13 is used in a simple positioning system as shown in Fig. P4.1. Assume that a difference amplifier, whose gain is 10, is used as the error detector and also supplies power to the control field.

a) What are the undamped natural frequency ω_n and the damping factor ζ?
b) What are the percent overshoot and time to peak resulting from the application of a unit step input?
c) Plot the error as a function of time on the application of a unit step input.

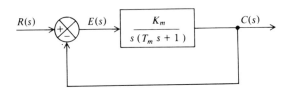

Figure P4.1

4.2 Repeat Problem 4.1 with the gain of the difference amplifier increased to 20. What conclusions can you draw from your result?

4.3 The two-phase ac induction motor and load, in conjunction with the gear train specified in Problem 3.19, is used in a simple positioning system as shown in Fig. P4.1. Assume that a difference amplifier, whose gain is 20, is used as the error detector and also supplies power to the control field.

a) What are the undamped natural frequency and damping factor ζ?
b) What are the percent overshoot and time to peak resulting from the application of a unit step input?
c) Plot the error as a function of time on the application of a unit step input.

4.4 Repeat Problem 4.3 with the gain of the difference amplifier increased to 40. What conclusions can you draw from your results?

4.5 Repeat Problem 4.3 with the gain of the difference amplifier decreased to 10. What conclusions can you draw from your result?

4.6 A typical aerodynamically controlled missile control system is synthesized by means of the appropriate application of moments to the airframe. These moments are generated by the deflection of control surfaces placed at large distances from the center of gravity. The result is that large moments are created with relatively small surface loads. The design of this type of control system requires a sufficiently high control loop gain in order to minimize the response time to input commands. In addition, it must not be so high as to cause high-frequency instabilities. Figure P4.6 illustrates an acceleration-control steering system of a missile. The command acceleration is compared with the output of an accelerometer to develop the basic error signal which drives the control system. The output of a rate gyro is utilized for damping [1].

a) Determine the transfer function, $C(s)/R(s)$, of this system.
b) Determine the undamped natural frequency ω_n of this system and the damping factor ζ,

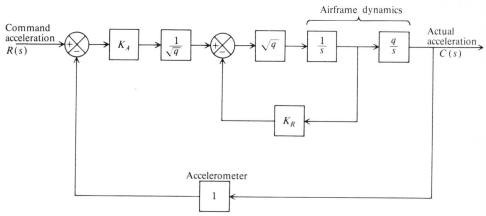

Figure P4.6

for the following set of parameters:

Amplifier gain $= K_A = 16,$ Aircraft gain factor $= q = 4,$ $K_R = 4.$

c) Determine the percent overshoot and time to peak resulting from an input command of a unit step of acceleration.

4.7 Figure P4.7 illustrates the control system of a map drive system used to display a portion of the map of an area to the commander of a task force. A main requirement of this control system is that a constant tension be maintained on the continuous sheet of paper between the wind-up and wind-off rolls in order that the paper does not tear. The system illustrated contains four rollers, and a spring which provides a restoring angular torque of $K_1\theta(t)$. As the radius R of the rollers varies, the tension changes and an adjustment in the wind-up motor speed is required. The map leaving the wind-off roll is assumed to have a velocity $v_1(t)$ and a tension $T_2(t)$; that approaching the wind-up roll is assumed to have a velocity $v_2(t)$ and a tension $T_1(t)$. A synchro-type sensing device, whose transfer function is K_e, is mounted on the pivot point of the tension arm and is used to sense angular deviations from $\theta = 0$. This error signal is amplified by a factor K_a by an amplifier. In addition, assume that the relationship for motor control voltage $E_c(t)$ caused by tension variations $T(t)$ is given by

$$E_c(t) = K_2 T(t).$$

The two-phase ac servomotor is assumed to have a transfer function given by Eq. (3.101). It also contains a gear reduction of $N:1$.

a) Determine the transfer function relating the change in tension $T(s)$ due to an input change velocity $v_1(s)$ of the wind-off roll in terms of the system parameters.

b) Determine the undamped natural frequency ω_n and the damping factor ζ in terms of the system parameters.

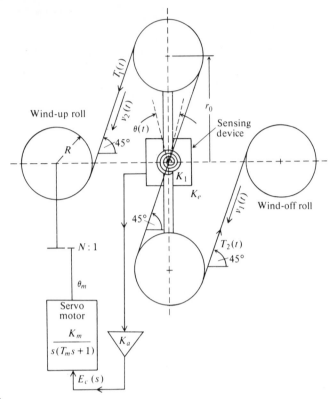

Figure P4.7

c) Calculate the damping factor ζ, for the case of a fully loaded wind-up roll, with the following parameters:

$$\text{Inertia of wind-up roller} = 0.101 \text{ oz in sec}^2$$
$$\text{Gear ratio} = N = 10.4\!:\!1$$
$$\text{Radius of wind-up roller} = R = 1.5 \text{ in}$$
$$\text{Motor inertia} = J_m = 4.43 \times 10^{-4} \text{ oz in sec}^2$$
$$\text{Motor constant} = K_m = 0.417 \text{ in oz/(rad/sec)}$$
$$\text{Synchro sensitivity} = K_e = 0.4 \text{ V/degree}$$
$$\text{Amplifier gain} = K_a = 1000$$
$$K_1 = 10.5 \text{ V/oz tension}$$
$$K_2 = 0.04 \text{ in oz/degree}$$

d) How can the damping factor be increased?

4.8 The Viking Mission conducted by the National Aeronautics and Space Administration included two launches in 1975 of a Viking Spacecraft by a Titan-Centaur launch vehicle consisting of an Orbiter and a Lander. The Orbiter had the capability of orbiting the planet Mars and of separating the Lander capsule, an automated laboratory in the search of signs of life, that entered the Martian atmosphere for soft landing on the surface

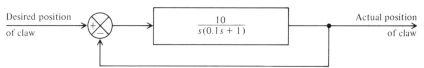

Fig. P4.8 (a) Lander capsule, Viking Project. (Official NASA photo) (b) Block diagram of the second-order control system for positioning the retractable claw.

of Mars. Looking over Mars from orbit, Viking cameras and other instruments aided in the confirmation of a suitable landing site. After this confirmation, the Lander separated from the Orbiter and began its descent to the Martian surface. The descent occurred when Mars was about 225 million miles from Earth and nearly on the other side of the sun. Therefore, this required a completely automated deorbit and landing operation because two-way communication at that distance was almost 45 minutes. Viking featured a series of scientific experiments in the areas of biology, geology, and meteorology. In order to collect material for these experiments, the Lander had a retractable claw with a 10-foot reach as shown in Fig. P4.8(a). It was used for scooping out soil samples and placing them in its automated chemical laboratory for analysis. For purposes of this example, assume that the control system which positions the retractable claw can be represented by the second-order control system shown in Fig. P4.8(b).

a) Determine the undamped natural frequency and the damping factor of this control system.
b) Determine the maximum percent overshoot and time to peak resulting from the application of a unit command signal.

4.9 A two-phase ac induction motor is used to position a device in a feedback configuration represented by Fig. P4.1. The time constant of the motor and load, T_m, is 0.5 seconds.

a) Determine the combined amplifier and motor constant gain, K_m, which will result in a damping factor of 0.5.
b) What is the resulting undamped, natural resonant frequency for the value of gain determined in part (a)?

REFERENCE

1. W. K. Waymeyer and R. W. Sporing, "Closed loop adaptation applied to missile control," in *Proceedings of the 1962 Joint Automatic Control Conference*, p. 18-3.

5 PERFORMANCE CRITERIA

5.1 INTRODUCTION

In the early days of control-system theory, engineers were generally less rigid in defining performance criteria. They were more apt to look on the feedback control system rather qualitatively and center attention primarily on stability and static accuracy. However, modern complex control systems have demanded the development of accurate criteria of performance.

The performance of a feedback control system is generally described in terms of stability, sensitivity, accuracy, transient response, and residual noise jitter. The exact specifications are usually dictated by the required system performance. Certain characteristics are more important in some systems than in others.

The great amount of literature that has appeared on the subject in recent years is evidence of the increasing importance that performance criteria have been given in feedback control system design. In order to keep pace with the requirements of modern feedback control systems, several new criteria of performance have been developed [1]. It is the purpose of this chapter to review and study several classical performance criteria together with more recent and sophisticated approaches. The control literature of the past decade abounds in various criteria of performance. Most significant are several performance criteria that have been postulated which are functions of time and error. After examining the literature, we find that the integral of time and error (ITAE) criterion for optimizing the transient response stands out as being useful, although somewhat limited; the integral square error (ISE) criterion is also quite popular. The concept of performance criteria, as presented in this chapter, is extended in Chapter 9, where optimal control theory is discussed.

5.2 STABILITY

A feedback control system must be stable even when the system is subjected to command signals, extraneous inputs anywhere within the loop, power supply variations, and changes in parameters of the feedback loop.

The qualitative statement that a system is stable, is meaningless. The question of how stable the system is must also be determined. In order to answer this question adequately, we must return to Eq. (2.117), which is repeated below:

$$\frac{C(s)}{R(s)} = \frac{G(s)}{1 + G(s)H(s)}.$$ (5.1)

As shown in Chapter 6, stability is determined by evaluating the denominator of this equation for $s = j\omega$. If $G(j\omega)H(j\omega) = -1$, the denominator would vanish and the system would oscillate indefinitely or the response would grow linearly with time. The margin by which $G(j\omega)H(j\omega)$ is shy of unity magnitude when its phase is 180° is known as the *gain margin*, and the phase by which it is shy of 180° when its magnitude is unity is known as the *phase margin*. These quantities indicate the degree to which the system is stable. They are used by the control engineer to determine how stable the feedback system is. Useful, qualitative, desirable design values are 30°–60° for the phase margin and 4–12 decibels for the gain margin. These numbers indicate that when $G(j\omega)H(j\omega)$ equals unity, its phase is 120°–150°, and when $G(s)H(s)$ has 180° phase shift, its magnitude is 0.25–0.63. In addition, the magnitude of the closed-loop resonant peak, M_p, is also a measure of the degree of stability. As shown in Chapter 6, a desirable value is between 1.0 and 1.4.

5.3 SENSITIVITY

Sensitivity is a measure of the dependence of a system's characteristics on those of a particular element. The differential sensitivity of a system's closed-loop transfer function T with respect to the characteristics of a given element K is defined as

$$S_K^T = \frac{d \ln T}{d \ln K}, \tag{5.2}$$

where

$$T = C(s)/R(s).$$

A more meaningful definition can be obtained by rewriting Eq. (5.2) as:

$$S_K^T = \frac{dT/T}{dK/K}. \tag{5.3}$$

Equation (5.3) states that the differential sensitivity of T with respect to K is the percentage change in T divided by that percentage change in K which has caused the change in T to occur. This definition is valid only for small changes. It is important to note that sensitivity is a function of frequency and an ideal system has zero sensitivity with respect to any parameter.

In order to illustrate the concept of sensitivity, consider the typical control system shown in Fig. 5.1(a). Here K_1 represents the transfer function of the input transducer, K_2 represents the transfer function of the feedback transducer, and G represents the combined transfer function of an amplifier, stabilizing network, motor, and gear train in the forward part of the feedback loop.

The overall system transfer function T is given by

$$T = \frac{C}{R} = \frac{K_1 G}{1 + K_2 G}. \tag{5.4}$$

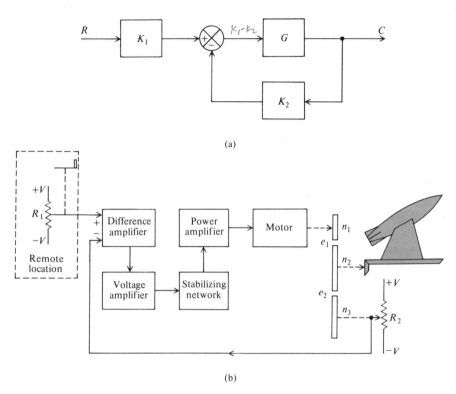

Fig. 5.1 (a) A representative control system. (b) An automatic positioning system for a missile launcher.

Let us now determine the sensitivity of the overall system transfer function with respect to changes in K_1, K_2, and G.

A. Sensitivity of T with respect to K_1 This is

$$u = 1+$$

$$S_{K_1}^T = \frac{dT/T}{dK_1/K_1} = \frac{K_1}{T}\frac{dT}{dK_1},$$

where

$$\frac{dT}{dK_1} = \frac{G}{1 + K_2 G} = \frac{T}{K_1}.$$

Therefore,

$$S_{K_1}^T = \frac{K_1}{T}\frac{T}{K_1} = 1. \tag{5.5}$$

B. Sensitivity of T with respect to K_2 This is

$$S_{K_2}^T = \frac{dT/T}{dK_2/K_2} = \frac{K_2}{T}\frac{dT}{dK_2},$$

where

$$\frac{dT}{dK_2} = \frac{0 - K_1 G^2}{(1 + K_2 G)^2} = \frac{-K_1^2 G^2}{K_1(1 + K_2 G)^2} .$$

Therefore,

$$S_{K_2}^T = \frac{K_2}{T} \frac{-K_1^2 G^2}{K_1(1 + K_2 G)^2} = \frac{-K_2}{T} \frac{T^2}{K_1} = \frac{-K_2 G}{1 + K_2 G} .$$

For cases where $K_2 G \gg 1$, this reduces to

$$S_{K_2}^T \simeq -1. \tag{5.6}$$

C. Sensitivity of T with respect to G

$$S_G^T = \frac{dT/T}{dG/G} = \frac{G}{T} \frac{dT}{dG} ,$$

where

$$\frac{dT}{dG} = \frac{(1 + K_2 G)K_1 - K_1 G K_2}{(1 + K_2 G)^2} = \frac{K_1}{(1 + K_2 G)^2} .$$

Therefore,

$$S_G^T = \frac{G}{T} \frac{K_1}{(1 + K_2 G)^2} = \frac{1}{1 + K_2 G} . \tag{5.7}$$

The results obtained in Eqs. (5.5), (5.6), and (5.7) are quite interesting. The symbols K_1 and K_2 represent input and feedback transducers, respectively, and Eqs. (5.5) and (5.6) illustrate that they are very critical. Any changes in their characteristics are directly reflected in an overall system transfer function change. Elements used for K_1 and K_2 must, therefore, possess precise and stable characteristics with temperature and time. Equation (5.7) shows that the sensitivity of the overall system transfer function with respect to G is divided by $1 + K_2 G$. From a sensitivity viewpoint, it appears desirable to design $K_2 G$ to have as large a value as possible. However, it need not be very precise.

It is also important to recognize that since sensitivity is a function of frequency, we should think of systems as being sensitive or insensitive only over certain frequency bands. In this example, $K_2 G$ is a function of frequency and will be large over only a limited range of frequencies. Therefore, T is insensitive to G only over a certain range of frequencies. This point is further illustrated in Problems 5.1 through 5.5.

Let us now try to extend the results derived in this section in order to determine qualitatively the requirements of the various elements shown in the simple missile launcher positioning device of Fig. 5.1(b). In this system R_1, R_2, $\pm V$, and the difference capability of the difference amplifier must all be precise. The gain characteristics of the difference amplifier, voltage amplifier, stabilizing network, power amplifier, and motor need not be precise. Any changes in the characteristics of these elements will be divided by $1 + K_2 G$. Let us now consider the gear train, which is composed of three gears n_1, n_2, n_3. It is assumed that gears n_2 and n_3 have the same number of teeth and each has 10 times as many teeth as gear n_1. The output is taken off gear n_2. The prime purpose of gear n_3 is to enable the output transducer to be coupled off another shaft. Gear meshes can result in system errors because the tooth space exceeds the thickness of an engaging tooth. This phenomenon is commonly referred to as backlash. It can be measured by holding one gear fast and observing the amount of motion in the other gear. Figure 5.1(b) denotes the backlash between gears n_1 and n_2 as e_1, and that between n_2 and n_3 as e_2. A sensitivity analysis shows that the error produced by backlash e_1 is reduced by $1 + K_2G$, while the error produced by backlash e_2 is added practically directly into the overall system transfer function. Therefore, there is a need for precision gearing for n_2 and n_3 but not for n_1. Chapter 8 illustrates that backlashes at e_1 and e_2, however, are very important from a stability viewpoint. From a sensitivity viewpoint, however, the backlash at e_1 is not as critical as that at e_2.

5.4 STATIC ACCURACY

Accuracy probably ranks as the next most important characteristic of a feedback control system. The designer always strives to design the system to minimize error for a certain anticipated class of inputs. This section considers techniques which are available for determining the system accuracy.

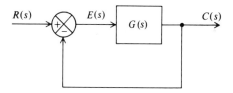

Fig. 5.2 A unity feedback system.

Theoretically, it is desirable for a control system to have the capability of responding to changes in position, velocity, acceleration, and changes in higher-order derivatives with zero error. Such a specification is very impractical and unrealistic. Fortunately, the requirements of practical systems are much less stringent. For example, let us consider the automatic positioning system of a missile launcher which is illustrated in Fig. 5.1(b). Its functioning is similar to the missile launcher positioning system in Fig. 1.8. Realistically, it would be desirable for this system to respond well to inputs of position and velocity, but not necessarily to those of acceleration. In addition, it probably would be desirable for this system to respond with zero error for positional-type inputs. However, a finite tracking error could probably be tolerated for inputs of velocity. In contrast to this system, where the stakes are quite high, let us consider a simpler positioning system which perhaps is only required to reproduce the angular position of a dial at some remote location. Such a control system would probably be only required to reproduce any positional inputs, but not any higher-order inputs such as those of velocity and acceleration.

A method for determining the steady-state performance of any control system is to apply the final-value theorem of the Laplace transform as given by Eq. 2.68. Let us consider the unity feedback system shown in Fig. 5.2. The relation between the resulting system error, $E(s)$, for a given input $R(s)$ is given by

$$\frac{E(s)}{R(s)} = \frac{1}{1 + G(s)}. \tag{5.8}$$

The steady-state error can be expressed as

$$e_{ss} = \lim_{t \to \infty} e(t) = \lim_{s \to 0} \frac{sR(s)}{1 + G(s)}. \tag{5.9}$$

The control engineer is usually interested in inputs of position, velocity, and acceleration. A step, ramp, and paraboloid are simple mathematical expressions which represent these physical quantities, respectively. They are defined in Eqs. (5.10) through (5.12), where the notation $U(t)$ means a unit step for $t > 0$.

$$\text{Position input: } r(t) = U(t), \qquad R(s) = 1/s, \tag{5.10}$$
$$\text{Ramp input: } r(t) = tU(t), \qquad R(s) = 1/s^2, \tag{5.11}$$
$$\text{Acceleration input: } r(t) = \tfrac{1}{2}t^2 U(t), \qquad R(s) = 1/s^3. \tag{5.12}$$

We next determine the steady-state error of several types of systems for each of these three inputs: the unit step, unit ramp, and unit paraboloid. It is assumed that the loop transfer function $G(s)$ has the general form

$$G(s) = \frac{K(1 + T_1 s)(1 + T_2 s) \cdots (1 + T_M s)}{s^n[(T_3 s)^2 + 2\zeta\omega_n s + 1](1 + T_4 s)(1 + T_5 s) \cdots (1 + T_N s)}, \tag{5.13}$$

where

$\quad s^n = $ a multiple pole at the origin of the complex plane,

$\quad K = $ gain factor of the expression.

1. Unit Step Input   The steady-state error can be obtained by substituting $R(s) = 1/s$ into Eq. (5.9):

$$e_{ss} = \lim_{s \to 0} \frac{s(1/s)}{1 + G(s)} = \frac{1}{1 + \lim_{s \to 0} G(s)}. \tag{5.14}$$

The quantity $\lim_{s \to 0} G(s)$ is defined as the position constant and is denoted by K_p:

$$K_p = \lim_{s \to 0} G(s). \tag{5.15}$$

Therefore the steady-state error in terms of the position constant is given by

$$e_{ss} = \frac{1}{1 + K_p} . \tag{5.16}$$

Equation (5.16) states that the steady-state tracking error of a feedback control system having a unit step input equals $1/(1 +$ the position constant). Table 5.1 summarizes the value of K_p as a function of the number of pure integrations of the open-loop transfer function $G(s)$.

Table 5.1

Pure integrations of $G(s)$	K_p
0	K
1	∞
2	∞
.	.
.	.
.	.
n, where $n > 0$	∞

Table 5.1 indicates that the position constant is infinite for all systems which contain one or more pure integration(s) in the open-loop transfer function $G(s)$. Therefore, Eq. (5.16) implies that all systems containing at least one pure integration result in a theoretical steady-state positional response error of zero. Table 5.1 indicates that the position constant is finite for a system containing no pure integrations and, therefore, the response error for a position input is nonzero.

2. Unit Ramp Input   The steady-state error can be obtained by substituting $R(s) = 1/s^2$ into Eq. (5.9):

$$e_{ss} = \lim_{s \to 0} \frac{s(1/s)^2}{1 + G(s)} = \frac{1}{\lim_{s \to 0} sG(s)} . \tag{5.17}$$

The quantity $\lim_{s \to 0} sG(s)$ is defined as the velocity constant and is denoted by K_v:

$$K_v = \lim_{s \to 0} sG(s). \tag{5.18}$$

Therefore, the steady-state error in terms of the velocity constant is given by

$$e_{ss} = 1/K_v. \tag{5.19}$$

Equation (5.19) states that the steady-state response error of a feedback control system having a unit ramp equals the reciprocal of the velocity constant. Table 5.2 summarizes the value of K_v as a function of the number of pure integrations of the open-loop transfer function $G(s)$.

Table 5.2

Pure integrations of $G(s)$	K_v
0	0
1	K
2	∞
.	.
.	.
.	.
n, where $n > 1$	∞

Table 5.2 indicates that the velocity constant is infinite for all systems which contain more than one pure integration in the open-loop transfer function $G(s)$. Therefore, Eq. (5.19) implies that all systems containing at least two pure integrations have a theoretical steady-state velocity response error of zero. Table 5.2 indicates that a system containing no pure integrations cannot follow a velocity input. Table 5.2 also indicates that a system containing one pure integration has a nonzero response error for a velocity input.

3. Unit Parabolic Input An expression for the steady-state error can be obtained by substituting $R(s) = 1/s^3$ into Eq. (5.9):

$$e_{ss} = \lim_{s \to 0} \frac{s(1/s^3)}{1 + G(s)} = \frac{1}{\lim_{s \to 0} s^2 G(s)}. \tag{5.20}$$

The quantity $\lim_{s \to 0} s^2 G(s)$ is defined as the acceleration constant and is denoted by K_a:

$$K_a = \lim_{s \to 0} s^2 G(s). \tag{5.21}$$

Therefore, the steady-state error in terms of the acceleration constant is

$$e_{ss} = 1/K_a. \tag{5.22}$$

Equation (5.22) states that the steady-state response error of a feedback control system having a unit parabolic input equals the reciprocal of the acceleration constant. Table 5.3 summarizes the value of K_a as a function of the number of pure integrations of the open-loop transfer function $G(s)$. Table 5.3 indicates that the acceleration constant is infinite for all systems which contain three or more pure integrations in the open-loop transfer function $G(s)$. Therefore, Eq. (5.22) implies that all systems containing at least three pure integrations have a theoretical steady-state acceleration response error of zero. Table 5.3 indicates that systems containing less than two pure integrations cannot follow an acceleration input. Table 5.3 also indicates that a system containing two pure integrations has a nonzero response error for an acceleration input.

Table 5.3

Pure integrations of $G(s)$	K_a
0	0
1	0
2	K
3	∞
.	.
.	.
.	.
n, where $n > 2$	∞

A summary of the results derived appears in Table 5.4. It is quite general and enables the reader to compare the capabilities of various types of systems. Notice from this table that the steady-state constants are zero, finite, or infinite. It is important to emphasize at this time that if the inputs are other than unit quantities the steady-state errors are proportionally increased. For example, should the input to a system containing one pure integration be a ramp whose value is B position units (ft/sec, yd/sec, etc.), then the steady-state error as given by Eq. (5.19) would be modified to read

$$e_{ss} = B/K_v .$$

It should be noted that the unit of the velocity constant is 1/sec and that of the acceleration constant is 1/sec². The position constant K_p has no dimensions.

Let us now consider an input composed of position, velocity, and acceleration which equal A ft, B ft/sec, and $C/2$ ft/sec², respectively. The form of the input can be represented as:

$$r(t) = A + Bt + \tfrac{1}{2}Ct^2.$$

The steady-state response of the system may be obtained by considering each component of the input separately, and then adding the results by means of superposition.

Table 5.4 Summary of steady state constants for various types of input

Number of pure integrations	Type of input		
	Unit step	Unit ramp	Unit paraboloid
0	K_p	0	0
1	∞	K_v	0
2	∞	∞	K_a
3	∞	∞	∞

The resulting steady-state error is of the following form:

$$e_{ss} = \frac{A}{1 + K_p} + \frac{B}{K_v} + \frac{C}{K_a}.$$

It is interesting to see how the various types of systems summarized in Table 5.4 would respond to this input.

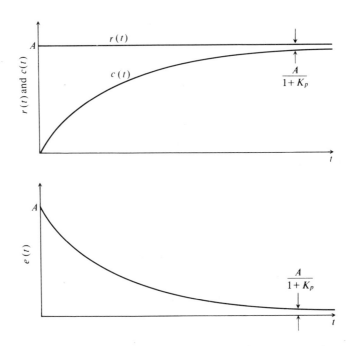

Fig. 5.3   Response of a system containing no pure integrations to a step input.

a) System Containing No Pure Integration.   The steady-state error is

$$e_{ss} = \frac{A}{1 + K_p} + \infty + \infty.$$

The result indicates that a system containing no pure integration will be able to follow the position input component of A ft, but not the velocity or acceleration inputs of B ft/sec and $C/2$ ft/sec², respectively.   Figure 5.3 illustrates a typical step response for this system (when there are no velocity or acceleration components of the input).

b) System Containing One Pure Integration.   The steady-state error is

$$e_{ss} = 0 + B/K_v + \infty.$$

The result indicates that a system containing one pure integration will follow the position input component with zero error and the velocity input component with a finite error of B/K_v. This system will not, however, be able to follow the acceleration input component. Note that the units of B/K_v are feet. This should be interpreted to mean that there is a fixed positional error due to the constant velocity input component. Figure 5.4 illustrates typical step and ramp responses of this system.

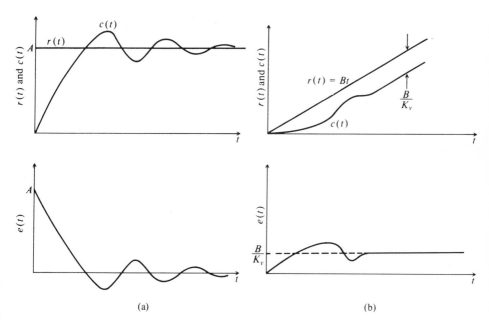

Fig. 5.4 Response of a system containing one pure integration to (a) step and (b) ramp inputs.

c) System Containing Two Pure Integrations. The steady-state error is

$$e_{ss} = 0 + 0 + C/K_a.$$

The result indicates that a system containing two pure integrations will follow the position and velocity input components with zero error and the acceleration input component with a finite error of C/K_a. Note that the units of C/K_a are feet. This should be interpreted to mean that there is a fixed positional error due to the constant acceleration input component. Figure 5.5 illustrates typical step and ramp responses of this system.

4. Relationships of Static Error Constants to Closed-Loop Poles and Zeros It is often important to relate the position, velocity, and acceleration constants to the *closed-loop* poles and zeros. Let us consider a general single-loop, unity-feedback system having a

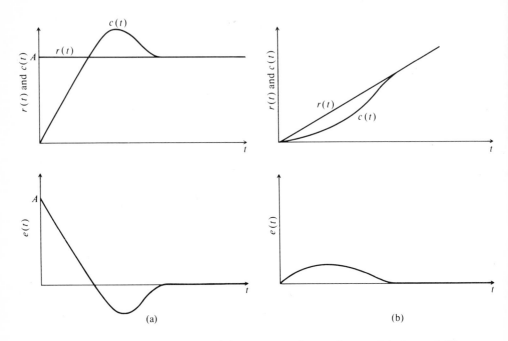

Fig. 5.5 Response of a system containing two pure integrations to (a) step and (b) ramp inputs.

forward transfer function, $G(s)$, where the closed-loop transfer function is given by

$$\frac{C(s)}{R(s)} = \frac{G(s)}{1 + G(s)}, \tag{5.23}$$

and the relationship between input and error is given by

$$\frac{E(s)}{R(s)} = \frac{1}{1 + G(s)}. \tag{5.24}$$

Since $E(s) = R(s) - C(s)$, it is evident that

$$\frac{C(s)}{R(s)} = 1 - \frac{E(s)}{R(s)}. \tag{5.25}$$

It is assumed that $C(s)/R(s)$ can be represented by the rational function

$$\frac{C(s)}{R(s)} = \frac{K(s + z_1)(s + z_2) \cdots (s + z_m)}{(s + p_1)(s + p_2) \cdots (s + p_n)} = \frac{K \prod_{j=1}^{m}(s + z_j)}{\prod_{j=1}^{n}(s + p_j)}. \tag{5.26a}$$

In addition, if $1/[1 + G(s)]$ in Eq. (5.24) is expanded as a power series in s, the error constants are defined in terms of the successive coefficients:

$$\frac{E(s)}{R(s)} = \frac{1}{1 + G(s)} = \frac{1}{1 + K_p} + \frac{1}{K_v} s + \frac{1}{K_a} s^2 + \cdots. \tag{5.26b}$$

Now the relationships between K_p, K_v, K_a and the closed-loop poles and zeros can be determined.

a) *Position Constant.* Letting s approach zero in Eq. (5.24), we have

$$\frac{E(0)}{R(0)} = \lim_{s \to 0} \frac{1}{1 + G(s)}. \tag{5.27}$$

Based on the definition of K_p in Eq. (5.15), we can rewrite Eq. (5.27) as

$$\frac{E(0)}{R(0)} = \frac{1}{1 + K_p}. \tag{5.28}$$

Substituting Eq. (5.28) into Eq. (5.25) with s set at zero, we obtain the following:

$$\frac{C(0)}{R(0)} = \frac{K_p}{1 + K_p}.$$

Solving for K_p in terms of $C(0)/R(0)$, we have

$$K_p = \frac{C(0)/R(0)}{1 - C(0)/R(0)}. \tag{5.29}$$

Using Eq. (5.26a) to represent $C(s)/R(s)$ and letting s approach zero in Eq. (5.26a), we have

$$\frac{C(0)}{R(0)} = \frac{K \prod_{j=1}^{m} z_j}{\prod_{j=1}^{n} p_j}, \tag{5.30}$$

where

$$\prod_{j=1}^{m} z_j = \text{product of zeros,}$$

$$\prod_{j=1}^{n} p_j = \text{product of poles.}$$

Substituting Eq. (5.30) into (5.29), the following expression for K_p in terms of the closed-loop poles and zeros is obtained:

$$K_p = \frac{K \prod_{j=1}^{m} z_j}{\prod_{j=1}^{n} p_j - K \prod_{j=1}^{m} z_j}. \tag{5.31}$$

b) *Velocity Constant.* In order to derive the velocity constant in terms of the closed-loop poles and zeros, let us substitute Eq. (5.26b) into Eq. (5.25):

$$\frac{C(s)}{R(s)} = 1 - \frac{1}{1 + K_p} - \frac{1}{K_v} s - \frac{1}{K_a} s^2 - \cdots.$$

Taking the derivative of this expression with respect to s, and then letting s equal zero, we obtain

$$\left[\frac{d}{ds}\left(\frac{C(s)}{R(s)}\right)\right]_{s=0} = -\frac{1}{K_v}. \tag{5.32}$$

In addition, we make use of the property that

$$\left[\frac{C(s)}{R(s)}\right]_{s=0} = \frac{C(0)}{R(0)} = 1 \tag{5.33}$$

in unity feedback systems containing one or more pure integrations. Equation (5.33) says that a closed-loop unity feedback system behaves with an ideal closed-loop transfer function of one at zero frequency. Dividing Eq. (5.32) by (5.33),

$$\frac{1}{K_v} = \frac{-\left[\frac{d}{ds}\left(\frac{C(s)}{R(s)}\right)\right]_{s=0}}{[C(s)/R(s)]_{s=0}} = -\left[\frac{d}{ds}\ln\frac{C(s)}{R(s)}\right]_{s=0}.$$

Substituting Eq. (5.26a) into this equation, we have the following:

$$\frac{1}{K_v} = -\left\{\frac{d}{ds}\left[\ln K + \ln(s + z_1) + \cdots + \ln(s + z_m)\right.\right.$$
$$\left.\left. - \ln(s + p_1) - \cdots - \ln(s + p_n)\right]\right\}_{s=0}.$$

This can also be written as

$$\frac{1}{K_v} = -\left(\frac{1}{s + z_1} + \cdots + \frac{1}{s + z_m} - \frac{1}{s + p_1} - \cdots - \frac{1}{s + p_n}\right)_{s=0}, \tag{5.34}$$

or

$$\frac{1}{K_v} = \sum_{j=1}^{n}\frac{1}{p_j} - \sum_{j=1}^{m}\frac{1}{z_j}. \tag{5.35}$$

Therefore, $1/K_v$ equals the sum of the reciprocals of the closed-loop poles minus the sum of the reciprocals of the closed-loop zeros.

c) *Acceleration Constant.* The acceleration constant in terms of the closed-loop poles and zeros can be derived in a similar manner. We know from Eq. 5.26(b) that

$$\frac{E(s)}{R(s)} = \frac{1}{1 + K_p} + \frac{1}{K_v}s + \frac{1}{K_a}s^2 + \cdots.$$

From Eq. (5.25), this can be rewritten as

$$\frac{C(s)}{R(s)} = 1 - \frac{1}{1 + K_p} - \frac{1}{K_v}s - \frac{1}{K_a}s^2 - \cdots.$$

It is obvious from this equation that $-2/K_a$ equals the zero-frequency value of the second derivative of $C(s)/R(s)$. Writing this in terms of the logarithmic derivative:

$$\frac{d^2}{ds^2}\left[\ln\frac{C(s)}{R(s)}\right] = \frac{(C/R)''}{C/R} - \left[\frac{(C/R)'}{C/R}\right]^2.$$

Setting $C(0)/R(0)$ equal to one, then

$$-\frac{2}{K_a} = \left\{\frac{d^2}{ds^2}\left[\ln\frac{C(s)}{R(s)}\right]\right\}_{s=0} + \frac{1}{K_v^2}.$$

Differentiating the right-hand side of Eq. (5.34) and letting s equal zero yields the following expression:

$$-\frac{2}{K_a} = \frac{1}{K_v^2} + \sum_{j=1}^{n}\frac{1}{p_j^2} - \sum_{j=1}^{m}\frac{1}{z_j^2}, \tag{5.36}$$

where K_v is defined by Eq. (5.35).

d) *An Example.* As an example, let us consider the second-order system illustrated in Fig. 4.1 whose characteristics are defined by Eqs. (4.1), (4.2), and (4.3):

$$\frac{C(s)}{R(s)} = \frac{K_m/T_m}{s^2 + (1/T_m)s + K_m/T_m},$$

or

$$\frac{C(s)}{R(s)} = \frac{\omega_n^2}{s^2 + 2\zeta\omega_n s + \omega_n^2},$$

where

$$\omega_n^2 = K_m/T_m \quad\text{and}\quad \zeta = 1/2\omega_n T_m.$$

This is representative of a wide class of control systems that was thoroughly analyzed in Section 4.2. The problem is to determine K_v in terms of the parameters ζ and ω_n. K_m is the system gain and T_m is the time constant of the open-loop transfer function. The velocity constant of the simple system illustrated in Fig. 4.1 is

$$K_v = K_m$$

by inspection. Let us relate this velocity constant to ζ and ω_n from the basic definitions of K_m and T_m in terms of ζ and ω_n:

$$K_v = K_m = \omega_n^2 T_m = \omega_n^2\left(\frac{1}{2\zeta\omega_n}\right) = \frac{\omega_n}{2\zeta}.$$

The fact that

$$K_v = \omega_n/2\zeta \tag{5.37}$$

is very important to remember for all second-order control systems that are characterized by a pair of complex-conjugate poles.

5.5 TRANSIENT RESPONSE

In addition to stability, sensitivity, and accuracy, we are always concerned with the transient response of a feedback system. Transient response characteristics are usually defined on the basis of a step input. The response of a second-order system, containing one pure integration, to a step input is quite useful for purposes of defining the various transient parameters. Should a problem arise where the system is higher than second order, a reasonably good approximation can be made by assuming that the system is second order if one pair of complex-conjugate roots dominate. This point is amplified during the discussion of the root locus in Chapters 6 and 7. For purposes of illustration, let us consider the second-order system analyzed in Section 4.2. There we had a unity feedback system whose closed-loop transfer function $C(s)/R(s)$ was given by

$$\frac{C(s)}{R(s)} = \frac{\omega_n^2}{s^2 + 2\zeta\omega_n s + \omega_n^2}. \tag{5.38}$$

Its response to a unit step input, for the case of $\zeta < 1$, was given by

$$c(t) = 1 + \frac{e^{-\zeta\omega_n t}}{\sqrt{1 - \zeta^2}} \sin(\omega_n t\sqrt{1 - \zeta^2} - \alpha), \tag{5.39}$$

where

$$\alpha = \sin^{-1}\sqrt{1 - \zeta^2}.$$

The input and output responses are illustrated in Fig. 5.6.

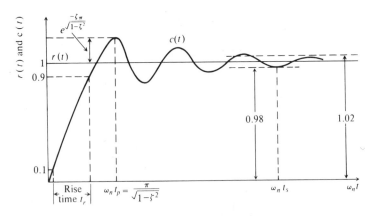

Fig. 5.6   Response of a second-order system to a unit step.

The time for the feedback system to reach its first overshoot is commonly referred to as the time to the first peak, t_p. As was derived in Section 4.2 (see Eq. 4.29) and is illustrated in Fig 5.6,

$$\omega_n t_p = \frac{\pi}{\sqrt{1 - \zeta^2}},$$

or

$$\text{time to first peak} = t_p = \frac{\pi}{\omega_n \sqrt{1 - \zeta^2}}. \tag{5.40}$$

The time required for the system to damp out all transients is commonly called the settling time, t_s. Theoretically, for a second-order system this is infinity. In practice, however, the transient is assumed to be over when the error is reduced below some minimum value. Typically, the minimum level is set at 2% of the final value. The settling time, which is approximately equal to four time constants of the envelope of the damped sinusoidal oscillation, is illustrated in Fig. 5.6 and is given by

$$t_s = 4/\zeta \omega_n. \tag{5.41}$$

The rise time t_r is defined as the time required for the response to rise from 10% to 90% of its final value. Figure 5.6 illustrates the rise time for the response shown.

Upon the application of a step input, the output of a feedback system will usually exceed the input. As is illustrated in Fig. 5.6, a second-order system may oscillate several times around the steady-state output, depending on the value of the damping ratio ζ. The first overshoot is of particular interest; the ratio of this overshoot's peak value to the steady-state settling value of the system is usually expressed as a percentage. The amount of overshoot allowable depends entirely on the particular problem. An overshoot of 10% is reasonable. Notice that the percentage overshoot of the system illustrated in Fig. 5.6 is $\exp(-\zeta\pi/\sqrt{1 - \zeta^2}) \times 100\%$. A 10% overshoot corresponds to a damping ratio of approximately 0.6 for this system.

Even though suitable and reasonable values for rise time, settling time, and the peak overshoot have been chosen, one is not sure whether a good system has been designed. For example, if the rise time is very small, invariably the peak value of the overshoot increases and so does the settling time. On the other hand, if the design is for minimum overshoot, the rise time increases. In order to resolve the conflict that exists between rise time, peak overshoot, and settling time, criteria have been proposed for synthesizing optimum transient performance. Most of these criteria consider that the error and/or the time at which the error has occurred during the transient is important [1, 2]. Several of these criteria are presented in the remainder of this chapter.

5.6 PERFORMANCE INDICES

Modern complex control systems usually require more sophisticated performance criteria than those presented so far. As seen in the previous section, error and the time at which it occurs are very important factors which usually must be considered simultaneously. A performance index is a single measure of a system's performance which emphasizes those characteristics of the response that are deemed to be important. The notion of a performance index is very important in optimal control theory where the system is designed to optimize this performance index given certain constraints. This subject will be discussed fully in Chapter 9.

References 1 and 2 discuss an entire class of performance indices which are various functions of error and/or time. This section reviews those which seem useful in the design of practical control systems.

A fairly useful performance index is the *integral of the absolute magnitude of the error* (IAE) criterion which is

$$S_1 = \int_0^\infty |e(t)|\ dt. \tag{5.42}$$

By utilizing the magnitude of the error, this integral expression increases for either positive or negative error, and results in a fairly good underdamped system. For a second-order system, S_1 has a minimum for a damping factor of approximately 0.7. The simplicity of this performance index makes it very applicable to analog-computer simulation.

Another useful performance index is the *integral of the square of the error* (ISE) criterion which is

$$S_2 = \int_0^\infty e^2(t)\ dt. \tag{5.43}$$

By focusing on the square of the error function, it penalizes both positive and negative values of the error. For a second-order system, S_2 has a minimum for a damping factor of approximately 0.5.

A very useful criterion which penalizes long-duration transients is known as the *integral of time multiplied by the absolute value of error* (ITAE). It is given by

$$S_3 = \int_0^\infty t\ |e|\ dt. \tag{5.44}$$

This performance index is much more selective than the IAE or the ISE: the minimum value of its integral is much more definable as the system parameters are varied. For a second-order system, S_3 has a minimum for a damping factor of 0.707.

Other figures of merit which have been proposed are the *integral of time multiplied by the squared error* (ITSE), *the integral of squared time multiplied by the absolute value of error* (ISTAE), and *the integral of squared time multiplied by square error*

(ISTSE). These performance indices are

$$\text{ITSE:} \qquad S_4 = \int_0^\infty t e^2 \, dt, \qquad\qquad (5.45)$$

$$\text{ISTAE:} \qquad S_5 = \int_0^\infty t^2 |e| \, dt, \qquad\qquad (5.46)$$

$$\text{ISTSE:} \qquad S_6 = \int_0^\infty t^2 e^2 \, dt. \qquad\qquad (5.47)$$

The performance indices S_4, S_5, and S_6 have not been applied to any great extent in practice due to the increased difficulty in handling them.

Note that all of the integrals proposed in Eqs. (5.42) through (5.47) converge only if $e(t) \to 0$ as $t \to \infty$.

A comparison of this array of performance indices is very interesting. The ISE criterion is not very sensitive to parameter variations since the minimum is usually broad [11]. In addition, the ISE criterion has the advantage of being easy to deal with mathematically. The IAE criterion gives a slightly better sensitivity than the ISE criterion. The ITAE criterion generally produces smaller overshoots and oscillations than the IAE and ISE criteria. In addition, it is the most sensitive of the three, and sometimes too sensitive—slight parameter variation degrades system performance [1, 2].

In practice, a relatively insensitive criterion may be more useful in those systems where the parameters may not be known very accurately. In addition, even though one tries to optimize a performance criterion, one may also have other performance characteristics in mind. Therefore, it is desirable in some applications to permit moderate deviation from the "optimum" setting of the parameters in order that these other performance characteristics can be achieved without appreciably increasing the performance index. Based on this logic, the ISE criterion may be the most desirable performance index in some practical applications and is considered further in Chapter 9 when optimal control theory is discussed. In addition, the reader is referred to Chapter 2 of Reference 11 for additional considerations of the ISE criterion.

The ITSE criterion has not been studied in great detail [1]. However, available data indicate that it will probably result in being a valuable performance criterion. Very limited information is available on the ISTAE performance criterion, and a judgment cannot be made here [1]. Data available on the ISTSE criterion indicate that it does result in good responses for systems containing one integration in the open-loop transfer function [1]. Information on the resulting response with more than one integration in the open-loop transfer function is not available.

A paper by Graham and Lathrop [2] created a great deal of interest in the ITAE criterion. An important aspect of this paper was the detailed discussion and presentation of results of a comparison for various performance criterion with the ITAE. It is more sensitive than the IAE and ISE criteria, and is useful in those practical applications which require a very sensitive criterion. Since the ITAE criterion has practical value and is interesting academically, this performance criterion is studied further. First, however, we are interested in determining the form of the closed-loop system transfer function in order that zero error results for various kinds of inputs. This is necessary in order to determine the relationship of the various coefficients in the numerator and denominator of the closed-loop system transfer function.

5.7 ZERO-ERROR SYSTEMS

The transfer function of the general feedback system, containing a forward transfer function $G(s)$ and a feedback transfer function $H(s)$, is given by the general expression

$$\frac{C(s)}{R(s)} = \frac{G(s)}{1 + G(s)H(s)} = \frac{A_1 s^l + A_2 s^{l-1} + A_3 s^{l-2} + \cdots + A_l s + A_{l+1}}{B_1 s^m + B_2 s^{m-1} + B_3 s^{m-2} + \cdots + B_m s + B_{m+1}}. \quad (5.48)$$

From Eq. (5.26b) of Section 5.4, the error for this system can be expressed as*

$$e(t) = \frac{r(t)}{1 + K_p} + \frac{\dot{r}(t)}{K_v} + \frac{\ddot{r}(t)}{K_a} + \cdots. \quad (5.49)$$

It can be shown [2] that $1 + K_p$ is a function of $B_{m+1} - A_{l+1}$. Therefore, it is necessary that $B_{m+1} = A_{l+1}$ for zero steady-state error when the input is a step function. In addition, this implies that the forward transfer function contains at least one pure integrator as discussed previously in Section 5.4. From the general system transfer function of Eq. (5.48), it can be seen that there are many possible forms of $C(s)/R(s)$ which will yield zero steady-state error with a step input. When the numerator consists of the constant B_{m+1}, the system is called a *zero steady-state step error system* [2]:

$$\frac{C(s)}{R(s)} = \frac{B_{m+1}}{B_1 s^m + B_2 s^{m-1} + B_3 s^{m-2} + \cdots + B_{m+1}}. \quad (5.50)$$

It can also be shown [2] that K_v is a function of $B_{m+1} - A_{l+1}$ and $B_m - A_l$. Therefore, for zero steady-state error with a ramp input, $B_m = A_l$ and $B_{m+1} = A_{l+1}$. In addition, this implies that the forward transfer function contains two or more pure integrators as discussed previously in Section 5.4. A *zero steady-state ramp input system* occurs when the system transfer function is given by

$$\frac{C(s)}{R(s)} = \frac{B_m s + B_{m+1}}{B_1 s^m + B_2 s^{m-1} + B_3 s^{m-2} + \cdots + B_m s + B_{m+1}}. \quad (5.51)$$

---

* It is assumed that the input and all of its derivatives do not have any discontinuities in the interval of interest.

5.8 THE ITAE PERFORMANCE CRITERION
FOR OPTIMIZING THE TRANSIENT RESPONSE

The ITAE performance index, as defined by Eq. (5.44), is considered further in this section. As discussed in Section 5.6, this performance index does not penalize large initial errors which are unavoidable. However, it does penalize long duration transients. We now consider the form that the system transfer function should take, for various order systems, in order to achieve zero steady-state step and ramp error systems and minimize the ITAE.

In the case of the zero steady-state step error system, Eq. (5.50) shows that the form of the system transfer function is given by

$$\frac{C(s)}{R(s)} = \frac{B_{m+1}}{B_1 s^m + B_2 s^{m-1} + B_3 s^{m-2} + \cdots + B_{m+1}}. \tag{5.52}$$

The procedure used to produce a table of standard system transfer functions of the form $C(s)/R(s)$ was to vary each coefficient in Eq. (5.52) separately until the integral of time multiplied by the absolute value of error became a minimum. Then the successive coefficients were varied in sequence to minimize the ITAE value.

If this criterion is applied to the second-order system described by Eq. (5.38), the optimum damping ratio is approximately 0.707. A table of system transfer functions, $C(s)/R(s)$, has been prepared by Graham and Lathrop [2]. They show the optimum form of the denominator, for systems whose transfer functions are of the form given by Eq. (5.52), which will minimize the integral of Eq. (5.44). For example, the optimum form for a second-order system is given by

$$\frac{C(s)}{R(s)} = \frac{\omega_n^2}{s^2 + 1.414\omega_n s + \omega_n^2}, \tag{5.53}$$

where $\zeta = 1.414/2 = 0.707$. The optimum form for a third-order system is given by

$$\frac{C(s)}{R(s)} = \frac{\omega_n^3}{s^3 + 1.75\omega_n s^2 + 2.15\omega_n^2 s + \omega_n^3}. \tag{5.54}$$

Table 5.5 The minimum ITAE standard forms for a zero-error-displacement system [2]

$$s + \omega_n$$
$$s^2 + 1.4\omega_n s + \omega_n^2$$
$$s^3 + 1.75\omega_n s^2 + 2.15\omega_n^2 s + \omega_n^3$$
$$s^4 + 2.1\omega_n s^3 + 3.4\omega_n^2 s^2 + 2.7\omega_n^3 s + \omega_n^4$$
$$s^5 + 2.8\omega_n s^4 + 5.0\omega_n^2 s^3 + 5.5\omega_n^3 s^2 + 3.4\omega_n^4 s + \omega_n^5$$
$$s^6 + 3.25\omega_n s^5 + 6.60\omega_n^2 s^4 + 8.60\omega_n^3 s^3 + 7.45\omega_n^4 s^2 + 3.95\omega_n^5 s + \omega_n^6$$
$$s^7 + 4.475\omega_n s^6 + 10.42\omega_n^2 s^5 + 15.08\omega_n^3 s^4 + 15.54\omega_n^4 s^3 + 10.64\omega_n^5 s^2 + 4.58\omega_n^6 s + \omega_n^7$$
$$s^8 + 5.20\omega_n s^7 + 12.80\omega_n^2 s^6 + 21.60\omega_n^3 s^5 + 25.75\omega_n^4 s^4 + 22.20\omega_n^5 s^3 + 13.30\omega_n^6 s^2 + 5.15\omega_n^7 s + \omega_n^8$$

Table 5.5, which has been obtained from Reference 2, shows the optimum denominator transfer function for systems through the eighth order which will minimize the integral of Eq. (5.44). These standard forms provide a quick and simple method for synthesizing an optimum dynamic response.

In the case of the zero steady-state ramp error systems, the system transfer function was shown in Eq. (5.51) to be given by:

$$\frac{C(s)}{R(s)} = \frac{B_m s + B_{m+1}}{B_1 s^m + B_2 s^{m-1} + B_3 s^{m-2} + \cdots + B_m s + B_{m+1}}. \tag{5.55}$$

The objective here also is to obtain a set of standard forms for the denominators of the system transfer functions given by Eq. 5.55. Table 5.6, obtained from Reference 2, was similarly obtained as Table 5.5.

Table 5.6 The minimum ITAE standard forms for a zero-error ramp system [2].

$$s^2 + 3.2\omega_n s + \omega_n^2$$

$$s^3 + 1.75\omega_n s^2 + 3.25\omega_n^2 s + \omega_n^3$$

$$s^4 + 2.41\omega_n s^3 + 4.93\omega_n^2 s^2 + 5.14\omega_n^3 s + \omega_n^4$$

$$s^5 + 2.19\omega_n s^4 + 6.50\omega_n^2 s^3 + 6.30\omega_n^3 s^2 + 5.24\omega_n^4 s + \omega_n^5$$

$$s^6 + 6.12\omega_n s^5 + 13.42\omega_n^2 s^4 + 17.16\omega_n^3 s^3 + 14.14\omega_n^4 s^2 + 6.76\omega_n^5 s + \omega_n^6$$

The ITAE criterion is applied to several problems in the problem section of this chapter (see Problems 5.17 through 5.21).

The ITAE criterion is a straightforward method for optimizing the transient response of a system when the transfer function is known. Generally, it produces smaller overshoots and oscillations than the other criteria presented. It should be emphasized, however, that the ITAE solution is very sensitive and may not be useful for certain systems where most of the system poles and zeros and gains may be specified initially. For the latter case, the designer does not have the flexibility for selecting as many of the parameters as he might wish.

5.9 OTHER PRACTICAL CONSIDERATIONS

The control engineer must concern himself with several other practical aspects before becoming able to state intelligently and completely the expected system performance. The concluding section of this chapter qualitatively discusses considerations of feedback system bandwidth, nonlinearities, size, weight, power consumption, and economics. Hopefully, this will aid in giving the reader a complete overall view of the problem.

Feedback system bandwidth is usually defined as the frequency at which the closed-loop magnitude equals unity. The bandwidth of a system is indicated by its particular application. Usually, the control engineer is interested in designing the system to respond to a certain spectrum of input signal frequencies and to suppress all inputs above a certain frequency. It is important to emphasize that we should not arbitrarily design for a large bandwidth. Although large bandwidths usually result in large error constants, with small resulting system error, they also result in a system that responds to extraneous noise inputs and has considerable jitter due to the noise. The desirable approach is to design the feedback system bandwidth to be just large enough to pass the desired input-signal frequency spectrum and then attenuate all higher frequency signals [3]. Feedback system bandwidth considerations are discussed in detail in Chapters 6 and 7, where methods for obtaining the frequency response are presented.

Nonlinearities are other factors which affect the performance of a control system. Primary concern is with backlash, stiction, and coulomb friction. Backlash, which was described in Section 5.3, is the amount of free motion of one gear while its mating gear is held fast. Stiction is the frictional force which prevents motion until the driving force exceeds some minimum value. Coulomb friction is a constant frictional drag which opposes motion, but has a magnitude that is independent of velocity. Each of these nonlinearities has an effect on the performance of feedback systems. More will be said regarding nonlinearities in Chapter 8.

Other factors of concern are size, weight, power consumption, and economics. The system must conform to certain specifications of size and weight. These are very important factors which usually dictate the design of the system. For example, these specifications may decide the type of power drive to be used. Power is another very important consideration. The system must usually perform within a certain allowable power limitation. Size, weight, and power consumption are usually very critical items for airborne and space applications. Last, but not least, is the question of economics. A basic fact of life is that most engineers work for organizations whose primary purpose is to make profit. Systems must therefore be designed as inexpensively as possible within the framework of good performance. A generally useful rule of thumb is that minimum bandwidth systems will consume the least power and be the most economical.

PROBLEMS

5.1 Assume that the system shown in Fig. P5.1 has the following characteristics:

$$K_1 = 10 \text{ V/rad},$$

$$K_2 = 10 \text{ V/rad},$$

$$G(s) = \frac{100}{s(s + 1)}.$$

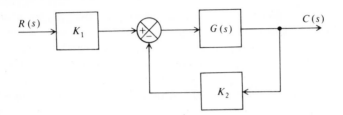

Figure P5.1

a) Determine the sensitivity of the system's transfer function T with respect to the input transducer, K_1.
b) Determine the sensitivity of the system's transfer function T with respect to the output transducer, K_2.
c) Determine the sensitivity of the system's transfer function T with respect to G.
d) Indicate qualitatively the frequency dependency of S_G^T.

5.2 For the system in Fig. P5.2, assume the following characteristics:

$$K_1 = 10 \text{ V/rad}, \qquad G_1 = \frac{10}{100 + s},$$

$$K_2 = 10 \text{ V/rad}, \qquad G_2 = \frac{20}{s(s + 2)},$$

$$K_3 = 2 \text{ V/rad}, \qquad H = 4s.$$

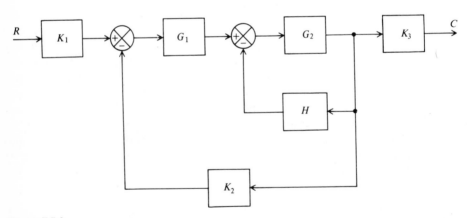

Figure P5.2

a) Determine the sensitivity of the system's transfer function T with respect to G_1 at $\omega = 1$ rad/sec.
b) Determine the sensitivity of the system's transfer function T with respect to G_2 at $\omega = 1$ rad/sec.
c) Determine the sensitivity of the system's transfer function T with respect to H at $\omega = 1$ rad/sec.

d) Determine the sensitivity of the system's transfer function T with respect to K_3 at $\omega = 1$ rad/sec.

e) List the answers of parts (a) through (d) in tabular form with the lowest value first and the largest value last in the vicinity of $\omega = 1$. Normalize your table with respect to the lowest sensitivity found. How could this table change for a different choice of frequency?

5.3 Figure P5.3(a) illustrates the block diagram of a computer-controlled machine tool which utilizes a position loop and a correction loop [4]. A practical problem in machine-tool

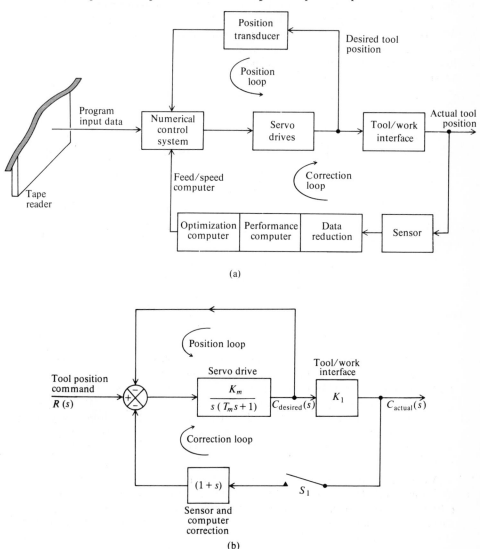

(a)

(b)

Figure P5.3

application is the fact that the desired tool position is ordinarily fed back by the position loop as shown, but this is not indicative of the condition and shape of the finished part. In reality, the finished part is ordinarily removed from the feedback effect due to the tool–work interface. Since these effects are complex and difficult to predict, the process output will not conform precisely to that desired by the programmer. By means of the correction loop, information is obtained from the process output by means of sensors coupled as closely as possible to the tool–work interface. These sensor signals, which represent variables such as torque, vibration, and temperature, are fed back to improve the performance of the machine tool control system. Figure P5.3(b) illustrates an equivalent block-diagram representation.

a) With switch S_1 open and the correction loop inoperative, determine the sensitivity of the system's transfer function T to variations of K_m, T_m, and K_1.

b) Repeat part (a) with switch S_1 closed and the correction loop operative.

c) What conclusions can you draw from your results?

5.4 For the system illustrated in Fig. P5.4 assume the transfer functions indicated.

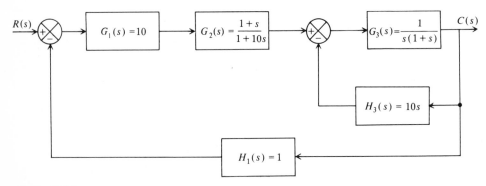

Figure P5.4

a) Determine the sensitivity of the system's transfer function with respect to $G_1(s)$.

b) Determine the sensitivity of the system's transfer function with respect to $G_3(s)$.

c) Determine the sensitivity of the system's transfer function with respect to $H_3(s)$.

d) Determine the sensitivity of the system's transfer function with respect to $H_1(s)$.

e) List the answers of parts (a) through (d) in tabular form with the lowest value of sensitivity first and the largest value last at a frequency of 1 rad/sec. Normalize your table with respect to the lowest sensitivity found. How could this table change for a different choice of frequency?

5.5 Automatic control theory can also be applied to automatic warehousing and inventory control systems [5]. Of particular importance in these systems is the smooth flow of material. Fig. P5.5 illustrates a block diagram that is representative of this class of control systems.

a) Determine the sensitivity of the system's transfer function $C(s)/R(s)$ with respect to production planning, $G_1(s)$.

b) Determine the sensitivity of the system's transfer function with respect to production, $G_2(s)$

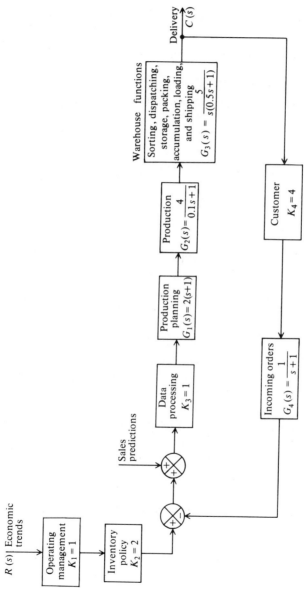

Figure P5.5

c) Determine the sensitivity of the system's transfer function with respect to the various warehouse functions, $G_3(s)$.

d) Determine the sensitivity of the system's transfer function with respect to incoming orders, $G_4(s)$.

e) Determine the sensitivity of the system's transfer function with respect to the inventory policy, K_2.

f) For which two parameters is the system most sensitive, and what conclusions can you reach from your results?

g) Indicate qualitatively the frequency dependency of the sensitivity functions.

5.6 The block diagram of a simple instrument servomechanism is shown in Fig. P5.6.

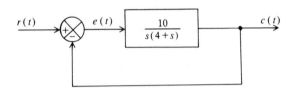

Figure P5.6

a) Determine the steady-state error resulting from the input of a ramp which may be represented by   4

$$r(t) = 10t.$$

b) Determine the steady-state error resulting from the following input:

$$r(t) = 4 + 6t + 3t^2.$$

5.7 Repeat Problem 5.6 with the transfer function of the system given by

$$G(s) = \frac{10}{s(1 + s)(1 + 10s)}.$$

5.8 For the system shown in Fig. P5.8, determine the following:

a) Steady-state error resulting from an input

$$r(t) = 10t. \quad \text{zero}$$

b) Steady-state error resulting from an input

$$r(t) = 4 + 6t + 3t^2. \quad 2.4$$

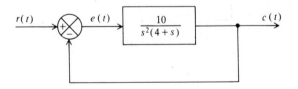

Figure P5.8

c) Steady-state error resulting from an input

$$r(t) = 4 + 6t + 3t^2 + 1.8t^3.$$

5.9 Repeat Problem 5.8 with the transfer function of the system given by

$$G(s) = \frac{10}{s^2(1 + s)(1 + 10s)}.$$

5.10 A common problem in the television industry is that of picture wobbling or jumping due to movement of the TV camera while a picture is being taken. The effect of this problem is easily understood by examining Figs. P5.10(a) and (b). When the camera is at rest (as illustrated in Fig. P5.10(a) a light-ray entering the camera lens impinges on point A within the camera. However, if the camera is jolted upward through an angle δ (as in Fig. P5.10b), the light ray is displaced from its original location at point A to point B. This can be corrected by means of the system illustrated in Fig. P5.10(b). The concept utilizes a device which changes the shape of a fluid lens such that the ray's impinging point does not move [6]. The front transparent plate is rotated in the vertical plane by a torque motor. The rear plate is rotated in the horizontal plane. Two rate gyros are mounted in the camera to detect any disturbances. Their output is fed to a servo amplifier which adjusts the driving current to the

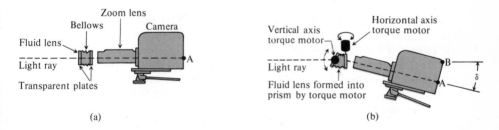

(a)   (b)

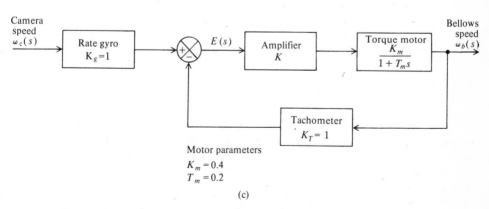

(c)

Fig. P5.10 (From *Control Engineering* published by The Reuben H. Donnelley Corp. Reproduced by permission.)

torque motors. Tachometers close the rate feedback loops. The equivalent block diagram of one such axis is illustrated in Fig. P5.10(c). Observe that the feedback loop uses speed from the rate gyro as the reference input and speed from the tachometer to indicate speed of the bellows.

a) Determine the required value of amplifier gain K in order that the steady-state error resulting from a camera scanning speed of 50°/sec is only 1°/sec.
b) What is the steady-state error of this system resulting from camera accelerations of any magnitude?

5.11 A servomechanism, shown in Fig. P5.11, is used to drive an inertia load through gearing of negligible inertia.

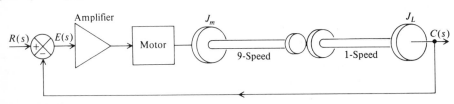

Figure P5.11

1. AC motor characteristics: Torque–speed slope $= 4.5 \times 10^{-6}$ lb ft/(rad/sec). Stall torque constant $= 8.5 \times 10^{-6}$ (lb ft)/V.
2. Load inertia $= J_L = 40 \times 10^{-6}$ lb ft sec² (Assume $J_m N^2 \ll J_L$)
3. Amplifier gain $= 10$
4. Gear ratio $= 9:1$ (steps motor speed down)
   a) What is the transfer function $G(s)$ relating $C(s)$ and $E(s)$ of the system?
   b) What are the undamped natural frequency ω_n and the damping factor ζ?
   c) What is the percent overshoot and time to peak resulting from the application of a unit step input?
   d) What is the steady-state error resulting from application of a unit step input?
   e) What is the steady-state error resulting from application of a unit ramp input?
   f) What is the steady-state error resulting from application of a parabolic input?

5.12 Repeat Problem 5.11 with the assumption that the motor inertia, J_m, is 1.11×10^{-6} lb ft sec².

5.13 Repeat Problem 5.11 with a motor inertia of 1.11×10^{-6} lb ft sec² and the amplifier's gain increased to 100.

5.14 Automatically controlled machine tools form an important aspect of control-system application. The major trend has been towards the use of automatic numerically controlled machine tools using tape inputs. The justification of using tape has been the elimination of costly contour templates and the reduction of the machine set-up procedure required. In addition, it eliminates the tedium of repetitive operations required of human operators, and the possibility of human error. Figure P5.14 illustrates the block diagram of an automatic numerically controlled machine-tool position control system, using a punched tape reader, to supply the reference signal.

a) What are the undamped natural frequency ω_n and damping factor ζ?
b) What are the percent overshoot and time to peak resulting from the application of a unit step input?

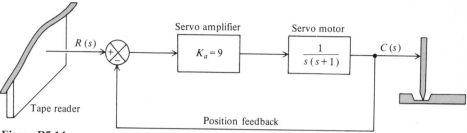

Figure P5.14

c) What is the steady-state error resulting from the application of a unit step input?

d) What is the steady-state error resulting from the application of a unit ramp input?

5.15 Modern ocean-going ships utilize stabilization techniques in order to minimize the effects of oscillations due to waves. By utilizing hydrofoils or fins, stabilizing torques can be generated in order to stabilize the ship [7]. Figure P5.15(a) illustrates this concept employing stabilizing fins. An equivalent block diagram of the system is illustrated in Fig. P5.15(b).

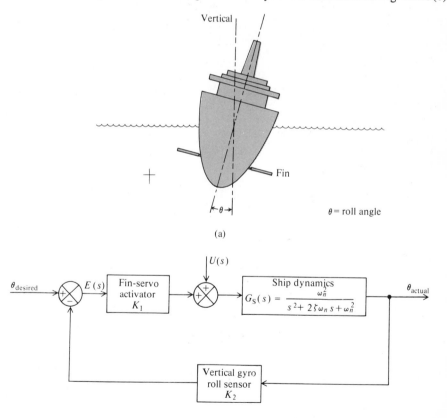

Figure P5.15   (b)

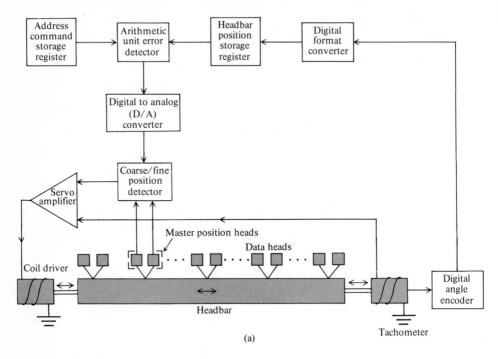

(a)

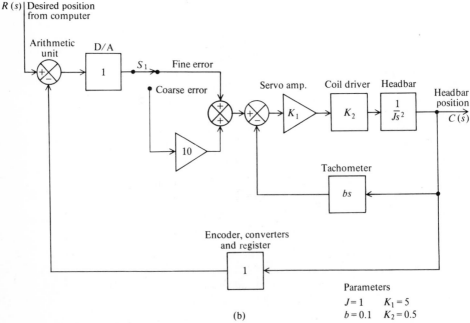

(b)

Parameters

$J = 1$ $K_1 = 5$

$b = 0.1$ $K_2 = 0.5$

Figure P5.16

The reference input signal, θ_{desired}, is normally set equal to zero. A vertical gyro, which senses deviations from $\theta = 0$, feeds back a correction signal which activates the fins in order to drive the error signal to zero. The disturbance signal, $U(s)$, represents the disturbance torque due to the waves.

a) Determine the effect of the disturbance torque $U(s)$ on system error $E(s)$ if it is assumed that $K_1 K_2 G_s(s) \gg 1$. What conclusions can you draw from your result?

b) Based on the approximation of part (a), what should K_1 be set at in order to reduce a disturbance torque input at $U(s)$ of $10°$ to an equivalent system error of $0.1°$?

5.16 As digital computers become more common in every phase of industry and the scientific communities, there is an ever increasing demand for the storage and speedy retrieval of data. The resulting control-system requirements imposed are usually quite stringent and usually require new techniques. Figure P5.16(a) illustrates a system used to control the position of the read/write heads in a random-access magnetic-drum memory system [8]. The particular application had 64 data heads mounted in vertical pairs at 2-inch intervals along a "headbar." Although the controlling action is obtained by utilizing digital components, an equivalent representation of the control system is given in Fig. P5.16(b).

The headbar is free to move longitudinally. It is located between two magnetic drums whose axes of rotation are parallel with the longitudinal axis of the headbar. Each of the data heads in this system has access to a total of 200 tracks. The track accessed depends on the placement of the headbar within the limits of its 2-inch travel. A fine and a coarse positioning loop are used to obtain the desired accuracy. An electronic switch S_1 switches the system so that it has a large amount of gain for large errors so that the system responds rapidly. For this condition, a large amount of overshoot is tolerated in order to achieve a fast rise time. For small errors, the fine system is activated which has a larger damping factor, smaller overshoot, and a somewhat longer response time. Assume that the system switches when $e(t) = 0.1$ unit.

a) Determine the undamped natural frequency, damping factor, overshoot, and time to peak for the "coarse" loop.

b) Repeat part (a) for the "fine" loop.

c) Determine the error of the "coarse" and "fine" loops to a unit step input.

d) Determine the error of the "coarse" and "fine" loops to a unit ramp input.

e) Determine the error of the "coarse" and "fine" loops to a unit paraboloid input.

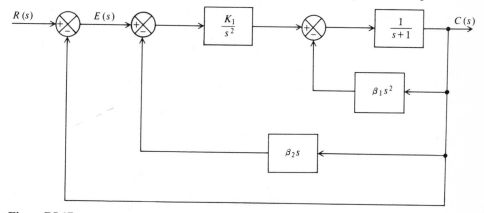

Figure P5.17

5.17 A fourth-order feedback system containing three feedback paths is illustrated in Fig. P5.17. In order to satisfy the ITAE criterion, determine the values of K_1, β_1, and β_2.

5.18 Utilizing an analog computer, determine the overshoot and rise time of the transient response for the system considered in Problem 5.17 to a step input as K_1, β_1, and β_2 are each varied $\pm 100\%$ [9, 10, 12].

5.19 Determine the values of K_1, K_2, and b, in the feedback system illustrated in Fig. P5.19, which will satisfy the ITAE criterion.

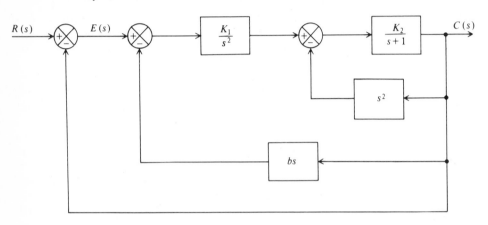

Figure P5.19

5.20 Repeat Problem 5.19 with the acceleration feedback element attenuated by a factor of 0.5. What conclusions can you draw from your result?

5.21 Repeat Problem 5.19 if the gain of the acceleration feedback element is doubled. What conclusions can you draw from your result?

5.22 Figure P5.22 illustrates an electronic pacemaker used to regulate the speed of the human heart. Assume that the transfer function of the pacemaker is given by $G_p(s) = K/(0.05s + 1)$ and assume that the heart acts as a pure integrater.

a) For optimum response, a closed-loop damping factor of 0.5 is desired. Determine the required gain, K, of the pacemaker in order to achieve this.
b) What is the sensitivity of the system transfer function, $C(s)/R(s)$, to small changes in K?
c) Determine this sensitivity at DC.
d) Find the magnitude of this sensitivity at the normal heart rate of 60 beats/minute.

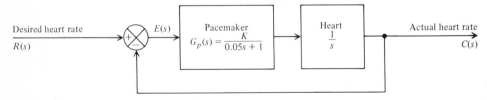

Figure P5.22

5.23 A control system used to position a load is shown in Fig. P5.23.

a) Determine the steady-state error for a step input of 10 units.

b) How should $G(s)$ be modified in order to reduce this steady-state error to zero?

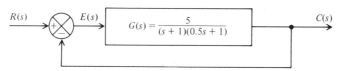

Figure P5.23

5.24 Repeat Problem 5.6 for

$$G(s) = \frac{10(s + 1)}{s^2(0.1s + 1)(s + 5)}.$$

REFERENCES

1. W. C. Schultz and C. V. Rideout, "Control system performance measures: Past, present, and future," *IRE Trans. Automatic Control* **AC-6,** 22 (1961).
2. D. Graham and R. C. Lathrop, "The synthesis of optimum transient response: Criteria and standard forms," *AIEE Trans.* **72,** 273 (1953).
3. S. M. Shinners, "Minimizing servo load resonance error with frequency selective feedback," *Control Eng.* **51,** 51–56 (January 1962).
4. R. M. Centner and J. M. Idelsohn, "Adaptive controller for a metal cutting process," in *Proceedings of the 1963 Joint Automatic Control Conference*, pp. 262–71.
5. R. Dallimonti, "Developments in automatic warehousing and inventory control," in *Proceedings of the 1965 Joint Automatic Control Conference*, pp. 281–5.
6. J. de la Cierva, "Rate servo keeps TV picture clear," *Control Eng.* **12,** 112 (May 1965).
7. J. Bell, "Control for ship stabilization," in *Proceedings of the 1st International Federation of Automatic Control Congress*, Butterworth, London, (1960), pp. 208–17.
8. R. Tickell, "A high performance position control," in *Proceedings of the 1966 Joint Automatic Control Conference*, pp. 230–42.
9. G. A. Korn and T. M. Korn, *Electronic Analog and Hybrid Computers*, McGraw-Hill, New York (1964).
10. J. R. Ashley, *Introduction to Analog Computation*, Wiley, New York (1963).
11. G. C. Newton, Jr., L. A. Gould, and J. F. Kaiser, *Analytical Design of Linear Feedback Controls*, Wiley, New York (1961).
12. S. M. Shinners, "Which computer-Analog, digital, or hybrid?" *Machine Design* **43,** 104–111 (January 21, 1971).

6 TECHNIQUES FOR DETERMINING CONTROL-SYSTEM STABILITY

6.1 INTRODUCTION

For proper controlling action, a feedback system must be stable. Previous chapters have indicated that feedback systems have the serious disadvantage that they may inadvertently act as oscillators. A feedback control system must maintain stability when the system is subjected to commands at its input, extraneous inputs anywhere within the feedback loop, power supply variations, and changes in the parameters of the elements comprising the feedback loop.

In the ensuing discussion, if, for every bounded input, the output is bounded, then the system is *stable*. In this chapter, analysis is limited to linear time-invariant systems, that is, systems for which the principle of superposition is valid and which may be described by an ordinary linear differential equation with constant coefficients. The analysis of nonlinear systems is presented in Chapter 8.

This chapter focuses attention on the stability of the general feedback system which was illustrated in Fig. 2.12. The closed-loop transfer function of this system, given by Eq. (2.117), is repeated below:

$$\frac{C(s)}{R(s)} = \frac{G(s)}{1 + G(s)H(s)}.$$ (6.1)

The characteristic equation for this generalized system can be obtained by setting the denominator of the system transfer function equal to zero:

$$1 + G(s)H(s) = 0.$$ (6.2)

In linear systems, stability is independent of the input excitation, and this is the equation that determines system stability. All the methods of stability analysis investigate this equation in some manner. One can show that if the roots of Eq. (6.2) lie in the left half-plane, the system is stable. However, the system is considered unstable if any of the roots of this equation have positive real parts or lie on the imaginary axis.

In general, the following two approaches exist for determining stability:

1. Calculating the exact roots of Eq. (6.2)

2. Determination of the region of system parameters which guarantees that the roots of Eq. (6.2) have negative real parts.

Using the first approach, the control engineer has at his disposal the following two methods:

a) Direct solution utilizing the classical approach
b) Root-locus method

Using the second approach, the control engineer has at his disposal the following criteria:

a) Routh-Hurwitz criterion
b) Nyquist criterion.

Clearly, calculation of the exact roots using the classical approach can be extremely tedious. Usually, the designer is interested in the root-locus method and the criteria of the second approach. This chapter presents each of these methods, except the classical approach, together with their relative merits. Additional graphical approaches based on the Nyquist criterion are also discussed. These include the use of the Bode diagram and Nichols chart. Application of these methods to actual design is deferred to Chapter 7.

6.2 STATE-SPACE DETERMINATION OF THE CHARACTERISTIC EQUATION

The characteristic equation can also be defined in terms of the state-space equation of the control system. We have stated in the previous section that stability is independent of the input. Therefore, the condition $\mathbf{x}(t) = 0$, where $\mathbf{x}(t)$ is the state vector, can be viewed as the equilibrium state of the system. Let us assume that a linear system is subjected to a disturbance at $t = 0$ resulting in an initial state, $\mathbf{x}(0)$, that is finite. If it returns to its equilibrium state as t approaches infinity, the system is considered to be stable. If it does not, in terms of our definition, it is considered to be unstable. These concepts of stability can be generalized. In the state-space approach [1, 2], a linear system is considered to be stable if, for a finite initial state $\mathbf{x}(0)$, there is a positive number A, that depends on $\mathbf{x}(0)$, where

$$\|\mathbf{x}(t)\| < A \quad \text{for} \quad t \geqslant 0, \tag{6.3}$$

$$\lim_{t \to \infty} \|\mathbf{x}(t)\| = 0. \tag{6.4}$$

The value $\|\mathbf{x}(t)\|$ denotes the norm of the state vector $\mathbf{x}(t)$. It is defined as

$$\|\mathbf{x}(t)\| = \left[\sum_{j=1}^{n} x_j(t)^2 \right]^{1/2}. \tag{6.5}$$

Equation (6.3) can be interpreted to mean that the transition of state for positive time, as represented by the norm of the vector $\mathbf{x}(t)$, is bounded. Equation (6.4) can be interpreted to mean that the system must reach its equilibrium point as t approaches infinity. Figure 6.1 illustrates the state-space stability criterion for a second-order system having states x_1 and x_2. Observe from this figure that a cylinder, whose radius is A, forms the bound for the trajectory as time increases. As time approaches infinity, the linear system reaches the equilibrium point $\mathbf{x}(t) = 0$. Note that, strictly

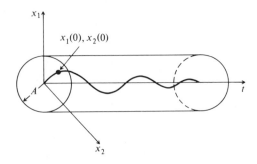

Fig. 6.1 State-space stability concept.

speaking, the above definition corresponds to asymptotic stability.* For simplicity, we will call the system stable if Eqs. (6.3) and (6.4) hold.

Let us next develop analytically, and apply, the state-space approach for determining the stability of a linear system. It will be shown that this method determines the location of the roots of the characteristic equation, and restricts the roots of the characteristic equation to the left half-plane for stability.

Consider the general differential equation of a linear system in scalar form:

$$A_n \frac{d^n c(t)}{dt^n} + A_{n-1} \frac{d^{n-1}c(t)}{dt^{n-1}} + \cdots + A_0 c(t) = r(t), \tag{6.6}$$

where all the coefficients are constants, $A_n \neq 0$, $r(t)$ represents the input to the system, and $c(t)$ represents the system output. Since the input does not affect stability, it is assumed to be zero. The Laplace transform of Eq. (6.6) is given by

$$C(s)[A_n s^n + A_{n-1}s^{n-1} + \cdots + A_0] = 0. \tag{6.7}$$

Stability can be determined from the characteristic equation of this system:

$$A_n s^n + A_{n-1}s^{n-1} + \cdots + A_0 = 0. \tag{6.8}$$

It remains now to determine the roots of this equation.

The nth-order differential equation, given by Eq. (6.6), may also be specified by n first-order differential equations. By defining

$$x_1 = c(t),$$

$$x_2 = \dot{x}_1 = \frac{dc(t)}{dt},$$

$$\cdots$$

$$\dot{x}_n = \frac{d^n c(t)}{dt^n}, \tag{6.9}$$

---

* According to Liapunov, this is asymptotic stability—the type of stability preferred by control system engineers. This is defined and discussed in detail in Section 8.17 on Liapunov's stability criterion.

the linear system can be specified by the following set of first-order differential equations:

$$\dot{x}_1 = A_{11}x_1 + \cdots + A_{1n}x_n + B_{11}r_1,$$
$$\cdots \tag{6.10}$$
$$\dot{x}_n = A_{n1}x_1 + \cdots + A_{nn}x_n + B_{n1}r_1,$$

which is equivalent to Eq. (6.6).

The set of equations (6.10) is solved for the case where the input $r(t)$ equals zero; assume an exponential solution of the form

$$x_i = f_i e^{st}, \tag{6.11}$$

where f_i is an unknown constant. To check the form of the assumed solution, Eq. (6.11) is differentiated and the result is substituted into the set of equations (6.10). Differentiating (6.11), we obtain

$$\dot{x}_i = f_i(se^{st}). \tag{6.12}$$

Substitution of Eqs. (6.11) and (6.12) into (6.10) results in

$$f_1 se^{st} = A_{11}f_1 e^{st} + \cdots + A_{1n}f_n e^{st},$$
$$\cdots \tag{6.13}$$
$$f_n se^{st} = A_{n1}f_1 e^{st} + \cdots + A_{nn}f_n e^{st}.$$

Equating coefficients of e^{st} and rearranging, we obtain the following:

$$(A_{11} - s)f_1 + \cdots + A_{1n}f_n = 0,$$
$$\cdots \tag{6.14}$$
$$A_{n1}f_1 + \cdots + (A_{nn} - s)f_n = 0.$$

This set of equations can be rewritten in the following matrix algebra form:

$$[\mathbf{A} - s\mathbf{I}]\mathbf{f} = \mathbf{0} \tag{6.15}$$

where $\mathbf{f}$ = column matrix, or vector, and $\mathbf{I}$ = identity matrix. Equation (6.15) will have a nontrivial solution only if

$$|\mathbf{A} - s\mathbf{I}| = 0. \tag{6.16}$$

By expanding the determinant $|\mathbf{A} - s\mathbf{I}|$ and solving for the roots of the equation $|\mathbf{A} - s\mathbf{I}| = 0$, the eigenvalues of the matrix $\mathbf{A}$ are obtained. Expanding the determinant

$$|\mathbf{A} - s\mathbf{I}| = 0 \tag{6.17}$$

results in the following expression:

$$s^n + B_1 s^{n-1} + \cdots + B_n = 0. \tag{6.18}$$

Notice that the B_i terms in Eq. (6.18) are equivalent to the A_i/A_n terms in Eq. (6.8). Therefore, the two equations are equivalent. The n values of s that satisfy Eqs. (6.17) are called the characteristic values, or eigenvalues, of the matrix.

As an example for comparing the stability determination of a system utilizing conventional Laplace transform and state-space techniques, consider the unity feed-back system shown in Fig. 6.2. Using the Laplace transform, we could easily obtain the closed-loop transfer function of the system as

$$\frac{C(s)}{R(s)} = \frac{G(s)}{1 + G(s)} = \frac{K}{T_1 T_2 s^3 + (T_1 + T_2)s^2 + s + K}. \tag{6.19}$$

The characteristic equation of this simple linear system is given by

$$T_1 T_2 s^3 + (T_1 + T_2)s^2 + s + K = 0. \tag{6.20}$$

System stability can be determined by locating the roots of this equation if the classical approach is to be used. The same problem will next be analyzed from the state-space viewpoint.

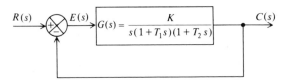

Fig. 6.2   Third-order feedback control system considered in Section 6.2.

From Eq. (6.19), we obtain

$$[T_1 T_2 s^3 + (T_1 + T_2)s^2 + s + K]C(s) = KR(s). \tag{6.21}$$

The time-domain expression equivalent to Eq. (6.21) is given by

$$T_1 T_2 \dddot{c}(t) + (T_1 + T_2)\ddot{c}(t) + \dot{c}(t) + Kc(t) = Kr(t). \tag{6.22}$$

As discussed previously, this third-order differential equation may be written as three first-order differential equations as follows:

$$c(t) = x_1$$
$$\dot{c}(t) = \dot{x}_1 = x_2$$
$$\ddot{c}(t) = \dot{x}_2 = x_3 \tag{6.23}$$
$$\dddot{c}(t) = \dot{x}_3 = \frac{1}{T_1 T_2}[-(T_1 + T_2)\ddot{c}(t) - \dot{c}(t) - Kc(t) + Kr(t)].$$

This can easily be transformed into matrix form:

$$\begin{bmatrix} \dot{x}_1 \\ \dot{x}_2 \\ \dot{x}_3 \end{bmatrix} = \begin{bmatrix} 0 & 1 & 0 \\ 0 & 0 & 1 \\ \dfrac{-K}{T_1 T_2} & \dfrac{-1}{T_1 T_2} & \dfrac{-(T_1 + T_2)}{T_1 T_2} \end{bmatrix} \begin{bmatrix} x_1 \\ x_2 \\ x_3 \end{bmatrix} + r(t) \begin{bmatrix} 0 \\ 0 \\ \dfrac{K}{T_1 T_2} \end{bmatrix}. \tag{6.24}$$

Using vector notation, the above equation becomes

$$\dot{\mathbf{x}} = \mathbf{A}\mathbf{x} + \mathbf{B}r(t),$$

where

$$\mathbf{x} = \begin{bmatrix} x_1 \\ x_2 \\ x_3 \end{bmatrix}, \quad \dot{\mathbf{x}} = \begin{bmatrix} \dot{x}_1 \\ \dot{x}_2 \\ \dot{x}_3 \end{bmatrix}, \quad \mathbf{A} = \begin{bmatrix} 0 & 1 & 0 \\ 0 & 0 & 1 \\ \dfrac{-K}{T_1 T_2} & \dfrac{-1}{T_1 T_2} & \dfrac{-(T_1 + T_2)}{T_1 T_2} \end{bmatrix}, \quad \mathbf{B} = \begin{bmatrix} 0 \\ 0 \\ \dfrac{K}{T_1 T_2} \end{bmatrix}. \quad (6.25)$$

In order to obtain the characteristic equation, the input will be set equal to zero and the solution x_i will be assumed equal to $f_i e^{st}$. As outlined previously (see Eq. 6.17), this procedure results in

$$|\mathbf{A} - s\mathbf{I}| = 0. \tag{6.26}$$

The resulting determinant is given by

$$\begin{vmatrix} -s & 1 & 0 \\ 0 & -s & 1 \\ \dfrac{-K}{T_1 T_2} & \dfrac{-1}{T_1 T_2} & \dfrac{-(T_1 + T_2)}{T_1 T_2} - s \end{vmatrix} = 0. \tag{6.27}$$

Expansion of the determinant by means of minors along the first row gives

$$-s \begin{vmatrix} -s & 1 \\ \dfrac{-1}{T_1 T_2} & \dfrac{-(T_1 + T_2)}{T_1 T_2} - s \end{vmatrix} -1 \begin{vmatrix} 0 & 1 \\ \dfrac{-K}{T_1 T_2} & \dfrac{-(T_1 + T_2)}{T_1 T_2} - s \end{vmatrix} = 0, \tag{6.28}$$

which reduces to

$$T_1 T_2 s^3 + (T_1 + T_2)s^2 + s + K = 0. \tag{6.29}$$

Equation (6.29), obtained utilizing the state-space approach, is the same characteristic equation as (6.20), which was obtained by using conventional Laplace transform techniques. Although the mathematics involved in obtaining Eq. (6.29) was more laborious than for Eq. (6.20) for this simple problem, many important features and characteristics of the state-space approach have been presented and applied. It is important to emphasize that for more complex problems involving multiple inputs and outputs, the state-space approach greatly simplifies the solution. In addition, it greatly facilitates automatic computation utilizing digital computers.

6.3 ROUTH-HURWITZ STABILITY CRITERION

The Routh-Hurwitz stability criterion is an algebraic procedure for determining whether a polynomial has any zeros in the right half-plane. It involves examining the signs and magnitudes of the coefficients of the characteristic equation without actually having to determine its roots. Although this method overcomes one of the disadvantages of the classical approach, it still does not indicate the relative degree of stability or instability.

Routh [3] and Hurwitz [4] independently determined the necessary and sufficient conditions for stability from the signs and magnitudes of the coefficients of the characteristic equation. Although their criteria differ somewhat, both furnish the same information. A useful form of their approach is described below.

Let us represent the general form of the characteristic equation by

$$B_1 s^m + B_2 s^{m-1} + B_3 s^{m-2} + \cdots + B_{m+1} = 0. \tag{6.30}$$

The coefficients of this equation are arranged in two rows as follows:

$$
\begin{array}{c|cccc}
s^m & B_1 & B_3 & B_5 & B_7 \cdots \\
s^{m-1} & B_2 & B_4 & B_6 & B_8 \cdots
\end{array}
\tag{6.31}
$$

All the coefficients are assumed to be real and, in addition, B_1 is assumed to be positive. Additional rows of coefficients are derived from these two in the following manner:

$$
\begin{array}{c|cccc}
s^m & B_1 & B_3 & B_5 & B_7 \cdots \\
s^{m-1} & B_2 & B_4 & B_6 & B_8 \cdots \\
s^{m-2} & U_1 & U_3 & U_5 & U_7 \cdots \\
s^{m-3} & U_2 & U_4 & U_6 & U_8 \cdots \\
s^{m-4} & V_1 & V_3 & V_5 & V_7 \cdots \\
s^{m-5} & V_2 & V_4 & V_6 & V_8 \cdots \\
\cdot & \cdot & \cdot & \cdot & \cdot \\
\cdot & \cdot & \cdot & \cdot & \cdot \\
\cdot & \cdot & \cdot & \cdot & \cdot \\
s^0 & Z_1 & & &
\end{array}
\tag{6.32}
$$

where

$$U_1 = \frac{B_2 B_3 - B_1 B_4}{B_2}, \qquad U_3 = \frac{B_2 B_5 - B_1 B_6}{B_2},$$

$$U_2 = \frac{U_1 B_4 - B_2 U_3}{U_1}, \qquad U_4 = \frac{U_1 B_6 - B_2 U_5}{U_1}, \tag{6.33}$$

$$V_1 = \frac{U_2 U_3 - U_1 U_4}{U_2}, \qquad V_3 = \frac{U_2 U_5 - U_1 U_6}{U_2}.$$

This pattern will continue until all the terms in a row are zero. The rows are indexed downwards, the first row being numbered m, the degree of the original polynomial; the last row being numbered 0. The number of rows obtained will be $m + 1$, where m

is the order of the characteristic equation. Note that there is one exceptional case where this will not be so and this is discussed later on. The criterion of stability is to check that all the terms in the left-hand column $(B_1, B_2, U_1, U_2, V_1, V_2, \ldots)$ have the same sign. If so, there are no roots in the right half-plane. If there are X changes of sign, then X roots exist in the right half-plane.

Let us illustrate the approach with a simple example. Consider the characteristic equation

$$1 + G(s)H(s) = s^3 + 4s^2 + 100s + 500 = 0. \tag{6.34}$$

Using the procedure described, the resulting array is

$$
\begin{array}{c|cc}
s^3 & 1 & 100 \\
s^2 & 4 & 500 \\
s & -25 & 0 \\
s^0 & 500 & 0
\end{array}
\tag{6.35}
$$

There are two changes of sign in the first column: 4 to -25 and -25 to 500; therefore, there are two roots in the right half-plane.

If the first term in any row is zero, and the other terms of the row are not zero, the array of Eq. (6.32) may be continued by replacing the first column zero by an arbitrary small positive constant ϵ. The process is then continued in the usual manner. Let us illustrate the procedure for this particular case with a simple example. Consider the following characteristic equation:

$$1 + G(s)H(s) = s^5 + s^4 + 4s^3 + 4s^2 + 2s + 1 = 0. \tag{6.36}$$

Using the procedure described, the resulting array is

$$
\begin{array}{c|l}
s^5 & \qquad\qquad\qquad\qquad 1 \quad 4 \quad 2 \\
s^4 & \qquad\qquad\qquad\qquad 1 \quad 4 \quad 1 \\
s^3 & \text{Replace this row} \rightarrow \quad (0 \quad 1 \quad 0) \\
    & \text{with} \rightarrow \qquad\qquad\quad \epsilon \quad 1 \quad 0 \\
s^2 & \qquad\qquad\qquad \dfrac{4\epsilon - 1}{\epsilon} \quad 1 \quad 0 \\
s & \qquad\qquad\quad \dfrac{-\epsilon^2 + 4\epsilon - 1}{4\epsilon - 1} \quad 0 \quad 0 \\
s^0 & \qquad\qquad\qquad\qquad 1 \quad 0 \quad 0
\end{array}
\tag{6.37}
$$

As ϵ goes to zero, the limiting value of the term in the left-hand column, fourth row, is negative. The limiting value of the term in the left-hand column, fifth row, is positive. Therefore there are two changes of sign, and two roots must lie in the right half-plane.

The exceptional case referred to above occurs when all the terms in a row are zero before the $(m + 1)$th row is reached. This means that there are pairs of real roots existing which are negatives of each other located on the real axis, pairs of conjugate roots on the imaginary axis, or quadruples of roots symmetrically located with respect to the origin. For this special case, the array of Eq. (6.32) can be completed by obtaining a subsidiary polynomial from the preceding row. The subsidiary polynomial is formed by constructing an odd or even polynomial whose coefficients are the coefficients of the last nonzero row. To determine the degree of the subsidiary polynomial, the rows are indexed downwards, the first row being numbered m, the degree of the original polynomial. Then the index of the last nonzero row is the degree of the subsidiary polynomial. This polynomial is then differentiated and the resulting coefficients are used to complete the array. The zeros of the subsidiary polynomial are actual roots of the characteristic equation. This procedure is illustrated with a simple example. Consider the following characteristic equation:

$$1 + G(s)H(s) = s^3 + 10s^2 + 16s + 160 = 0. \tag{6.38}$$

Using the procedure described, the resulting array is

$$\begin{array}{c|cc} s^3 & 1 & 16 \\ s^2 & 10 & 160 \\ s & 0 & 0 \end{array} \tag{6.39}$$

The presence of zeros in the third row indicates the exceptional case. Using the coefficients of the second row for the subsidiary equation, we obtain

$$F(s) = 10s^2 + 160 = 0. \tag{6.40}$$

In order to complete the array, Eq. (6.40) is differentiated and the resulting coefficients are then inserted into the array as follows:

$$\begin{array}{c|cc} s^3 & 1 & 16 \\ s^2 & 10 & 160 \\ s & 20 & 0 \\ s^0 & 160 & 0 \end{array} \tag{6.41}$$

No roots lie in the right half-plane since there are no changes of sign in the left-hand column. The roots which are negatives of each other can be obtained from Eq. (6.40) as $\pm j4$, indicating a pair of imaginary roots.

Although the Routh-Hurwitz criterion gives a relatively quick determination of absolute stability, it does not show how to improve the design. In addition, it does not give an indication of relative system performance. Its main attribute is to serve as a check of other design criteria. The greatest difficulty with using the Routh-Hurwitz criterion, however, is that it assumes that the characteristic polynomial is known.

6.4 NYQUIST STABILITY CRITERION

The Nyquist stability criterion [5] is a very valuable tool which determines the degree of stability, or instability, of a feedback control system. In addition, it is the basis for other methods which are used to improve both the steady-state and the transient response of a feedback control system. Application of the Nyquist stability criterion requires a polar plot of the open-loop transfer function, $G(j\omega)H(j\omega)$, which is usually referred to as the Nyquist diagram.

The Nyquist criterion determines the number of roots of the characteristic equation which have positive real parts from a polar plot of the open-loop transfer function, $G(j\omega)H(j\omega)$, in the complex plane. Let us consider the characteristic equation,

$$F(s) = 1 + G(s)H(s) = 0. \tag{6.42}$$

System stability can be determined from Eq. (6.42) by locating its roots in the complex plane. Assuming that $G(s)$ and $H(s)$, in their general form, are functions of s which are given by

$$G(s) = \frac{N_A(s)}{D_A(s)} \tag{6.43}$$

and

$$H(s) = \frac{N_B(s)}{D_B(s)}, \tag{6.44}$$

then we can say that

$$G(s)H(s) = \frac{N_A(s)N_B(s)}{D_A(s)D_B(s)}. \tag{6.45}$$

Substituting Eq. (6.45) into Eq. (6.42), we obtain the following equivalent expression for $F(s)$:

$$F(s) = 1 + G(s)H(s) = 1 + \frac{N_A(s)N_B(s)}{D_A(s)D_B(s)} = \frac{D_A(s)D_B(s) + N_A(s)N_B(s)}{D_A(s)D_B(s)}. \tag{6.46}$$

In terms of factors, we may rewrite Eq. (6.46) as

$$F(s) = \frac{(s + s_1)(s + s_2)(s + s_3)\cdots}{(s + s_A)(s + s_B)(s + s_C)\cdots}. \tag{6.47}$$

The factors $s + s_1, s + s_2, \ldots$ are called the zero factors of $F(s)$. This terminology is due to the fact that $F(s)$ vanishes when $s = -s_1, -s_2$, and so on. The factors $s + s_A, s + s_B, \ldots$ are called the pole factors of $F(s)$. This terminology is due to the fact that $F(s)$ is infinite when $s = -s_A, -s_B$, and so on.

Since $F(s)$ is the denominator of the closed-loop system transfer function given by Eq. (6.1), we see that the zeros of $F(s)$ are the poles of Eq. (6.1). Therefore, for a stable system, it is necessary that the zeros $s_1, s_2, s_3, \ldots$ have negative real parts. The roots s_A, s_B, s_C, have no real restrictions on them. As we shall shortly see, however, if we try to determine stability based on $G(s)H(s)$ above, then a knowledge of the roots $s_A, s_B, s_C, \ldots$ is also required. The only limitations of the Nyquist criterion are that the system be describable by a linear differential equation having constant coefficients.

The Nyquist diagram is a polar plot in the complex plane of $G(s)H(s)$ as s follows the contour shown in Fig. 6.3. Notice that the locus of s avoids poles of $G(s)H(s)$ which lie anywhere on the imaginary axis by small semicircular paths passing to the right. These semicircular paths are assumed to have radii of infinitesimal magnitude. Any roots of the characteristic equation having positive real parts will lie within the contour shown in Fig. 6.3.

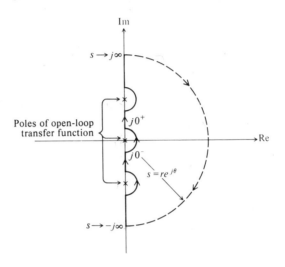

Fig. 6.3   Locus of s in the complex plane for determining the Nyquist diagram.

Use is made of Cauchy's [6] *principle of the argument*, which states that if a function $F(s)$ is analytic on and within a closed contour, except for a finite number of poles and zeros within the contour, then the number of times the origin of $F(s)$ in the F-plane is encircled as s traverses the closed contour once is equal to the number of zeros minus the number of poles of $F(s)$, the poles and zeros being counted according to their multiplicity. In our particular case, the function $F(s)$ equals $1 + G(s)H(s)$. Observe that for this case the origin of $1 + G(s)H(s)$ is given by

$$G(s)H(s) = -1. \qquad (6.48)$$

Therefore, if $G(s)H(s)$ is sketched for the contour defined by Fig. 6.3, the number of times that $G(s)H(s)$ encircles the point $-1 + j0$ equals the number of zeros minus the number of poles of $1 + G(s)H(s)$ for s in the right half-plane.

Figure 6.4 illustrates a polar plot, corresponding to the contour defined by s in the complex plane, for a system whose characteristic equation is

$$1 + G(s)H(s) = 1 + \frac{K}{s(1 + T_1 s)(1 + T_2 s)} = 0. \qquad (6.49)$$

To obtain the direct polar plot of this function, we substitute $s = j\omega$ into Eq. (6.49) as follows:

$$1 + G(j\omega)H(j\omega) = 1 + \frac{K}{j\omega\,(1 + T_1 j\omega)(1 + T_2 j\omega)}. \qquad (6.50)$$

Figure 6.4 represents the polar plot of $1 + G(s)H(s)$ for values of s along the imaginary axis where $-j\infty < j\omega < j\infty$.* The dotted portion of the curve denotes the negative-frequency portion. Notice that the polar plot for negative frequencies is the conjugate of the positive portion. Figure 6.5 illustrates the same polar plot with the origin shifted to the point $-1 + j0$. This simplifies the drawing of the polar plot of $1 + G(s)H(s)$, since we need now only sketch $G(s)H(s)$. The resulting curve will correspond to $1 + G(s)H(s)$ as is indicated on the new set of coordinates. In practice, this is very advantageous since $G(s)H(s)$ is a function which is much more readily available than $1 + G(s)H(s)$.

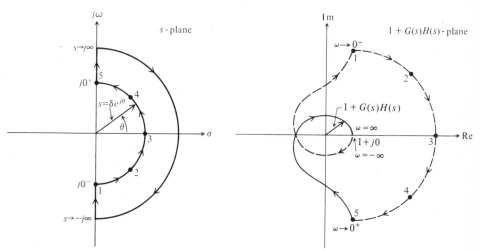

Fig. 6.4   Locus of s in the complex plane and polar plot of $1 + G(s)H(s)$.

---

* The path along the imaginary axis in the vicinity of $\omega = 0$ is modified to be a semicircle of infinitesimal radius δ in the positive half-plane in order to avoid passing through this pole on the imaginary axis at the origin (see Fig. 6.3). For this semicircular portion of the path, from $j0^-$ to $j0^+$, $s = \delta e^{j\theta}$ where $\delta \to 0$ and $-\pi/2 \leqslant \theta \leqslant \pi/2$. Therefore, for $s \to 0$, Eq. (6.49) becomes

$$1 + G(s)H(s) = 1 + K/s = K/\delta e^{j\theta} = (K/\delta)e^{-j\theta} = (K/\delta)e^{j\alpha}.$$

Observe that the magnitude of $K/\delta \to \infty$ as $\delta \to 0$, and $\alpha = -\theta$ goes from $\pi/2$ to $-\pi/2$ as the directed segment s goes from $\delta \lfloor -\pi/2$ to $\delta \lfloor \pi/2$. This implies that the end points from $\omega \to 0^-$ and $\omega \to 0^+$ in the polar plot are joined by a semicircle of infinite radius in the first and fourth quadrants as indicated. Numerals 1–5 are used for clarity in correlating the locus of s in the complex plane and the polar plot.

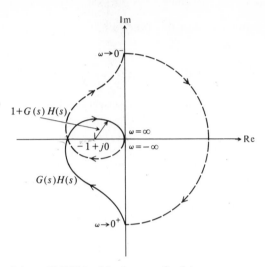

Fig. 6.5   Polar plot of $1 + G(s)H(s)$ with change of origin.

With the Nyquist diagram sketched on a set of axes where the origin is shifted to the $-1 + j0$ point, the Nyquist stability criterion can be stated algebraically as*

$$N = Z - P \tag{6.51}$$

where N is the number of clockwise encirclements of the $-1 + j0$ point by the Nyquist locus, P is the number of poles of the open-loop transfer function $G(s)H(s)$ having positive real parts, and Z is the number of roots of the characteristic equation having positive real parts (Z must equal zero for stability). In most practical cases, the open-loop transfer function is in itself stable and P would equal zero. Since Z must equal zero for stability, N must equal zero for these practical systems.

Figure 6.6 illustrates several examples of application of the Nyquist stability criterion using the relationship given by Eq. (6.51). In all cases, the values of the open-loop transfer function $G(s)H(s)$ are shown together with the values of N, P, and Z. Notice that those systems whose Nyquist diagrams are illustrated in parts (a) through (d) are open-loop stable while those in parts (e) and (f) are open-loop unstable. The systems illustrated in parts (a), (d), and (e) are closed-loop stable, while the systems illustrated in parts (b), (c), and (f) are closed-loop unstable.

It is quite clear that a good picture of a system's margin of stability can be obtained from the Nyquist diagram. The proximity of the $G(s)H(s)$ locus to the $-1 + j0$ point is an indication of relative stability. The farther away the locus is from this point, the greater the margin of stability. One conventional measure of the relative degree of stability is the distance two points on the $G(s)H(s)$ locus are from the point $-1 + j0$. This is illustrated in Fig. 6.7 by the points A and B.

---

* The proof of the Nyquist stability criterion is contained in Appendix B.

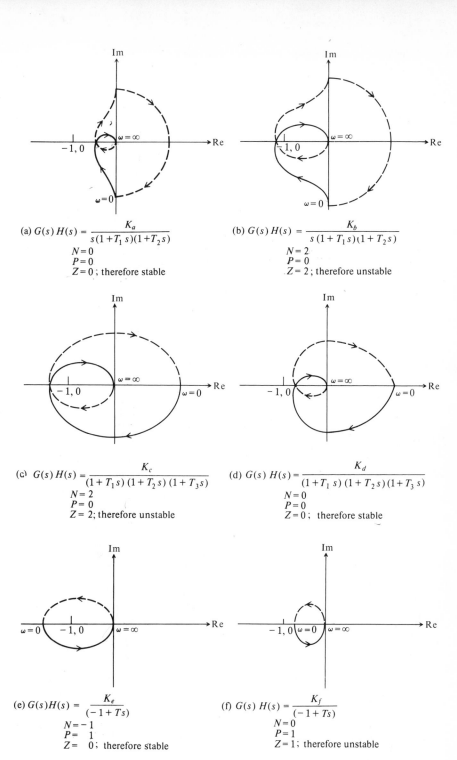

(a) $G(s)H(s) = \dfrac{K_a}{s(1+T_1 s)(1+T_2 s)}$

$N = 0$
$P = 0$
$Z = 0$; therefore stable

(b) $G(s)H(s) = \dfrac{K_b}{s(1+T_1 s)(1+T_2 s)}$

$N = 2$
$P = 0$
$Z = 2$; therefore unstable

(c) $G(s)H(s) = \dfrac{K_c}{(1+T_1 s)(1+T_2 s)(1+T_3 s)}$

$N = 2$
$P = 0$
$Z = 2$; therefore unstable

(d) $G(s)H(s) = \dfrac{K_d}{(1+T_1 s)(1+T_2 s)(1+T_3 s)}$

$N = 0$
$P = 0$
$Z = 0$; therefore stable

(e) $G(s)H(s) = \dfrac{K_e}{(-1+Ts)}$

$N = -1$
$P = 1$
$Z = 0$; therefore stable

(f) $G(s)H(s) = \dfrac{K_f}{(-1+Ts)}$

$N = 0$
$P = 1$
$Z = 1$; therefore unstable

Fig. 6.6 Examples of typical Nyquist diagrams.

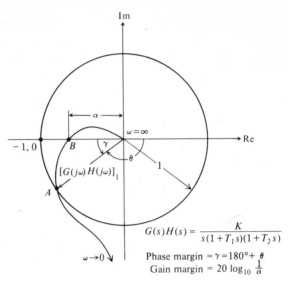

Fig. 6.7 Definition of phase and gain margin.

The point A is defined by the intersection of the $G(s)H(s)$ locus and the unit circle. Obviously, the magnitude of $G(s)H(s)$ at point A is unity and is denoted by $[G(j\omega)H(j\omega)]_1$ in Fig. 6.7. The angle of $[G(j\omega)H(j\omega)]_1$ with respect to the positive real axis is defined as being positive in the counterclockwise sense and is designated as θ. The phase margin γ is defined as the angle which $[G(j\omega)H(j\omega)]_1$ makes with respect to the negative real axis. It is related to θ by the equation

$$\gamma = 180° + \theta. \tag{6.52}$$

A positive value of phase margin tends to indicate stability and a negative value of phase margin tends to indicate instability although there are exceptions.* A zero value of phase margin indicates that the $G(s)H(s)$ locus passes through the $-1 + j0$ point. The magnitude of $+\gamma$ indicates the relative degree of stability. Usually, a desirable value of γ is between 30° and 60°.

The point B is defined by the intersection of the $G(s)H(s)$ locus and the negative real axis. The gain margin is defined as the reciprocal of the magnitude of the $G(s)H(s)$ locus at point B. For the configuration illustrated in Fig. 6.7, the gain margin equals $1/\alpha$. The significance of the gain margin is that the system gain could be increased by a factor of $1/\alpha$ before the $G(s)H(s)$ locus would intersect the $-1 + j0$ point. A value of gain margin greater than unity tends to indicate stability; a gain margin smaller than unity tends to indicate instability. The magnitude of the gain

---

* The exceptions refer to conditionally stable systems which are considered shortly.

margin indicates the relative degree of stability. Gain margin is usually expressed in decibels as follows:

$$\text{gain margin in db} = 20 \log_{10}(1/\alpha). \tag{6.53}$$

Usually, a desirable value of gain margin is between 4 and 12 db.

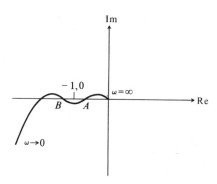

Fig. 6.8   A conditionally stable system.

One important issue to be considered is that of conditionally stable systems. If Fig. 6.7 is examined, the erroneous impression can be obtained that an increase in gain will always cause a system to become unstable. This is not always true, as illustrated in Fig. 6.8. Here, if the gain is increased, then the point A will move to enclose the $-1 + j0$ point and the system will become unstable. However, if the gain is decreased, point B will move to enclose the $-1 + j0$ point and the system will become unstable. Although they do occur in practice, conditionally stable systems should be avoided if possible.

6.5   BODE-DIAGRAM APPROACH

The Bode-diagram approach [8] is one of the most commonly used methods for the analysis and synthesis of linear feedback control systems. This method, which is basically an extension of the Nyquist stability criterion, has the same limitations and uses as the Nyquist diagram. The presentation of information in the Bode-diagram approach, however, is modified to permit relatively quick determinations of the effects of changes in system response without the laborious calculations associated with the Nyquist diagram.

The Nyquist diagram gives the amplitude and phase of the open-loop transfer function $G(s)H(s)$ as s traverses a contour that encloses the right half-plane. As s traverses the positive imaginary axis, it has the value of real frequency ω, and the plot corresponds to $G(j\omega)H(j\omega)$. We can illustrate the same amount of information by means of two diagrams which have ω as a common axis. These two diagrams,

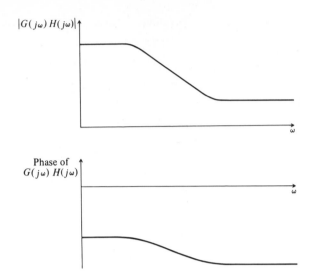

Fig. 6.9   Typical pair of Bode diagrams.

illustrated in Fig. 6.9, are usually referred to as Bode diagrams. It is important to emphasize that the Bode diagrams provide information only of s corresponding to the imaginary axis and therefore represent the frequency response.

Let us consider the system illustrated in Fig. 6.10. It is assumed that the input, $r(t)$, is a sinusoidal waveform of frequency ω and unity amplitude given by

$$r(t) = \sin \omega t. \tag{6.54}$$

Assuming a linear system, the output would have the general form

$$c(t) = a \sin (\omega t - \phi), \tag{6.55}$$

where $a =$ gain of system, and $\phi =$ phase shift of the system. Let us represent the system transfer function as

$$T(j\omega) = \frac{C(j\omega)}{R(j\omega)} = a(j\omega)e^{-j\phi(j\omega)}. \tag{6.56}$$

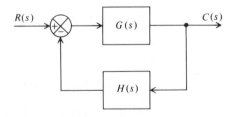

Fig. 6.10   Representative feedback control system.

Here $T(j\omega)$ is a complex function which can be represented by an amplitude $a(j\omega)$ and a phase shift of $\phi(j\omega)$. In the Bode-diagram approach, the amplitude is expressed in decibels and is plotted versus frequency on semilogarithmic graph paper. The amplitude, in decibels, is given by

$$\text{amplitude in db} = 20 \log_{10} a(j\omega). \tag{6.57}$$

The phase shift $\phi(j\omega)$ is conventionally expressed in degrees.

By introducing the logarithm concept, the tedious process of multiplying two complex numbers is simplified to one of addition. For example, let us consider two complex numbers: $G_a(j\omega)$ and $G_b(j\omega)$. The logarithm of the product of two complex quantities

$$G_a(j\omega) = a_a(j\omega)e^{-j\phi_a(j\omega)} \tag{6.58}$$

and

$$G_b(j\omega) = a_b(j\omega)e^{-j\phi_b(j\omega)} \tag{6.59}$$

is given by

$$\log_{10} G_a(j\omega)G_b(j\omega) = \log_{10} a_a(j\omega) + \log_{10} a_b(j\omega)$$
$$+ 0.434j[\phi_a(j\omega) + \phi_b(j\omega)]. \tag{6.60}$$

Equation (6.60) illustrates that the logarithm of the product of two complex numbers is the sum of the logarithms of the magnitude components plus $0.434j$ times the sum of the phase-angle components. In addition, Eq. (6.60) shows that the logarithm of the magnitude and phase components, respectively, are separate functions of the common parameter ω and can be sketched separately as was shown in Fig. 6.9. By convention, the factor 0.434 is not considered, and only the phase angle itself is used.

Bode's theorems are presented in Chapter 7 when we discuss the Bode-diagram approach for the design of feedback control systems in order to meet certain specifications. In this chapter, however, it is important to emphasize the fact that the Bode diagram approach applies primarily only to *minimum-phase* networks. The basic definition of a minimum-phase network [7] is a network whose phase shift is the minimum possible for the number of energy storage elements in the network. This definition restricts the zeros of minimum-phase networks to the left half of the complex plane. A little thought indicates that when we specify either the amplitude or phase of a minimum-phase network, we have also automatically specified the other. This concept is the basis of one of Bode's theorems. However, the Bode diagram can be applied to nonminimum phase systems if we know where the nonminimum phase characteristics come from.

Understanding the basic concepts of the Bode diagram, let us now gain the facility for constructing them. The laborious procedure of plotting the amplitude and phase of $G(j\omega)H(j\omega)$ by means of substituting several values of $j\omega$ is not necessary when drawing the Bode diagram, because we can use several short cuts. These short cuts are based on simplifying approximations which allow us to represent the exact, smooth plots with straight-line asymptotes. The difference between actual amplitude characteristics and the asymptotic approximations is only a few decibels. Now we shall demonstrate the application of this approximating technique to seven common, representative transfer functions: a constant, a pure integration, a pure differentiation, a simple phase-lag network, a simple phase-lead network, a quadratic phase-lag network, and a time-delay factor. The basic concepts illustrated are then used to draw the Bode diagram of the transfer function for representative systems.

A. Bode Diagram of a Constant Using the definition of Eq. (6-57), the logarithm of a constant K, or $1/K$, where $K > 1$, is given by

$$(K)_{db} = 20 \log_{10} K, \tag{6.61}$$

$$\left(\frac{1}{K}\right)_{db} = -20 \log_{10} K. \tag{6.62}$$

The corresponding phase angle of a constant is $0°$ or $180°$, depending on whether the constant is positive or negative, respectively. Figure 6.11 illustrates the Bode diagram of a constant.

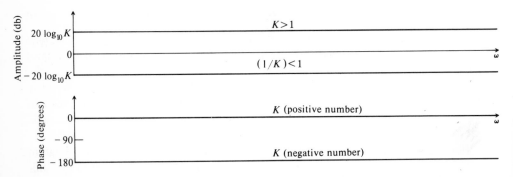

Fig. 6.11 Bode diagram of a constant.

B. Bode Diagram of a Pure Integration The Bode diagram of a pure integration

$$G(j\omega) = 1/j\omega \tag{6.63}$$

can be obtained by taking the logarithm, yielding

$$20 \log_{10} \frac{1}{j\omega} = -20 \log_{10} \omega - j\frac{\pi}{2}. \tag{6.64}$$

Figure 6.12 illustrates the Bode diagram of a pure integration. Notice that the resulting amplitude curve is linear when the amplitude is plotted on a linear scale and the frequency is plotted on a logarithmic scale. The slope of the amplitude characteristic is a constant and equals -20 db/decade. A slope of -20 db/decade also corresponds to a slope of -6 db/octave. The phase characteristic for the pure integration is constant and equals $-90°$.

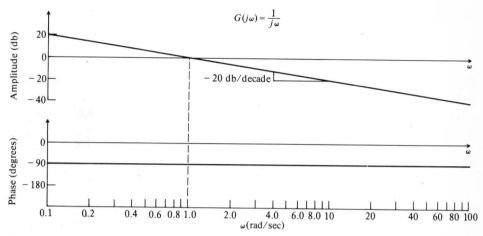

Fig. 6.12   Bode diagram of a pure integration.

C. Bode Diagram of a Pure Differentiation The Bode diagram of a pure differentiation

$$G(j\omega) = j\omega \tag{6.65}$$

can be obtained in a manner similar to that used for the pure integration. The major differences are that the amplitude characteristics now have a positive slope and the phase characteristic is positive. Figure 6.13 illustrates the Bode diagram of a pure differentiation.

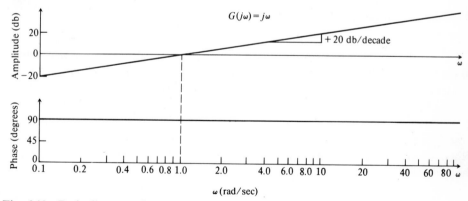

Fig. 6.13   Bode diagram of a pure differentiation.

D. Bode Diagram of a Simple Phase-lag Network A phase-lag network produces a phase lag which is a function of frequency. The transfer function of such a network is given by

$$G(j\omega) = \frac{a}{j\omega + a} = \left(j\frac{\omega}{a} + 1\right)^{-1}. \tag{6.66}$$

The Bode diagram for this transfer function can be obtained as follows:

$$20 \log_{10} \left(j \frac{\omega}{a} + 1 \right)^{-1} = -20 \log_{10} \left(\frac{\omega^2}{a^2} + 1 \right)^{1/2} - j \tan^{-1} \frac{\omega}{a}. \qquad (6.67)$$

Figure 6.14 illustrates the Bode diagram of a simple phase-lag network. The dashed portion of the amplitude characteristic represents the exact plot, and the heavy line segments represent the straight-line asymptotic approximation. They differ by a maximum of 3 db at $\omega/a = 1$.

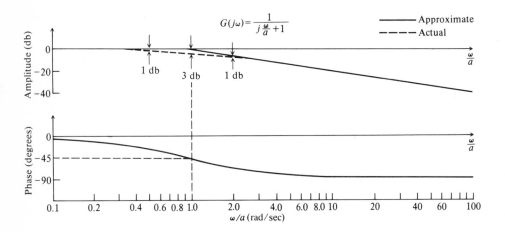

Fig. 6.14 Bode diagram of a simple phase-lag network.

The asymptotic approximations can be drawn quite easily. For example, when ω/a is much less than unity, the imaginary component is very much smaller than the real component (unity), and the imaginary component can be neglected. Therefore,

$$20 \log_{10} 1 = 0 \text{ db} \qquad \text{for} \qquad \frac{\omega}{a} \ll 1. \qquad (6.68)$$

When ω/a is much greater than unity, the imaginary component is much greater than the real component (unity) and the real component may be neglected. Therefore

$$20 \log_{10} \left(j \frac{\omega}{a} \right)^{-1} = -20 \log_{10} \frac{\omega}{a} \text{ db} \qquad \text{for} \qquad \frac{\omega}{a} \gg 1. \qquad (6.69)$$

Equation (6.69) is very similar to the amplitude characteristic of a pure integrator, given by Eq. (6.64), and has a slope of -20 db/decade. At $\omega/a = 1$, the two asymptotes join each other. This frequency ($\omega = a$) is referred to as the break frequency. Ordinarily, the straight-line asymptotic approximation is accurate for most applications. If further correction is needed, the exact curve can be obtained from the approximate curve by using the following corrections: -1 db at $\omega/a = 0.5$ and 2; -3 db at $\omega/a = 1$.

The phase shift produced by the simple phase-lag network can be obtained from the expression

$$\text{phase lag} = -\tan^{-1}\frac{\omega}{a}. \tag{6.70}$$

The phase shift at the break frequency is $-45°$; at $\omega = 0$ it is $0°$; at $\omega = \infty$ it is $-90°$. Notice that this phase-lag network is a minimum-phase network since it has the minimum phase shift possible for the number of energy storage elements in the network (one).

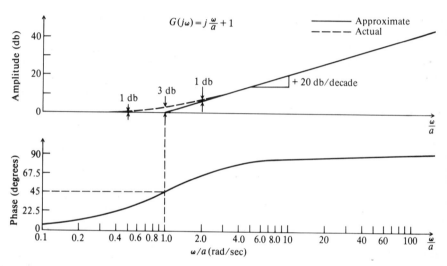

Fig. 6.15 Bode diagram of a simple phase-lead network.

E. Bode Diagram of a Simple Phase-lead Network The Bode diagram of a simple phase-lead network

$$G(j\omega) = j\frac{\omega}{a} + 1 \tag{6.71}$$

can be obtained in a manner similar to that used for the simple phase-lag network. The major differences are that the amplitude characteristic has a positive slope, and the phase characteristic has a positive (phase lead) value. Figure 6.15 illustrates the Bode diagram of a simple phase-lead network.

F. Bode Diagram of a Quadratic Phase-lag Network Let us consider the second-order quadratic, phase-lag transfer function given by

$$G(j\omega) = \frac{\omega_n^2}{(j\omega)^2 + 2\zeta\omega_n(j\omega) + \omega_n^2}, \tag{6.72}$$

or

$$G(j\omega) = \left[\left(j\frac{\omega}{\omega_n} \right)^2 + 2\zeta\left(j\frac{\omega}{\omega_n} \right) + 1 \right]^{-1}. \tag{6.73}$$

The Bode diagram for this transfer function can be obtained as follows:

$$20 \log_{10}\left[\left(j\frac{\omega}{\omega_n} \right)^2 + 2\zeta\left(j\frac{\omega}{\omega_n} \right) + 1 \right]^{-1}$$

$$= -20 \log_{10}\left[\left(\frac{2\zeta\omega}{\omega_n} \right)^2 + \left(1 - \frac{\omega^2}{\omega_n^2} \right)^2 \right]^{1/2} - j0.434 \tan^{-1}\frac{2\zeta\omega_n\omega}{\omega_n^2 - \omega^2}. \tag{6.74}$$

Figures 6.16 and 6.17 illustrate the Bode diagram of the quadratic phase-lag network for various values of damping factor, ζ. In Fig. 6.16, the dashed portion of the amplitude characteristic represents the straight-line asymptotic approximation and the heavy-line portions represent the exact plot. For $\zeta = 1$, the two curves differ by a maximum of 6.2 db at ω/ω_n equal to unity.

The asymptotic approximations can be drawn quite simply. For example, when ω/ω_n is much less than unity, the imaginary component is very much smaller than the real component, and the imaginary component can be neglected. Therefore,

$$20 \log_{10}\sqrt{1} = 0 \text{ db/decade} \quad \text{for} \quad \frac{\omega}{\omega_n} \ll 1. \tag{6.75}$$

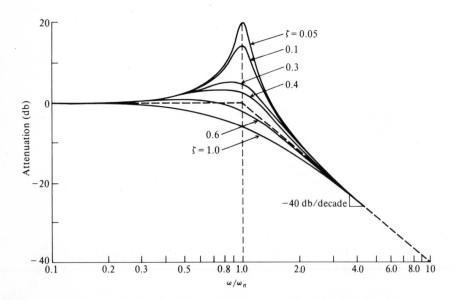

Fig. 6.16 Amplitude portion of Bode diagram for a quadratic phase-lag network.

When ω/ω_n is much greater than unity, the dominant term is

$$20 \log_{10} \left[\left(j \frac{\omega}{\omega_n} \right)^2 \right]^{-1} = -20 \log_{10} \left[\left(-\frac{\omega^2}{\omega_n^2} \right)^2 \right]^{1/2}$$

$$= -40 \log_{10} \frac{\omega}{\omega_n} \quad \text{for} \quad \frac{\omega}{\omega_n} \gg 1. \qquad (6.76)$$

This indicates that the slope of the straight-line asymptotic approximation for $\omega/\omega_n \gg 1$ is twice that of a phase-lag network. The difference between the approximate curve and the exact curve depends on the damping factor, ζ. Similar analysis indicates the phase shift possible is twice that of the simple phase-lag network.

G. Bode Diagram of a Time-delay Factor The time-delay factor [9] occurs in systems which are characterized by the movement of mass that requires a finite time to pass from one point to another. The transfer function of a pure time-delay factor, without attenuation, is given by

$$G(j\omega) = e^{-j\omega T}. \qquad (6.77)$$

The delay factor $e^{-j\omega T}$ results in a phase shift

$$\phi(\omega) = -\omega T \, \text{rad}, \qquad (6.78)$$

which has to be added to the phase shift resulting from the rest of the system. Observe that the magnitude of the time-delay factor is always unity and, therefore, does not affect the magnitude characteristics. Figure 6.18 illustrates the Bode diagram of the pure time-delay factor on a logarithmic scale.

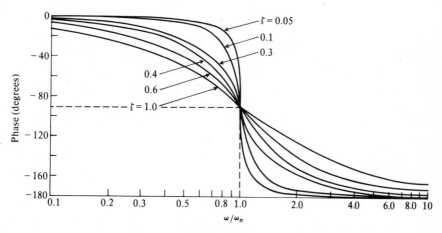

Fig. 6.17 Phase portion of Bode diagram for a quadratic phase-lag network.

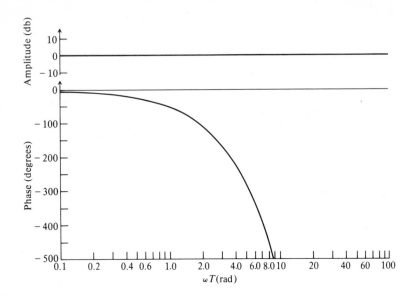

Fig. 6.18 Bode diagram of a pure time-delay factor.

H. Bode Diagram of a Composite Transfer Function—Example 1 Let us next apply the basic notions developed in this section for the transfer function of a representative system. Consider a unity feedback system whose open-loop transfer function is given by

$$G(j\omega) = \frac{775}{j\omega} \frac{0.1j\omega + 1}{0.5j\omega + 1} \frac{0.2j\omega + 1}{j\omega + 1}$$

$$\times \; [0.655 \times 10^{-4}(j\omega)^2 + 6.55 \times 10^{-3}j\omega + 1]^{-1}. \qquad (6.79)$$

This transfer function contains a pure integration, two phase lags, two phase leads, and a quadratic phase-lag network. Each of these characteristics has previously been considered individually. The simplest procedure for plotting the amplitude characteristic for the entire transfer function is to start by locating one point, say $\omega = 1$. Using the straight-line asymptotic approximation technique, the assumption is made that either the real or the imaginary component (whichever is larger at $\omega = 1$) predominates. (If they are both equal, either component may be used for the straight-line asymptote.) At the same time, the slope at $\omega = 1$ is determined. For the problem at hand, at $\omega = 1$,

$$[G(j\omega)]_{\omega=1} = \left(\frac{775}{j1}\right)\left(\frac{1}{1}\right)\left(\frac{1}{j1 \;\; \text{or} \;\; 1}\right)\left(\frac{1}{1}\right). \qquad (6.80)$$

Equation (6.80) indicates that the gain of the straight-line asymptotic curve is 775 (57.5 db) and the slope is changing from -20 db/decade to -40 db/decade at $\omega = 1$.

The next step is to locate all the break frequencies (frequencies at which the slopes change because a real or imaginary component starts or stops dominating). The break frequencies for this transfer function are at $\omega = 1$ (−20 db/decade to −40 db/decade); $\omega = 2$ (−40 db/decade to −60 db/decade); $\omega = 5$ (−60 db/decade to −40 db/decade); $\omega = 10$ (−40 db/decade to −20 db/decade); $\omega = 123.8$ (−20 db/decade to −60 db/decade). The corresponding phase characteristics can most easily be determined using superposition of the individual phase characteristics of each component and utilizing the scale shown in Fig. 6.19. This scale, which can be easily derived from the tangent relationship of Eq. (6.70), illustrates the effect on the resultant phase shift occurring at frequency ω for a first-order factor, due to a break frequency occurring at $\omega = \omega_0$. For example, at $\omega/\omega_0 = 5$, the phase shift contributed by the break occurring at $\omega/\omega_0 = 1$ is 78.7°, while at $\omega/\omega_0 = 0.2$ the phase shift contributed by this break is only 11.3°.

Figure 6.20 illustrates the composite Bode diagram for the transfer function given by Eq. (6.79). Notice that the peaking due to the quadratic phase-lag component occurs at $\omega_n = 123.8$, and corresponds to $\zeta = 0.405$.

I. Relationship Between Bode and Nyquist Diagrams   At this point in the development of the Bode diagram, it seems appropriate to relate the Nyquist stability criterion to the Bode amplitude and phase diagrams. When we discussed the Nyquist diagram, the degree of stability was defined in terms of the phase and gain margins present (see Fig. 6.7). This can easily be related to the Bode diagrams by making use of the following two facts.

1. The unit circle of the Nyquist diagram transforms into the unity or 0 db line of the amplitude plot for all frequencies.

2. The negative real axis of the Nyquist diagram transforms into a negative 180° phase line for all frequencies.

Therefore, the phase margin can be determined on the Bode diagram by determining the phase shift present when $G(j\omega)H(j\omega)$ crosses the 0 db line (this is commonly referred to as the crossover frequency). For the example illustrated in Fig. 6.20, the phase shift is −126° (this corresponds to the θ of Fig. 6.7) and the resulting phase margin, γ, is 54°. The gain margin can be obtained on the Bode diagram by determining the gain present when $G(j\omega)H(j\omega)$ crosses the −180° line. For the example

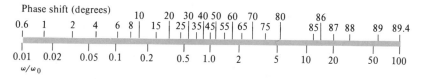

Fig. 6.19   Phase-shift scale.

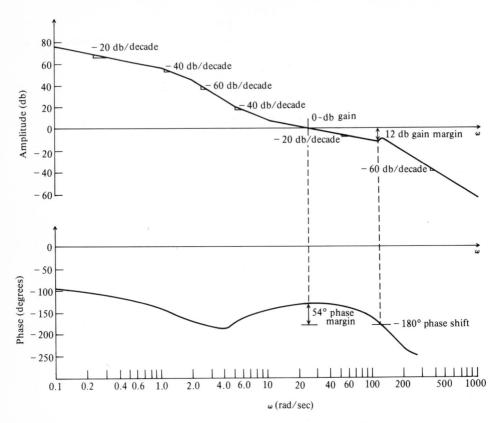

Fig. 6.20 Composite Bode diagram for a unity feedback system where

$$G(j\omega) = \frac{775}{j\omega} \frac{0.1j\omega + 1}{0.5j\omega + 1} \frac{0.2j\omega + 1}{j\omega + 1} [0.655 \times 10^{-4}(j\omega)^2 + 6.55 \times 10^{-3}j\omega + 1]^{-1}.$$

illustrated in Fig. 6.20, the gain is -12 db when the phase shift is $-180°$ (this corresponds to $20 \log_{10} \alpha$ of Fig. 6.7) and the resulting gain margin, $20 \log_{10}(1/\alpha)$, is 12 db.

J. Bode Diagram of a Composite Transfer Function—Example 2 As a second example of a Bode diagram, let us consider a unity feedback system whose open-loop transfer function is given by

$$G(j\omega) = \frac{20(1 + 0.2j\omega)}{j\omega(1 + 0.5j\omega)} e^{-0.1j\omega}. \tag{6.81}$$

This transfer function contains a pure integration, one phase lag, one phase lead, and a time-delay factor. This represents the transfer function in a steel mill and the time-delay factor is caused by the finite time it takes the steel to move from one point

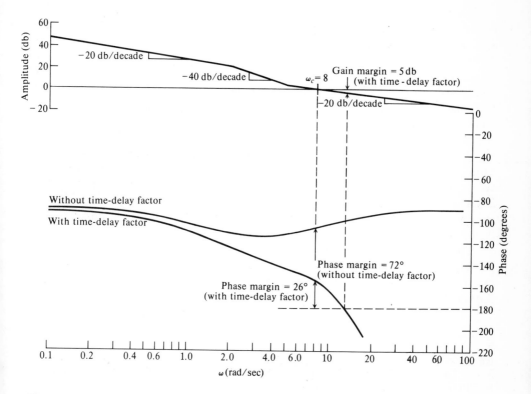

Fig. 6.21 Composite Bode diagram for a unity feedback system where

$$G(j\omega) = \frac{20(1 + 0.2j\omega)}{j\omega(1 + 0.5j\omega)} e^{-0.1j\omega}.$$

to another. In this system, a motor adjusts the separation, l, of two rolls so that the thickness error of the steel is minimized. If the steel is traveling at a velocity v of 10 ft/sec, and the nominal separation l is 1 ft, then the time delay between the roll thickness measurement and thickness adjustment is given by

$$T = \frac{l}{v} = \frac{1 \text{ ft}}{10 \text{ ft/sec}} = 0.1 \text{ sec.}$$

The overall transfer function $G(j\omega)$, given by Eq. (6.81), relates the transfer function of the system error, defined as system error = desired thickness − actual thickness and the output which represents the actual thickness.

Each of the other characteristics in Eq. (6.81) has been previously considered separately. Again, our procedure for plotting the amplitude characteristics is to start locating the point for $\omega = 1$. For this problem, the amplitude of G at $\omega = 1$ is

$$[G(j\omega)]_{\omega=1} = \frac{20}{j1} \frac{1}{1} 1. \tag{6.82}$$

Equation (6.82) indicates that the gain is 20 (26 db) and the slope is -20 db/decade at $\omega = 1$. The break frequencies occur at $\omega = 2$ where the slope changes from -20 db/decade to -40 db/decade, and $\omega = 5$ when the slope changes from -40 db/ decade to -20 db/decade. The corresponding phase characteristic can be readily determined using superposition of the individual phase characteristics of each component and utilizing the scale of Fig. 6.19. Figure 6.21 illustrates the composite Bode diagram for the transfer function given by Eq. (6.81). For interest, the phase characteristic is illustrated with and without the time-delay factor. Observe from this curve that the crossover frequency is 8 rad/sec and the phase margin of the system without the time-delay factor is 72°. However, with the time-delay factor in the system, the phase margin drops to 26°, the gain margin becomes 5 db, and the system has a smaller margin of stability than before. The control system engineer should always give particular attention to time-delay factors and avoid them if possible since they produce a phase lag and decrease the phase margin.

The Bode-diagram method has the very practical virtue that the amplitude and phase frequency responses can easily be measured in the laboratory. The relative ease of synthesizing feedback control systems with the Bode-diagram approach is further demonstrated in Chapter 7.

6.6 DIGITAL COMPUTER TECHNIQUES FOR OBTAINING THE OPEN-LOOP AND CLOSED-LOOP FREQUENCY RESPONSE, AND THE TIME-DOMAIN RESPONSE*

As well as making possible computations and analyses which are not practicable by other means, the digital computer provides valuable assistance in obtaining solutions to relatively common, "standard" problems. It is a very valuable tool for computing the gain and phase characteristics as a function of frequency. Several program languages can be used to perform this computation. Perhaps the simplest languages are BASIC (Beginner's All-purpose Symbolic Instruction Code) [22, 23, 26] and FORTRAN (FORmula TRANslator) [24, 26, 27]. Several problems are solved in this chapter and in Chapter 8 utilizing these languages.

Consider a unity feedback control system where

$$G(s) = \frac{99(1 + 0.1s)}{(1 + 0.01s)^2(1 + 0.2s)(1 + 1.5s)}. \tag{6.83}$$

It is desired to determine the phase and gain margins of this linear control system, and a BASIC program will be utilized in this problem [22, 23, 26]. The coding

---

* The reader should consult Refs. 34 and 35 for a review of digital computer techniques.

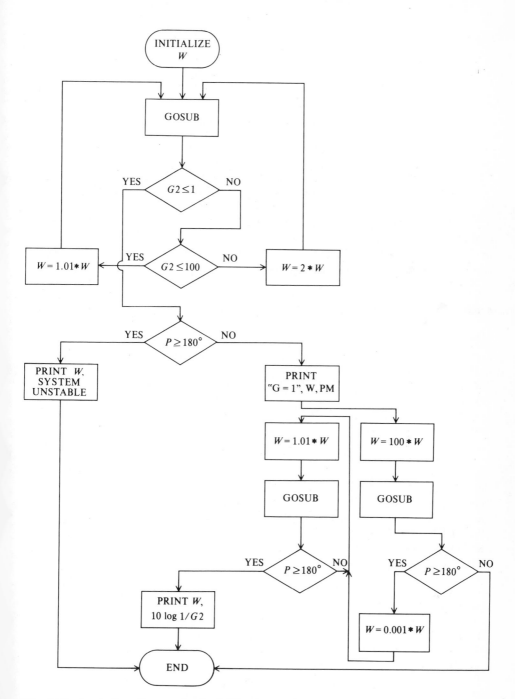

Fig. 6.22 Logic flow diagram for Bode-diagram analysis.

symbols used are as follows:

$$W = \omega$$

$$G2 = |G(s)|^2$$

$$P = \text{phase of } G(s)$$

$$PM = \text{phase margin}$$

$$GOSUB \rightarrow \text{Compute G2, P}$$

Figure 6.22 illustrates the logic flow diagram for developing the program that determines the phase and gain margins. Table 6.1 illustrates the actual program. Figure 6.22 and Table 6.1 should be compared in order to obtain a thorough understanding of the method. Table 6.2 illustrates the computer's answer for phase and gain margin. It indicates that unity gain occurs at $\omega = 31.2$ rad/sec where the phase shift is 132.1°. Therefore, the phase margin is $180 - 132.1 = 47.9°$. It also indicates that at $\omega = 96.1$ rad/sec, the phase shift is 180°, and the gain margin is 14.9 db. To

Table 6.1 Computer program for Bode-diagram analysis (BASIC program)

```
LIST
  1  GEN PHASE AND GAIN MARGIN COMPUTATION FOR GENERAL
       RESPONSE
 10  LET W = 0.01
 20  GOSUB 170
 30  IF G2 < = 1 THEN 90
 40  IF G2 < = 100 THEN 70
 50  LET W = 2*W
 60  GO TO 20
 70  LET W = 1.01*W
 80  GO TO 20
 90  IF P > = 180 THEN 210
100  PRINT "UNITY GAIN," "W =" W, "P =" P
110  LET W = 1.01*W
120  GOSUB 170
130  IF P > = 180 THEN 150
140  GO TO 110
150  PRINT "W =" W, "GAIN MARGIN =" 4.3429448*LOG(1/G2)
160  GO TO 900
170  LET P =57.29578*(2*ATN(0.01*W)+ATN(0.2*W)+ATN(1.5*W)−ATN(0.1*W))
180  LET X = W*W
190  LET G2 = 99*99*(1+0.01*X)/((1+0.0001*X)↑2*(1+0.04*X)*(1+2.25*X))
200  RETURN
210  PRINT "W =" W, "SYSTEM UNSTABLE"
900  END
```

Table 6.2 Results of computer analysis for $G(s) = \dfrac{99(1 + 0.1s)}{(1 + 0.01s)^2(1 + 0.2s)(1 + 1.5s)}$

RUN
UNITY GAIN   W = 31.21   P = 132.107
W = 96.0747   GAIN MARGIN = 14.9268

RUNNING TIME: 01.4 SECS
READY
BYE

utilize the conventional asymptotic Bode-diagram technique as a check, Fig. 6.23 is drawn. It indicates a phase margin of 48° at a crossover of 31 rad/sec, which is in good agreement. The frequency where 180° phase shift occurs is at 97 rad/sec (compared with 96.1 in the computer run). The gain margin using the straight-line asymptotic curve is only 10 db and is quite far from the computer run (approximately 15 db) since a double break occurs at 100 rad/sec. However, if we correct each of these breaks by the 3 db error factor, we should obtain a gain margin of approximately 16 db which is fairly close to the computer solution of 14.9 db.

Notice the preciseness and simplicity of the digital computer's solution. In addition, the speed of the computer's solution is most interesting. Its usefulness as an aid to the control system engineer is quite evident.

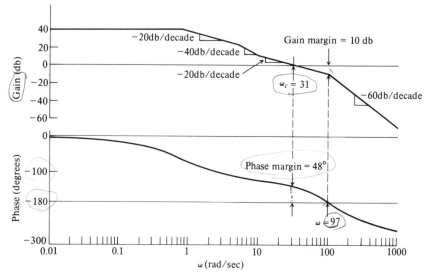

Fig. 6.23 Bode-diagram analysis of

$$G(s) = \frac{99(1 + 0.1s)}{(1 + 0.01s)^2(1 + 0.2s)(1 + 1.5s)}.$$

Let us next extend the digital computer as a tool for also determining the closed-loop frequency response, and the time-domain response to a number of standard input signals. Consider a unity feedback control system where

$$G(s) = \frac{60(1 + 0.5s)}{s(1 + 5s)}. \tag{6.84}$$

A FORTRAN [24, 26, 27] program will be utilized in this example in order to determine the open- and closed-loop frequency response, and the BASIC [22, 23, 26] program will be utilized to determine the time-domain response.

Figure 6.24 shows the logic flow chart of a FORTRAN program, called FRECOM, which computes the following relations for various frequencies:

a) $20 \log |G(s)H(s)|$ } for open-loop Bode diagram
b) Phase of $G(s)H(s)$ }

c) $20 \log \left| \dfrac{C(s)}{R(s)} = \dfrac{G(s)}{1 + G(s)H(s)} \right|$ } for closed-loop frequency response

d) Phase of $\dfrac{C(s)}{R(s)}$ }

The FORTRAN program for computing these quantities is shown in Table 6.3 and should be compared with the flow chart of Fig. 6.24 to obtain a thorough understanding of the technique. The program utilizes statements available from the Rapidata FORTRAN IV language [23]. Note how simple the complex specification statements become in this language. One should also observe the use of complex function subprograms $G(s)$ and $H(s)$. Once written, the programs can now be revised for different systems merely by replacing the two lines containing the arithmetic statements for $G(j\omega)$ and $H(j\omega)$.

The actual use of the program is extremely simple. During program execution, one enters the starting frequency value and the number of decades of frequency range required. Table 6.4 shows the computer run for the open and closed-loop frequency response of this system. Figure 6.25 is a plot of these results. It indicates a phase margin of 74.5° at a crossover frequency of 6.4 rad/sec and a closed-loop peaking of 1.8 db at 2.5 rad/sec. It is immediately evident from examination of the open-loop frequency characteristics that the system is always stable since the phase shift never exceeds $-144.87°$.

In order to obtain the time-domain response, the differential equations which describe the system must be derived from the given frequency-domain description. Since, in the running example,

$$G(s) = \frac{60(1 + 0.5s)}{s(1 + 5s)}$$

and

$$H(s) = 1$$

Table 6.3 Computer program FRECOM

FRECOM
PROGRAM FRECOM

```
1$NDM
100   COMPLEX S,TFRFCT,LOOPG
101   COMPLEX G,H
102   EXTERNAL G,H
105   COMPLEX TEMP
106   PRINT, "INPUT INITIAL FREQUENCY AND # OF DECADES,"
108   READ, WMIN,JM
110   PRINT, "FREQUENCY    LOOP GAIN            RESPONSE"
120   PRINT, "RAD/SEC      DB      DEGREES    DB      DEGREES"
125   PRINT, "---------    ----------    ----------"
221   FAC1=20./ALOG(10.)
222   FAC2=180./FPI(1.)
231   WINC=WMIN
240   DO 1 J=1,JM
241   IF(J.GT.1)  WINC=WMIN*(10.**(J-1))
250   DO 1 K=1,9
260   W=K*WINC
270   S=CMPLX(O.,W)
280   TEMP=G(S)
290   LOOPG=TEMP*H(S)
300   TFRFCT=TEMP/(1.+LOOPG)
310   LOOPG=CLOG(LOOPG)
320   TFRFCT=CLOG(TFRFCT)
330   V1=REAL(LOOPG)*FAC1
340   V2=AIMAG(LOOPG)*FAC2
350   V3=REAL(TFRFCT)*FAC1
360   V4=AIMAG(TFRFCT)*FAC2
370   PRINT 900,W,V1,V2,V3,V4
380   900 FORMAT(F7.2,2X,2F10.2,1X,2F10.2)
390   1 CONTINUE
400   STOP
410   END
500   COMPLEX FUNCTION G(S)
510   COMPLEX S
520   G = 60.*(1.+0.5*S)/(S*(1.+5.*S))
530   RETURN
540   END
550   COMPLEX FUNCTION H(S)
560   COMPLEX S
570   H=1.
580   RETURN
590   END
```

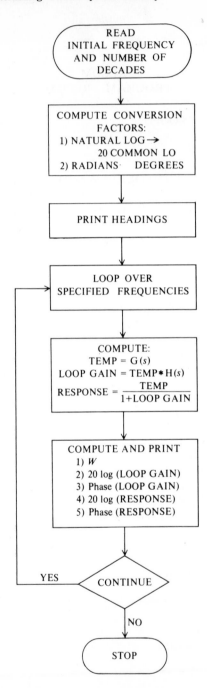

Fig. 6.24 Flow diagram FRECOM for computing open and closed-loop frequency responses.

Table 6.4 Open and closed-loop frequency response of $G(s) = \dfrac{60(1 + 0.5s)}{s(1 + 5s)}$

READY
RUNNH
INPUT INITIAL FREQUENCY AND # OF DECADES 0.1, 4

| FREQUENCY | LOOP GAIN | | RESPONSE | |
RAD/SEC	DB	DEGREES	DB	DEGREES
0.10	54.60	−113.70	0.01	−0.10
0.20	46.58	−129.29	0.03	−0.21
0.30	41.00	−137.78	0.06	−0.35
0.40	36.70	−142.13	0.10	−0.52
0.50	33.24	−144.16	0.15	−0.74
0.60	30.37	−144.87	0.22	−1.02
0.70	27.94	−144.76	0.29	−1.37
0.80	25.84	−144.16	0.36	−1.79
0.90	24.01	−143.24	0.44	−2.28
1.00	22.38	−142.13	0.53	−2.84
2.00	12.51	−129.29	1.21	−12.17
3.00	7.60	−119.88	1.20	−24.53
4.00	4.48	−113.70	0.57	−35.73
5.00	2.22	−109.51	−0.34	−44.55
6.00	0.45	−106.53	−1.34	−51.26
7.00	−1.00	−104.31	−2.32	−56.39
8.00	−2.24	−102.60	−3.24	−60.39
9.00	−3.31	−101.26	−4.11	−63.57
10.00	−4.27	−100.16	−4.91	−66.14
20.00	−10.41	−95.14	−10.57	−77.99
30.00	−13.96	−93.43	−14.03	−81.98
40.00	−16.47	−92.58	−16.51	−83.99
50.00	−18.41	−92.06	−18.43	−85.19
60.00	−20.00	−91.72	−20.01	−85.99
70.00	−21.34	−91.47	−21.35	−86.56
80.00	−22.50	−91.29	−22.51	−86.99
90.00	−23.52	−91.15	−23.53	−87.33
100.00	−24.44	−91.03	−24.44	−87.59
200.00	−30.46	−90.52	−30.46	−88.80
300.00	−33.98	−90.34	−33.98	−89.20
400.00	−36.48	−90.26	−36.48	−89.40
500.00	−38.42	−90.21	−38.42	−89.52
600.00	−40.00	−90.17	−40.00	−89.60
700.00	−41.34	−90.15	−41.34	−89.66
800.00	−42.50	−90.13	−42.50	−89.70
900.00	−43.52	−90.11	−43.52	−89.73

STOP
USED: CPU 2.3 I/0 3.8
READY

then

$$\frac{C(s)}{R(s)} = \frac{G(s)}{1 + G(s)}.$$

Upon substitution of $G(s)$ into the expression for $C(s)/R(s)$, the following is obtained:

$$\frac{C(s)}{R(s)} = \frac{30s + 60}{5s^2 + 31s + 60}$$

or

$$C(s)[5s^2 + 31s + 60] = R(s)[30s + 60].$$

This is equivalent to the following differential equation:

$$5\frac{d^2c(\tau)}{d\tau^2} + 31\frac{dc(\tau)}{d\tau} + 60c(\tau) = 60r(\tau) + 30\frac{dr(\tau)}{d\tau}. \qquad (6.85)$$

In order to obtain the first-order dynamic equations, let

$$x_1(\tau) = c(\tau),$$

$$x_2(\tau) = \frac{dc(\tau)}{d\tau},$$

$$y(\tau) = 12r(\tau) + 6\frac{dr(\tau)}{d\tau}.$$

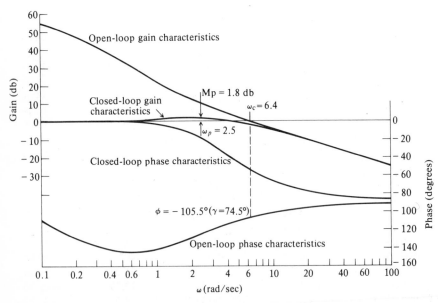

Fig. 6.25   Open and closed-loop frequency response of

$$G(s) = \frac{60(1 + 0.5s)}{s(1 + 5s)}.$$

Table 6.5 Coding of a BASIC program for solving two simultaneous differential equations

STEP 1	Read in $x_1(\tau_0)$, $x_2(\tau_0)$, τ_0, $\Delta\tau$, τ_f
	Set $n = 0$, $N_{max} = \dfrac{(\tau_f - \tau_0)}{\Delta\tau}$
STEP 2	Compute
	$f = f(x_1(\tau_n), x_2(\tau_n), \tau_n)$
	$g = g(x_1(\tau_n), x_2(\tau_n), \tau_n)$
STEP 3	Compute
	$x_1' = x_1(\tau_n) + f * \Delta\tau$
	$x_2' = x_2(\tau_n) + g * \Delta\tau$
	$\tau_{n+1} = \tau_n + \Delta\tau$
STEP 4	Compute
	$f' = f(x_1', x_2', \tau')$
	$g' = g(x_1', x_2', \tau')$
STEP 5	Compute
	$x(\tau_{n+1}) = (x_1' + x_1)/2 + \Delta\tau * f'/2$
	$y(\tau_{n+1}) = (x_2' + x_2)/2 + \Delta\tau * g'/2$
STEP 6	If the final time has not yet been reached, set $n = n + 1$ and go back to Step 2. Otherwise stop.

Therefore, Eq. (6.85) is equivalent to the following two first-order differential equations:

$$\frac{dx_1(\tau)}{d\tau} = x_2(\tau), \tag{6.86}$$

$$\frac{dx_2(\tau)}{d\tau} = -12x_1(\tau) - \tfrac{3}{5}\tfrac{1}{}x_2(\tau) + y(\tau). \tag{6.87}$$

Table 6.5 illustrates the coding of a BASIC program called INTER 2, which applies a second-order Runge-Kutta numerical integration method to the solution of two simultaneous differential equations [24, 25, 32, 33]:

$$\dot{x}_1 = f(x_1, x_2, \tau)$$
$$\dot{x}_2 = g(x_1, x_2, \tau).$$

Table 6.5 shows how the values of $x_1(\tau)$ and $x_2(\tau)$ are obtained for $\tau = \tau_0 + n\,\Delta\tau$, for $n = 1, 2, \ldots$. A flow chart of the program is shown in Fig. 6.26, and the BASIC language program is shown in Table 6.6.

In order to use this program, one must define the specific functions $f(x_1, x_2, \tau)$, $g(x_1, x_2, \tau)$, and $y(\tau)$ on the specified lines. One must then enter the initial values of τ_0, $x_1(\tau_0)$, $x_2(\tau_0)$, and the time increment $\Delta\tau$. The tabulation step size is a multiple of $\Delta\tau$ for which a print out of results is required (this cuts down on a voluminous output).

Table 6.6 Computer program INTER 2

```
READY
LISTNH
 1   REM      THIS ROUTINE APPLIES THE RUNGE-KUTTA METHOD WITH
 2   REM      SECOND-ORDER ACCURACY TO THE SOLUTION OF THE
 3   REM      SYSTEM OF DIFFERENTIAL EQUATIONS X' = F(X,Y,T), Y' =
 4   REM      G(X,Y,T) WITH THE INITIAL CONDITIONS X(T0) = X0, Y(T0) = Y0.
 5   REM      THE INTEGRATION STEP IS H AND THE SOLUTIONS X(T) AND
 6   REM      Y(T) ARE TABULATED ON THE INTERVAL TO < = T < = B IN
 7   REM      STEPS OF SIZE L.
 8   REM          THE FUNCTIONS F(X,Y,T), G(X,Y,T) ARE ENTERED IN LINES
 9   REM      509 AND 510 RESPECTIVELY AS FOLLOWS:

10   REM                          509 LET F = F(X,Y,T)
11   REM                          510 LET G = G(X,Y,T)

12   REM          THE NUMBERS T0, X0, Y0, B, L, and H ARE ENTERED AS
13   REM      DATA IN LINE 900.
14   REM
15   REM
100  READ T,X,Y,B,L,H
110  LET M = INT(L/H)
120  LET N = INT((B−T)/L)
130  PRINT "VALUE OF T,"
         "VALUE OF X," "VALUE OF Y"
140  PRINT
150  PRINT
160  PRINT T,X,Y
170  FOR J = 1 TO N
180  FOR I = 1 TO M
190  LET X1 = X
200  LET Y1 = Y
210  GOSUB 500
220  LET X = X+H*F
230  LET Y = Y+H*G
240  LET T = T+H
250  GOSUB 500
260  LET X = (X1+X)/2 + 0.5*H*F
270  LET Y = (Y1+Y)/2 + 0.5*H*G
280  NEXT I
290  PRINT, T,X,Y
300  NEXT J
310  STOP
500  LET Y9 = 12
509  LET F = Y
510  LET G = −(31/5)*Y−12*X+Y9
520  RETURN
900  DATA 0,0,1,10,1,0.01
999  END
```

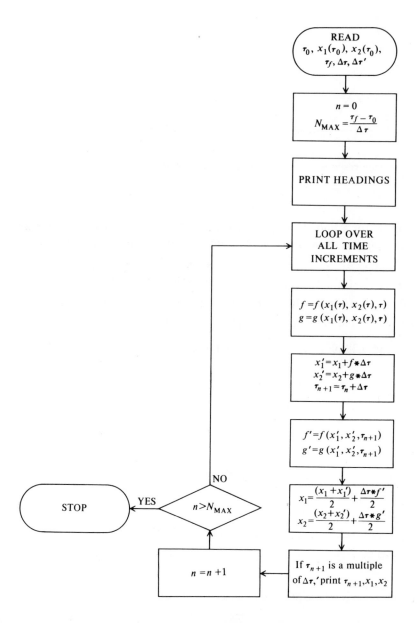

Fig. 6.26   Flow diagram INTER 2 for computing the time-domain response.

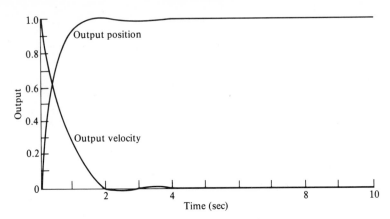

Fig. 6.27 Unit step response of unity feedback system where

$$G(s) = \frac{60(1 + 0.5s)}{s(1 + 5s)}.$$

The solution for a unit step input is shown in Table 6.7. Similarly, the solution for a unit ramp input is shown in Table 6.8. For both cases, the initial input parameters were:

$\tau_0 = 0$	(initial time)	$\tau_f = 10$	(final time)
$x_1(\tau_0) = 0$	(initial value of $x_1(\tau)$)	$\Delta\tau = 0.01$	(time increment)
$x_2(\tau_0) = 1$	(initial value of $x_2(\tau)$)	$\Delta\tau' = 0.5$	(tabulation step size)

Figures 6.27 and 6.28 plot the results of the tabulation in Tables 6.7 and 6.8.

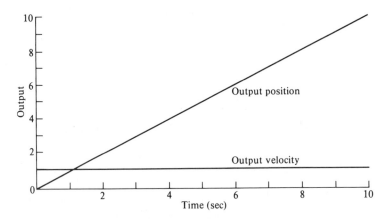

Fig. 6.28 Unit ramp response of unity feedback system where

$$G(s) = \frac{60(1 + 0.5s)}{s(1 + 5s)}.$$

Table 6.7 Unit step response of unity feedback system where $G(s) = \dfrac{60(1 + 0.5s)}{s(1 + 5s)}$

```
11   USERS     10:30
510 LET G = -(31/5)*Y-12*X+Y9
RUNNH
```

VALUE OF T	VALUE OF X	VALUE OF Y
0	0	1
1.	0.937669	0.260442
2.	1.00189	$-1.43078E\text{-}3$
3.	1.00013	$-5.31968E\text{-}4$
4.	0.999996	$1.68118E\text{-}6$
5.	1.	$1.08411E\text{-}6$
6.	1.	$-4.45157E\text{-}10$
7.	1.	$-1.49581E\text{-}9$
8.	1.	$3.54907E\text{-}10$
9.	1.	$3.56945E\text{-}10$
10.	1.	$3.56945E\text{-}10$

```
USED:  CPU  3.2  I/0  1.2
```

Table 6.8 Unit ramp response of unity feedback system where $G(s) = \dfrac{60(1 + 0.5s)}{s(1 + 5s)}$

```
READY
500 LET Y9 = 12*T+6
RUNNH
```

VALUE OF T	VALUE OF X	VALUE OF Y
0	0	1
1.	0.984858	0.994173
2.	1.9833	0.999987
3.	2.98333	1.00001
4.	3.98333	1.
5.	4.98333	1.
6.	5.98333	1.
7.	6.98333	1.
8.	7.98333	1.
9.	8.98333	1.
10.	9.98333	1.

```
USED:   CPU  3.0  I/0  1.2
```

6.7 THE NICHOLS CHART

The Nichols chart [8, 10] is a very useful technique for determining stability and the closed-loop frequency response of a feedback system. Stability is determined from a plot of the open-loop gain versus phase characteristics. At the same time, the closed-loop frequency response of the system is determined by utilizing contours of constant closed-loop amplitude and phase shift which are overlaid on the gain–phase plot.

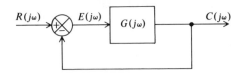

Fig. 6.29   General block diagram of a simple feedback control system.

In order to derive the basic Nichols-chart relationships, let us consider the unity feedback system illustrated in Fig. 6.29. The closed-loop transfer function is given by

$$\frac{C(j\omega)}{R(j\omega)} = \frac{G(j\omega)}{1 + G(j\omega)}, \tag{6.88}$$

or

$$\frac{C(j\omega)}{R(j\omega)} = M(\omega)e^{j\alpha(\omega)}, \tag{6.89}$$

where $M(\omega)$ represents the amplitude component of the transfer function and $\alpha(\omega)$ represents the phase component of the transfer function. The radian frequency at which the maximum value of $C(j\omega)/R(j\omega)$ occurs is called the resonant frequency of the system, ω_p, and the maximum value of $C(j\omega)/R(j\omega)$ is denoted by M_p. For the system illustrated in Fig. 6.29, we would expect a typical closed-loop frequency response to have the general form shown in Figure 6.30.

From Section 4.2 we know that a small margin of stability would mean a relatively small value for ζ and a relatively large value for M_p. This can be further clarified if we

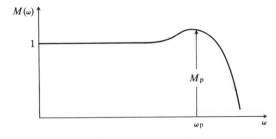

Fig. 6.30   A typical closed-loop frequency-response curve for the system shown in Fig. 6.29.

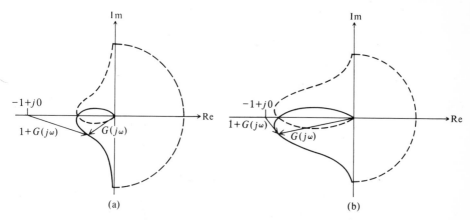

Fig. 6.31 Qualitative stability comparison of two systems illustrating corresponding values of $M(\omega)$. (a) Stable system with large degree of stability:

$$\left|\frac{C(j\omega)}{R(j\omega)}\right| = \left|\frac{G(j\omega)}{1 + G(j\omega)}\right| = M(\omega) \ll 1.$$

(b) Stable system with very small degree of stability: $M(\omega) \gg 1$.

examine the Nyquist diagrams illustrated in Fig. 6.31. Notice that $M(\omega)$ is much less than unity for the case of a system having a large degree of stability, as shown in part (a) of the figure. This situation would correspond to $\zeta \gg 1$. In contrast, $M(\omega)$ is much greater than unity for the case of a system having a very small degree of stability, as shown in part (b) of the figure. This situation would correspond to $\zeta \ll 1$. We next develop the Nichols-chart contours in the complex plane in order to be able to determine M_p and ω_p quantitatively for a feedback control system.

Reconsidering the basic relationships of the unity feedback control system illustrated in Fig. 6.29, let us represent the complex vector $G(j\omega)$ by an amplitude and phase as follows:

$$G(j\omega) = |G(j\omega)|e^{j\theta}. \tag{6.90}$$

Substituting Eq. (6.90) into the system transfer function, as given by Eq. (6.88), we obtain

$$\frac{C(j\omega)}{R(j\omega)} = \frac{G(j\omega)}{1 + G(j\omega)}$$

$$= \frac{|G(j\omega)| e^{j\theta}}{1 + |G(j\omega)| e^{j\theta}}. \tag{6.91}$$

Dividing through by $|G(j\omega)|e^{j\theta}$, we obtain

$$\frac{C(j\omega)}{R(j\omega)} = \left[\frac{e^{-j\theta}}{|G(j\omega)|} + 1\right]^{-1}. \tag{6.92}$$

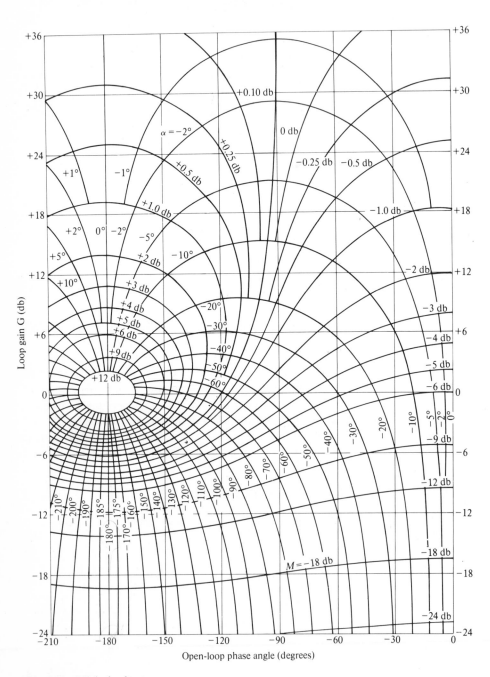

Fig. 6.32   Nichols chart.

Using the trigonometric relationship for the exponential term, we obtain the expression

$$\frac{C(j\omega)}{R(j\omega)} = \left[\frac{\cos\theta}{|G(j\omega)|} - \frac{j\sin\theta}{|G(j\omega)|} + 1\right]^{-1}. \tag{6.93}$$

The expression of Eq. (6.93) has a magnitude M and a phase angle α which are given by

$$M(\omega) = \left\{\left[1 + \frac{1}{|G(j\omega)|^2} + \frac{2\cos\theta}{|G(j\omega)|}\right]^{1/2}\right\}^{-1}, \tag{6.94}$$

$$\alpha(\omega) = -\tan^{-1}\frac{\sin\theta}{\cos\theta + |G(j\omega)|}. \tag{6.95}$$

A detailed plot of G in decibels versus θ, with M and α as parameters, is illustrated in Fig. 6.32. Notice the symmetry of these curves about the $-180°$ line. The stability criterion for the Nichols chart is quite simple since the $-1 + j0$ point of the complex plane corresponds to the 0 db, $-180°$ point of Fig. 6.32. Therefore, for minimum-phase systems, the phase margin can be determined and the feedback control system is stable if a plot of $G(j\omega)H(j\omega)$ on the Nichols chart lies to the right of the $-180°$ line when crossing the 0 db line.

The Nichols chart can be used for the analysis and/or synthesis of a feedback control system. For analysis, we can use the Nichols chart to obtain the closed-loop frequency response from the Bode diagram and determine the maximum value of peaking, M_p, and the frequency at which it occurs, ω_p. From a synthesis viewpoint, we can use the Nichols chart to meet certain requirements as to the values of M_p and ω_p. We demonstrate the use of this method as a tool for analysis in this section. Its value as a tool for synthesis is illustrated in Chapter 7.

Let us determine the closed-loop frequency response for the third-order feedback control system which is illustrated in Fig. 6.33. Specifically, we are now interested in obtaining the values of M_p and ω_p using the Nichols chart. They can be obtained by first drawing the Bode diagram as shown in Fig. 6.34. Then, for each value of ω, the magnitude and phase of $G(j\omega)$ are then plotted onto a Nichols chart as shown in Fig. 6.35. This figure indicates a phase margin of 30° which agrees with the value

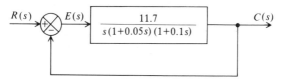

Fig. 6.33 A third-order feedback system.

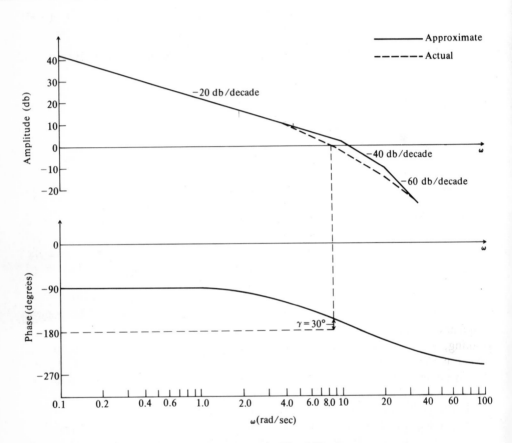

Fig. 6.34   Bode diagram for the system shown in Fig. 6.33 where

$$G(s)H(s) = \frac{11.7}{s(1 + 0.05s)(1 + 0.1s)}.$$

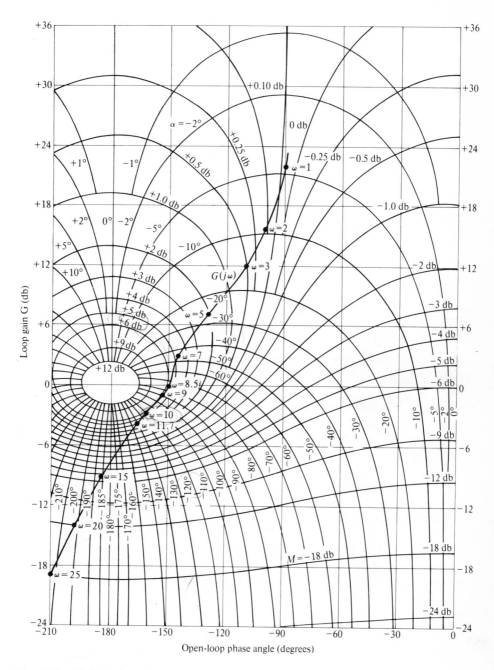

Fig. 6.35 Nichols chart for the system shown in Fig. 6.33.

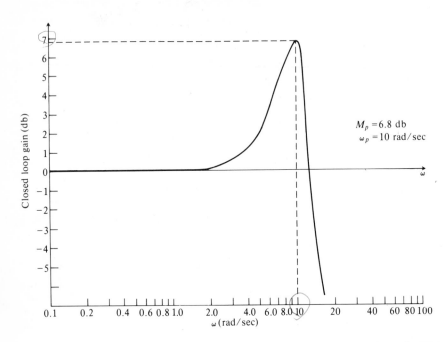

Fig. 6.36 Closed-loop frequency response of the system shown in Fig. 6.33 from the Nichols chart of Fig. 6.35.

obtained from the Bode diagram (using the corrected curve of Fig. 6.34). The intersections of $G(j\omega)$ with $M(\omega)$ on the Nichols chart give the closed-loop response from which M_p and ω_p can be obtained. The resulting response, shown in Fig. 6.36, indicates a value of $M_p = 6.8$ db (2.2) and $\omega_p = 10$ rad/sec.

In practice, a value of $M_p = 2.2$ is a little too high. Usually, M_p is chosen somewhere between 1.0 and 1.4 (see Section 6.8). The technique of compensation of this system utilizing the Nichols chart is illustrated in Chapter 7. Let us here, however, indicate how the constant M loci on the Nichols chart may be used to limit M_p to some maximum value, say 2.2 db (1.3). Since the interior of the $M = 2.2$ db locus consists entirely of constant-magnitude loci which represent M greater than 2.2 db, we must confine the Nichols locus to the exterior of the $M = 2.2$ db locus. The system illustrated in Fig. 6.33 is compensated, to meet certain maximum requirements of M_p, in Chapter 7 utilizing the Nichols chart.

6.8 RELATIONSHIP BETWEEN CLOSED-LOOP FREQUENCY RESPONSE AND THE TIME-DOMAIN RESPONSE

Section 6.7 has illustrated how the closed-loop frequency-domain response may be obtained from the open-loop transfer function. The next logical question to ask is how to determine the relationship between M_p and the peak overshoot one obtains in the time domain. In Chapter 4, we defined the time at which the peak overshoot occurs as t_p in terms of ζ and ω_n (see Eq. 4.28). For example, does an M_p of 1.3 mean a 30% transient overshoot in the time domain?

This problem has been analyzed for the general, unity-feedback system of Fig. 6.29 [21]. After determining the Fourier transform of the input, $R(\omega)$, and that of the output, $C(\omega)$, the value of $C(\omega)$ was then related to the transient response in the time domain. The resulting approximate general relationship between M_p and the peak overshoot of the transient in the time domain has been shown [21] to be given by

$$c(t)_{max} \leqslant 1.18 M_p. \tag{6.96}$$

Therefore, the overshoot in the time domain is, in general, related to M_p by some factor equal to or less than 18%. This approximation is extremely important since it permits the control-system engineer to correlate peak overshoots in the frequency and time domains.

For second-order systems, exact relationships can be obtained for M_p and ω_p in terms of the damping factor ζ. In order to derive this expression, let us reconsider the closed-loop transfer function of a second-order system in the following form:

$$\frac{C(j\omega)}{R(j\omega)} = \frac{1}{(1 - \omega^2/\omega_n^2) + j2\zeta(\omega/\omega_n)} = M(\omega)e^{j\alpha(\omega)}. \tag{6.97}$$

Therefore, the magnitude of the closed-loop response is given by

$$M(\omega) = \frac{1}{[(1 - \omega^2/\omega_n^2)^2 + 4\zeta^2(\omega^2/\omega_n^2)]^{1/2}}. \tag{6.98}$$

In order to find the maximum value M_p of M, and the frequency at which it occurs, ω_p, Eq. (6.98) is differentiated with respect to frequency and set equal to zero. The frequency ω_p at which M_p occurs is found to be given by

$$\omega_p = \omega_n\sqrt{1 - 2\zeta^2}. \tag{6.99}$$

If this value of ω_p is substituted into Eq. (6.98), the value of M_p is found to be given by

$$M_p = \frac{1}{2\zeta\sqrt{1 - \zeta^2}}. \tag{6.100}$$

2nd order

A plot of M_p versus ζ is illustrated in Fig. 6.37. Observe from this figure that M_p increases very rapidly for $\zeta < 0.4$. The resulting transient oscillatory response is excessively large in this region and is undesirable from practical considerations. Therefore, systems having a $\zeta < 0.4$ are not normally desired. In mechanical systems, values of M_p in the range of

$$1 < M_p < 1.4$$

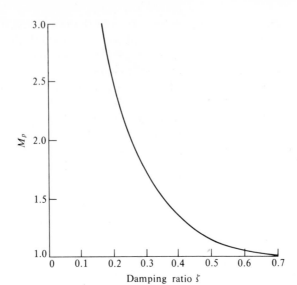

Fig. 6.37 M_p versus the damping ratio.

are usually specified, which requires $\zeta > 0.4$ if the system is a pure second-order system. In industrial process control, values of M_p up to 2.0 are often used.

Let us carry the analysis of the second-order system one step further by correlating its frequency and time-domain response. In Section 4.2, we found that the peak value of $c(t)$ for the step response of a second-order system is given by

$$c(t)_{\max} = 1 + e^{-\zeta\pi/\sqrt{1-\zeta^2}}. \tag{6.101}$$

To check the correspondence between M_p (from Eq. 6.100) and $c(t)_{\max}$ (from Eq. 6.101), consider the case where $\zeta = 0.4$. Substituting into Eqs. (6.100) and (6.101), we find that

$$M_p = 1.364, \tag{6.102}$$

$$c(t)_{\max} = 1.254, \quad \le 1.18 \times 1.364 = 1.6 \tag{6.103}$$

and they are within 18% of each other as stated in Eq. (6.96). Actually, the difference is only 8.76%. As the damping factor ζ increases, the correspondence gets better. For example, if $\zeta = 0.6$, then $M_p = 1.04$ and $c(t)_{\max} = 1.09$; a difference of only 4.8%. In general, when $\zeta > 0.4$, there is a close correspondence between M_p and $c(t)_{\max}$. For values of $\zeta < 0.4$, the correspondence between M_p and $c(t)_{\max}$ is only qualitative. For example, for the second-order system and at $\zeta = 0$, we find that $M_p = \infty$ and $c(t)_{\max} = 2$. However, since practical second-order systems usually do not use a $\zeta < 0.4$, the 18% relationship given in Eq. 6.96 is quite accurate for systems of interest.

$$1.09 \le 1.18 \times 1.04 = 1.23$$

6.9 THE ROOT-LOCUS METHOD FOR NEGATIVE-FEEDBACK SYSTEMS

The root-locus method is a graphical technique for determining the roots of the closed-loop characteristic equation of a system as a function of the static gain. This method is based on the relationship which exists between the poles of the closed-loop transfer function and the poles and zeros of the open-loop transfer function. The root-locus method, which was conceived by Evans [11–13], has several distinct advantages. A complete, detailed, and very accurate transient and steady-state solution can be obtained since the closed-loop poles can be obtained directly from the root loci. Alternatively, approximate solutions may be obtained, with a considerable reduction of labor, if very accurate solutions are not required. This and the following sections present the methods of constructing the root locus and of interpreting the results for negative and positive feedback systems, respectively. The technique for synthesizing a system utilizing the root-locus method is discussed in Chapter 7. The method is a useful one and should be part of the designer's bag of tricks.

Let us consider the general feedback control system illustrated in Fig. 6.38. In order to find the poles of the closed-loop transfer function, we require that

$$1 + G(s)H(s) = 0 \qquad (6.104)$$

$$G(s)H(s) = -1 = 1\underline{/(2n + 1)\pi} \qquad (6.105)$$

where $n = 0, \pm1, \pm2, \ldots$ Equation (6.105) specifies two conditions which must be satisfied for the existence of a closed-loop pole.

1. The angle of $G(s)H(s)$ must be an odd multiple of π:

$$\text{angle of } G(s)H(s) = (2n + 1)\pi \qquad (6.106)$$

   where $n = 0, \pm1, \pm2, \ldots$
2. The magnitude of $G(s)H(s)$ must be unity:

$$|G(s)H(s)| = 1. \qquad (6.107)$$

The construction of the root locus for a particular system can start by locating the open-loop poles and zeros in the complex plane. Other points on the locus can be obtained by choosing various test points and determining whether they satisfy

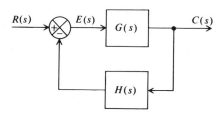

Fig. 6.38   A nonunity feedback control system.

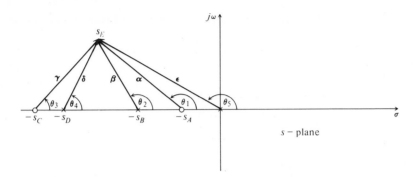

Fig. 6.39 Vector representation of

$$G(s_E)H(s_E) = \frac{K(s_E + s_A)(s_E + s_C)}{s_E(s_E + s_B)(s_E + s_D)}$$

at the exploratory point, s_E.

Eq. (6.106). The angle of $G(s)H(s)$ can be easily determined at any test point in the complex plane by measuring the angles contributed to it by the various poles and zeros. For example, consider a feedback control system where

$$G(s)H(s) = \frac{K(s + s_A)(s + s_C)}{s(s + s_B)(s + s_D)}. \tag{6.108}$$

At some exploratory point s_E, $G(s)H(s)$ has the value

$$G(s_E)H(s_E) = \frac{K(s_E + s_A)(s_E + s_C)}{s_E(s_E + s_B)(s_E + s_D)}. \tag{6.109}$$

Pictorially, Eq. (6.109) can be represented by Fig. 6.39, where the vectors

$$\alpha = s_E + s_A \tag{6.110}$$
$$\beta = s_E + s_B \tag{6.111}$$
$$\gamma = s_E + s_C \tag{6.112}$$
$$\delta = s_E + s_D \tag{6.113}$$
$$\epsilon = s_E. \tag{6.114}$$

The angle of $G(s_E)H(s_E)$ is the sum of the angles θ_1, θ_2, θ_3, θ_4, and θ_5, determined by the vectors α, β, γ, δ, and ϵ, respectively:

Angle of $G(s_E)H(s_E) = \sum$ angles of vectors α, β, γ, δ, and ϵ to point s_E

$$= \theta_1 - \theta_2 + \theta_3 - \theta_4 - \theta_5. \tag{6.115}$$

If this angle equals $(2n + 1)\pi$ where $n = 0, \pm 1, \ldots$, then s_E lies on the root locus. If it does not, the point s_E does not lie on the locus and a new point must be tried.

When a point is found which does satisfy Eq. (6.106), the vector magnitudes are determined and are substituted into Eq. (6.107) in order to find the value of the gain constant K at the exploratory point s_E:

$$|G(s_E)H(s_E)| = \frac{K\,|(s_E + s_A)|\,|(s_E + s_C)|}{|s_E|\,|(s_E + s_B)|\,|(s_E + s_D)|} = 1. \tag{6.116}$$

Fortunately, the actual construction of a root locus does not entail an infinite search through the complex plane. Since the zeros of the characteristic equation are continuous functions of the coefficients, the root locus is a continuous curve. Therefore, the root locus must have certain general patterns which are governed by the location and number of open-loop zeros and poles. Once these governing rules are established, the drawing of a root locus is not a tedious and lengthy trial-and-error procedure. We next present eleven basic rules which aid in determining the approximate location of the root locus.

Rule 1 The number of branches of the locus equals the order of the characteristic equation. This is true since there are as many roots (and branches) of the root locus as the order of the characteristic equation. Each segment, or branch, of the root locus describes the motion of a particular pole of the closed-loop system as the gain is varied.

Rule 2 The open-loop poles define the start of the root locus ($K = 0$), and the open-loop zeros define the termination of the root locus ($K = \infty$). This can easily be shown by considering Eq. (6.116). At open-loop zeros, K must equal infinity since there is a zero in the expression due to either s_A or s_C. However, K must equal zero when open-loop poles occur since there is a zero in the expression due to s_B or s_D. When the order of the denominator of $G(s)H(s)$ is greater than the numerator, the root locus ends at infinity, whereas if the order of the numerator is greater than the denominator, the root locus starts at infinity.

Rule 3 Complex portions of the root locus always occur as complex-conjugate pairs if the characteristic equation is a rational function of s having real coefficients.

Rule 4 Sections of the real axis are part of the root locus if the number of poles and zeros to the right of an exploratory point along the real axis is odd. This is easily demonstrated from Eq. (6.115). Since the angular contribution along the real axis due to complex-conjugate poles cancels, the total angle of $G(s)H(s)$ is due only to the contributions of the real poles and zeros. Therefore, at any exploratory point along the real axis, the angular contribution due to a pole or zero to its right is 180°, while that due to a pole or zero to its left is zero.

Rule 5 Angles of asymptotes to the root locus α_m are given by

$$\alpha_m = \pm \frac{(2m + 1)\pi}{p - z}, \tag{6.117}$$

where

$$p = \text{number of open-loop poles,}$$

$$z = \text{number of open-loop zeros,}$$

$$m = 0, 1, 2, \dots, \text{up to } m = p - z \text{ (exclusively).}$$

This can be shown by considering the root locus as it is mapped far away from the group of open-loop poles and zeros. In this area, all the poles and zeros contribute about the same angular component. Since the total angular component must add up to ± 180° or some odd multiple, Eq. (6.117) follows. Figure 6.40 illustrates the

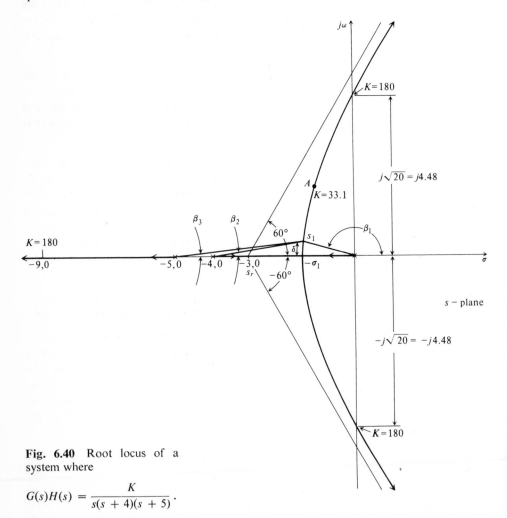

Fig. 6.40 Root locus of a system where

$$G(s)H(s) = \frac{K}{s(s + 4)(s + 5)}.$$

asymptotes of a third-order system where

$$G(s)H(s) = \frac{K}{s(s + 4)(s + 5)}. \tag{6.118}$$

The asymptotic angles for the root locus illustrated in Fig. 6.40 are

$$\alpha_0 = \pm\pi/3 = \pm 60°,$$
$$\alpha_1 = \pm 3\pi/3 = \pm 180°.$$

Rule 6 The intersection of the asymptotes and the real axis occurs along the real axis at s_r, where

$$s_r = \frac{\sum_{\text{poles}} - \sum_{\text{zeros}}}{\text{no. of finite poles} - \text{no. of finite zeros}}. \tag{6.119}$$

The value of s_r is the centroid of the open-loop pole and zero configuration. The intersection of the asymptotes for the root locus illustrated in Fig. 6.40 is given by

$$s_r = \frac{(-4 - 5) - 0}{3 - 0} = -3.$$

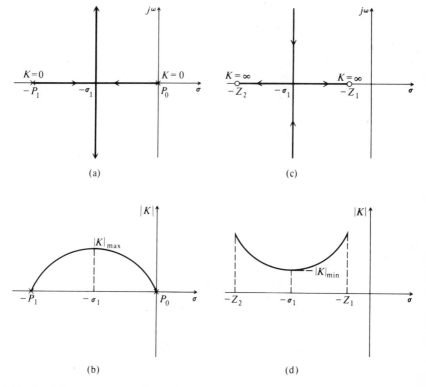

(a)

(c)

(b)

(d)

Fig. 6.41 Loci for two consecutive poles and zeros on the real axis and the corresponding relation of $|K|$ versus σ.

Rule 7 Consider the simple case where the locus has a branch on the real axis between two poles as shown in Fig. 6.41(a). A point must exist where the two branches *break-away* from the real axis and enter the complex region of the *s*-plane in order to approach zeros which are finite or are located at infinity. Since K has a value of zero at the two poles and increases in value as the locus moves along the real axis away from the poles, the K's for the two branches simultaneously reach a maximum value at the breakaway point. A plot of $|K|$ versus σ is shown in Fig. 6.41(b). For the case where the locus has branches on the real axis between the two zeros as illustrated in Fig. 6.41(c), branches come from poles in the complex region and *break-in* onto the real axis. The variation in value of $|K|$ along the real-axis locus between these two zeros is shown in Fig. 6.41(d). The two poles that enter the real axis and then move to zeros on the real axis will enter simultaneously with a value of K which is a minimum since the gain is increasing continuously as the loci approach the zeros. Thus the breakaway and break-in points can be evaluated from the magnitude condition for $\text{Re } s = \sigma$ and solving for $K(\sigma)$. This can be accomplished graphically or by utilizing

$$\frac{dK(\sigma)}{d\sigma} = 0 \qquad (6.120)$$

to find all the maxima and minima of $K(\sigma)$ and their locations. The most straightforward method for isolating the factor K is to rearrange the characteristic equation. An example will illustrate the procedure.

For the system analyzed in Fig. 6.40,

$$G(s)H(s) = \frac{K}{s(s + 4)(s + 5)}.$$

The characteristic equation of this system is given by

$$1 + G(s)H(s) = 1 + \frac{K}{s(s + 4)(s + 5)} = 0.$$

Alternatively, this can be written as

$$K = -s(s + 4)(s + 5).$$

For the case of $\text{Re } s = \sigma$, we obtain

$$K(\sigma) = -\sigma(\sigma + 4)(\sigma + 5).$$

Multiplying the factors together, we have

$$K(\sigma) = -\sigma^3 - 9\sigma^2 - 20\sigma.$$

Taking the derivative of this function and setting it equal to zero, we can determine the breakaway point:

$$\frac{dK}{d\sigma} = -3\sigma^2 - 18\sigma - 20 = 0.$$

The roots are

$$\sigma_1 = -1.47,$$
$$\sigma_2 = -4.53.$$

The breakaway point of σ_1 is indicated in Fig. 6.40; the value σ_2 is not a possible solution in this negative feedback example since the root locus does not exist on the real axis at this point. It is interesting to note that the point -4.53 is a breakaway point for this system when there is positive feedback present (see Section 6.10).

Points of breakaway of the root locus from the real axis can also be obtained by considering the transition from the real axis to a point s_1 which is a small distance δ off the axis. The basis of this method is that the transition from the real axis to s_1 must result in a zero net change of the angle of $G(s)H(s)$. This is illustrated for the root locus considered in Fig. 6.40. For this example,

$$-(\beta_1 + \beta_2 + \beta_3) = (2n + 1)\pi. \tag{6.121}$$

The very small angles we are considering are equal to their tangents, in radians, as follows:

$$-\left(\pi - \frac{\delta}{\sigma_1}\right) - \frac{\delta}{4 - \sigma_1} - \frac{\delta}{5 - \sigma_1} = -\pi. \tag{6.122}$$

This can be rewritten as

$$-\frac{\delta}{\sigma_1} + \frac{\delta}{4 - \sigma_1} + \frac{\delta}{5 - \sigma_1} = 0. \tag{6.123}$$

Canceling δ, and simplifying, results in the equation

$$-\frac{1}{\sigma_1} + \frac{1}{4 - \sigma_1} + \frac{1}{5 - \sigma_1} = 0. \tag{6.124}$$

Solution of Eq. (6.124) yields values of σ_1 equal to 1.47 and 4.53. As indicated before, the value of 4.53 is impossible for negative feedback.

Both techniques presented for determining the breakaway and break-in points are utilized in this book. They are referred to as the *maximization* (or *minimization*) *of* $K(\sigma)$ and the *transition from the real-axis to the complex-plane* methods.

Rule 8 The intersection of the root locus and the imaginary axis can be determined by applying the Routh-Hurwitz stability criterion to the characteristic equation. The characteristic equation for the system illustrated in Fig. 6.40 is given by

$$s^3 + 9s^2 + 20s + K = 0.$$

The resulting Routh-Hurwitz array is given by

$$
\begin{array}{c|cc}
s^3 & 1 & 20 \\
s^2 & 9 & K \\
s & \dfrac{180 - K}{9} & \\
s^0 & K &
\end{array}
$$

For this simple array, a zero in the third row indicates a pair of complex-conjugate poles crossing the imaginary axis. The corresponding value of gain and the value of s at which this occurs can be obtained as follows: For the third row to equal zero,

$$\frac{180 - K}{9} = 0$$

or

$$K = 180. \tag{6.125}$$

Therefore, this system is stable for all gains up to a value of 180. The corresponding value of s occurring at the crossing of the imaginary axis can be obtained from the expression

$$9s^2 + 180 = 0,$$

or

$$s = \pm j\sqrt{20} = \pm j4.48. \tag{6.126}$$

These values are illustrated in Fig. 6.40.

Rule 9 The angles made by the root locus leaving a complex pole can be evaluated by applying the principle of Eq. (6.106). This is illustrated by considering the system shown in Fig. 6.42. Let us calculate the angle that the root locus makes with the complex pole located at $-2 + 2j$. An exploratory point, s_E, will be assumed slightly displaced from this pole. The angle made by the root locus leaving the pole at $-2 + 2j$ to the point s_E is assumed to be $-\theta$ as illustrated in Fig. 6.42. The angles contributed to the point s_E, due to various open-loop poles of the system, are given by (see Fig. 6.42)

$$-135° - 90° + \theta = -180°,$$

or

$$\theta = 45°.$$

Therefore, the angle contributed by the branch of the root locus leaving the pole at $-2 + 2j$ must be sufficient to satisfy the basic relationship given by Eq. (6.106). The negative sign appears before each of the angles in this angular equation, since they are in the denominator of the expression given by Eq. (6.106). If zeros were present, however, they would contribute positive angles since they are in the numerator of the expression given by Eq. (6.106).

Rule 10 In order to derive a useful relation between the poles of the open-loop transfer function and the roots of the characteristic equation, consider the following form of the open-loop transfer function for the system illustrated in Fig. 6.38, where z_i represents the open-loop zeros and p_a represents the open-loop poles excluding those at the origin:

$$G(s)H(s) = \frac{K \prod_{i=1}^{x} (s - z_i)}{s^n \prod_{q=1}^{y} (s - p_a)}. \tag{6.127}$$

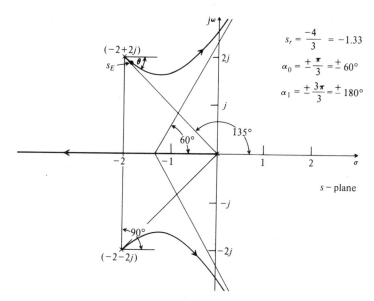

$$s_r = \frac{-4}{3} = -1.33$$

$$\alpha_0 = \pm \frac{\pi}{3} = \pm 60°$$

$$\alpha_1 = \pm \frac{3\pi}{3} = \pm 180°$$

Fig. 6.42   Root locus of system where

$$G(s)H(s) = \frac{K}{s(s^2 + 4s + 8)}.$$

In practical physical systems

$$n + y > x$$

and the denominator of $C(s)/R(s)$ is of the form

$$1 + G(s)H(s) = \frac{\prod_{j=1}^{m} (s - r_j)}{s^n \prod_{a=1}^{y} (s - p_a)}, \tag{6.128}$$

where

$$m = n + y,$$

$$r_j = \text{roots described by the root locus.}$$

Substituting Eq. (6.127) into Eq. (6.128), and equating the expressions on each side of the resulting equation, results in the following:

$$s^n \prod_{a=1}^{y} (s - p_a) + K \prod_{i=1}^{x} (s - z_i) = \prod_{j=1}^{m} (s - r_j).$$

Expanding the product terms of this equation yields

$$\left(s^m - \sum_{a=1}^{y} p_a s^{m-1} + \cdots \right) + K \left(s^x - \sum_{i=1}^{x} z_i s^{x-1} + \cdots \right)$$

$$= s^m - \sum_{j=1}^{m} r_j s^{m-1} + \cdots . \tag{6.129}$$

For those open-loop transfer functions where the denominator of $G(s)H(s)$ is at least of degree two higher than that of the numerator (which is often the case in practice),

$$x \leqslant m - 2$$

and the following is obtained by equating the coefficients of s^{m-1} in Eq. (6.129):

$$\sum_{a=1}^{u} p_a = \sum_{j=1}^{m} r_j.$$

By defining p_j to represent all of the open-loop poles, including those at the origin, this equation can be written as

$$\sum_{j=1}^{m} p_j = \sum_{j=1}^{m} r_j. \tag{6.130}$$

Equation (6.130), known as Grant's rule, indicates that the sum of the system roots is a constant as the gain is varied from zero to infinity. Therefore, the sum of the system roots is conserved and is independent of gain. This rule, sometimes also referred to as the conservation of the sum of the roots, aids in drawing the root locus, since it implies that as certain loci turn to the right, others must turn to the left in order that the sum of the closed-loop poles may be constant. In addition, this rule, as described by Eq. (6.130), aids in determining the gain along the root locus. For interest, we can determine the location of the third root of the system illustrated in Fig. 6.40 when the root locus crosses the imaginary axis as follows:

$$\sum (-4 - 5) = \sum (+j4.48 - j4.48 + r).$$

Therefore, at $r = -9$, the gain is also 180.

Rule 11 The gain along the root locus can be determined in a number of ways. One of the most fundamental rules of the root locus, Eq. (6.107), can be used to determine this as indicated previously in Fig. 6.39 and Eq. (6.116). Basically, for any point along the root locus, the control-system engineer can substitute the distance of the various poles and zeros to the point into Eq. (6.107) and solve for K. As an example, let us reconsider the root locus illustrated in Fig. 6.40. We have already determined that the gain when the root locus intersects the imaginary axis is 180 (using rule 8). Now, let us determine the value of gain at point A. To do this, we need to solve the following equation:

$$\left| \frac{K}{s_A(s_A + 4)(s_A + 5)} \right| = 1. \tag{6.131}$$

Measuring the distances $|s_A|$, $|s_A + 4|$, and $|s_A + 5|$ to be 2.2, 3.5, and 4.3, respectively, and substituting these values into Eq. (6.131), we obtain

$$\frac{K}{(2.2)(3.5)(4.3)} = 1, \tag{6.132}$$

$$K = 33.1.$$

Gains along the rest of the root locus can be similarly determined.

After the root locus has been sketched by using the eleven rules presented, the graphical accuracy may be improved by determining the exact location of a few points. This can easily be performed by applying the relationship of Eq. (6.106).

In general, constructing a detailed root locus is a very tedious, time-consuming method. Several mechanical construction aids are available. The most commonly used device is known as the Spirule, shown in Fig. 6.43 [14]. This clear-plastic device, which consists of a disk and arm, functions as an angle summer and locus calibrator. A logarithmic spiral curve on the arm portion enables the logarithm of a length to be calibrated as an angle. Therefore, the addition of angles reduces to a process of adding logarithms. Analog and digital computers are other very important tools used in plotting the root locus. These approaches are discussed in detail in Section 6.11.

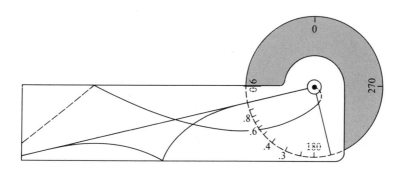

Fig. 6.43   A Spirule.

This section will conclude with illustrative examples of the root-locus procedure. The technique of stabilizing systems utilizing the root-locus method are illustrated in Chapter 7.

As the first example, consider a unity negative feedback system whose open-loop transfer function is given by

$$G(s) = \frac{K}{(s + 1)(s - 1)(s + 4)^2} \,. \tag{6.133}$$

This system has four poles: three on the negative real axis and one on the positive real axis. In addition, it has four zeros at infinity. The root locus of this system, illustrated in Fig. 6.44, can be drawn on the basis of the eleven rules, as follows.

Rule 1   There are four separate loci since the characteristic equation, $1 + G(s)H(s)$, is a fourth-order equation.

Rule 2   The root locus starts ($K = 0$) from the poles located at 1, -1, and a double pole at -4. All loci terminate ($K = \infty$) at zeros which are located at infinity for this problem.

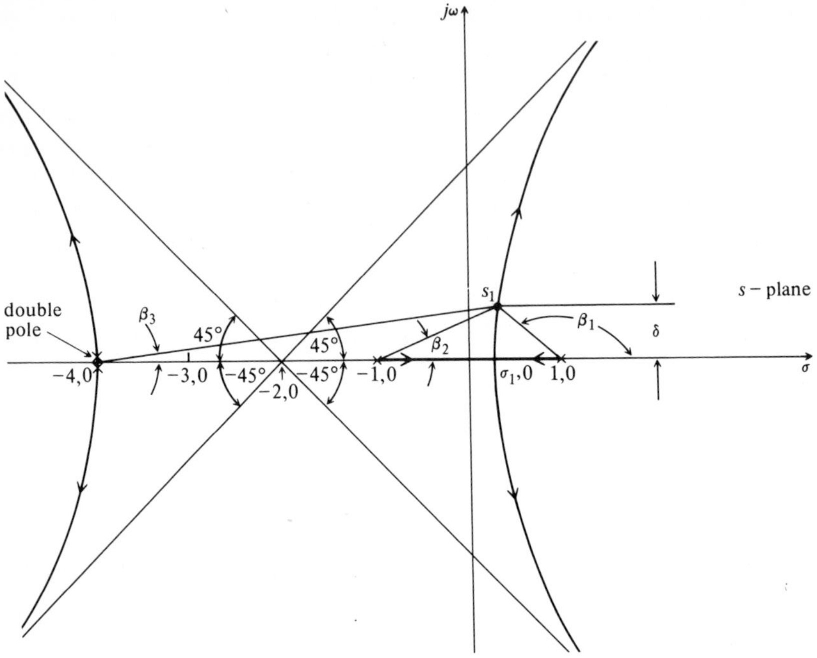

Fig. 6.44 Root locus of system where

$$G(s)H(s) = \frac{K}{(s + 1)(s - 1)(s + 4)^2}.$$

Rule 3 Complex portions of the root locus occur in complex-conjugate pairs.

Rule 4 The portions of the real axis between -1 and 1 are part of the root locus.

Rule 5 The four loci approach infinity as K becomes large at angles given by

$$\alpha_0 = \pm\pi/4 = \pm45°$$

and

$$\alpha_1 = \pm3\pi/4 = \pm135°.$$

Rule 6 The intersections of the asymptotic lines and real axis occur at

$$s_r = \frac{-8 - 0}{4 - 0} = -2.$$

Rule 7 The point of breakaway from the real axis is determined using the two techniques presented: maximization of $K(\sigma)$ and the transition from the real-axis to the complex-plane methods.

Maximization of $K(\sigma)$ Method. From the equation

$$1 + G(s)H(s) = 1 + \frac{K}{(s + 1)(s - 1)(s + 4)^2} = 0,$$

we obtain

$$K = -(s + 1)(s - 1)(s + 4)^2,$$

$$K(\sigma) = -(\sigma^2 - 1)(\sigma + 4)^2,$$

$$\frac{dK(\sigma)}{d\sigma} = -2(\sigma + 4)(2\sigma^2 + 4\sigma - 1) = 0.$$

The roots are

$$\sigma_1 = 0.22, \qquad \sigma_2 = -2.22, \qquad \sigma_3 = -4.$$

Of these, σ_1 represents the breakaway point from the positive real axis, σ_3 represents the breakaway from the double pole at -4, 0, and σ_2 represents the breakaway point for positive feedback (see Problem 6.33).

Transition from the Real-axis to the Complex-plane Method. The point of breakaway from the real axis occurring between -1 and 1 will be assumed to lie along the positive real axis at σ_1. The angles contributed from the various poles to a point s_1 that lies a small distance δ off the positive real axis are

$$-(\beta_1 + \beta_2 + 2\beta_3) = (2n + 1)\pi$$

$$-\left(\pi - \frac{\delta}{1 - \sigma_1}\right) - \frac{\delta}{1 + \sigma_1} - \frac{2\delta}{4 + \sigma_1} = -\pi,$$

or

$$\frac{-\delta}{1 - \sigma_1} + \frac{\delta}{1 + \sigma_1} + \frac{2\delta}{4 + \sigma_1} = 0.$$

Solving, we obtain roots at 0.22 and -2.22. The interpretation of these roots is the same as with the first method.

Rule 8 This particular root locus intersects the imaginary axis only at the origin.

Rule 9 This rule does not apply to this problem.

Rule 10 This rule shows that as certain of the loci turn to the right, others turn to the left to ensure that the sum of the roots is a constant.

Rule 11 This rule does not apply to this problem since the problem does not require the calculation of gains along the locus.

It is interesting to observe from the root locus illustrated in Fig. 6.44 that the system is always unstable since at least one root of the characteristic equation always lies in the right half-plane. We illustrate in Chapter 7 the stabilization of this system by means of a lead network.

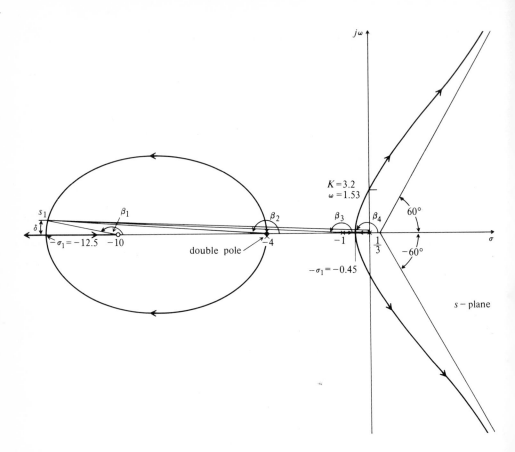

Fig. 6.45 Root locus of negative-feedback system where

$$G(s)H(s) = \frac{1.6K(s + 10)}{s(s + 1)(s + 4)^2}.$$

As a second example, consider a unity negative-feedback system where

$$G(s) = \frac{K(1 + 0.1s)}{s(1 + s)(1 + 0.25s)^2} = \frac{1.6K(s + 10)}{s(s + 1)(s + 4)^2}. \qquad (6.134)$$

This system has four poles (two being a double pole) and one zero, all on the negative real axis. In addition, it has three zeros at infinity. The root locus of this system, illustrated in Fig. 6.45, can be drawn on the basis of the eleven rules presented as follows:

Rule 1 There are four separate loci since the characteristic equation, $1 + G(s)H(s)$, is a fourth-order equation.

Rule 2 The root locus starts ($K = 0$) from the poles located at zero, -1, and a double pole located at -4. One pole terminates ($K = \infty$) at the zero located at -10 and three loci terminate at zeros which are located at infinity.

Rule 3 Complex portions of the root locus occur in complex-conjugate pairs.

Rule 4 The portions of the real axis between the origin and -1, the double poles at -4, and between -10 and $-\infty$ are part of the root locus.

Rule 5 The loci approach infinity as K becomes large at angles given by

$$\alpha_0 = \pm \frac{\pi}{4 - 1} = \pm 60°$$

and

$$\alpha_1 = \pm \frac{3\pi}{4 - 1} = \pm 180°.$$

Rule 6 The intersection of the asymptotic lines and the real axis occur at

$$s_r = \frac{-9 - (-10)}{4 - 1} = 0.33.$$

Rule 7 The point of breakaway from the real axis is determined using the maximization of $K(\sigma)$ and the transition from the real-axis to the complex-plane method.

Maximization of $K(\sigma)$ Method. From the relation

$$1 + G(s)H(s) = 1 + \frac{1.6K(s + 10)}{s(s + 1)(s + 4)^2} = 0,$$

we have

$$K = - \frac{s(s + 1)(s + 4)^2}{1.6(s + 10)},$$

$$K(\sigma) = - \frac{\sigma(\sigma + 1)(\sigma + 4)^2}{1.6(\sigma + 10)}.$$

Taking the derivative of $K(\sigma)$ with respect to σ and solving, we obtain roots at -0.45, -2.25, and -12.5. The root at -2.25 is impossible for the negative-feedback case, since the root locus doesn't lie here. In the following section we shall find that -2.25 is the breakaway point for a positive-feedback system. As mentioned previously, the third breakaway point occurs at the location of the double pole at -4.

Transition from the Real-axis to the Complex-plane Method. The points of breakaway from the real axis occurring between the origin and -1, and -10 and $-\infty$, are evaluated by summing the angles contributed from the various poles and zero to a point s_1, located a small distance δ off the negative real axis. In addition, there must be a breakaway point at the double pole at -4 since sections of the negative real axis on both sides of this double pole are not part of the root locus. The point chosen, as illustrated in Fig. 6.45, is to the left of $-10, 0$. Actually, it doesn't matter if the point is chosen on this segment of the root locus or between the $-1, 0$ point and the origin. The resulting equation, in either case, will result in the other points of breakaway:

$$[\beta_1 - 2\beta_2 - \beta_3 - \beta_4] = (2n + 1)\pi,$$

$$\left(\pi - \frac{\delta}{\sigma - 10}\right) - 2\left(\pi - \frac{\delta}{\sigma - 4}\right) - \left(\pi - \frac{\delta}{\sigma - 1}\right) - \left(\pi - \frac{\delta}{\sigma}\right) = -3\pi.$$

Solving, we obtain values of σ_1 at 0.45, 2.25, and 12.5. The interpretation of these roots is the same as with the first method.

Rule 8 The intersection of the root locus and the imaginary axis can be determined by applying the Routh-Hurwitz stability criterion to the characteristic equation:

$$s^4 + 9s^3 + 24s^2 + (16 + 1.6K)s + 16K = 0.$$

A simpler technique in this problem, since the characteristic equation is a fourth-order equation, is first to solve for the frequencies where the locus intersects the imaginary axis, and then obtain the maximum values of gain from this expression. Letting $s = j\omega$, the characteristic equation becomes

$$\omega^4 - j9\omega^3 - 24\omega^2 + j\omega(16 + 1.6K) + 16K = 0. \tag{6.135}$$

The frequencies where the locus crosses the imaginary axis can be calculated from

$$\omega^4 - 24\omega^2 + 16K = 0 \tag{6.136}$$

or

$$\omega = \sqrt{\frac{16 + 1.6K}{9}}. \tag{6.137}$$

Substituting this value into Eq. (6.136), we obtain

$$\left(\frac{16 + 1.6K}{9}\right)^2 - 24\left(\frac{16 + 1.6K}{9}\right) + 16K = 0.$$

Therefore,

$$K_{max} = 3.2$$

and the system is stable for $0 < K < 3.2$. Substituting $K_{max} = 3.2$ into Eq. 6.137, we obtain

$$\omega = \sqrt{\frac{16 + 1.6(3.2)}{9}} = 1.53 \text{ rad/sec}$$

as the frequency of crossover.

Rule 9 This rule does not apply to this problem.

Rule 10 This rule shows that as certain of the loci turn to the right, others turn to the left to ensure that the sum of the roots is a constant.

Rule 11 This rule does not apply to this problem since the problem does not require the calculation of gains along the locus.

6.10 THE ROOT-LOCUS METHOD FOR POSITIVE-FEEDBACK SYSTEMS

The rules presented in the previous section for constructing the root locus were directed towards negative feedback systems. For positive feedback systems, several of these rules must be modified. The purpose of this section is to indicate the changes to the rules and apply it to the last problem considered in Section 6.9 for positive, instead of negative, feedback.

For positive feedback, Eq. (6.104) becomes:

$$1 - G(s)H(s) = 0 \qquad (6.138)$$

or

$$G(s)H(s) = 1 = 1\underline{/2n\pi} \qquad (6.139)$$

where $n = 0, \pm 1, \pm 2, \pm 3, \dots$. Equation (6.139) specifies two conditions which must be satisfied for the existence of a closed-loop pole in positive-feedback systems.

1. The angle of $G(s)H(s)$ must be an even multiple of π:

$$\text{angle of } G(s)H(s) = 2n\pi, \qquad (6.140)$$

where $n = 0, \pm 1, \pm 2, \pm 3, \dots$.
2. The magnitude of $G(s)H(s)$ must be unity:

$$|G(s)H(s)| = 1. \qquad (6.141)$$

Based on Eqs. (6.140) and (6.141), it is necessary to modify Rules 4, 5, 7, and 9 given in Section 6.9 for construction of the root locus with negative feedback as follows:

Rule 4 This rule is modified for positive feedback so that sections of the real axis are part of the root locus if the number of poles and zeros to the right of an exploratory point along the real axis is even.

Rule 5 For positive feedback, the numerator in Eq. (6.117) is changed to $2m\pi$.

Rules 7 and 9 The transition from the real-axis to the complex-plane method of Rules 7 and 9 is modified in both cases so that the sum of the angles in the calculations is $2n\pi$ instead of $(2n + 1)\pi$. The procedure of Rule 7 using the maximization of $K(\sigma)$ method is exactly the same for negative- and positive-feedback systems.

These changes will now be considered in view of the last example presented in Section 6.9 with positive instead of negative feedback. The open-loop transfer

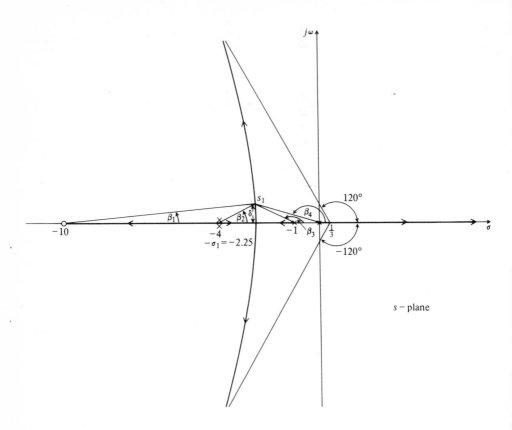

Fig. 6.46 Root locus of a positive-feedback system where

$$G(s)H(s) = \frac{1.6K(s + 10)}{s(s + 1)(s + 4)^2}.$$

function of the unity feedback linear system with positive feedback is given by (see Eq. (6.134))

$$G(s) = \frac{1.6K(s + 10)}{s(s + 1)(s + 4)^2}. \tag{6.142}$$

Let us construct the root locus for this system by reconsidering the changes required to Rules 4, 5, 7, and 9.

Rule 4 The portions of the real axis between $+\infty$ and the origin, and between -1 and -10, are part of the root locus as indicated in Fig. 6.46.

Rule 5 The loci approach infinity as K becomes large at angles given by

$$\alpha_0 = 0$$

and

$$\alpha_1 = \pm \frac{2\pi}{4 - 1} = \pm 120°.$$

Rule 7 The point of breakaway from the real axis is determined using the maximization of $K(\sigma)$ and the transition from the real-axis to the complex-plane methods.

Maximization of $K(\sigma)$ Method. The procedure and resulting equations are exactly the same as for the negative-feedback case presented in Section 6.9. The only difference is in the interpretation of the resulting roots. For the positive-feedback case only the root at -2.25 is possible; the roots at -0.45 and -12.5 are impossible.

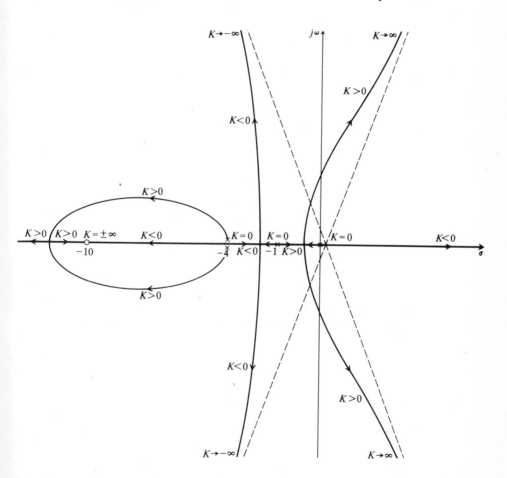

Fig. 6.47 Root locus of a feedback system where

$$G(s)H(s) = \frac{1.6K(s + 10)}{s(s + 1)(s + 4)^2}.$$

for $-\infty < K < \infty$.

Transition from the Real-axis to the Complex-plane Method. The point of breakaway from the real axis can be computed from the following equation:

$$[\beta_1 - 2\beta_2 - \beta_3 - \beta_4] = 2n\pi, \qquad (6.143)$$

$$\left[\frac{\delta}{10 - \sigma_1} - \frac{2\delta}{4 - \sigma_1} - \left(\pi - \frac{\delta}{\sigma_1 - 1}\right) - \left(\pi - \frac{\delta}{\sigma_1}\right)\right] = -2\pi. \qquad (6.144)$$

This equation is similar to the one obtained for the negative-feedback system case. Solving, we obtain the values of σ_1 as before: 0.45, 2.25, and 12.5. In this case, however, only the root at -2.25 is possible. The other solutions are impossible since the root locus doesn't exist along the real axis there.

Rule 9 This rule does not apply to this problem.

The complete root locus for this system is illustrated in Fig. 6.46. It is interesting to observe from this figure that the system is unstable for all values of gain, as compared to the negative feedback case (see Fig. 6.45), where we found that the system was stable for $0 < K < 3.2$.

A useful interpretation is possible if we combine the analyses of Figs. (6.45) and (6.46) and study the behavior of the feedback system where

$$G(s)H(s) = \frac{1.6K(s + 10)}{s(s + 1)(s + 4)^2}$$

for

$$-\infty < K < \infty.$$

Figure 6.47 illustrates this combination of negative- and positive-feedback behavior. Note that the root locus is a continuous curve in going from negative to positive feedback.

6.11 COMPUTER TECHNIQUES FOR PLOTTING THE ROOT LOCUS

The root locus can be plotted automatically using a variety of methods [2]. A special-purpose analog computer which satisfies the necessary angle condition of Eq. (6.106) (or (6.140)) as s is varied can be constructed using servo multipliers [15]. Another special-purpose analog computer method utilizes two-dimensional electric potential distribution techniques in order to plot the complex variables which satisfy Eqs. (6.106) and (6.107) (or Eqs. 6.140 and 6.141) [16, 17]. In addition the root locus can be determined utilizing a digital computer [18–20]. Since the analog computer techniques are less satisfactory and not as commonly used today as the digital computer method, we will focus our attention on the digital computer technique.

In our discussions, we will only consider the case of negative feedback. The procedures are similar for positive feedback.

The digital computer is a very versatile and flexible tool that is easily adaptable for automatically determining the roots in the complex plane. The method presented, based on the material of References 18, 19, and 20, starts at the poles of the control system and searches for the overall root locus in a segmented manner.

Reference 18 discusses this conceptual algorithm for obtaining the root locus using a digital computer. It presents the logic that can be used to code a digital computer in order to determine the locus of points which satisfy Eqs. (6.106), (6.107), and (6.120). The techniques presented are conceptual and are not designed for a particular machine or computer language. In addition, specific problems associated with computer coding are not presented in the following discussion, but must be considered when determining the root locus of a specific system utilizing a specific digital computer.

The computer logic flow diagrams of this technique are illustrated in Fig. 6.48(a) through (d), and (f). The basis of the conceptual algorithm is a convergent trial-and-error procedure for progressing from a known point on the root locus to a

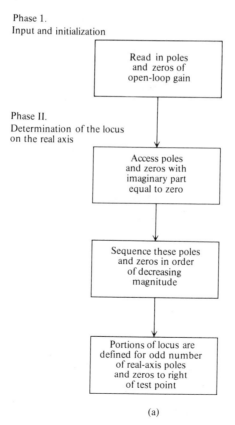

Phase 1.
Input and initialization

Read in poles
and zeros of
open-loop gain

Phase II.
Determination of the locus
on the real axis

Access poles
and zeros with
imaginary part
equal to zero

Sequence these poles
and zeros in order
of decreasing
magnitude

Portions of locus are
defined for odd number
of real-axis poles
and zeros to right
of test point

(a)

Fig. 6.48   (a) Digital computer logic flow chart, phases 1 and 2.

succeeding point. Initial points for the algorithm can be determined by realizing that the locus begins at the poles where the loop gain equals zero. For poles located on the real axis, the locus will remain on it until some breakaway point is reached. For these loci, it will be necessary to determine the breakaway points and begin a search in the complex plane from these points. Terminating points of the root locus are zeros that are located at some finite value or at infinity. Therefore, it is necessary to test and determine if a locus point is near a zero or if it appears to be heading away from the origin of the complex plane.

The procedure determines the root locus in segments which are ultimately joined together. The process is composed of the following phases:

a) Input and initialization—Fig. 6.48(a).

b) Determine locus segments on real axis—Fig. 6.48(a).

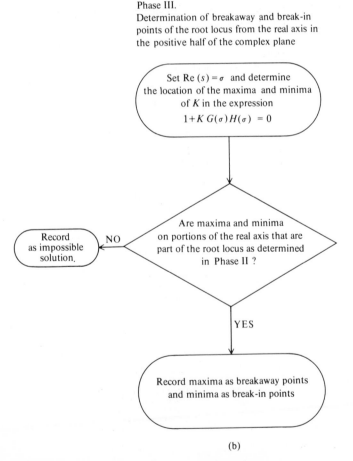

Phase III.
Determination of breakaway and break-in points of the root locus from the real axis in the positive half of the complex plane

Set Re $(s) = \sigma$ and determine the location of the maxima and minima of K in the expression

$1 + K\, G(\sigma)H(\sigma) = 0$

Are maxima and minima on portions of the real axis that are part of the root locus as determined in Phase II ?

Record as impossible solution.   NO

YES

Record maxima as breakaway points and minima as break-in points

(b)

Fig. 6.48   (b) Digital computer logic flow chart, phase 3.

Phase IV.
Search for root loci which begin at poles in the positive half of the complex plane

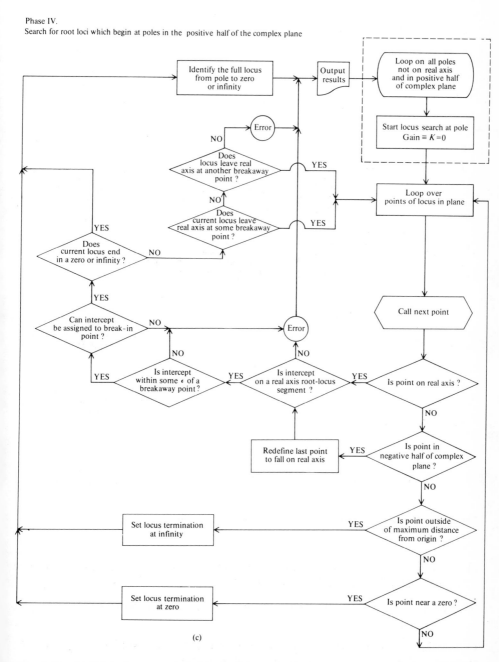

(c)

Fig. 6.48   (c) Digital computer logic flow chart, phase 4.

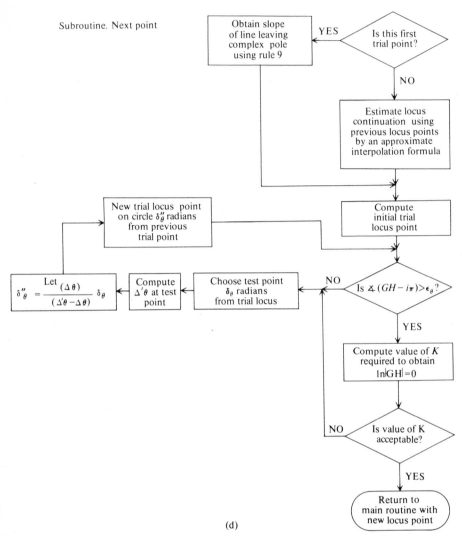

(d)

Fig. 6.48   (d) Digital computer logic flow chart, subroutine.

c) Determine breakaway and break-in points of root locus from real axis in positive (upper) half of complex plane—Fig. 6.48(b).

d) Search for root loci which begin at poles in positive (upper) half of complex plane—Fig. 6.48(c).

e) Search for root loci which begin at poles on real axis and subsequently enter the positive (upper) half of complex plane—Fig. 6.48(f).

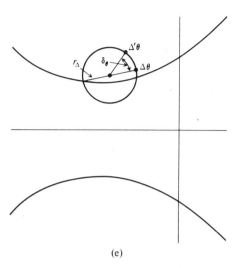

(e)

Fig. 6.48   (e) Estimated root locus point.

f)   Reflect the root-locus segments of the positive (upper) half-plane into the negative (lower) half-plane—Fig. 6.48(f).

g)   Produce remaining root loci from unassigned segments—Fig. 6.48(f).

Figure 6.48(d) illustrates a computer subroutine associated with Fig. 6.48(c) and denoted as "next point." Let us next consider some of the details associated with each of these computer steps.

The *input and initialization phase*, illustrated in Fig. 6.48(a), is concerned with reading in poles and zeros of the open-loop gain.

The second phase of the procedure, denoted as the *determination of the locus on the real axis* in Fig. 6.48(a), consists of three sequences. Poles and zeros with imaginary parts equal to zero are determined and are then sequenced in order of decreasing magnitude. Portions of the locus are then defined for an odd number of real-axis poles and zeros which exist to the right of a test point.

The third phase, illustrated in Fig. 6.48(b), is concerned with the *determination of breakaway and break-in points of the root locus from the real axis in the positive half of the complex plane*. The procedure involves determining the location of the maxima and minima of the function $K(\sigma)$ as discussed in Rule 7 for construction of the root locus.

Search for root loci which begin at poles in the positive half of the complex plane, the fourth phase, is illustrated in Fig. 6.48(c). Successive points of the root locus are determined and the search for additional points are terminated if the following conditions occur:

a) A zero is recognized as the termination of the current locus;

b) The termination of the current locus occurs at infinity;

c) The current locus intercepts the real axis.

If the last condition occurs, then it is necessary to determine the succeeding behavior of the root locus. In general, the following three situations can occur:

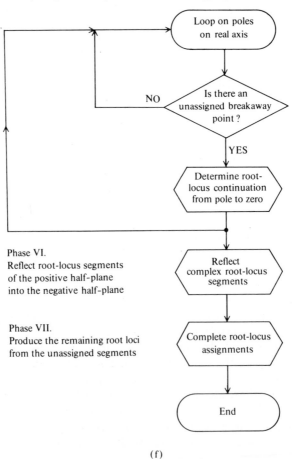

Phase V.
Search for root loci which begin at poles on the real axis and subsequently enter the positive-half of the complex plane

Phase VI.
Reflect root-locus segments of the positive half-plane into the negative half-plane

Phase VII.
Produce the remaining root loci from the unassigned segments

(f)

Fig. 6.48 (f) Digital computer logic flow chart, phases 5, 6, and 7 [18].

d) The locus continues onto the real axis and terminates at a zero on the real axis which is located at a finite value or at infinity;

e) The locus continues on the real axis until a breakaway point occurs and it then reenters the full complex plane;

f) The locus immediately reenters the imaginary portion of the complex plane.

As each new locus point is obtained, it is examined to determine whether conditions (a), (b), or (c) occur.

Each real-axis intercept is tested to determine if it lies between the bounds of real-axis root-locus segments determined in the second phase. If it does not, an error flag is set, the search for the current root locus is discontinued, and the search for a new root locus is started. Similarly, the real-axis intercepts are compared with the breakaway points determined in the third phase. If the intercept is not within some small error, then an error condition is recognized. If more than one break-in point exists at the breakaway point, tests are made to determine which of them has the same slope as the current root locus. If no correspondence is obtained, then an error condition is recognized.

The root-locus continuation is examined next. A test is made to determine whether the current root locus can be ended at a nearby zero or a zero located at infinity. If it can, the full locus from start to completion is recorded and displayed. If the root locus reenters the imaginary portion of the complex plane, it will occur at the nearest unassigned breakaway point. The repeated use of "call next point" is then applied as indicated.

The "next point" flow diagram is illustrated in Fig. 6.48(d). The real and imaginary parts of the point in the complex plane are treated in a real two-dimensional space. An estimate is made of the continuation of the root locus. The initial trial locus point is obtained by computing the intersection of the estimated root-locus continuation with a circle having a small radius r_Δ, that is centered at the last known locus point as illustrated in Fig. 6.48(e). There will be two intersecting points and the computer logic must be able to differentiate between them. The net angle contribution for all poles and zeros is computed at the trial locus point. The difference $\Delta\theta$ is computed and compared with the maximum permissible angular deviation of GH from $i\pi$, ϵ_θ. If $\Delta\theta \leqslant \epsilon_\theta$, then the angle criterion is satisfied and the value of K required to obtain $|GH| = 1$ is determined. However, if $\Delta\theta > \epsilon_\theta$, a new trial locus point must be found. The procedure used is to search for an acceptable solution which falls on the previously defined circle illustrated in Fig. 6.48(e). In the figure, a test point δ_θ is chosen. The difference of the angular contribution of the poles and zeros from $i\pi$, evaluated at the new test point, is denoted as $\Delta'\theta$ in Fig. 6.48(e). The change in the angular error is used to determine a new trial locus point on the circle. After the location of the point and the value of gain are determined, the computer returns to the main routine.

The fifth phase, illustrated in Fig. 6.48(f), is to *search for root loci which begin at poles on the real axis and subsequently enter the positive half of the complex plane.* Only those, as yet unassigned, breakaway points in the same segments as the poles which lead into the right half-plane and which are also unassigned, are examined. If an assignment can be made, the continuation of the current root locus is determined. The coding in the box labeled "determine root-locus continuation from pole to zero" is exactly the same as that used in the fourth phase (see Fig. 6.48(c)) which exists outside the dotted lines.

The sixth phase, illustrated in Fig. 6.48(f), is concerned with *reflecting the root-locus segments of the positive half-plane into the negative half-plane.* The fourth and fifth phases of the computer routine are sufficient to have determined all of the root-locus segments in the positive half-plane. Similar segments are also part of the root locus in the negative half-plane.

Producing the remaining root loci from the unassigned segments, the seventh and final computer phase, is illustrated in Fig. 6.48(f). The complete root loci in the negative half-plane is not completely symmetrical with the root loci previously determined. It is the function of this phase to join all unconnected segments remaining in the negative half-plane and on the origin to form a complete root locus. This process can be accomplished by successively accessing unassigned poles, and then adjoining unassigned segments which are continuous and end in the unassigned zeros.

PROBLEMS

6.1 The differential equation expressing the output $y(t)$ of a second-order system in terms of its input $x(t)$ is given by

$$A \frac{d^2y(t)}{dt^2} + B \frac{dy(t)}{dt} + Cy(t) = x(t).$$

a) Find the ratio of $Y(j\omega)/X(j\omega)$ for a sinusoidal input motion in terms of A, B, and C.
b) Plot the ratio $Y(j\omega)/X(j\omega)$ in the complex plane when $A = 1$, $B = 4$, $C = 16$.

6.2 The field of fluidics is a relatively new field with a great potential [2]. The use of fluids as a power source is not new, but the discovery that this energy can be manipulated and utilized in much the same way as electricity, and without the need for moving parts, has sparked the new technology of fluidics. The term *fluidics* refers to the field of technology which is concerned with the use of either liquid or gaseous fluids in motion to perform functions such as amplification, sensing, switching, computation, and control. These systems have the advantages of high reliability, operation under extreme environmental conditions, resistance to radiation, and low cost. Fluidic control systems are especially applicable to fluid-flow systems, such as those using a turbine. Figure P6.2 illustrates the equivalent block diagram of a fluidic speed control system. Actual speed, derived from a tuning-fork device, is compared with the desired speed. Any difference is amplified by a fluidic amplifier which then activates a valve used to control the turbine's speed.

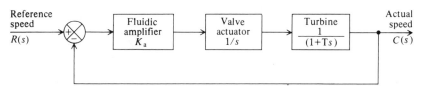

Figure P6.2

a) Determine the state-variable signal-flow diagram and the vector differential equation of this system when $K_a = 10$ and $T = 100$.
b) Determine the characteristic equation of this system from the relation $|\mathbf{P} - s\mathbf{I}| = 0$.
c) Utilizing the Routh-Hurwitz criterion, determine whether the system is stable.

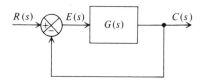

Figure P6.3

6.3 Using the Routh-Hurwitz stability criterion, determine if the feedback control system shown in Fig. P6.3 is stable for the following transfer functions:

a) $G(s) = \dfrac{100}{s(s^2 + 8s + 24)}$

b) $G(s) = \dfrac{3s + 1}{s^2(300s^2 + 600s + 50)}$

c) $G(s) = \dfrac{24}{s(s + 2)(s + 4)}$

d) $G(s) = \dfrac{0.2(s + 2)}{s(s + 0.5)(s + 0.8)(s + 3)}$

6.4 Figure P6.4(a) illustrates the submarine depth control problem. The object is to adjust the actual depth C to equal a desired depth R. The control system depends on a pressure transducer which is used to measure the actual depth. Any difference between the actual and desired depths is amplified and is used to drive the stern plane actuator through an appropriate angle θ until the actual depth equals the desired depth. An equivalent block diagram of such a system is illustrated in Fig. P6.4(b). Utilizing the Routh-Hurwitz criterion, determine whether this system is stable for the parameters indicated.

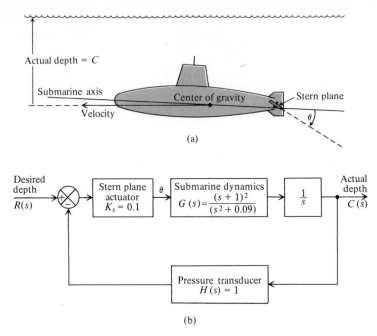

(a)

(b)

Figure P6.4

6.5 Using the Routh-Hurwitz stability criterion, determine if the feedback control system illustrated in Fig. P6.5 is stable.

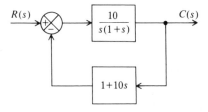

Figure P6.5

6.6 Automatic control systems are being used to an ever-increasing degree in the railroad transportation field [28]. A very widely acclaimed high-speed rail transportation system is in operation in Japan [29]. Figure P6.6(a) is a photograph of the Tokyo-to-Osaka railroad which is commonly referred to as the Tokaido line. Figure P6.6(b) illustrates an equivalent block diagram for the automatic braking system used to regulate this class of high-speed trains.

a) Determine the signal-flow diagram and the characteristic equation of this system.
b) Using the Routh-Hurwitz criterion, determine the allowable values of amplifier gain K_a

Fig. P6.6 (a) Tokaido Line. (Courtesy of Japanese National Railways)

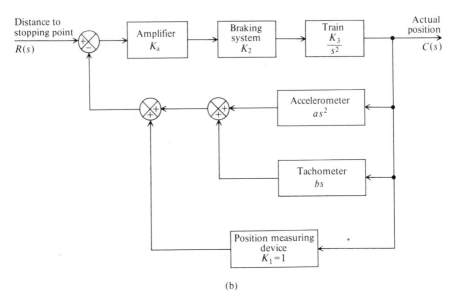

(b)

Figure P6.6   (b)

for system stability. Assume the following parameters:

$$K_1 = 1, \qquad K_2 = 1000, \qquad K_3 = 0.001;$$
$$a = 0.1, \qquad b = 0.1.$$

6.7 Using the Routh-Hurwitz stability criterion, determine if the feedback control system illustrated in Fig. P6.7 is stable.

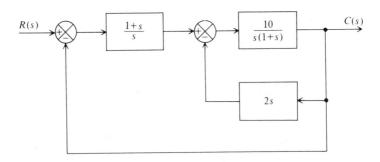

Figure P6.7

6.8 The tool of mathematical modeling can be used to study economic problems of a much larger scope than that found in a single business organization. The economics concerned with national income, government policy on spending, private business investment, business production, taxes, and consumer spending may be represented [2] by the block diagram of

Fig. P6.8. Although business production lags available funds according to a pure time delay, the representation of $G_2(s)$ by the lag factor $1/(1 + Ts)$, is adequate. Assuming that $E(s)$ and $U(s)$ are related by

$$U(s) = -(A + Bs)E(s)$$

and government policy is represented by

$$G_1(s) = C + Ds,$$

determine the requirements on C and D in terms of A and B for system stability.

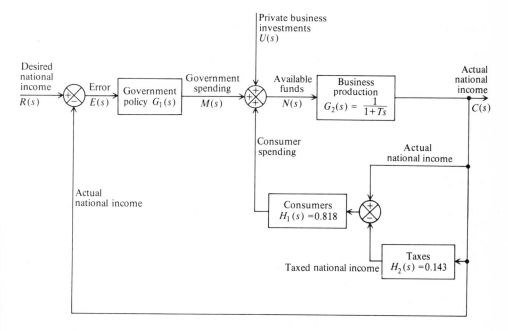

Figure P6.8

6.9 By means of complex-plane plots, determine whether feedback systems represented by the following values of $G(j\omega)H(j\omega)$ are stable.

a) $\qquad G(j\omega)H(j\omega) = \dfrac{10}{(1 + j\omega)(1 + 2j\omega)(1 + 3j\omega)}$

b) $\qquad G(j\omega)H(j\omega) = \dfrac{10}{j\omega(1 + j\omega)(1 + 10j\omega)}$

c) $\qquad G(j\omega)H(j\omega) = \dfrac{10}{(j\omega)^2(1 + 0.1j\omega)(1 + 0.2j\omega)}$

d) $\qquad G(j\omega)H(j\omega) = \dfrac{2}{(j\omega)^2(1 + 0.1j\omega)(1 + 10j\omega)} \cdot$

Do not attempt to plot the exact values of $G(Hj\omega)(j\omega)$ for all values of frequency. It should only be necessary to determine a few values of frequency exactly.

6.10 By means of a complex-plane plot, determine whether the system illustrated in Fig. P6.10 is stable.

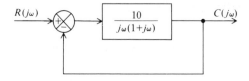

Figure P6.10

6.11 Determine whether the system illustrated in Fig. P6.11 is stable by means of a complex-plane plot.

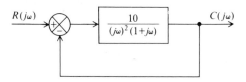

Figure P6.11

6.12 The positive-frequency portions of the complex-plane plots for several transfer functions are shown in Fig. P6.12. Complete the Nyquist diagram and determine stability, assuming that $G(j\omega)$ has no poles in the right half-plane.

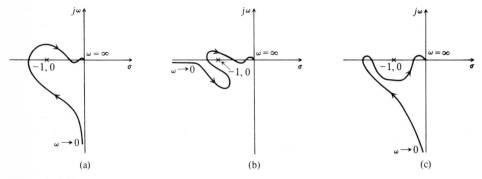

Figure P6.12

6.13 The design of automatic control systems for electronic activation of human limb movements is an interesting example of the application of control theory [30]. By means of electrical pulses, a paralyzed limb can be made to contract, with the result that functional movements of the extremity are performed. Figure P6.13(a) illustrates the concept implemented using conventional control-system techniques. The electronic controller $G(s)$ feeds electrical signals to the contracting muscle (agonists) which stretch the opposing muscles

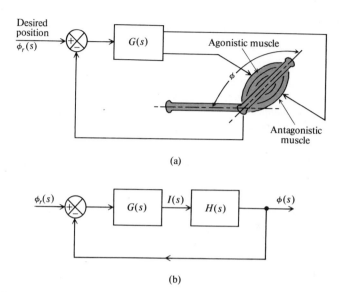

(a)

(b)

Figure P6.13

(antagonists). Figure P6.13(b) illustrates the equivalent block diagram where the following transfer functions can be assumed:

$$G(s) = A,$$

$$H(s) = \frac{\phi(s)}{I(s)} = \frac{K'e^{-\tau_0 s}}{s(Js + C)}.$$

Typical values of the parameters are as follows:

$$J = 1, \quad C = 20, \quad \tau_0 = 0.1, \quad K' = 1.$$

a) Utilizing the Nyquist diagram, determine the phase margin when the electronic gain, A, is set at 2.

b) Repeat part (a) when the electronic gain is set at 100.

6.14 Draw the straight-line attenuation diagrams, showing the magnitude in decibels, and the phase characteristics, showing the phase angle in degrees, as a function of frequency, for the following transfer functions:

a) $\quad G_A = \dfrac{20}{s(1 + 0.5s)(1 + 0.1s)}$,

b) $\quad G_B = \dfrac{2s^2}{(1 + 0.4s)(1 + 0.04s)}$,

c) $\quad G_C = \dfrac{50(0.6s + 1)}{s^2(4s + 1)}$,

d) $\quad G_D = \dfrac{7.5(0.2s + 1)(s + 1)}{s(s^2 + 16s + 100)}$.

Assuming that G_A and G_C represent the open-loop transfer functions of unity-gain feedback systems, determine the phase and gain margins in parts (a) and (c).

6.15 A feedback control system has the configuration shown in Fig. P6.15.
a) Draw the Nyquist diagram of the loop gain function.
b) Draw the Bode diagram showing the magnitude, in decibels, and phase angle, in degrees, as a function of frequency. $_{120}$ and $_{14 dB}$
c) Find the phase margin and gain margin of the system. Illustrate these points on the graphs for parts (a) and (b).
d) Find the values of K_p, K_v, and K_a.
e) What is the steady-state velocity-lag error for a velocity input of 5 rad/sec?

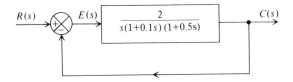

Figure P6.15

6.16 Repeat Problem 6.15 for the transfer function

$$G(s) = \frac{10}{s(1 + 0.1s)(1 + 0.01s)} .$$

6.17 A feedback control system has the configuration shown in Fig. P6.17.
a) Plot the Bode diagram for this system. 3.75 $-9°$
b) What is the gain crossover frequency? What is the phase margin at gain crossover? Is the system stable?
c) Find the values of K_p, K_v, K_a.
d) What is the steady-state velocity-lag error for a velocity input of 40 rad/sec?

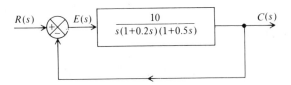

Figure P6.17

6.18 Repeat Problem 6.17 for the transfer function

$$G(s) = \frac{20}{s(1 + 0.5s)(1 + 0.001s)} .$$

6.19 A tank level control is shown in Fig. P6.19. It is desired to hold the tank level C within limits even though the outlet flow rate V is varied. If the level is not correct, an error voltage E_n is developed which is amplified and applied to a servomotor. This in turn adjusts a valve through appropriate gearing N and thereby restores balance by adjusting the inlet rate, M_2.

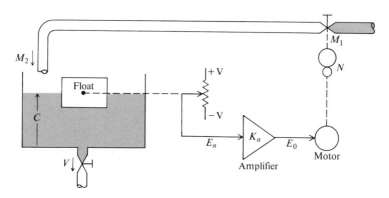

Figure P6.19

The following relations are valid from this figure:

$E_n = R - C$ = error in level (in),
$E_0 = K_a E_n$ = voltage applied to motor (volts),
$M_1 = G_1 E_0$ = valve position (rad),
$M_2 = K_v M_1$ = tank feed flow (ft³/sec),
$C = G_2 M_2$ = tank level (in).

The pertinent constants and transfer functions are as follows:

$$G_1 = \frac{1}{s(0.1s + 1)} \text{ radians of valve motion per volt,}$$

$G_2 = 0.5/s$ inches of level per ft³/sec,
$K_v = 0.1$ ft³/sec per radian of valve motion,
K_a = amplifier gain to be set,
Error detector = 1 volt per inch of error.

a) Draw the system block diagram showing all transfer functions.
b) Using a Bode diagram for the solution, determine the amplifier gain K_a required to meet a required gain crossover frequency of 1.5 rad/sec.
c) With the gain set at the value determined in part (b), what are the resulting phase and gain margins?

6.20   Repeat Problem 6.19 with an error-detector sensitivity of 10 volts per inch of error.

6.21   By utilizing automatic ship steering systems, a ship can maintain a desired heading much more accurately than if it depended on a helmsman correcting the heading at infrequent intervals. Accurate ship heading is a particularly crucial problem for minesweepers who must sweep desired, prescribed, straight paths with very little allowable deviations. Figure P6-21 illustrates the overall control problem where δ represents the deviation of the mine sweeper from the desired heading and θ represents the angle of deflection of the steering rudder. The transfer function relating $\theta(s)$ and $\delta(s)$ for a 180-ft minesweeper moving at 13 ft/sec is given by [31]

$$\frac{\delta(s)}{\theta(s)} = \frac{0.164(s + 0.2)(-s + 32)}{s^2(s + 0.25)(s - 0.009)}.$$

Determine the Bode diagram for this transfer function.

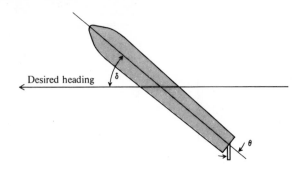

Desired heading

Figure P6.21

6.22 A feedback control system has the configuration shown in Fig. P6.22, where $U(s)$ represents an extraneous signal appearing at the input to the plant.
a) Assuming that $G_1(s) = 1$, plot the decibel–log frequency diagram for this system.
b) It is desired that the steady-state error resulting from an extraneous unit step input signal at $U(s)$ shall be 0.1 unit. Assuming $G_1(s) = K$, determine K to meet this specification.

$K = 9.9$

c) Determine the crossover frequency and phase margin resulting from part (b).

15 rad/sec.   ~32°

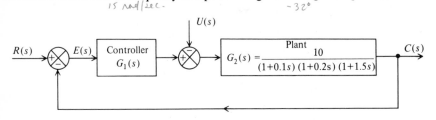

Figure P6.22

6.23 Repeat Problem 6.22 with the transfer function of $G_2(s)$ given by

$$G_2(s) = \frac{2}{s(1 + s)(1 + 0.1s)}.$$

6.24 A second-order servomechanism has a forward transfer function given by

$$G(s) = \frac{16}{s(2 + s)}.$$

The feedback function is unity.
a) Draw the Bode diagram showing the magnitude and the phase characteristics as a function of frequency.
b) Using the Nichols chart, plot a curve of frequency response of the closed loop system.
c) What are ω_p and M_p?
d) Can this system ever be unstable no matter how large the forward gain is made? Explain.

6.25 Repeat Problem 6.24 for

$$G(s) = \frac{60(1 + 0.5s)}{s(1 + 5s)}.$$

6.26 Repeat Problem 6.24 for

$$G(s) = \frac{60(1 + s)}{s^2(1 + 0.1s)} .$$

6.27 A feedback control system has a block diagram as shown in Fig. P6.27.
a) Determine the required gain K for a steady-state velocity-lag error of 30° with an input
velocity of 10 rad/sec. 1.91
b) What are the values of M_p and ω_p?

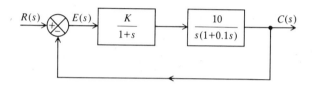

$R(s)$ $E(s)$ $\dfrac{K}{1+s}$ $\dfrac{10}{s(1+0.1s)}$ $C(s)$

Figure P6.27

6.28 A unity feedback control system has a forward transfer function given by

$$G(s) = \frac{10}{s(1 + 0.1s)(1 + 0.05s)} .$$

a) Plot $G(j\omega)$ on the Nichols chart.
b) What are the values of M_p and ω_p?
6.29 Repeat Problem 6.28 for

$$G(s) = \frac{2}{s(1 + s)(1 + 10s)} .$$

6.30 A unity negative feedback control system has a forward transfer function given by

$$G(s) = \frac{K}{s(1 + 0.1s)(1 + s)} .$$

a) Sketch the root locus giving all pertinent characteristics of the locus.
b) At what value of gain does the system become unstable?
6.31 Repeat Problem 6.30 for the following transfer functions:

a) $G(s) = \dfrac{K}{s^2} ,$

b) $G(s) = \dfrac{K(1 + s)}{s^2(1 + 0.1s)} ,$

c) $G(s) = \dfrac{K(s + 1)}{s(s^2 + 8s + 16)} ,$

d) $G(s) = \dfrac{K(s + 0.1)^2}{s^2(s^2 + 9s + 20)} .$

6.32 Sketch the root locus for a feedback control system having the following forward and feedback transfer functions:

$$G(s) = \frac{K(s + 0.1)}{s^2(s + 0.01)}, \qquad H(s) = 1 + 0.6s.$$

6.33 Sketch the root locus of a unity, positive, feedback system whose transfer function is given by

$$G(s) = \frac{K}{(s + 1)(s - 1)(s + 4)^2}.$$

6.34 Sketch the root locus for a negative feedback control system having the following forward and feedback transfer functions:

$$G(s) = \frac{K(s + 1)}{s^2(s + 2)(s + 4)}, \qquad H(s) = 1.$$

6.35 Repeat Problem 6.34 for positive feedback. What conclusions can you draw from your result?

6.36 Draw the root locus of a positive feedback system where

$$G(s) = \frac{K}{(s + 1)^2(s + 4)^2}, \qquad H(s) = 1.$$

6.37 Determine the root locus for a feedback-system whose open-loop transfer function is given by

$$G(s)H(s) = \frac{K(s + 2)}{s(s + 4)(s + 8)(s^2 + 2s + 5)}$$

for $-\infty < K < \infty$. Indicate all pertinent values on the root locus.

6.38 A negative feedback-system has an open-loop transfer function given by

$$G(s)H(s) = \frac{K(s + 4)(s + 40)}{s^3(s + 200)(s + 900)}.$$

a) Draw the root locus and label all pertinent values on the root locus.
b) For what range of values of gain is the system stable?
c) Draw the Bode diagram of the system and correlate the region of stability and instability with the root-locus results.

6.39 Repeat Problem 6.36 for the case of negative feedback.

6.40 Reconsider the ecological model of rabbits and rabbit-eating foxes presented in Problem 2.51. The state equations of the process were given by

$$\dot{x}_1 = Ax_1 - Bx_2,$$
$$\dot{x}_2 = -Cx_2 + Dx_1,$$

where x_1 represented the number of rabbits and x_2 represented the fox population. Determine the requirements on A, B, C, and D for a stable ecological system. What occurs if A is greater than C?

6.41 Sketch the root locus for the following negative feedback control system, and determine the range of K for which the system is stable:

$$G(s) = \frac{K(s + 6)}{s(s + 1)(s + 4)}, \qquad H(s) = 1.$$

6.42 Reconsider the model depicting the development of unindustrialized nations discussed previously in Problem 2.47. For the coefficients listed in part (a) of Problem 2.47,
determine whether the system is stable. Stability in this problem means that the states
$x_1(t)$ and $x_2(t)$ decrease to zero as time increases.

6.43 The stability of the feedback control system shown in Fig. P6.43 is to be determined.

a) Using the Routh-Hurwitz stability criterion, determine whether the system is stable for
the following transfer functions:

$$G_A(s) = \frac{45}{s + 2},$$

$$G_B(s) = \frac{2}{(s + 3(s + 5)},$$

$$H(s) = 1.$$

b) Repeat part (a) for the following transfer functions:

$$G_A(s) = \frac{45}{s + 2},$$

$$G_B(s) = \frac{2}{s + 3},$$

$$H(s) = \frac{1}{s + 5}.$$

c) What conclusions concerning stability can you draw from your results in parts (a) and
(b)?
d) Will the outputs, $c(t)$, in parts (a) and (b), differ in response to the same input? Discuss
your results.

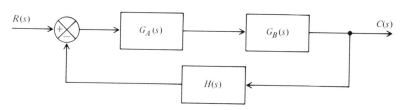

Figure P6.43

6.44 Determine the range of positive real gain K that will result in a stable system for the system shown in Fig. P6.44 for the following conditions:

a) The system has negative feedback.
b) The system has positive feedback.

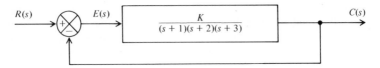

$$G(s) = \frac{K}{(s + 1)(s + 2)(s + 3)}$$

with $R(s)$, $E(s)$, $C(s)$ labels and summing junction.

Figure P6.44

6.45 The transfer function of an unknown element in a control system can be determined by measuring its frequency response using a sinusoidal input. For a particular element, the frequency and amplitude data shown in Table P6.45 was obtained.

Table P6.45   Experimental data

ω	$G(j\omega)$ in db
0.1	66
0.2	60
0.4	54
0.7	51
1.0	49
2.0	47
4.0	46
7.0	45
10.0	43
20.0	39
40.0	34
70.0	28
100.0	23
200.0	13
400.0	2
700.0	-8
1000.0	-14

The form of the transfer function is given by

$$G(s) = \frac{K[1 + (s/\omega_a)]}{s[1 + (s/\omega_b)][1 + (s/\omega_c)]}.$$

Determine the values of K, ω_a, ω_b, and ω_c.

6.46 The transfer function of an element used in a feedback system is not known explicitly, but its frequency response has been measured experimentally and is given in Fig.

P6.46. Based on your knowledge of Bode diagram characteristics, determine the transfer function of this element.

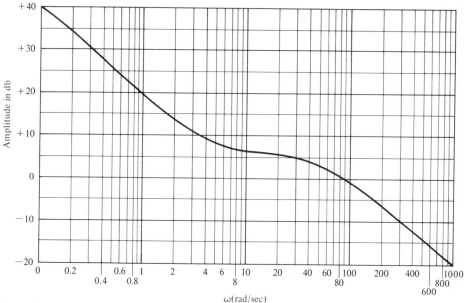

Figure P6.46

6.47 Electromechanical nose-wheel steering systems have been developed that can supply general aviation short-haul aircraft with needed maneuverability and durability [36]. These intermediate-sized aircraft must be able to utilize small airfields, many of which do not have elaborate service and maintenance facilities. During takeoff from such landing strips, the task of keeping the aircraft on the proper heading is achieved by utilizing nose-wheel power steering provided by this device. The block diagram of a nose-wheel steering system is illustrated in Fig. P6.47. Pilot command signals are compared with a feedback signal representing the nose-wheel's heading. The resulting error signal is amplified and applied to a magnetic particle clutch which activates the rotation of the wheel heading. Assuming that $L/R = 0.1$, $K_c = 1$, $J = 1$, $C_v = 9$, and $K_e = 9$, utilize the Bode diagram to determine the gain required to achieve a phase margin of $40°$.

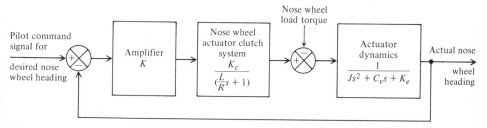

Figure P6.47

6.48 The proper design of man-machine control systems requires as much knowledge of the human element as that of the machine. Therefore, the determination of the human transfer function is very essential in order that the performance of man-machine systems can be evaluated. This problem can be better understood by referring to Fig. P6.48, which depicts the classical compensatory manual tracking problem. In this configuration, an operator attempts to maintain a moveable follower coincident with a stationary reference point that represents the target. A very general and useful form of the human transfer function, $G_H(s)$, which can be applied to manual tracking problems, was provided in Problem 3.26. It is given by

$$G_H(s) = \frac{K(1 + T_A s)\, e^{-Ds}}{(1 + T_L s)(1 + T_N s)},$$

where D represents the operator's transportation lag, T_A represents the operator's anticipation time constant, T_L represents the operator's error-smoothing lag time constant, and T_N represents the operator's short neuromuscular delay. Representative values for the elements are as follows:

$$D = 0.2 \text{ sec} \pm 20\%,$$

$$T_A = 0 \text{ to } 2.5 \text{ sec (variable)},$$

$$T_L = 0 \text{ to } 20 \text{ sec (variable)},$$

$$T_N = 0.1 \text{ sec} \pm 20\%,$$

$$K = 1 \text{ to } 100 \text{ (variable)}.$$

The gain K and the time constants T_A and T_L are considered to be variable according to the control task being performed. This transfer function has met with reasonable success for predicting manual tracking control system response where the bandwidth is relatively low. For this problem, assume that the human transfer function is given by the following expression:

$$G_H(s) = \frac{10(1 + s)\, e^{-0.2s}}{(1 + 10s)(1 + 0.1s)}.$$

Assume that the controlled system has a transfer function given by

$$G_S(s) = \frac{2}{s}.$$

Draw the Bode diagram and determine the phase and gain margin of the resulting system.

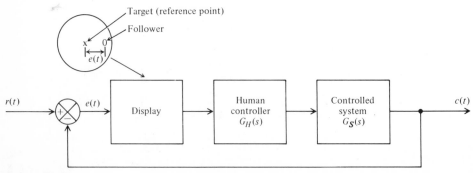

Figure P6.48

6.49 The pitch attitude control system for a booster rocket containing attitude and rate gyros is shown in Fig. P6.49. Sketch the root locus and determine the maximum value of K that would permit stable operation.

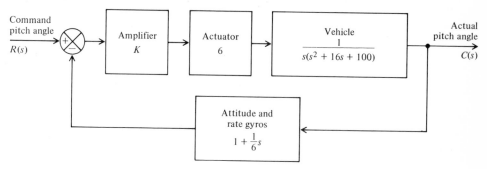

Figure P6.49

6.50 A unity feedback control system has the following forward transfer function:

$$G(s) = \frac{K}{s(s + 1)(s + 5)}.$$

a) If the gain is set at 1.1 and one of the closed-loop poles is known to be located at $s_1 = -0.5$, is the system stable?
b) If the gain is increased to 30 and one of the closed-loop poles is known to be located at $s_1 = -6$, is the system stable?

REFERENCES

1. R. J. Schwarz and B. Friedland, *Linear Systems*, McGraw-Hill, New York (1965).
2. S. M. Shinners, *Techniques of System Engineering*, McGraw-Hill, New York (1967).
3. E. Routh, *Advanced Dynamics of a System of Rigid Bodies*, Macmillan, London (1905).
4. A. Hurwitz, "Uber die Bedingungen, unter welchen eine Gleichung nur Wurzeln mit negativen realen Theilen besitzt," *Math. Ann.* **46**, 273 (1895).
5. H. Nyquist, "Regeneration theory," *Bell System Tech. J.* **11**, 126 (1932).
6. W. T. Thompson, *Laplace Transformation*, Prentice-Hall, Englewood Cliffs, N.J. (1950).
7. S. M. Shinners, "Minimizing servo load resonance error with frequency selective feedback," *Control Eng.* **51**, 51–56 (January 1962).
8. H. Chestnut and R. W. Mayer, *Servomechanisms and Regulating System Design*, (2nd Edn.), Vol. I, Wiley, New York (1959).
9. R. C. Dorf, *Modern Control Systems*, Addison-Wesley, Reading, Mass. (1967).
10. H. M. James, N. B. Nichols, and R. S. Phillips, *Theory of Servomechanisms*, McGraw-Hill, New York (1947).
11. W. R. Evans, "Graphical analysis of control systems," *Trans. AIEE* **67**, 547 (1948).
12. W. R. Evans, "Control system synthesis by root locus method," *Trans. AIEE* **69**, 66 (1950).

13. W. R. Evans, *Control System Dynamics*, McGraw-Hill, New York (1954).

14. *The Spirule*, available from the Spirule Co., Whittier, Calif.

15. F. E. Liethen, *An Automatic Root Locus Plotter Using an Electronic Analog Computer*, M.S. Thesis, Air Force Institute of Technology, Wright-Patterson AFB, Ohio, August 1959.

16. M. L. Morgan, "Algebraic function calculation using potential analog pairs," *Proc. IRE* **49**, 276–82 (1961).

17. Information supplied by Electro Scientific Industries, Inc., 13900 N.W. Science Park Drive, Portland, Oregon.

18. J. Lipow, *A Computer Algorithm for Obtaining the Root Locus*, National Biscuit Co., New York (1962).

19. M. J. Remec, "Saddle-points of a complete root locus and an algorithm for their easy location on the complex frequency plane," *Proc. Natl. Electronics Conf.* **21**, 605–8 (1965).

20. H. M. Paskin, *Automatic Computation of Root Loci Using a Digital Computer*, M.S. Thesis, Air Force Institute of Technology, Dayton, Ohio (March 1962).

21. A. Papoulis, *The Fourier Integral and Its Application*, McGraw-Hill, New York (1962).

22. *"BASIC" Language*, Reference Manual, General Electric Co., Information Systems Division, Pheonix, Arizona (June 1965, Rev. January 1967).

23. Information supplied by Rapidata, The Empire State Building, 350 Fifth Avenue, New York, N.Y. 10001.

24. D. D. McCracken and W. J. Dorn, *Numerical Methods and FORTRAN Programming: With Applications in Engineering and Science*, Wiley, New York (1964).

25. S. Gill, "A process for the step-by-step integration of differential equations in an automatic digital computing machine," *Proc. Roy. Soc. (London)* **A193**, 407–33 (1948).

26. S. A. Hovanessian and L. A. Pipes, *Digital Computer Methods in Engineering*, McGraw-Hill, New York (1969).

27. E. I. Organick, *A FORTRAN IV Primer*, Addison-Wesley, Reading, Mass. (1966).

28. B. Blake, "Four views on train control," *Control Eng.* **11**, 62–68 (December 1964).

29. I. Nakamura and S. Yamazaki, "On the centralized system for train operation and traffic control—Including signaling and routing information," *Railway Technical Research Institute* **5**, 9–11 (1964).

30. W. Crochetiere, L. Vovovnik, and J. B. Resnick, "The design of control systems for electronic activation of human limb movements," in *Proceedings of the 1967 Joint Automatic Control Conference*, pp. 51–57.

31. J. Goclowski and A. Gelb, "Dynamics of an automatic ship steering system," in *Proceedings of the 1966 Joint Automatic Control Conference*, pp. 294–304.

32. H. H. Rosenbrick and C. Storey, *Computational Techniques for Chemical Engineers*, Pergamon, Oxford (1966).

33. R. W. Hamming, *Numerical Methods for Scientists and Engineers*, McGraw-Hill, New York (1962).

34. T. C. Bartee, *Digital Computer Fundamentals* (2nd Edn.), McGraw-Hill, New York (1966).

35. S. M. Shinners, "Which computer—Analog, digital or hybrid?" *Machine Design* **43**, 104–111 (January 21, 1971).

36. J. Camp and M. J. Campbell, "Aircraft Power Steering," *Sperry Rand Engineering Review* **24**, 37–40 (1971).

7 LINEAR FEEDBACK SYSTEM DESIGN

7.1 INTRODUCTION

After the stability of a feedback control system has been analyzed, by using any of the tools presented in Chapter 6, it will often be found that system performance is not satisfactory and needs to be modified. It is necessary to ensure that the open-loop gain is adequate for accuracy, and that the transient response is desirable for the particular application. In order for the system to meet the requirements of stability, accuracy, and transient response, certain types of equipment must be added to the basic feedback control system. We use the term *design* to encompass the entire process of basic system modification in order to meet the specifications of stability, accuracy, and transient response. The term *stabilization* is usually used to indicate the process of achieving the requirements of stability alone; the term *compensation* is usually used to indicate the process of increasing accuracy and speeding up the response.

The compensating (or stabilizing) device may be inserted into the system either in cascade with the forward portion of the loop (cascade compensation) as shown in Fig. 7.1, or as part of a minor feedback loop (feedback compensation) as shown in Fig. 7.2 [1–2]. The cascade-compensation technique is usually concerned with the addition of phase-lag, phase-lead, and phase-lag–lead devices. The feedback-compensation technique is primarily concerned with the addition of rate or acceleration feedback. The type of compensation chosen usually depends on the nonlinearities and the location of the noises in the loop.

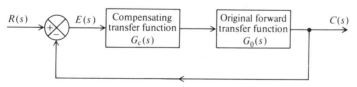

Fig. 7.1 Illustration of cascade compensation.

This chapter focuses attention on the tools presented in Chapter 6 which are of practical and useful interest to the control engineer. Since not all the stability criteria presented are useful with both cascade and feedback compensation, we will focus our attention only on the application of the techniques to those particular design problems

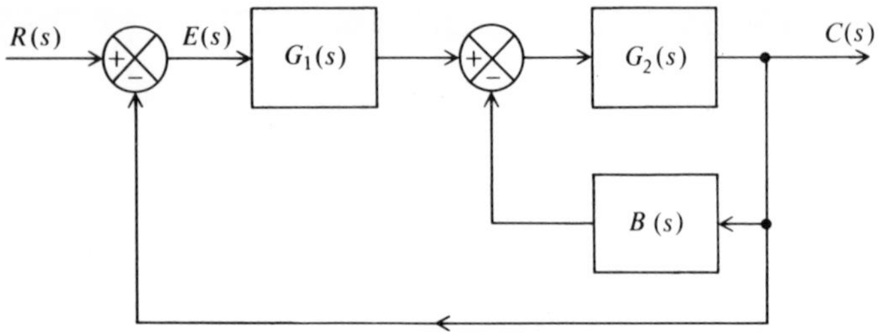

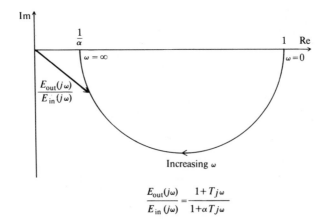

Fig. 7.2 Illustration of minor-loop feedback compensation.

they are most suited to solve. Chapter 9 discusses the design of linear feedback control systems from the point of view of modern optimal control theory.

7.2 CASCADE-COMPENSATION TECHNIQUES

Let us consider the system of Fig. 7.1 as our basic starting point in order to analyze the effects of cascade compensation. The compensating transfer function, $G_c(s)$, is designed in order to provide additional phase lag, phase lead, or a combination of both, in order to achieve certain specifications regarding stability and accuracy. We will illustrate and derive the transfer functions for representative compensating networks [1–6].

A *phase-lag network* is a device which shifts the phase of the control signal in order that the phase of the output lags the phase of the input. An electrical network

$$\frac{E_{\text{out}}(j\omega)}{E_{\text{in}}(j\omega)} = \frac{1+Tj\omega}{1+\alpha Tj\omega}$$

Fig. 7.3 A complex-plane plot for a phase-lag network.

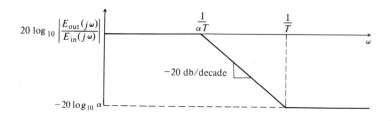

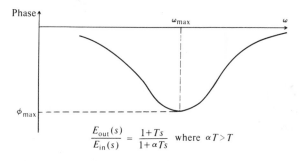

Fig. 7.4   Bode diagram of a phase-lag network.

performing this function was illustrated in Table 2.2 as item 4. Its transfer function was

$$\frac{E_{out}(s)}{E_{in}(s)} = \frac{1 + R_2 C_2 s}{1 + (R_1 + R_2)C_2 s}. \tag{7.1}$$

This can be written in the following more useful form:

$$\frac{E_{out}(s)}{E_{in}(s)} = \frac{1 + Ts}{1 + \alpha Ts}, \tag{7.2}$$

where

$$T = R_2 C_2, \qquad \alpha = 1 + R_1/R_2.$$

Observe that $\alpha T > T$. A complex-plane plot of this network as a function of frequency is shown in Fig. 7.3. Notice that the output voltage lags the input in phase angle for all positive frequencies. In addition, observe that the magnitude of E_{out}/E_{in} decreases from unity at $\omega = 0$ to $1/\alpha$ at $\omega = \infty$. A corresponding Bode diagram for the phase-lag network is illustrated in Fig. 7.4. The frequency at which the maximum phase lag occurs, ω_{max}, and the maximum phase lag, ϕ_{max}, can be easily derived. The results are

$$\omega_{max} = 1/(T\sqrt{\alpha}) \tag{7.3}$$

and

$$\phi_{max} = \sin^{-1}\frac{1 - \alpha}{1 + \alpha}. \tag{7.4}$$

Values of ϕ_{max} for certain values of α, which are useful for design purposes, are listed in Table 7.1.

Table 7.1 ϕ_{max} as a function of α

α	ϕ_{max} (degrees)
1	0
2	−19.4
4	−36.9
8	−51.0
10	−55.0
20	−64.8

A *phase-lead network* is a device which shifts the phase of the control signal in order that the phase of the output leads the phase of the input. An electrical network performing this function was illustrated in Table 2.2, as item 3. Its transfer function was as follows:

$$\frac{E_{out}(s)}{E_{in}(s)} = \frac{R_2}{R_1 + R_2} \frac{1 + R_1 C_1 s}{1 + [R_2/(R_1 + R_2)]R_1 C_1 s}. \tag{7.5}$$

This can be written in the following more useful form:

$$\frac{E_{out}(s)}{E_{in}(s)} = \frac{1}{\alpha}\left(\frac{1 + \alpha T s}{1 + T s}\right) \tag{7.6}$$

where

$$T = \frac{R_1 R_2}{R_1 + R_2} C_1, \qquad \alpha = 1 + \frac{R_1}{R_2}.$$

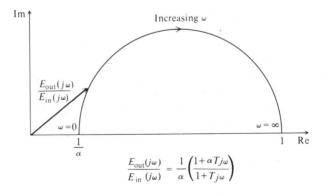

$$\frac{E_{out}(j\omega)}{E_{in}(j\omega)} = \frac{1}{\alpha}\left(\frac{1 + \alpha T j\omega}{1 + T j\omega}\right)$$

Fig. 7.5 A complex-plane plot for a phase-lead network.

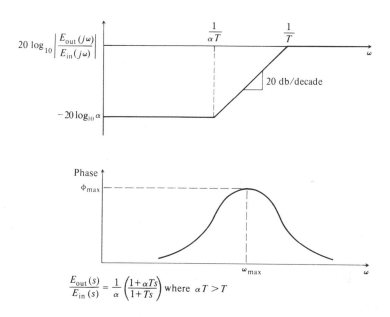

Fig. 7.6 Bode diagram of a phase-lead network.

Observe that $\alpha T > T$. A complex-plane plot of this network as a function of frequency is shown in Fig. 7.5. Notice that the output voltage leads the input in phase angle for all positive frequencies. In addition, notice that the magnitude of E_{out}/E_{in} increases from $1/\alpha$ at $\omega = 0$ to unity at $\omega = \infty$. A corresponding Bode diagram for the phase-lead network is illustrated in Fig. 7.6. The corresponding values of ω_{max} and ϕ_{max} for the phase-lead network are

$$\omega_{max} = 1/(T\sqrt{\alpha}) \tag{7.7}$$

and

$$\phi_{max} = \sin^{-1}\frac{\alpha - 1}{\alpha + 1}. \tag{7.8}$$

The values shown in Table 7.1 are also true for the phase-lead case except for the sign. An important practical point to emphasize is that the control engineer would never in practice use any ratio of $\alpha > 10$ since the lead network acts as an approximate differentiator and emphasizes high-frequency noise, which leads to saturation. In addition, the $1/\alpha$ term acts as an attenuation which must be made up for somewhere in the feedback control system, with an amplification whose ratio is α.

A *phase-lag–lead network* is a device which shifts the phase of a control signal in order that the phase of the output lags at low frequencies and leads at high frequencies relative to the input. An electrical network performing this function was illustrated in Table 2.2 as item 5. Its transfer function was as follows:

$$\frac{E_{out}(s)}{E_{in}(s)} = \frac{(1 + R_1C_1s)(1 + R_2C_2s)}{R_1R_2C_1C_2s^2 + (R_1C_1 + R_2C_2 + R_1C_2)s + 1}. \tag{7.9}$$

Defining

$$T_1 = R_1C_1, \qquad T_2 = R_2C_2, \qquad T_{21} = R_1C_2$$

we can rewrite Eq. (7.9) as

$$\frac{E_{out}(s)}{E_{in}(s)} = \frac{T_1T_2s^2 + (T_1 + T_2)s + 1}{T_1T_2s^2 + (T_1 + T_2 + T_{21})s + 1}.$$

A complex-plane plot of this network as a function of frequency is shown in Fig. 7.7. Notice that the output voltage lags the input in phase angle for low frequencies and leads in phase angle for high frequencies. In addition, notice that the magnitude of E_{out}/E_{in} decreases at an intermediate range of frequencies and increases to unity as ω approaches 0 and ∞. A corresponding Bode diagram for the phase-lag–lead network is illustrated in Fig. 7.8.

The stabilizing effect of cascaded, phase-shifting networks can easily be demonstrated for a simple second-order system. For example, let us consider the configuration illustrated in Fig. 7.1, where the original forward transfer function, $G_0(s)$, is given by

$$G_0(s) = \frac{\omega_n^2}{s(s + 2\zeta\omega_n)}. \tag{7.10}$$

If this system were uncompensated $(G_c(s) = 1)$, then the transfer function $G_0(s)$ would result in the familiar second-order system response which was discussed at great length in Section 4.2. The resulting damping factor of the system would be given

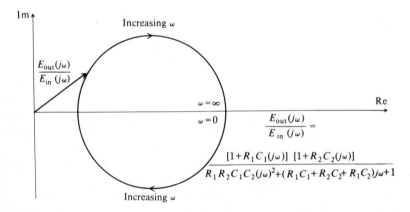

Fig. 7.7 A complex-plane plot for a phase-lag–lead network.

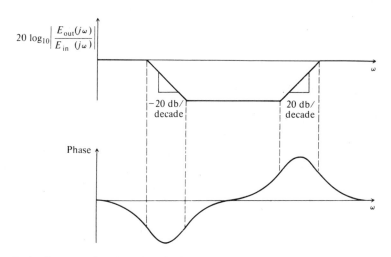

Fig. 7.8   Bode diagram of a phase-lag–lead network.

by ζ and its natural frequency by ω_n. Let us now assume that we add a lead network to this system whose transfer function is given by

$$G_c(s) = \frac{1 + \alpha Ts}{1 + Ts}. \tag{7.11}$$

It is assumed that the attenuation factor of $1/\alpha$ is negated with an amplification increase of α, and the system will maintain the same static error. Let us consider the case where $T \ll \alpha T$ and, therefore, $G_c(s)$ can be approximated by a zero factor:

$$G_c(s) \approx 1 + \alpha Ts. \tag{7.12}$$

The form of Eq. (7.12) suggests that this lead network is equivalent to a proportional plus derivative controller. The resulting system transfer function with the lead network is given by

$$\frac{C(s)}{R(s)} \approx \frac{\omega_n^2(1 + \alpha Ts)}{s^2 + (2\zeta\omega_n + \alpha T\omega_n^2)s + \omega_n^2}. \tag{7.13}$$

Comparing the denominators of Eqs. (7.13) and (4.3), we observe that it is still of second order and ω_n remains the same, but ζ is greater due to the increase in the coefficient of s in the denominator. The equivalent damping factor with $G_c(s)$ can be obtained as follows:

$$2\zeta\omega_n + \alpha T\omega_n^2 = 2\zeta_{eq}\omega_n \tag{7.14}$$

where

ζ_{eq} = an equivalent damping factor with phase-lead compensation.

Solving for ζ_{eq}, we obtain

$$\zeta_{eq} = \zeta + \alpha T\omega_n/2. \tag{7.15}$$

Therefore, we can conclude that the addition of a zero factor in $G_c(s)$ has increased the damping factor from ζ to ζ_{eq} by an amount equal to $\alpha T \omega_n / 2$. This assumes that T is positive, or the zero of the factor $(1 + \alpha Ts)$ is in the left half of the s-plane.

The next question is what is the steady-state error resulting from cascade compensation? To answer this, we must find the steady-state errors resulting from the application of a unit ramp input for the cases of no compensation and compare them with those resulting from cascade compensation. We choose a unit ramp as our input since it is the only input which results in a finite response error for a system with a pole at the origin. The transfer function relating error to input for the system shown in Fig. 7.1 is given by

$$\frac{E(s)}{R(s)} = \frac{1}{1 + G_c(s)G_0(s)}. \tag{7.16}$$

Assuming that

$$G_c(s) = 1 \quad \text{(no cascade compensation)},$$

$$G_0(s) = \frac{\omega_n^2}{s(s + 2\zeta\omega_n)},$$

and

$$R(s) = 1/s^2 \quad \text{(a unit ramp input)},$$

we find that

$$E(s) = \frac{s + 2\zeta\omega_n}{s(s^2 + 2\zeta\omega_n s + \omega_n^2)}. \tag{7.17}$$

Applying the final-value theorem to Eq. (7.17), we find the steady-state error to be

$$e_{ss(\text{ramp input})} = \lim_{s \to 0} sE(s) = \frac{2\zeta\omega_n}{\omega_n^2} = \frac{2\zeta}{\omega_n}. \tag{7.18}$$

For the case with cascade compensation, a similar analysis yields the following result:

$$G_c(s) = \frac{1 + \alpha Ts}{1 + Ts}, \quad G_0(s) = \frac{\omega_n^2}{s(s + 2\zeta\omega_n)}, \quad R(s) = \frac{1}{s^2}.$$

Therefore,

$$E(s) = \frac{1}{s} \frac{(s + 2\zeta\omega_n)(1 + Ts)}{[s(s + 2\zeta\omega_n)(1 + Ts) + \omega_n^2(1 + \alpha Ts)]}. \tag{7.19}$$

Applying the final-value theorem to Eq. (7.19), the steady-state error is found to be

$$e_{ss(\text{ramp input})} = \lim_{s \to 0} sE(s) = \frac{2\zeta\omega_n}{\omega_n^2} = \frac{2\zeta}{\omega_n}. \tag{7.20}$$

Comparing the results of Eqs. (7.18) and (7.20), we conclude that the addition of the cascade lead network as given by Eq. (7.11) does not increase or decrease the steady-state response error of the system.

It is important to emphasize that the relationships derived in this analysis apply only to the simple system considered. For example, if a zero factor were contained in the numerator of Eq. (7.10), then these relationships are modified (see Problems 7.5 and 7.6).

If we attempt to extend this analysis of a second-order system to the case of phase-lag compensation, the characteristic equation becomes third order and difficult to factor. For example, if $G_c(s)$ were only to represent the pole factor of the phase-lag network, then

$$G_c(s) = \frac{1}{1 + \alpha T s}. \tag{7.21}$$

In a similar manner, the closed-loop system transfer function can be found to be given by

$$\frac{C(s)}{R(s)} = \frac{\omega_n^2}{\alpha T s^3 + (2\zeta\omega_n\alpha T + 1)s^2 + 2\zeta\omega_n s + \omega_n^2}. \tag{7.22}$$

The factorization of this characteristic equation is not trivial and a similar analysis to that performed for the lead-network case is not possible.

7.3 MINOR-LOOP FEEDBACK COMPENSATION TECHNIQUES

Let us next consider the general system illustrated in Fig. 7.2. The compensating element in this case is the transfer function $B(s)$. In order to have a basis of comparison, we will follow an analysis for minor-loop feedback compensation similar to that performed for the case of lead-network cascade compensation.

The minor-loop feedback element $B(s)$ usually represents rate feedback or acceleration feedback. In general, phase-lag, lead, and/or lag–lead networks may also be cascaded with $B(s)$.

The stabilizing effect of minor-loop feedback compensation can easily be demonstrated for a simple second-order system. We assume that the system illustrated in Fig. 7.2 consists of simple rate feedback. The specific transfer functions for the system are

$$G_1(s) = 1, \tag{7.23}$$

$$G_2(s) = \frac{\omega_n^2}{s(s + 2\zeta\omega_n)}, \tag{7.24}$$

$$B(s) = bs. \tag{7.25}$$

Without any rate feedback, the configuration represents a simple second-order system whose damping factor is ζ and natural frequency is ω_n. The resulting system transfer function with rate-feedback compensation is given by

$$\frac{C(s)}{R(s)} = \frac{\omega_n^2}{s^2 + (2\zeta\omega_n + \omega_n^2 b)s + \omega_n^2}. \tag{7.26}$$

Comparing the denominator of Eq. (7.26) with that of Eq. (4.3), we observe that it is still of second order and ω_n remains the same, but ζ is greater due to the increase in the coefficient of s in the denominator. The equivalent damping factor with rate feedback added can be obtained by setting the coefficients of the s terms equal to each other, as follows:

$$2\zeta\omega_n + \omega_n^2 b = 2\zeta_{eq}\omega_n, \tag{7.27}$$

where ζ_{eq} = an equivalent damping factor with rate feedback added. Solving for ζ_{eq}, we obtain

$$\zeta_{eq} = \zeta + \frac{\omega_n b}{2}. \tag{7.28}$$

Therefore, we can conclude that the addition of the minor-loop using rate feedback has increased the damping factor from ζ to ζ_{eq} by an amount equal to $\omega_n b/2$. This assumes that b is positive (negative feedback).

It is important at this point to compare Eqs. (7.15) and (7.28). Note that they are very similar, and they imply that

$$\alpha T = b. \tag{7.29}$$

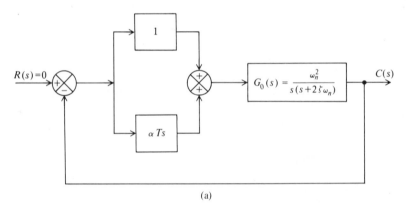

(a)

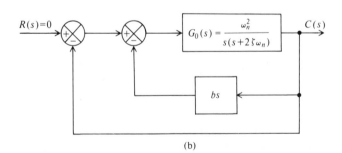

(b)

Fig. 7.9 The stabilizing effects of the systems illustrated are equivalent. (a) Cascade compensation—proportional plus derivative controller. (b) Feedback compensation.

The fact that rate feedback behaves as the approximated lead network, as defined by Eq. (7.12) (proportional plus derivative controller), can be easily demonstrated from Figs. 7.9(a) and (b). Let us assume that there is zero input to both systems, since we are concerned only with the system poles. Clearly, in both cases, there are two negative feedback paths in parallel around $G_0(s)$. In the cascade-compensation case, the total feedback around $G_0(s)$ is $1 + \alpha Ts$; in the rate-feedback-compensation case, the total feedback around $G_0(s)$ is $1 + bs$. Therefore, the stabilizing effects of αT and b are equivalent.

Let us next determine the steady-state error resulting from the use of minor-loop rate-feedback compensation. We assume that the input to this system is a unit ramp in order to have a finite steady-state response error and a basis for comparison. From our discussion of cascade compensation in Section 7.2 we know from Eq. (7.18) that the resulting steady-state error of this system without any compensation ($b = 0$) is $2\zeta/\omega_n$. For the case of minor-loop rate-feedback compensation, the resulting expression for $E(s)$ is given by

$$E(s) = \frac{1}{s}\left[\frac{s + 2\zeta\omega_n + b\omega_n^2}{s(s + 2\zeta\omega_n + b\omega_n^2) + \omega_n^2}\right]. \qquad (7.30)$$

Applying the final-value theorem, the steady-state error is found to be

$$e_{ss(\text{ramp input})} = \lim_{s \to 0} sE(s) = \frac{2\zeta\omega_n + b\omega_n^2}{\omega_n^2} = \frac{2\zeta}{\omega_n} + b. \qquad (7.31)$$

Therefore, the steady-state response error of the system with minor-loop rate-feedback compensation has increased by a factor of b. This unfavorable result can easily be remedied by placing a high-pass filter in cascade with the rate signal. Such a filter would block the steady-state value of the rate output. This technique is illustrated in Fig. 7.10.

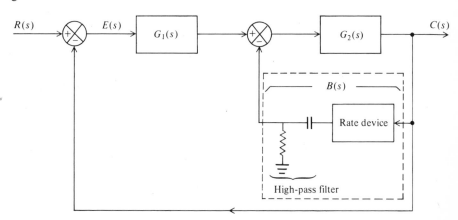

Fig. 7.10 Illustration of minor-loop feedback compensation using a rate device in cascade with a high-pass filter.

As in the preceding section, it is important to emphasize that the relationships derived apply only to the simple system considered. Problems 7.5 and 7.6 illustrate how these relationships change if a zero factor is added to the basic system transfer function considered.

7.4 AN EXAMPLE OF THE DESIGN OF A LINEAR FEEDBACK CONTROL SYSTEM

In this section we consider the design and resulting performance of the second-order system by means of cascade and minor-loop rate-feedback techniques. This problem is useful in unifying concepts which were introduced in Chapters 4, 5, and 6, together with the design techniques to be illustrated later in this chapter.

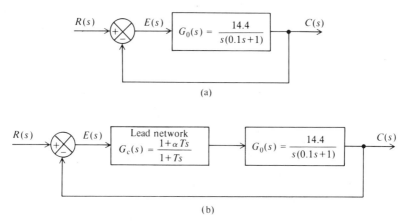

Fig. 7.11 Design of a second-order system.

Let us consider the second-order system illustrated in Fig. 7.11(a). We will assume that the original forward-loop transfer function $G_0(s)$ is given by

$$G_0(s) = \frac{14.4}{s(0.1s + 1)}. \tag{7.32}$$

The closed-loop transfer function, $C(s)/R(s)$, is given by

$$\frac{C(s)}{R(s)} = \frac{G_0(s)}{1 + G_0(s)} = \frac{14.4/[s(0.1s + 1)]}{1 + 14.4/[s(0.1s + 1)]} \tag{7.33}$$

or

$$\frac{C(s)}{R(s)} = \frac{144}{s^2 + 10s + 144}. \tag{7.34}$$

Comparing Eqs. (4.3) and (7.34), we observe that the natural frequency ω_n and damping factor ζ of the system are given by

$$\omega_n = 12 \ \text{rad/sec} \tag{7.35}$$

and

$$\zeta = 0.417. \tag{7.36}$$

If the system is subjected to a unit step input, the transient response will have the form shown in Fig. 4.4 (interpolate between $\zeta = 0.4$ and $\zeta = 0.6$). The maximum percent overshoot can be obtained from Eq. (4.32) and is found to be 23.5%.

Let us assume, for this application, that it is desired to have a critically damped system ($\zeta = 1$). We will demonstrate how this can be achieved using a cascaded network and minor-loop rate feedback.

Figure 7.11(b) illustrates the form that the system illustrated in Figure 7.11(a) would have if a cascaded network were used. We attempt to achieve a damping factor equal to 1 using a lead network, where

$$G_c(s) = \frac{1 + \alpha T s}{1 + T s}. \tag{7.37}$$

As was assumed previously, in Section 7.2, we assume that $T \ll \alpha T$, and therefore $G_c(s)$ can be approximated by

$$G_c(s) \approx 1 + \alpha T s. \tag{7.38}$$

The closed-loop system transfer function for this case was derived in Section 7.2 (see Eq. 7.13) as

$$\frac{C(s)}{R(s)} \approx \frac{\omega_n^2 + \alpha T \omega_n^2 s}{s^2 + (2\zeta\omega_n + \alpha T \omega_n^2)s + \omega_n^2}. \tag{7.39}$$

The equivalent damping factor for this situation was derived in Section 7.2 (see Eq. 7.15) as

$$\zeta_{eq} = \zeta + \frac{\alpha T \omega_n}{2}. \tag{7.40}$$

The object in this problem is to design $\zeta_{eq} = 1$ for the case where

$$\zeta = 0.417 \quad \text{and} \quad \omega_n = 12.$$

Substituting the values into Eq. (7.40), we find that

$$\alpha T = 0.0972.$$

We know that the resulting system will be stable and critically damped, and will have a steady-state response error for a unit step input of zero. The steady-state response error of the system to a unit ramp input was derived (see Eq. 7.18) as

$$e_{ss(\text{ramp input})} = \frac{2\zeta}{\omega_n} = \frac{2(0.417)}{12} = 0.0695 \ \text{units}. \tag{7.41}$$

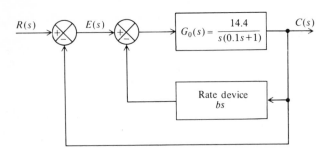

Fig. 7.12   Minor-loop feedback added to the system shown in Fig. 7.11(a).

Let us next attempt to achieve the same type of performance using minor-loop rate feedback. Figure 7.12 illustrates the form that the system illustrated in Fig. 7.11(a) would have if minor loop feedback were used. Our goal is to achieve a damping factor of $\zeta = 1$. The closed-loop system transfer function for this case was derived, previously, in Section 7.2 (see Eq. 7.26) as

$$\frac{C(s)}{R(s)} = \frac{\omega_n^2}{s^2 + (2\zeta\omega_n + \omega_n^2 b)s + \omega_n^2}. \tag{7.42}$$

The equivalent damping factor for this configuration was derived in Section 7.2 (see Eq. 7.28) as

$$\zeta_{eq} = \zeta + \omega_n b/2. \tag{7.43}$$

The object in this problem is to design $\zeta_{eq} = 1$ for the case where

$$\zeta = 0.417 \tag{7.44}$$

and

$$\omega_n = 12. \tag{7.45}$$

Substituting these values into Eq. (7.43), we find that

$$b = 0.0972. \tag{7.46}$$

We know that the resulting system will be stable and critically damped, and will have a steady-state response error of zero for a unit step input. The steady-state response error of the system to a unit ramp input was derived (see Eq. 7.31) as

$$e_{ss(\text{ramp input})} = \frac{2\zeta}{\omega_n} + b = 0.0695 + 0.0972 = 0.1667 \text{ units.} \tag{7.47}$$

This increase has been accounted for in Section 7.3; by using a high-pass filter in the feedback path to block a steady-state output from the rate feedback, the steady-state error can be reduced to 0.0695.

7.5 DESIGN UTILIZING THE BODE-DIAGRAM APPROACH

The techniques necessary to construct and analyze the open-loop frequency response of a feedback control system utilizing the Bode-diagram approach were presented in Section 6.5. This section illustrates how the Bode diagram can be used for designing a feedback control system in order to meet certain specifications regarding relative stability, transient response, and accuracy. It is important to emphasize that the Bode-diagram approach is used very frequently by the practicing control engineer. Its use is due to the fact that the anticipated theoretical results may be relatively simply checked with actual performance in the laboratory just by opening the feedback loop and obtaining an open-loop frequency response of the system.

Bode's primary contribution to the control art is summarized in two theorems [7]. We introduce the concepts embodied in these theorems first in a qualitative manner, and then the mathematical statements are given.

A. Bode's Theorems *Bode's first theorem* essentially states that the slopes of the asymptotic amplitude–log frequency curve implies a certain corresponding phase shift. For example, in Section 6.5 it was shown that a slope of $20n$ db/decade (or $6n$ db/octave) corresponded to a phase shift of $90n°$ for $n = 0, \pm 1, \pm 2, \ldots$. Furthermore, this theorem states that the slope at crossover (where the attenuation–log frequency curve crosses the 0 db line) is weighted more heavily towards determining system stability than a slope further removed from this frequency. This results in a rather complex weighting factor which is a measure of relative importance toward determining system stability.

From what has been presented so far, this theorem is intuitively seen to be valid. The crossover frequency is one of the two points that is checked to determine the degree of stability when using the Bode diagram. Specifically, the phase shift is measured at this particular frequency in order to determine the phase margin. A feedback control system whose slope at crossover is -20 db/decade, and whose other slope sections are relatively far away from crossover in accordance with the relative weighting function, implies a phase shift of approximately $-90°$ in the vicinity of crossover and a corresponding phase margin of about $90°$. This value of phase margin certainly implies a stable system. A system, however, whose slope at crossover is -40 db/decade, and whose other slope sections are relatively far away from crossover in accordance with the relative weighting function, implies a phase shift of approximately $-180°$ and a corresponding phase margin of about $0°$. This value of phase margin implies a system which is on the verge of being unstable and would probably be so when actually tested. Steeper slopes would indicate negative phase margins and definitely unstable systems. Therefore, one strives to maintain the slope of the amplitude–log frequency curve in the vicinity of crossover at a slope of -20 db/decade. Notice that the system, illustrated in Fig. 6.20, has slopes of -60 db/decade that are relatively far removed from the crossover frequency. Therefore, this system has a fairly respectable phase margin of $54°$ by maintaining the -20 db/decade slope for about an octave below, and about 2 octaves above crossover.

Bode's second theorem essentially states that the amplitude and phase characteristics of linear, minimum-phase shift systems are uniquely related. When we specify the slope of the amplitude–log frequency curve over a certain frequency interval, we have also specified the corresponding phase shift characteristics over that frequency interval. Conversely, if we specify the phase shift over a certain frequency interval, we have also specified the corresponding amplitude–log frequency characteristic over that frequency interval. The theorem emphasizes the fact that we can specify the amplitude–log frequency characteristic over a certain interval of frequencies together with the phase shift–log frequency characteristics over the remaining frequencies. It should be emphasized that these conclusions apply only if the transfer function is minimum phase.

The second theorem may appear quite trivial at first glance. Its implications however, are quite important. We will make further use of this theorem when designing feedback control systems using the Bode-diagram approach.

The formal *mathematical statement of Bode's first theorem* is given by the following expression

$$\phi(\omega_d) = \frac{\pi}{2} \left| \frac{dG}{dn} \right|_d + \frac{1}{\pi} \int_{-\infty}^{\infty} \left[\left| \frac{dG}{dn} \right| - \left| \frac{dG}{dn} \right|_d \right] \ln \coth \left| \frac{n}{2} \right| dn, \qquad (7.48)$$

where $\phi(\omega_d)$ represents the phase shift of the system in radians at the desired frequency ω_d, G represents the gain in nepers (1 neper $= \ln |e|$), $n = \ln (\omega/\omega_d)$, $|dG/dn|$ represents the slope of the amplitude–log frequency curve in nepers per unit change of n (note that 1 neper/unit change of n is equivalent to 20 db/decade), $|dG/dn|_d$ represents the slope of the amplitude–log frequency curve at the reference frequency ω_d, and $\ln \coth |n/2|$ represents the weighting function which is plotted in Fig. 7.13. The first

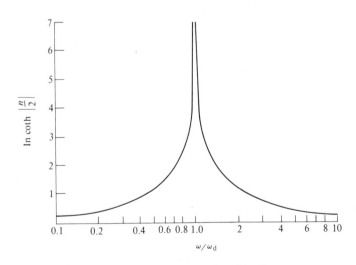

Fig. 7.13   Plot of the weighting function used in Bode's first theorem.

term of Eq. (7.48) represents the phase shift contributed by the slope of the amplitude–log frequency curve at the reference frequency ω_d. For example, it yields a phase shift of 90° for every neper per unit of n (20 db/decade). The second term of Eq. (7.48) is proportional to the integral of the product of the weighting function and the difference in slope of the amplitude–log frequency curve at a frequency ω as compared to its value at the reference frequency, ω_d. Attention is drawn to the fact that it is the weighting function that determines the phase shift contribution at ω_d due to the amplitude–log frequency curve which exists at some frequency ω. Since the second term of Eq. (7.48) is zero for large values of n and where $n = 0$, the value of the integral will be relatively small compared with the first term if the slope of dG/dn is constant over a relatively wide range of frequencies about ω_d. Therefore, under these conditions, the phase shift would be determined primarily by the first term of Eq. (7.48). Following this line of reasoning, the slope of the amplitude–log frequency curve should be less than -2 nepers per unit of n (-40 db/decade) over a relatively wide range of frequencies at crossover in order to ensure stability.

The formal *mathematical statement of Bode's second theorem* is given by the following expression:

$$\int_0^{\omega_s} \frac{G\,d\omega}{\sqrt{\omega_s^2 - \omega^2}\,(\omega^2 - \omega_d^2)} + \int_{\omega_s}^{\infty} \frac{\phi\,d\omega}{\sqrt{\omega^2 - \omega_s^2}\,(\omega^2 - \omega_d^2)}$$

$$= \begin{cases} \dfrac{\pi}{2}\dfrac{\phi(\omega_d)}{\omega_d\sqrt{\omega_s^2 - \omega_d^2}} & \text{(for } \omega_d < \omega_s\text{)} \\[4mm] -\dfrac{\pi}{2}\dfrac{G(\omega_d)}{\omega_d\sqrt{\omega_d^2 - \omega_s^2}} & \text{(for } \omega_d > \omega_s\text{)} \end{cases} \qquad (7.49)$$

where ω_s represents the frequency in radians per second below which the amplitude–log frequency characteristic is specified and above which the phase characteristic is specified. This theorem emphasizes the interdependence of amplitude and phase shift over the entire range of positive frequencies. In addition, notice that although it is possible to specify amplitude or phase in one range of frequencies, and the other quantity in the remaining frequencies, these quantities reflect their presence back into the other range of frequencies. Therefore the integration with respect to frequency is performed over the entire range of positive frequencies.

The design of several systems using the Bode-diagram approach is considered next. We shall illustrate a method that determines steady-state accuracy from the Bode diagram as well as meeting relative stability requirements.

B. Example of Lead and Lag Compensation Let us first consider the third-order system illustrated in Fig. 7.14. Its open-loop transfer function, $G_0(s)$, is given by

$$G_0(s) = \frac{80}{s(1 + 0.02s)(1 + 0.05s)}. \qquad (7.50)$$

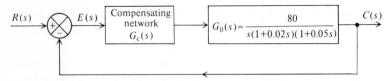

Fig. 7.14   A third-order system which is to be compensated.

The Bode diagram for the uncompensated system $(G_c(s) = 1)$ is illustrated in Fig. 7.15. It indicates a crossover frequency ω_c of 38 rad/sec, a phase margin γ of $-17°$, and a gain margin of -6 db. These results indicate that the uncompensated system is unstable. Let us next attempt to compensate this system with a simple lag and then a lead network. To achieve an acceptable transient response for the specific application of this system, the phase margin should be approximately 30° and the gain margin about 7 db (see Section 6.4). Also, it is assumed that a sinusoidal disturbance at 1 rad/ sec is present, and a gain of 38 db is required at this frequency to nullify its effect.

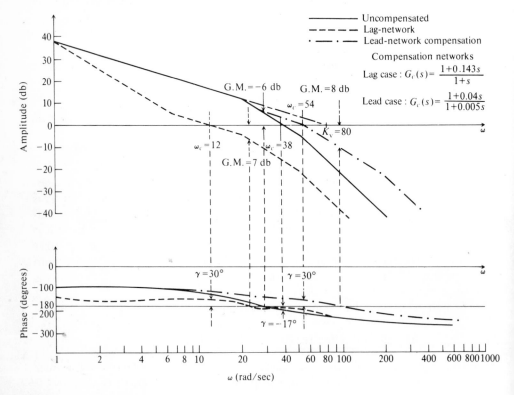

Fig. 7.15   Compensation of a third-order system where

$$G_0(s) = \frac{80}{s(1 + 0.02s)(1 + 0.05s)}.$$

Let us first consider the lag-network compensation case. Applying Bode's theorems in order to achieve the specified phase and gain margins, we would expect that the -20 db/decade slope in the vicinity of the new crossover frequency should not extend over too wide a range of frequencies since the relative stability that is desired is rather moderate. The lag network is of the form:

$$G_c(s) = \frac{1 + Ts}{1 + \alpha Ts},$$ (7.51)

where $\alpha T > T$. In Section 7.2 we studied the characteristics of the phase-lag network. In particular, Fig. 7.4 illustrated the Bode diagram of a general phase-lag network. Notice that this type of network is of such a nature that it attenuated all high frequency components above $\omega = 1/T$ by a factor of $1/\alpha$. From the Bode diagram viewpoint, this attenuating characteristic can be used for stabilization purposes by reshaping the uncompensated amplitude characteristics, so that the initial -20 db/decade is made to cross over the 0 db line rather than the -40 db/decade segment. In other words, one would attempt to stabilize this system with a phase-lag network by placing the frequencies $1/\alpha T$ and $1/T$ in the range of frequencies below about 10 rad/sec. It would be desirable that the -20 db/decade segment start at least around 10 rad/sec and cross over the 0 db line before 20 rad/sec where the amplitude–log frequency characteristic changes to a slope of -40 db/decade. In addition, we would not want $1/\alpha T$ to occur at less than 1 rad/sec, since an open-loop gain of 38 db has been specified at $\omega = 1$ rad/sec. The final phase of the solution is by means of iteration. However, the procedure converges quite rapidly. Usually, two or three iterations should prove sufficient. For the requirements specified, a phase-lag network given by

$$G_c(s) = \frac{1 + 0.143s}{1 + s}$$ (7.52)

results in a phase margin of 30° and a gain margin of 7 db.

Let us next consider the lead-network compensation case. The lead network has the form

$$G_c(s) = \frac{1}{\alpha}\left(\frac{1 + \alpha Ts}{1 + Ts}\right),$$ (7.53)

where $\alpha T > T$. In Section 7.2 we studied the characteristics of the phase-lead network. We assume that any low-frequency attenuation, which is due to the value of $1/\alpha$, is made up for by increasing the gain of the feedback control system by a like amount. The Bode diagram of a general phase-lead network was shown in Fig. 7.6. From the Bode-diagram viewpoint, this type of characteristic can be used for stabilization purposes by reshaping the uncompensated amplitude characteristics so that a -40 db/decade slope is made to cross over the 0 db line along a synthesized -20 db/decade slope rather than the -40 db/decade segment. The range of frequencies where one can place the frequencies $1/\alpha T$ and $1/T$ is quite limited in this particular problem. The value of $1/\alpha T$ can be placed between 20 and 38 rad/sec. The closer it is to 20 rad/sec, the greater will be its stabilizing effect. The further away the break at $1/T$ is from 38 rad/sec, the greater will be the stabilizing effect of the lead network. It can be seen that this particular solution does not modify the open-loop characteristics in the vicinity of 1 rad/sec and the accuracy specification of 38 db at 1 rad/sec will easily be achieved. After some trial and error, a phase-lead network given by

$$G_c(s) = \frac{1 + 0.04s}{1 + 0.005s} \tag{7.54}$$

results in a phase margin of 27° and a gain margin of 8 db. This is close enough to the desired result for all practical purposes. Observe the fact that although we are actually crossing the 0 db line at an asymptotic slope of -40 db/decade, the system is still stable. This can clearly be understood from Bode's first theorem and the weighting diagram shown in Fig. 7.13. Notice that the resulting lead network has a low-frequency attenuation of 0.005/0.04 which must be made up by increasing the gain by a factor of 0.04/0.005.

If we have a choice of using the phase-lag or phase-lead network, the phase-lag network solution would be preferable since it meets the required specifications with a narrower bandwidth than the lead network case ($\omega_c = 12$ for the former; $\omega_c = 54$ for the latter). A feedback control system having a narrower bandwidth will reject a greater amount of noise than one having a wider bandwidth, as well as requiring less power and cost. In addition, the phase-lead network has the disadvantage of requiring a greater amount of amplification within the control system than the lag network.

C. Obtaining Steady-State Error Coefficients from the Bode Diagram The steady-state error coefficients can be determined from the Bode diagram. The definition and importance of these error coefficients have been discussed in Section 5.4. For a system having one pure integration, the velocity constant K_v can be obtained by extending the initial -20 db/decade slope until it intersects the 0 db line. The frequency at which it intersects this line is equal to the velocity constant. We recall from our discussion in Section 5.4 that K_v was obtained by letting s approach zero when utilizing the final-value theorem (see Eq. 5.18). Therefore, the time-constant factor terms having the form $(1 + Ts)$ or $[(Ts)^2 + 2\zeta Ts + 1]$ all approach unity. This permits one to obtain K_v directly by considering only the initial slope of the Bode diagram. For the Bode diagram shown in Fig. 7.15, the value K_v obtained graphically is 80. As a check, using the definition given by Eq. (5.18), we obtain

$$K_v = \lim_{s \to 0} sG(s)H(s). \tag{7.55}$$

Since

$$G(s) = \frac{80}{s(1 + 0.02s)(1 + 0.05s)}, \tag{7.56}$$

$$H(s) = 1, \tag{7.57}$$

then,

$$K_v = \lim_{s \to 0} \frac{s(80)}{s(1 + 0.02s)(1 + 0.05s)} = 80. \tag{7.58}$$

It is also interesting to note that the velocity constant is the same for the uncompensated and compensated systems.

For a system which has a double pole at the origin, the acceleration constant K_a can be obtained in a similar manner. The initial -40 db/decade slope is extended until it intersects the 0 db line. The square of the frequency at which it intersects this line is equal to the acceleration constant.

D. Application of Bode Diagram to a System Containing a Disturbance The next system we consider is illustrated in Fig. 7.16. This consists of a third-order system which has an unwanted external input $U(s)$. The open-loop transfer function for the uncompensated system, $G_0(s)$, is given by

$$G_0(s) = \frac{2.2}{(1 + 0.1s)(1 + 0.4s)(1 + 1.2s)}. \tag{7.59}$$

It is desired that the steady-state error resulting from an unwanted, external step input signal at $U(s)$, should not exceed 0.1 unit. The compensation device, $G_c(s)$, is to contain amplification which will meet this accuracy requirement together with a phase-lead or phase-lag network which will provide a phase margin of about 30° and a gain margin of 5 db.

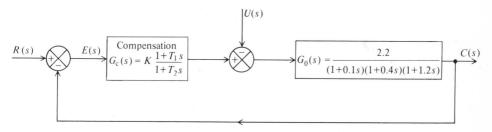

Fig. 7.16 A third-order system having an unwanted external input.

The value of gain required for $G_c(s)$ will be computed first. For this calculation, $U(s)$ is assumed to be the input and $E(s)$ is assumed to be the output. The transfer function between these two points is given by

$$\frac{E(s)}{U(s)} = \frac{2.2(1 + T_2s)}{(1 + T_2s)(1 + 0.1s)(1 + 0.4s)(1 + 1.2s) + 2.2K(1 + T_1s)}. \tag{7.60}$$

Setting $U(s) = 1/s$, we obtain the expression for the Laplace transform of the error, $E(s)$, as

$$E(s) = \frac{2.2(1 + T_2s)}{s[(1 + T_2s)(1 + 0.1s)(1 + 0.4s)(1 + 1.2s) + 2.2K(1 + T_1s)]}. \tag{7.61}$$

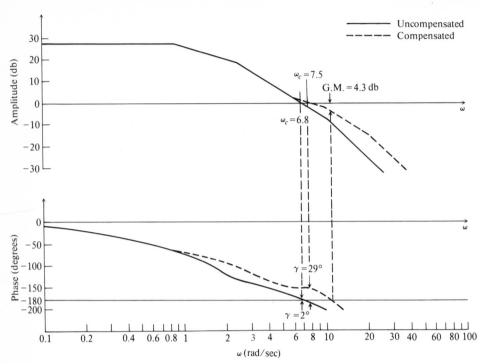

Fig. 7.17   Compensation of third-order system where

$$KG_0(s) = \frac{21}{(1 + 0.1s)(1 + 0.4s)(1 + 1.2s)} \quad \text{and} \quad G_c(s) = \frac{1 + 0.167s}{1 + 0.05s}.$$

The required value of K can be obtained by applying the final-value theorem to $E(s)$ and setting the result equal to 0.1 unit. We find

$$K = 9.55. \tag{7.62}$$

Let us next determine the compensating network required to achieve a phase margin of 30° and a gain margin of 5 db. The transfer function of the uncompensated system, with $K = 9.55$, is given by

$$KG_0(s) = \frac{21}{(1 + 0.1s)(1 + 0.4s)(1 + 1.2s)}. \tag{7.63}$$

Its Bode diagram, which is drawn in Fig. 7.17, indicates a phase margin of 2° and a gain margin of 1.7 db. We wish to increase the phase margin by about 30°. From Table 7.1, it can be seen that a time-constant ratio of 4 will result in a phase shift of approximately 37° at a frequency which was given by Eq. (7.7):

$$\omega_{\max} = 1/(T\sqrt{\alpha}). \tag{7.64}$$

We must be careful when using this approach, since we actually desire to have this phase shift occur at the crossover frequency in order to achieve the specified phase margin. However, specifying α, T, and ω_{max} does not ensure that ω_{max} will occur at the crossover frequency. Therefore, a trial-and-error procedure is followed. A phase-lead network corresponding to

$$\frac{1 + 0.167s}{1 + 0.05s} \tag{7.65}$$

results in a phase margin of 29° and a gain margin of 4.3 db. This is close enough to the desired result for all practical purposes. It is interesting to note that this lead network gives its maximum phase lead of about 32° at a frequency of 11 rad/sec. This accounts for the fact that we only achieved a phase margin of 29° at a crossover frequency of 7.5 rad/sec.

E. Application of the Bode Diagram to a Two-Loop System The concluding problem we consider using the Bode-diagram approach consists of designing the feedback control system illustrated in Fig. 7.18(a). For this particular system, we desire that the steady-state error resulting from a velocity input of 110 rad/sec be equal to 0.25 rad. The uncompensated open-loop transfer function $G_0(s)$ is given by

$$G_0(s) = K_v/(s(1 + T_m s)), \tag{7.66}$$

where

$$K_v = \text{velocity constant,}$$

$$T_m = \text{motor time constant} = 0.025 \text{ sec.}$$

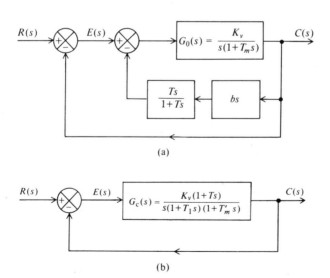

(a)

(b)

Fig. 7.18 (a) Use of rate feedback in cascade with a high-pass filter in order to compensate a feedback control system. (b) Equivalent block diagram for the system shown in Fig. 7.18(a).

This transfer function consists of an amplifier, positioning motor, gear train, and load. In order to achieve the required accuracy, K_v must equal

$$K_v = \frac{\omega}{\text{error}} = \frac{110 \text{ rad/sec}}{0.25 \text{ rad}} = 440/\text{sec}. \tag{7.67}$$

We want a phase margin of approximately 55° for this system. This will be achieved by means of minor-loop rate feedback compensation which is cascaded with a simple RC high-pass filter (lead network) in order not to increase the steady-state response error for velocity inputs.

The open-loop frequency response we must plot on the Bode diagram is that obtained with the minor rate loop closed and the outer position loop opened. Therefore, we are interested in obtaining the equivalent transfer function between $E(s)$ and $C(s)$. This is easily found to be

$$\frac{C(s)}{E(s)} = \frac{K_v(1 + Ts)}{s[(1 + T_m s)(1 + Ts) + K_v bTs]}. \tag{7.68}$$

Expanding the denominator of Eq. (7.68), we obtain the expression

$$s[T_m Ts^2 + (T_m + T + K_v bT)s + 1]. \tag{7.69}$$

This expression can be put into the form

$$s[(1 + T'_m s)(1 + T_1 s)]. \tag{7.70}$$

by defining the time constants T'_m and T_1 as

$$T'_m = \frac{T_m}{T_1} T \tag{7.71}$$

and

$$T_1 = -T'_m + T_m + (1 + K_v b)T. \tag{7.72}$$

Therefore, we may redraw Fig. 7.18(a) as shown in Fig. 7.18(b). For any set of values for T_m, T, K_v, and b, we can derive T'_m and T_1 by solving the simultaneous equations (7.71) and (7.72). Another approach is to choose T and T_1 from the Bode diagram which meets the specified phase margin and solve for the required rate feedback constant b.

The procedure we follow when compensating this system is to draw the Bode diagram for the uncompensated system in accordance with Eq. (7.66) and then fit the characteristics of the compensated system in accordance with

$$G_c(s) = \frac{C(s)}{E(s)} = \frac{K_v(1 + Ts)}{s(1 + T'_m s)(1 + T_1 s)} \tag{7.73}$$

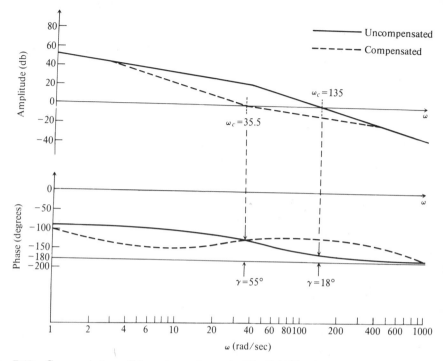

Fig. 7.19 Compensation of the system shown in Fig. 7.18(b) where

$$G_0(s) = \frac{440}{s(1 + 0.025s)} \quad \text{and} \quad G_c(s) = \frac{440(1 + 0.033s)}{s(1 + 0.33s)(1 + 0.0025s)}.$$

until a phase margin of 55° is achieved. The compensated characteristics will then determine T and T_1, from which T'_m and the rate feedback constant b can be determined. Figure 7.19 illustrates the Bode diagram of the uncompensated and compensated systems. Values of

$$T_1 = 0.33, \tag{7.74}$$
$$T = 0.033, \tag{7.75}$$

and

$$T'_m = \frac{T_m}{T_1} T = 0.0025 \quad \text{[from Eq. (7.71)]} \tag{7.76}$$

results in a phase margin of 55°. From Eq. (7.72), the corresponding value of rate feedback constant b is 0.0186.

The type of compensation just illustrated is used quite frequently in practice. In order to really understand what is actually happening, it is important to examine the Bode diagram of Fig. 7.19 closely. The net effect of the minor-loop rate feedback has been to move the equivalent motor break frequency from $1/T_m$ to $1/T'_m$ by the ratio given in Eq. (7.71). This technique is used quite frequently to compensate power servos. The net effect of the phase-lead network, in the minor-loop feedback path, is to appear as a phase lag, for the equivalent open-loop characteristics of Fig. 7.18(b). This can be easily understood since we effectively see the reciprocal of the feedback element when looking into a closed-loop system which has an open-loop gain much greater than unity. This is a very important fact that can be utilized to approximate the Bode diagram in preliminary designs. This point is now expanded upon in the following section.

7.6 APPROXIMATE METHODS FOR PRELIMINARY COMPENSATION DESIGN UTILIZING THE BODE DIAGRAM

Having presented detailed compensation methods, let us next focus our attention on approximate methods for obtaining a first cut at compensation utilizing the Bode diagram [9–11]. Although the procedures presented in this section are approximate, it is generally adequate for the preliminary stage of design. Before the system design is completed, however, the exact attenuation and phase curves should be drawn as indicated previously. The practice of utilizing approximate methods for preliminary design is generally employed as a convenience in obtaining significant time constants, gains, and phase characteristics required for a design.

A. Approximate Closed-Loop Response from the Bode Diagram In order to develop the concept of this approach, let us consider the feedback system illustrated in Fig. 7.20. The closed-loop transfer function of this system is given by

$$\frac{C(s)}{R(s)} = \frac{G(s)}{1 + G(s)H(s)}. \tag{7.77}$$

Let us modify this equation into the following more convenient form:

$$\frac{C(s)}{R(s)} = \frac{1}{H(s)}\left[\frac{G(s)H(s)}{1 + G(s)H(s)}\right]. \tag{7.78}$$

As shown previously, the magnitude (in db) of $G(s)H(s)$ can be approximated by straight lines of constant slope when plotted against frequency on a log scale. Therefore, it appears reasonable that the term $1 + G(s)H(s)$ in Eq. (7.78) may also be

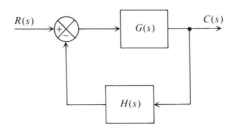

Fig. 7.20   A feedback system.

dealt with in a similar approximate manner as follows:

$$1 + G(s)H(s) \approx 1 \qquad \text{for} \qquad |G(s)H(s)| < 1, \qquad (7.79)$$

$$1 + G(s)H(s) \approx G(s)H(s) \qquad \text{for} \qquad |G(s)H(s)| > 1. \qquad (7.80)$$

Substituting these approximations into Eq. (7.78), we find that

$$\frac{C(s)}{R(s)} \approx G(s) \qquad \text{for} \qquad |G(s)H(s)| < 1, \qquad (7.81)$$

$$\frac{C(s)}{R(s)} \approx \frac{1}{H(s)} \qquad \text{for} \qquad |G(s)H(s)| > 1. \qquad (7.82)$$

The approximations of Eqs. (7.81) and (7.82) are very useful as a first cut in the preliminary design of a control system. However, it is important to point out that they are approximate relationships, and are subject to error particularly at those frequencies where $G(s)H(s) = 1$. However, the amount by which the approximations are in error can be calculated, and this correction can then be applied to correct the approximate value. The resultant corrected value is then an exact solution.

The technique of applying the approximation of Eqs. (7.81) and (7.82) can best be illustrated through an example. Consider the feedback system of Fig. 7.20, where

$$G(s) = \frac{K}{1 + Ts} \qquad (7.83)$$

and

$$H(s) = 1. \qquad (7.84)$$

This example will illustrate how feedback, around an element containing a time constant, reduces the time constant of that element. Substituting Eqs. (7.83) and (7.84) into Eq. (7.77), the resultant system transfer function is given by

$$\frac{C(s)}{R(s)} = \frac{K/(1 + Ts)}{1 + K/(1 + Ts)} = \frac{K}{1 + K} \frac{1}{1 + Ts/(1 + K)}. \qquad (7.85)$$

For purposes of illustration, it is assumed that

$$K = 10, \qquad (7.86)$$

$$T = 1 \qquad (7.87)$$

in this example. For these values, Fig. 7.21 illustrates the straight-line approximations to the equations for $C(s)/R(s)$, $G(s)$, and $G(s)H(s)$. The dotted curve representing $G(s)$ (and $G(s)H(s)$ since $H(s) = 1$) has a gain of 10 (20 db) for frequencies lower than $\omega = 1/T = 1$ rad/sec. At frequencies higher than $\omega = 1$ rad/sec, $G(s)$ and $G(s)H(s)$ have an attenuation of 20 db/decade. At $\omega = \omega_n = K/T = \frac{10}{1} = 10$, $G(s)$ and $G(s)H(s)$ approximately equal 0 db. Equation (7.82) indicates that for frequencies lower than $\omega_n = 10$, the system transfer function $C(s)/R(s)$ equals

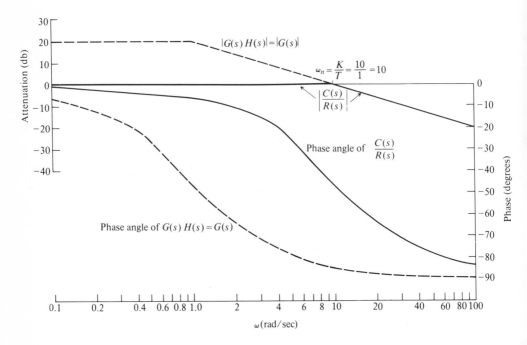

Fig. 7.21 Open- and closed-loop gain and phase characteristics for the system of Fig. 7.20 where $G(s) = 10/(1 + s)$ and $H(s) = 1$.

$1/H(s) = 1$ and is shown by the solid line at 0 db. For frequencies greater than ω_n, Eq. (7.81) indicates that $C(s)/R(s)$ is equal to $G(s)$ as shown. Figure 7.21 illustrates very clearly that the use of unity gain feedback around a single time-constant element results in a reduction in its time constant. This admirable characteristic is achieved at the expense of a loss of gain.

It is important to recognize that the results obtained for the closed-loop transfer function are approximate. For example, Eq. (7.85) indicates that the gain is actually $K/(1 + K) = 10/(1 + 10) = 0.909$ instead of 1 for $\omega < \omega_n$, and the closed-loop time constant is $T/(1 + K) = 1/(1 + 10) = 0.0909$ instead of $T/K = \frac{1}{10} = 0.1$. Note that these errors decrease as the gain K increases. Generally, K is quite large and the error between the approximate and exact curves is very small.

The form of Eq. (7.78) is very well suited for determining the value of $C(s)/R(s)$ when $|G(s)H(s)| > 1$, with the bracketed term representing the difference between the approximate and exact curves. In order to determine a general analytic expression for finding the difference between the approximate and exact curves when $|G(s)H(s)| < 1$, let us reconsider Eq. (7.77) and rewrite it as follows:

$$\frac{C(s)}{R(s)} = G(s) \frac{1}{1 + G(s)H(s)}. \tag{7.88}$$

Therefore,

$$\frac{C(s)}{R(s)} = G(s)\left[\frac{1/[G(s)H(s)]}{1 + 1/[G(s)H(s)]}\right]. \tag{7.89}$$

For those frequencies where $|G(s)H(s)| < 1$, the bracketed term of Eq. (7.89) represents the error between the approximate solution

$$\frac{C(s)}{R(s)} \approx G(s) \qquad \text{for} \qquad |G(s)H(s)| < 1 \tag{7.90}$$

and the exact solution. Another way of looking at this is that the bracketed term represents a correction factor which can be used to correct the results of the approximation given by Eq. (7.81).

B. The Straight-Line Phase-Shift Approximation For preliminary design purposes, it is usually sufficient to obtain quantitative information regarding phase shift without resorting to the exact but tedious method of Bode's phase integral. We have found it convenient to represent the amplitude characteristics by a straight-line approximation, and can utilize a similar technique for the phase-shift function. In order to introduce the method, let us consider the phase shift due to the transfer function given by

$$G(j\omega) = 1 + j\left(\frac{\omega}{\omega_1}\right). \tag{7.91}$$

Figure 7.22(a) illustrates the straight-line amplitude approximation and Fig. 7.22(b) illustrates the exact phase shift and its straight-line approximation. The straight-line approximate phase shift has been constructed by drawing a straight line tangent to the actual curve at $\omega = \omega_1$.

In order to construct these straight-line phase-shift approximations by inspection, the dependence of ω_A and ω_B on ω_1 must be known. This can be obtained by first considering the slope of the actual curve at $\omega = \omega_1$ as follows:

$$\phi = \tan^{-1}\frac{\omega}{\omega_1}. \tag{7.92}$$

The slope at $\omega = \omega_1$ can be obtained by differentiating this expression:

$$\frac{d\phi}{d(\ln\omega)}\bigg]_{\omega=\omega_1} = \frac{d\phi}{d\omega}\frac{d\omega}{d(\ln\omega)}\bigg]_{\omega=\omega_1} = \frac{\omega/\omega_1}{1 + (\omega/\omega_1)^2}\bigg]_{\omega=\omega_1} = \frac{1}{2}. \tag{7.93}$$

Knowing the slope at $\omega = \omega_1$, the intercepts at ω_A and ω_B can be obtained from

$$\frac{\pi/4}{\ln\omega_1 - \ln\omega_A} = \frac{1}{2}, \tag{7.94}$$

which gives

$$\frac{\pi}{4} = \tfrac{1}{2}\ln\frac{\omega_1}{\omega_A}. \tag{7.95}$$

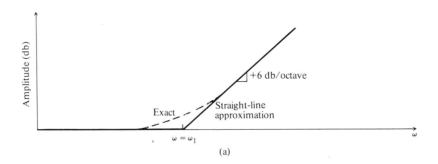

(a)

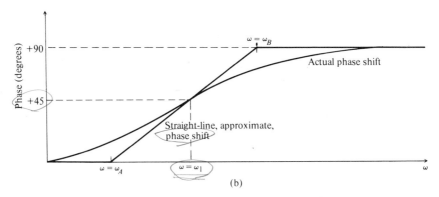

(b)

Fig. 7.22   Straight-line, approximate, phase-shift method for a lead term.

Therefore,

$$\frac{\omega_1}{\omega_A} = \frac{\omega_B}{\omega_1} = e^{\pi/2} = 4.81. \tag{7.96}$$

It is interesting to note that if a number other than e were chosen for the logarithmic base, the slope in Eq. (7.93) would change, but the frequency ratio given by Eq. (7.96) would remain the same.

Based on this result, complicated approximate straight-line phase-shift curves can be obtained. Figure 7.23 illustrates the application of this approach for a transfer function given by

$$G(s) = \frac{(1 + 10s)(1 + 0.01s)^2}{(1 + s)^2(1 + 0.001s)} \tag{7.97}$$

and compares the approximate straight-line phase-shift curve with the exact phase-shift curve. Observe that the errors obtained by using the approximate phase curve are relatively larger than the corresponding errors of the amplitude approximation. However, the use of this approximation greatly aids the control-system engineer in obtaining a first cut at preliminary design.

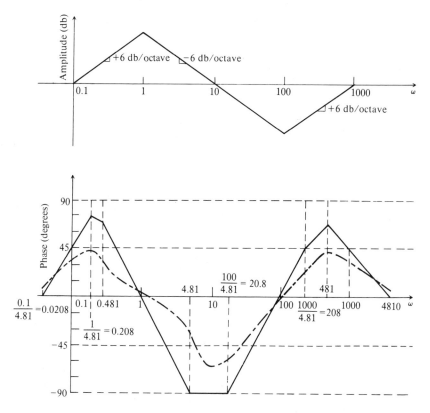

Fig. 7.23 Approximate straight-line phase-shift (——) and exact phase-shift (— — —) characteristics of

$$G(s) = \frac{(1 + 10s)(1 + 0.01s)^2}{(1 + s)^2(1 + 0.001s)}.$$

The example just presented contained roots that were all real. What happens if the transfer function contains underdamped quadratic factors? In order to analyze this situation, let us consider the following normalized quadratic term:

$$G(j\omega) = \frac{1}{(j\omega/\omega_n)^2 + 2\zeta(j\omega/\omega_n) + 1}. \qquad (7.98)$$

It was shown in Section 6.5 that the amplitude and phase-shift terms were given respectively by (see Eq. 6.74):

$$-20 \log_{10}\left[\left(\frac{2\zeta\omega}{\omega_n}\right)^2 + \left(1 - \frac{\omega^2}{\omega_n^2}\right)^2\right]^{1/2}, \qquad (7.99)$$

$$\phi = -\tan^{-1}\frac{2\zeta\omega_n\omega}{\omega_n^2 - \omega^2}. \qquad (7.100)$$

Let us focus attention on the phase-shift term. Differentiation of Eq. (7.100) results in the slope at $\omega = \omega_n$ being given by

$$\text{slope} = \frac{d\phi}{d(\ln \omega)} = \frac{1}{\zeta}. \tag{7.101}$$

As before, the two intercepts ω_A and ω_B can be found as follows:

$$\frac{\pi/2}{\ln \omega_n - \ln \omega_A} = \frac{1}{\zeta}. \tag{7.102}$$

Therefore,

$$\frac{\omega_n}{\omega_A} = \frac{\omega_B}{\omega_n} = e^{(\pi/2)\zeta} = 4.81^{\zeta}. \tag{7.103}$$

Exact phase characteristics corresponding to Eq. (7.102) and its straight-line approximation, based on the relationship of Eq. (7.103), are illustrated in Fig. 7.24. In addition, the exact amplitude characteristics of Eq. (7.99) and its straight-line approximation are also illustrated. Observe from Eq. (7.103) that as ζ approaches zero, the two frequency ratios decrease and approach unity in the limits. This

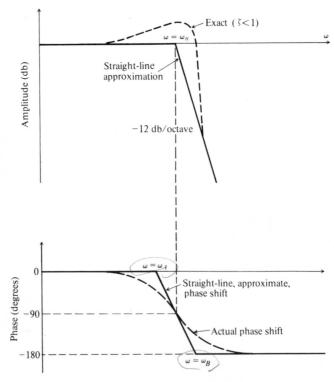

Fig. 7.24 Straight-line, approximate, phase shift method for a quadratic lag factor ($\zeta < 1$).

certainly agrees with the exact phase characteristics of a quadratic phase lag illustrated in Fig. 6.17. On the other hand, when $\zeta = 1$, the roots become real and Eq. (7.103) reduces to Eq. (7.96).

The reader is again reminded that the straight-line phase-shift approximation is not exact. It should only be used in order to obtain a first cut for preliminary design work. It should also be noted that the errors of the straight-line phase shift approximation are generally larger than the corresponding errors of the amplitude approximation. For this reason, the phase-shift approximation is not too widely used. In the final design, the actual phase-shift characteristics must be employed as previously illustrated.

7.7 DESIGN UTILIZING THE NICHOLS CHART

The Nichols chart method has been developed in Section 6.7. We demonstrated in that section how one could obtain the closed-loop frequency response of a feedback control system by superimposing the open-loop gain–phase characteristics onto the Nichols chart. Specifically, we obtained the closed-loop frequency response of the system shown in Fig. 6.33. The intersections of the open-loop gain–phase characteristics and the Nichols closed-loop gain characteristics were shown in Fig. 6.35. The resulting closed-loop frequency response was illustrated in Fig. 6.36. It indicated a maximum value of peaking, M_p, of 6.8 db (2.2) and the frequency at which it occurred, ω_p, was 10 rad/sec. We indicated in Sections 6.7 and 6.8 that a value of $M_p = 6.8$ db (2.2) does not represent a good design. This section demonstrates how the control engineer may use the Nichols chart in order to achieve a specified performance.

Let us assume for this problem that an acceptable value of M_p is 1.3 (2.3 db). This may be achieved by adding a phase-lag or phase-lead network in cascade with the forward-loop transfer function $G(s)$. A phase-lag network is used for this problem although a solution can be found as easily using a phase-lead network. We shall demonstrate that for an M_p of 2.3 db the object is to modify the gain–phase characteristics on the Nichols chart so that it is just tangent to the 2.3 db locus and does not enter it. By restricting the gain–phase characteristics to areas external to the $M = 2.3$ db locus, we will have limited M_p to 2.3 db, since the interior of this locus represents values of M greater than 2.3 db.

Studying the characteristics of Fig. 6.35, we see that relatively large magnitudes of $G(j\omega)$ exist for $\omega < 3$ rad/sec. Therefore, it is not desirable to shift these magnitudes inside the $M_p = 2.3$ db curve. In addition, it is desirable to attenuate $G(j\omega)$ by a factor of about 3 in the range of frequencies of $\omega = 5$ to 12 rad/sec. A phase-lag network, $(1 + Ts)/(1 + \alpha Ts)$, whose factor α equals 3 will achieve this if $\omega_{\max} = 1/(T\sqrt{\alpha})$ is chosen at about 1 rad/sec. Solving for αT and T, we get $\alpha T = 1.74$ and $T = 0.58$.

In order to obtain the gain–phase characteristics of the open-loop system, the Bode diagram is first drawn as indicated in Fig. 7.25. Then for each value of ω the

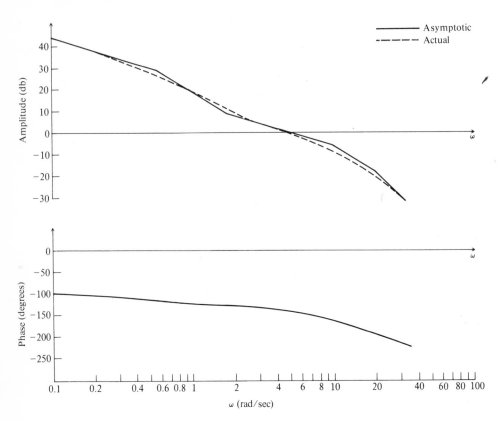

Fig. 7.25 Bode diagram for the compensated system of Fig. 6.33 where

$$G(s)G_c(s)H(s) = \frac{11.7(1 + 0.58s)}{s(1 + 0.05s)(1 + 0.1s)(1 + 1.74s)}.$$

magnitude and phase of the open-loop compensated characteristics are plotted onto a Nichols chart as shown in Figure 7.26. Since the open-loop gain–phase characteristics are just tangent to the constant-magnitude locus corresponding to $M = 2.3$ db (1.3), we have achieved our goal. Notice that we have shifted $\omega = 1$ rad/sec by about $-35°$, but this does not increase M_p.

7.8 DESIGN UTILIZING THE ROOT LOCUS

The root-locus technique has been developed previously in Sections 6.9 and 6.10. It is a very helpful tool which the control engineer can use in order to study the variation of gain, system parameters, and effect of compensation. We demonstrated in Section 6.9 the migration of poles in the complex plane as the gain of the system was varied from zero to infinity. Specifically, we obtained the root locus for a

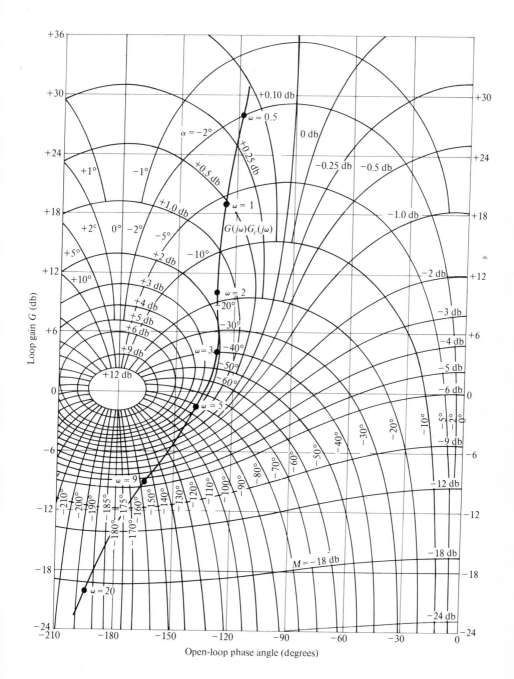

Fig. 7.26 Compensation of the system shown in Fig. 6.33 for $M = 2.3$ db (1.3).

feedback system where

$$G(s)H(s) = \frac{K}{(s + 1)(s - 1)(s + 4)^2}. \qquad (7.104)$$

This was illustrated in Fig. 6.44. An analysis of the root locus for this system indicated that it was always unstable, since at least one of the roots of the characteristic equation always occurred in the right half-plane. This section demonstrates how this system may be compensated by means of a lead network. This problem is followed by considering lag-network compensation for the system illustrated in Fig. 6.40. In addition, we shall demonstrate how the control engineer may determine the transient response of the compensated systems in order to meet certain specifications.

A. Lead Network Compensation Let us attempt to stabilize the system of Fig. 6.44 by means of a phase-lead network in cascade with the forward-loop transfer function. The form of its transfer function is given by

$$G_c(s) = \frac{s + \alpha}{s + \beta}. \qquad (7.105)$$

We assume that the effect of the pole introduced by the phase-lead network has a negligible effect compared with its zero. Therefore, we assume that the transfer function of the cascaded phase-lead network can be approximated by the simple expression:

$$G_c(s) \approx s + \alpha. \qquad (7.106)$$

The resulting value of $G_c(s)G(s)H(s)$, which is to be examined on the root locus, is given by

$$G_c(s)G(s)H(s) = \frac{K(s + \alpha)}{(s + 1)(s - 1)(s + 4)^2}. \qquad (7.107)$$

We wish to investigate the effect on stability of a variation in α as follows:

Case A: $\alpha = 0.5$ Case D: $\alpha = 4$

Case B: $\alpha = 1$ Case E: $\alpha = 6$

Case C: $\alpha = 2$

The resulting root loci for all these cases are presented next. It is important to emphasize at this point that, although we make use of most of the analytic tools developed in Section 6.9, we do not use all of them. This omission is due to the fact that some of the analytic techniques developed are too complex to use for higher-order systems. For example, it is very tedious to determine the value of the gain K along the root loci utilizing the relationship given by Eq. (6.107). Fortunately, this can be obtained much more easily with the Spirule. The approach we take in presenting the resulting root loci for this problem is to outline the results of the eleven rules developed previously in Section 6.9 and use the Spirule wherever it is helpful. In addition, the

values of gain obtained using the Spirule which are pertinent for an intelligent evaluation of the problem will be indicated. It is our feeling that this dual approach is the
best procedure when using the root-locus method.

Case A. $\alpha = 0.5$. See Fig. 7.27 for the root-locus sketch.

Rule 1. There are four separate loci since the characteristic equation,

$$1 + G_c(s)G(s)H(s) = 0,$$

is a fourth-order equation.

Rule 2. The root locus starts ($K = 0$) from the poles located at 1, -1, and a double
pole at -4. One branch terminates ($K = \infty$) at the zero located at $-\alpha$ and three
branches terminate at zeros located at infinity.

Rule 3. Complex portions of the root locus occur in complex-conjugate pairs.

Rule 4. The portions of the real axis between 1 to $-\alpha$; -1 to -4; and -4 to ∞
are part of the root locus.

Rule 5. The branches approach infinity as K becomes large at angles given by

$$\alpha_0 = \pm \frac{\pi}{3} = \pm 60°,$$

and

$$\alpha_1 = \pm \frac{3\pi}{3} = \pm 180°.$$

Rule 6. The intersection of the asymptotic lines and the real axis occur at

$$s_r = \frac{-8 - (-0.5)}{4 - 1} = -2.5.$$

Rule 7. Using the Spirule, we found the point of breakaway from the real axis to
occur at approximately -1.75.

Rule 8. Using the Spirule, we found the intersection of the root locus and the
imaginary axis to occur at approximately $s = \pm j3.3$, where the gain is 104, and at the
origin, where the gain is 32.

Rule 9. This rule does not apply here.

Rule 10. This rule shows that as certain of the loci turn to the right, others turn to the
left to ensure that the sum of the roots are a constant.

Rule 11. A Spirule was used to obtain the gains along the root locus.

The resulting root locus indicates that the system is stable when $32 < K < 104$.

*Case B.** $\alpha = 1$. See Fig. 7.27 for the root-locus sketch. A zero at $\alpha = -1$ cancels

---

* Only the results of applying the eleven rules for constructing the root locus are indicated
for Cases B–E.

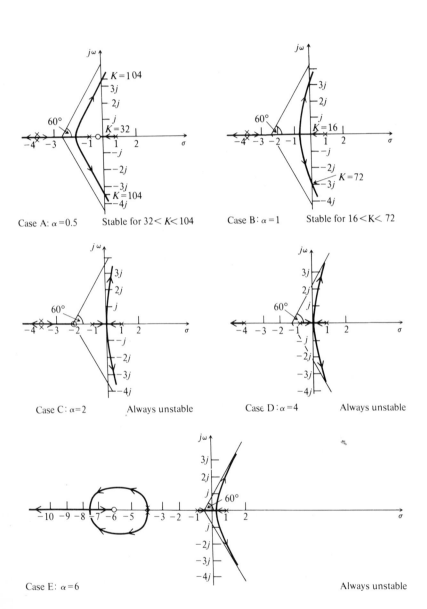

Case A: $\alpha = 0.5$ Stable for $32 < K < 104$

Case B: $\alpha = 1$ Stable for $16 < K < 72$

Case C: $\alpha = 2$ Always unstable

Case D: $\alpha = 4$ Always unstable

Case E: $\alpha = 6$ Always unstable

Fig. 7.27 Compensation of the root locus shown in Fig. 6.44 using a cascaded phase-lead network where $G_c(s) = (s + \alpha)$.

the pole at -1. The resulting root-locus sketch indicates that this system is stable when $16 < K < 72$.

Case C. $\alpha = 2$. See Fig. 7.27 for the root-locus sketch. The resulting root-locus sketch indicates that the system is always unstable since at least one of the roots of the characteristic equation always occurs in the right half-plane except for the condition where two poles exist at the origin.

Case D. $\alpha = 4$. See Fig. 7.27 for the root-locus sketch. A zero at $\alpha = 4$ cancels one of the poles located at -4. The resulting root-locus sketch indicates that the system is always unstable since at least one of the roots of the characteristic equation always occurs in the right half-plane.

Case E. $\alpha = 6$. See Fig. 7.27 for the root-locus sketch. The resulting root-locus sketch indicates that the system is always unstable since at least one of the roots of the characteristic equation always occurs in the right half-plane.

The interpretation of Fig. 7.27 is quite interesting and revealing. It indicates that the exact location of the zero is very important from a stability viewpoint. Cases A and B were the only configurations which had regions of stability. As a matter of fact, the closer the zero lies to the imaginary axis, the greater is its stabilizing effect. This point is very important. Since Case A resulted in larger values of gain, it would result in a more accurate system and is, therefore, preferred to Case B.

B. Determination of the Transient Response The transient response of the system can also be obtained by reasoning along these lines: The transient performance is often dominated by the pair of complex-conjugate poles located closest to the origin. This occurs when the other poles are far to the left of the dominant poles, or the other poles are near a zero. The resulting transient components due to these other poles are small under these conditions and diminish rapidly. For this case, the poles closest to the origin are conventionally referred to as the dominant poles. From the discussion of Section 4.2, the expression associated with these complex poles can be given by the following expression (see Eq. 4.3):

$$\frac{C(s)}{R(s)} = \frac{\omega_n^2}{s^2 + 2\zeta\omega_n s + \omega_n^2} \qquad (7.108)$$

where $\zeta =$ damping factor, and $\omega_n =$ natural resonant frequency. We found in Section 4.2 that the transient response to a unit step input, for $\zeta < 1$, is given by the following expression (see Eq. 4.24):

$$c(t) = 1 + \frac{e^{-\zeta\omega_n t}}{\sqrt{1 - \zeta^2}} \sin(\omega_n\sqrt{1 - \zeta^2}\, t - \alpha), \qquad (7.109)$$

where

$$\alpha = \cos^{-1}(-\zeta).$$

Figure 4.3 illustrated the complex-plane location of these dominant poles. The values derived for the time to the first peak (see Eq. 4.28) and maximum percent overshoot (see Eq. 4.32) are specifically for a second-order system whose closed-loop transfer function is given by Eq. (7.108). These quantities change if other closed-loop poles and zeros exist in addition to the dominant complex pair. However, if the other poles and zeros are at least twice as far from the origin as the dominant pair, the approximation gives reasonable results. Expressions for time to the first peak and percent overshoot, which consider other poles and zeros and give more accurate results, can be derived [6]. These expressions assume that

a) Other poles are far to the left of the dominant poles, so that the amplitude of transients due to these other poles is small.

b) Poles which are not far to the left of the dominant poles are near a zero so that the transient amplitude due to such poles is small.

The approximate expressions, for unity feedback systems, are given by

$$[t_p]_{\text{modified}} = \frac{1}{\sqrt{1 - \zeta^2}\,\omega_n}\left[\frac{\pi}{2} - \sum \phi_z + \sum \phi_p\right] \qquad (7.110)$$

where

$$\sum \phi_z = \text{sum of the angles from the zeros of } C/R \text{ to one of the dominant poles,}$$

$$\sum \phi_p = \text{sum of the angles from the poles of } C/R \text{ to one of the dominant poles,}$$

and maximum percent overshoot

$$= \frac{\begin{bmatrix}\text{product of distances from all}\\ \text{poles of } C/R \text{ to origin,}\\ \text{excluding distances of two}\\ \text{dominant poles from origin}\end{bmatrix}\begin{bmatrix}\text{product of distances from}\\ \text{all zeros of } C/R \text{ to}\\ \text{dominant pole } P_0\end{bmatrix}}{\begin{bmatrix}\text{product of distances from all}\\ \text{other poles of } C/R \text{ to}\\ \text{dominant pole } P_0 \text{ excluding}\\ \text{distance between dominant}\\ \text{poles}\end{bmatrix}\begin{bmatrix}\text{product of distances from}\\ \text{all zeros of } C/R \text{ to}\\ \text{origin}\end{bmatrix}}\, e^{-\zeta\omega_n t_p} \times 100\%.$$

The expression for maximum percent overshoot can be stated symbolically as

$$\text{max. \% overshoot} = \left[\left(\frac{P_1}{|P_1 - P_0|}\right)\left(\frac{P_2}{|P_2 - P_0|}\right)\left(\frac{P_3}{|P_3 - P_0|}\right)\cdots\right]$$
$$\times \left[\left(\frac{|Z_1 - P_0|}{Z_1}\right)\left(\frac{|Z_2 - P_0|}{Z_2}\right)\left(\frac{|Z_3 - P_0|}{Z_3}\right)\cdots\right]e^{-\zeta\omega_n t_p} \times 100\%, \quad (7.111)$$

where the first set of brackets represents the product of the ratios of the values of s at which poles occur to their absolute distances from the dominant pole. The second set of brackets represents the product of the ratios of the absolute distances of the zeros from the dominant pole and the values of s at which the zeros occur. Let us next apply these expressions in the following design problem.

C. Lag Network Compensation and Overall System Performance The concluding design problem we consider using the root locus consists of employing cascaded phase-lag compensation in order to improve the steady-state performance of a feedback control system. The object is to increase its gain while maintaining a good dynamic response. Specifically, we consider the system whose root locus was illustrated in Fig. 6.40. For this system

$$G(s)H(s) = \frac{K}{s(s + 4)(s + 5)}. \tag{7.112}$$

The root locus of Fig. 6.40 indicated that the system was stable when $0 < K < 180$. Let us assume that a damping factor of 0.707 achieves a desirable dynamic response for this system. In addition, we must maintain a velocity constant K_v of 30 in order to meet specified accuracy requirements. Analyzing this problem, by means of the root locus, we can find the value of K which will give the required damping factor. For example, the redrawn version of Fig. 6.40 shown in Fig. 7.28 indicates that a $K = 23.6$ will result in a damping factor of 0.707. This value of gain, obtained using the Spirule, does not maintain the required velocity constant of 30. The actual value of K_v resulting from $K = 23.6$ is

$$K_v = \lim_{s \to 0} sG(s)H(s) = \lim_{s \to 0} \frac{s(23.6)}{s(s + 4)(s + 5)} = 1.18/\text{sec.} \tag{7.113}$$

It is therefore clear that we cannot just increase the gain K to a value that produces the required velocity constant, since this would decrease the damping factor and adversely affect the transient response or cause the system to become unstable. Using the root locus for a solution, we show how these two conflicting factors can be resolved.

Let us assume that the phase-lag compensator is of the form

$$G_c(s) = \frac{s + n\alpha}{s + \alpha}, \tag{7.114}$$

where n = ratio of the break frequencies. Equation (7.114) indicates that this compensator provides a low-frequency gain in addition to the phase lag. The open-loop transfer function of the compensated system is given by

$$G_c(s)G(s)H(s) = \frac{K}{s(s + 4)(s + 5)}\left(\frac{s + n\alpha}{s + \alpha}\right). \tag{7.115}$$

In general, the distances of $n\alpha$ and α from the origin in the s-plane are chosen to be small compared with the distances of the other zeros and poles of the uncompensated open-loop transfer function, so that the added pole and zero of the compensator will not contribute significant phase lag in the vicinity of the closed-loop bandwidth (crossover frequency). This result is quite clear from a study of the Bode diagram.

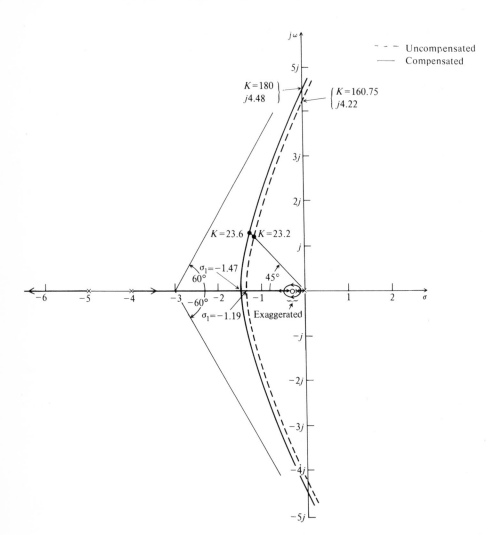

Fig. 7.28 Compensation of the root locus shown in Fig. 6.40 using a cascaded phase-lag network.

Certainly we do not wish to add the phase-lag contribution in the vicinity of the crossover frequency. Therefore, the combination of pole and zero will appear quite close together on the root locus and very close to the origin. This combination is usually called a dipole.

In order to complete the design, α will be chosen as 0.01 and n will be chosen, using the following derivation, which will achieve a $K_v = 30$:

$$K_v = 30 = \lim_{s \to 0} sG_c(s)G(s)H(s). \tag{7.116}$$

Substituting Eq. (7.115) into Eq. (7.116), we obtain

$$30 = \lim_{s \to 0} s \left(\frac{K}{s(s+4)(s+5)} \right) \left(\frac{s + n\alpha}{s + \alpha} \right), \tag{7.117}$$

or

$$30 = \frac{Kn}{(4)(5)}. \tag{7.118}$$

Since we desire that $K = 23.6$ from a transient viewpoint, we must have

$$n = \frac{30(4)(5)}{23.6} = 25.4. \tag{7.119}$$

The completed root locus for the compensated system whose open-loop transfer function is given by

$$G_c(s)G(s)H(s) = \frac{K}{s(s+4)(s+5)} \left(\frac{s + 0.254}{s + 0.01} \right)$$

must now be determined. Since the dipole is added near the origin, the original root locus is not changed significantly, because the two poles and the zero near the origin tend to merge into a single pole.

Let us next determine the new resulting root locus and analyze the effect of the dipole on it. Specifically, we wish to know whether the new root locus will indeed have a $K_v = 30$. In addition, we would like to determine the transient response of the compensated system. Each of the eleven rules for constructing the root locus will be considered.

Rule 1. There are four separate branches since the characteristic equation,

$$1 + G(s)H(s) = 0,$$

is a fourth-order equation.

Rule 2. The root locus starts ($K = 0$) from the poles located at the origin, -0.01, -4, and -5. One branch terminates ($K = \infty$) at the zero located at -0.254 and the other three branches terminate at zeros which are located at infinity.

Rule 3. Complex portions of the root locus occur in complex-conjugate pairs.

Rule 4. The portions of the real axis between the origin and -0.01, -0.254 and -4, and -5 to $-\infty$ are part of the root locus.

Rule 5. The four branches approach infinity as K becomes large at angles given by

$$\alpha_0 = \pm \frac{\pi}{4 - 1} = \pm 60°,$$

$$\alpha_1 = \pm \frac{3\pi}{4 - 1} = \pm 180°.$$

Rule 6. The intersection of the asymptotic lines and the real axis occurs at

$$s_r = \frac{-9.01 - (-0.254)}{4 - 1} = \frac{-8.756}{3} = -2.92.$$

Rule 7. The point of breakaway from the real axis can be computed from the following equation:

$$(\beta_1 - \beta_2 - \beta_3 - \beta_4 - \beta_5) = (2n + 1)\pi,$$

where

β_1 = angle from the zero at -0.254 to the point s_1 that is located a small distance δ off the positive real axis,

β_2 = angle from the pole at the origin to the point s_1,

β_3 = angle from the pole at -0.01 to the point s_1,

β_4 = angle from the pole at -4 to the point s_1,

β_5 = angle from the pole at -5 to the point s_1.

The equation of the transition of the root locus from the real axis to a point s_1 which is a small distance δ off the axis is given by

$$\left[\left(\pi - \frac{\delta}{\sigma_1 - 0.254}\right) - \left(\pi - \frac{\delta}{\sigma_1}\right) - \left(\pi - \frac{\delta}{\sigma_1 - 0.01}\right)\right.$$

$$\left. - \left(\frac{\delta}{4 - \sigma_1}\right) - \left(\frac{\delta}{5 - \sigma_1}\right)\right] = -\pi.$$

Solving, we obtain $\sigma_1 = 1.19$, which compares with a value of 1.47 obtained from Eq. (6.122) for the uncompensated system.

Rule 8. The intersection of the root locus and the imaginary axis can be determined by applying the Routh-Hurwitz stability criterion to the characteristic equation

$$s(s + 0.01)(s + 4)(s + 5) + K(s + 0.254) = 0,$$

which becomes

$$s^4 + 9.01s^3 + 20.09s^2 + (K + 0.2)s + 0.254K = 0.$$

The resulting Routh-Hurwitz array is given by

s^4	1	20.09	0.254K
s^3	9.01	K + 0.2	
s^2	$20.068 - 0.111K$	0.254K	
s	$\dfrac{-0.111K^2 + 17.8K + 4}{20.068 - 0.111K}$		
s^0	0.254K		

An interesting situation occurs in this Routh-Hurwitz array, since the first terms of the third and fourth rows can go to zero for certain values of gain, K. When the equation

$$20.068 - 0.111K = 0$$

is satisfied, then a possible solution is

$$K_{max} = 186.3.$$

When the equation

$$\frac{-0.111K^2 + 17.8K + 4}{20.068 - 0.111K} = 0$$

is satisfied, then a possible solution is

$$K_{max} = 160.75.$$

Therefore, in order to find which is (or are) valid, let us substitute $s = j\omega$ into the characteristic equation and find out where the root locus crosses the imaginary axis. The result can be separated into a real and imaginary part and written in the following form:

$$\omega^4 - 20.09\omega^2 + 0.254K + j\omega[-9.01\omega^2 + K + 0.2] = 0. \qquad (7.120)$$

For K to be real, the imaginary part of this equation must equal zero. Therefore,

$$j\omega[-9.01\omega^2 + K + 0.2] = 0,$$

$$\omega = \pm\sqrt{\frac{K + 0.2}{9.01}}.$$

Now, to find the value of K which corresponds to this value of crossing of the imaginary axis, let us substitute this value of ω into the real part of Eq. (7.120). The result is the following equation:

$$K^2 - 160.5K - 36.1 = 0.$$

Therefore, we find that

$$K = 160.75$$

is the only possible real value of gain when the root locus crosses the imaginary axis. This analysis indicates that the maximum value of gain, before the system becomes unstable, is 160.75. In Eq. (6.125), we found that the uncompensated system had a maximum allowable gain of 180. Therefore, this result indicates that the dipole has an effect on the maximum allowed gain. The corresponding value of s occurring at the crossing of the imaginary axis is found to be $\pm j4.22$ by substituting K_{max} into the equation for ω. This compares with a value of $s = j4.48$ obtained previously for the uncompensated case (see Eq. 6.126).

Rule 9. This rule does not apply to this problem.

Rule 10. This rule shows that as certain of the loci turn to the right, others turn to the left to ensure that the sum of the roots is a constant.

Rule 11. This rule is quite important to us in this case since we want to determine the value of gain when $\zeta = 0.707$ (the intersection of a line making an angle of $+45°$ with the negative real axis and the dashed curve). For the uncompensated case, we found that $K = 23.6$. The new value can be obtained from the following expression:

$$\left| \frac{K(s + 0.254)}{s(s + 4)(s + 5)(s + 0.01)} \right| = 1.$$

Measuring the distance from the various poles and zero to the point of interest, we obtain

$$\left| \frac{K(1.45)}{(1.7)(3)(3.85)(1.65)} \right| = 1,$$

$$K = 23.2.$$

Therefore, the gain has decreased slightly from 23.6 to 23.2. This will result in a slight reduction of K_v from 30 to 29.5. For this reason, the value of n calculated in Eq. (7.119) should always be increased by about 5 percent to allow for a margin of safety for the inherent reduction in gain due to the addition of the dipole. In this problem, an n of 25.9 will achieve a K_v of 30.

To conclude this problem, we can calculate the value of the time to the first peak and the maximum per cent overshoot from Eqs. (7.110) and (7.111), respectively. These results can then be compared with the values obtained from Eqs. (4.28) and (4.32), which assume that the transient response is completely controlled only by the pair of complex-conjugate poles located closest to the imaginary axis. The time to the first peak of the compensated system can be calculated from Eq. (7.110). In order to use this equation, the location of the other two roots must be determined. Using a Spirule, these were found to be located at -6.45 and -0.26. Therefore,

$$t_p = \frac{1}{(\sqrt{1 - \zeta^2}\, \omega_n)} \left(\frac{\pi}{2} - \Sigma\, \phi_z + \Sigma\, \phi_p \right)$$

where

$\zeta = 0.707,$

$\omega_n = 1.63$ (distance from the origin of the complex plane to the two dominant poles located at $1.15 \pm j1.15$—dashed curve of Fig. 7.28),

$\Sigma\, \phi_z = 141° = 2.47$ rad,

$\Sigma\, \phi_p = 136° + 90° + 14° = 240° = 4.19$ rad.

Substituting these values into the equation for t_p, we obtain

$$t_p = \frac{1}{(\sqrt{1 - 0.707^2}\, 1.63)} (1.57 - 2.47 + 4.19) = 2.86 \text{ sec.}$$

Therefore, the time to the first peak is 2.86 sec. If one simply assumes that the transient response is governed by the pair of complex-conjugate poles located at $1.15 \pm j1.15$ and uses Eq. (4.28) to determine the time to the first peak, then the following is obtained:

$$t_p = \frac{\pi}{\omega_n\sqrt{1 - \zeta^2}} \qquad \text{[from Eq. (4.28)]},$$

$$t_p = \frac{\pi}{1.63\sqrt{1 - 0.707^2}} = \frac{3.14}{1.15} = 2.73 \text{ sec.}$$

Therefore, we see the improved accuracy obtained using Eq. (7.110).

A similar analysis of the uncompensated system, utilizing Eq. (7.110), results in a time to the first peak of 2.81 sec. For this case, the two complex-conjugate poles are located at $-1.2 \pm j1.2$. The third root can be determined analytically from rule 10:

$$\sum p_i = \sum r_j,$$
$$-4 - 5 = -1.2 + j1.2 - 1.2 - j1.2 + r,$$
$$r = -6.6.$$

Therefore,

$$t_p = \frac{1}{(\sqrt{1 - \zeta^2}\,\omega_n)}\left(\frac{\pi}{2} - \sum \phi_z + \sum \phi_p\right),$$

where

$$\zeta = 0.707,$$

$\omega_n = 1.7$ (distance from the origin of the complex plane to the two dominant poles—solid curve of Fig. 7.28),

$$\sum \phi_z = 0,$$
$$\sum \phi_p = 90° + 13° = 103° = 1.8 \text{ rad.}$$

Hence,

$$t_p = \frac{1}{\sqrt{1 - 0.707^2}\,(1.7)}(1.57 + 1.8) = 2.81 \text{ sec.}$$

Observe that the dipole compensation increases the time to the first peak.

The maximum percent overshoot of the compensated system can be obtained from Eq. (7.111). To use this equation, we have to determine the location of the other two roots. As mentioned previously, these were found to be at -6.45 and -0.26. Therefore,

$$\text{maximum percent overshoot} = \frac{P_1 P_2}{(|P_1 - P_0|)(|P_2 - P_0|)}\frac{|Z_1 - P_0|}{Z_1}e^{-\zeta\omega_n t_p} \times 100\%$$

$$= \frac{(0.26)(6.45)}{(1.45)(4.9)}\frac{(1.46)}{(0.254)}e^{-0.707(1.63)2.86} \times 100\%$$

$$= 4.95\%.$$

Therefore, the resulting maximum percent overshoot is only 4.95%. If one simply assumes that the transient response is governed by the pair of complex-conjugate poles located at $1.15 \pm j1.15$ and uses Eq. (4.32) to determine the maximum percent overshoot, then the following is obtained:

$$\text{maximum percent overshoot} = (e^{-\zeta\pi/\sqrt{1-\zeta^2}}) \times 100\%$$

$$= (e^{-0.707\pi/\sqrt{1-0.707^2}}) \times 100\% = 4.3\%.$$

Therefore, we see the increased accuracy obtained using Eq. (7.111).

A similar analysis of the uncompensated system utilizing Eq. (7.111) results in a maximum percent overshoot of 4.48%; for this case, the third root is located at -6.6 as illustrated before:

$$\text{maximum percent overshoot} = \frac{P_1}{|P_1 - P_0|} e^{-\zeta\omega_n t_p} \times 100\%$$

$$= \frac{6.6}{5} e^{-0.707(1.7)(2.81)} \times 100\% = 4.48\%.$$

Observe that the dipole compensation increases the maximum percent overshoot.

The results of the transient analysis indicate that the effect of the dipole is to increase the time to the first peak slightly and to increase the maximum overshoot slightly. Of most importance, the dipole increases the velocity constant greatly, from 1.18 to 30.

One should note that the dominant pole concept is very useful here, the results coming quite close to the more accurate calculation. In fact, due to the ever-present uncertainty of the parameters of the actual system, it is rarely, if ever, necessary to carry out the detailed calculations indicated.

7.9 THE CONCEPT OF LINEAR-STATE-VARIABLE FEEDBACK

Having presented methods for designing linear control systems using classical techniques, let us now look at the problem from the viewpoint of state-variable feedback [8]. In order to do this, let us first look at the basic feedback problem illustrated in Fig. 7.29. This figure illustrates the concept of feeding back the states of

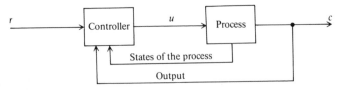

Fig. 7.29 General feedback system problem illustrating feedback of the output and the states of the process.

the process in addition to that of the output. Since a linear process can be characterized by the equations

$$\dot{\mathbf{x}} = \mathbf{P}\mathbf{x} + \mathbf{b}u, \tag{7.121}$$

$$c = \mathbf{L}\mathbf{x}, \tag{7.122}$$

let us consider the configuration of Fig. 7.30. It is important to observe from this
figure that the control signal is generated from a knowledge of the reference input r
and the state variables $\mathbf{x}$. Note that r, u, and c represent scalars.

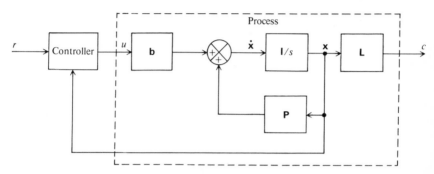

Fig. 7.30 General feedback system with state-variable feedback.

In general, the control input u can be represented as

$$u = f(\mathbf{x}, r). \tag{7.123}$$

Rather than considering the controller in such a broad sense, let us consider the
specific condition of linear state-variable feedback where the controller weights the
sum of the state variables in a linear manner. In addition, it is assumed that
the controller provides a linear gain K which multiplies the difference between the
reference input and the linear weighted sum of state variables fed back. Therefore, u
can be represented as

$$u = K[r - (h_1 x_1 + h_2 x_2 + h_3 x_3 + \cdots + h_n x_n)], \tag{7.124}$$

where h_i is defined as the ith feedback coefficient. In matrix form, u can be represented
as:

$$u = K[r - \mathbf{h}\mathbf{x}], \tag{7.125}$$

where

$$\mathbf{h} = [h_1 \quad h_2 \quad h_3 \cdots h_n], \tag{7.126}$$

$$\mathbf{x} = \begin{bmatrix} x_1 \\ x_2 \\ x_3 \\ \cdot \\ \cdot \\ \cdot \\ x_n \end{bmatrix}. \tag{7.127}$$

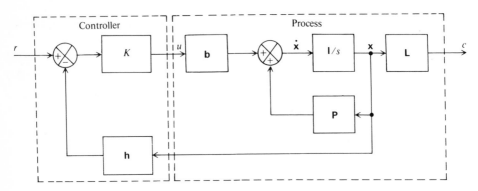

Fig. 7.31 Linear-state-variable feedback representation.

Figure 7.31 presents a matrix representation of the concept of linear-state-variable feedback, and Fig. 7.32 is a physical representation of a typical system as implied by Fig. 7.31. In the following discussion, it is assumed that all state variables are directly available for measurement and control. In practice, this is not always possible, and techniques for modifying and extending the design procedure presented, to the case where all the state variables are not available, are also discussed.

How does linear feedback of the state variables affect the behavior of the process given by Eqs. (7.121) and (7.122)? This can easily be determined by substituting Eq. (7.125) into Eq. (7.121):

$$\dot{\mathbf{x}} = \mathbf{P}\mathbf{x} + \mathbf{b}[K(r - \mathbf{h}\mathbf{x})]. \tag{7.128}$$

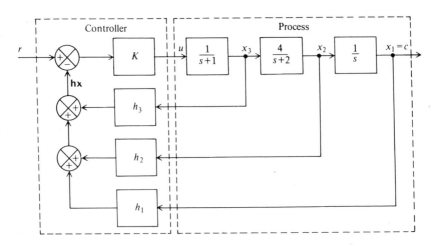

Fig. 7.32 Example of a linear-state-variable feedback system.

Simplifying Eq. (7.128), and incorporating Eq. (7.122), we obtain the closed-loop equations

$$\dot{\mathbf{x}} = \mathbf{P}_h\mathbf{x} + K\mathbf{b}r, \tag{7.129}$$

$$c = \mathbf{L}\mathbf{x}, \tag{7.130}$$

where

$$\mathbf{P}_h = \mathbf{P} - K\mathbf{bh} \tag{7.131}$$

is the closed-loop-system matrix. Comparing Eqs. (7.121) and (7.122) with (7.129) and (7.130), we observe that they are identical except that the **P**-matrix has been replaced by $\mathbf{P}_h$ and u becomes Kr.

How can we relate the closed-loop-system matrix, $\mathbf{P}_h$, to the closed-loop transfer function, $C(s)/R(s)$? This can be accomplished by taking the Laplace transform of Eqs. (7.129) and (7.130). Since the results will be used to find a transfer function, all initial conditions are assumed to be zero:

$$s\mathbf{X}(s) = \mathbf{P}_h\mathbf{X}(s) + K\mathbf{b}R(s), \tag{7.132}$$

$$C(s) = \mathbf{L}\mathbf{X}(s). \tag{7.133}$$

Solving for $\mathbf{X}(s)$ from Eq. (7.132), we get

$$\mathbf{X}(s) = K[s\mathbf{I} - \mathbf{P}_h]^{-1}\mathbf{b}R(s). \tag{7.134}$$

The inverse matrix $[s\mathbf{I} - \mathbf{P}_h]^{-1}$ is defined as the closed-loop resolvent matrix, $\mathbf{\Phi}_h(s)$, where

$$\mathbf{\Phi}_h(s) = [s\mathbf{I} - \mathbf{P}_h]^{-1}. \tag{7.135}$$

Therefore, Eq. (7.134) may be rewritten as

$$\mathbf{X}(s) = K\mathbf{\Phi}_h(s)\mathbf{b}R(s). \tag{7.136}$$

Substituting Eq. (7.136) into Eq. (7.133), we obtain a relation between $C(s)$ and $R(s)$:

$$C(s) = K\mathbf{L}\mathbf{\Phi}_h(s)\mathbf{b}R(s). \tag{7.137}$$

Therefore, the closed-loop transfer function in terms of the closed-loop resolvent matrix is given by

$$\frac{C(s)}{R(s)} = K\mathbf{L}\mathbf{\Phi}_h(s)\mathbf{b}. \tag{7.138}$$

In addition, the characteristic equation in terms of the closed-loop system matrix can also easily be determined, simply by substituting the numerator and denominator portions of the inverse matrix, $\mathbf{\Phi}_h(s)$:

$$\frac{C(s)}{R(s)} = \frac{K\mathbf{L}[\text{adj}\,(s\mathbf{I} - \mathbf{P}_h)]\mathbf{b}}{\det\,(s\mathbf{I} - \mathbf{P}_h)}. \tag{7.139}$$

The corresponding characteristic equation of the closed-loop system in terms of the closed-loop system matrix is given by

$$\det\,(s\mathbf{I} - \mathbf{P}_h) = 0. \tag{7.140}$$

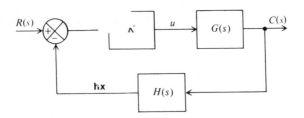

Fig. 7.33 An equivalent model of Fig. 7.31.

Since we are concerned with synthesizing control systems in terms of linear-state-variable feedback concepts, we would like to force the system illustrated in Fig. 7.31 into the generalized form illustrated in Fig. 7.33, and study its properties. Let us first consider the derivation of $H(s)$. From Fig. 7.33, we observe that

$$H(s) = \frac{\mathbf{h}\mathbf{X}(s)}{C(s)}. \tag{7.141}$$

Substituting Eq. (7.133) into Eq. (7.141), we obtain

$$H(s) = \frac{\mathbf{h}\mathbf{X}(s)}{\mathbf{L}\mathbf{X}(s)}. \tag{7.142}$$

After substitution of Eq. (7.136) for $\mathbf{X}(s)$, Eq. (7.142) becomes

$$H(s) = \frac{\mathbf{h}\mathbf{\Phi}_h(s)\mathbf{b}}{\mathbf{L}\mathbf{\Phi}_h(s)\mathbf{b}}. \tag{7.143}$$

The term $G(s)$ can also be derived in terms of $\mathbf{\Phi}_h(s)$. The closed-loop transfer function of the system illustrated in Fig. 7.33 is given by

$$\frac{C(s)}{R(s)} = \frac{KG(s)}{1 + KG(s)H(s)}. \tag{7.144}$$

Substituting Eqs. (7.138) and (7.143) into Eq. (7.144), we obtain the expression

$$G(s) = \frac{\mathbf{L}\mathbf{\Phi}_h(s)\mathbf{b}}{1 - K\mathbf{h}\mathbf{\Phi}_h(s)\mathbf{b}}. \tag{7.145}$$

Combining Eqs. (7.143) and (7.145), the open-loop transfer function is found to be given by

$$KG(s)H(s) = \frac{K\mathbf{h}\mathbf{\Phi}_h(s)\mathbf{b}}{1 - K\mathbf{h}\mathbf{\Phi}_h(s)\mathbf{b}}. \tag{7.146}$$

Let us compare Eqs. (7.138), (7.143)—(7.145), and (7.146) in order to draw conclusions regarding $G(s)$, $H(s)$, the open-loop transfer function $KG(s)H(s)$, and the closed-loop transfer function $C(s)/R(s)$. These characteristics will be important for designing systems with linear-state-variable feedback techniques in the following section. Based on these five equations, we can state the following properties:

1. The poles of $KG(s)H(s)$ are the poles of $G(s)$.

2. The zeros of $C(s)/R(s)$ are the zeros of $G(s)$.

3. The pole–zero excess of $C(s)/R(s)$ must be equal to the pole–zero excess of $G(s)$.

With these properties as a basis, we consider in the following section the design of control systems from the viewpoint of linear-state-variable feedback.

7.10 CONTROL-SYSTEM DESIGN WITH LINEAR-STATE-VARIABLE FEEDBACK

The preceding section has indicated several important relationships between open-loop and closed-loop transfer functions. This is very important in the design of control systems for the case where the closed-loop transfer function is specified and it is desired to determine the open-loop transfer function. A typical problem might specify the desired velocity constant; then use is made of Eq. (5.35) in Section 5.4 which gave the velocity constant in terms of the closed-loop poles and zeros. The problem is to determine the resulting linear-state-variable feedback system.

Let us illustrate the procedure by considering the following problem. It is desired that the closed-loop characteristics of a unity feedback control system be given by the following parameters:

$$\omega_n = 50 \text{ rad/sec}, \qquad K_v = 35/\text{sec}, \qquad \zeta = 0.707.$$

What form of closed-loop transfer function will satisfy these requirements? Let us first try a simple quadratic control system having a pair of complex-conjugate poles. From Eq. (5.37), such a system has a velocity constant given by

$$K_v = \frac{\omega_n}{2\zeta} = \frac{50}{2(0.707)} = 35.7/\text{sec}.$$

Therefore, a simple quadratic control system having a pair of complex-conjugate poles will satisfy these specifications. From Eq. (4.18),

$$\cos \alpha = -\zeta. \tag{7.147}$$

For a damping factor of 0.707, $\alpha = 45°$ and the relations among the complex-conjugate poles, ω_n and ζ are illustrated in Fig. 7.34. Therefore, the closed-loop control system is given by

$$\frac{C(s)}{R(s)} = \frac{\omega_n^2}{s^2 + 2\zeta\omega_n s + \omega_n^2}. \tag{7.148}$$

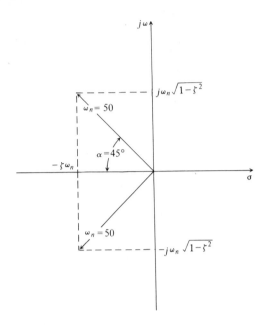

Fig. 7.34 Closed-loop poles.

By substituting $\zeta = 0.707$ and $\omega_n = 50$ into Eq. (7.148), we obtain the following desired closed-loop transfer function:

$$\frac{C(s)}{R(s)} = \frac{2500}{s^2 + 70.7s + 2500}. \tag{7.149}$$

Let us assume that the open-loop process that is being controlled is illustrated in Fig. 7.35. The corresponding state-variable representation is readily found to be

$$\dot{\mathbf{x}} = \begin{bmatrix} 0 & 1 \\ 0 & -70 \end{bmatrix} \mathbf{x} + \begin{bmatrix} 0 \\ 1 \end{bmatrix} u, \tag{7.150}$$

$$c = [1 \quad 0]\mathbf{x}, \tag{7.151}$$

where

$$\mathbf{x} = \begin{bmatrix} x_1 \\ x_2 \end{bmatrix} = \begin{bmatrix} c \\ \dot{c} \end{bmatrix}.$$

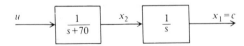

Fig. 7.35 Open-loop process to be controlled.

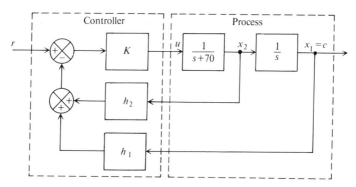

Fig. 7.36 State-variable feedback representation of system.

The resulting linear-state-variable feedback representation is illustrated in Fig. 7.36. This feedback representation can be simplified by the configuration illustrated in Fig. 7.37. The resulting closed-loop transfer function is given by

$$\frac{C(s)}{R(s)} = \frac{K/[s(s + 70)]}{1 + (h_1 + h_2 s)K/[s(s + 70)]}, \tag{7.152}$$

which can be reduced to the following expression:

$$\frac{C(s)}{(Rs)} = \frac{K}{s^2 + (70 + h_2 K)s + Kh_1}. \tag{7.153}$$

The values of $K_1 h_1$, and h_2 can be found from Eqs. (7.149) and (7.153). The following set of simultaneous equations result:

$$K = 2500, \tag{7.154}$$

$$70 + h_2 K = 70.7, \tag{7.155}$$

$$Kh_1 = 2500. \tag{7.156}$$

We have three equations and three unknowns. Solving, we find that $h_1 = 1$, $K = 2500$, and $h_2 = 2.8 \times 10^{-4}$. The final step is to draw the root locus and examine the relative stability, and the sensitivity as a function of slight gain variations. For this simple system, the final step is not necessary.

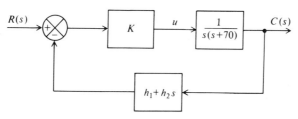

Fig. 7.37 Equivalent configuration of Fig. 7.36.

'Although this example has been solved using block diagrams and transfer functions, it could also have been solved using the matrix-algebra approach. For example, using Eq. (7.138), we could have also found $C(s)/R(s)$ from Eq. (7.153) as follows.

From Eq. (7.138), we have

$$\frac{C(s)}{R(s)} = K\mathbf{L}\boldsymbol{\Phi}_h(s)\mathbf{b}. \tag{7.157}$$

From Eq. (7.151), $\mathbf{L}$ is given by

$$\mathbf{L} = [1 \quad 0]. \tag{7.158}$$

From Eq. (7.150), $\mathbf{b}$ is given by

$$\mathbf{b} = \begin{bmatrix} 0 \\ 1 \end{bmatrix}. \tag{7.159}$$

We can evaluate $\boldsymbol{\Phi}_h(s)$ from Eq. (7.135), and it is given by

$$\boldsymbol{\Phi}_h(s) = [s\mathbf{I} - \mathbf{P}_h]^{-1} = \frac{\text{adj}\,(s\mathbf{I} - \mathbf{P}_h)}{\det\,(s\mathbf{I} - \mathbf{P}_h)}, \tag{7.160}$$

and the value of $\mathbf{P}_h$ can be evaluated from Eq. (7.131), using the value of $\mathbf{P}$ from Eq. (7.150):

$$\mathbf{P} = \begin{bmatrix} 0 & 1 \\ 0 & -70 \end{bmatrix}. \tag{7.161}$$

In this simple example, $\mathbf{h}$ is given by (this is needed in Eq. (7.131))

$$\mathbf{h} = [h_1 \quad h_2]. \tag{7.162}$$

From these matrix equations, we can determine $C(s)/R(s)$. The matrix approach is generally preferable, but in this simple example, the block diagram/transfer function method is adequate.

With this fundamental example as a basis, the general design procedure can be formulated as follows:

1. Determine the desired closed-loop transfer function based on the discussion of Section 5.4.

2. Determine the representation of the process to be controlled.

3. Represent the closed-loop system in terms of an equivalent linear-state-variable-feedback configuration.

4. Determine the closed-loop transfer function $C(s)/R(s)$ from the equivalent model in terms of K and $\mathbf{h}$.

5. Equate the $C(s)/R(s)$ expressions from Steps 1 and 4 and determine K and $\mathbf{h}$.*

6. Plot the resulting root locus of $KG(s)H(s)$ and evaluate the relative stability and sensitivity as a function of gain variations.

---

* This assumes that all of the states are measurable.

Let us apply this procedure next to the following more complex example. The problem concerns the control of a process in a unity feedback closed-loop system whose transfer function is given by

$$G(s) = \frac{1}{s(s + 1)(s + 10)}. \tag{7.163}$$

It is assumed that the transient response of the system is governed by a pair of dominant complex-conjugate poles, and that the following parameters are desired:

$$K_v = 0.93,$$
$$\zeta = 0.707,$$
$$\omega_n = 1 \text{ rad/sec.}$$

What should the closed-loop transfer function be? From Eq. (5.37), a pair of complex-conjugate poles in the denominator would only have a velocity constant given by

$$K_v = \frac{\omega_n}{2\zeta} = 0.707. \tag{7.164}$$

Therefore, a simple pair of complex-conjugate poles is inadequate to meet the velocity constant requirement of 0.93. By examining Eq. (5.35), we conclude that a zero Z must be added to the closed-loop transfer function. How many poles should the closed-loop system have? Since

$$(N_{P_c} - N_{Z_c})_{C/R} = (N_{P_0} - N_{Z_0})_G, \tag{7.165}$$

where

$$N_{P_c} = \text{number of closed-loop poles} = ?$$
$$N_{Z_c} = \text{number of closed-loop zeros} = 1$$
$$N_{P_0} = \text{number of open-loop poles} = 3$$
$$N_{Z_0} = \text{number of open-loop zeros} = 0.$$

Therefore,

$$(N_{P_c} - 1) = (3 - 0), \tag{7.166}$$

and

$$N_{P_c} = 4. \tag{7.167}$$

Since the resulting unity-feedback, closed-loop transfer has to have one zero and 4 poles, it has the following general form:

$$\frac{C(s)}{R(s)} = \frac{\omega_n^2 P_3 P_4}{Z} \frac{(s + Z)}{(s^2 + 2\zeta\omega_n s + \omega_n^2)(s + P_3)(s + P_4)}. \tag{7.168}$$

The value of the zero Z can be found from Eq. (5.35) as follows:

$$\frac{1}{K_v} = \frac{2\zeta}{\omega_n} + \frac{1}{P_3} + \frac{1}{P_4} - \frac{1}{Z}.$$

Due to external, overall system, factors in which this feedback system is to operate, it is assumed that the poles at P_3 and P_4 are specified to occur at 9 and 16, respectively. Therefore,

$$\frac{1}{0.93} = \frac{2(0.707)}{1} + \frac{1}{9} + \frac{1}{16} - \frac{1}{Z},$$

so that $Z = 2$, and the desired closed-loop transfer function is given by

$$\frac{C(s)}{R(s)} = \frac{72(s + 2)}{(s^2 + 1.414s + 1)(s + 9)(s + 16)} \tag{7.169}$$

or

$$\frac{C(s)}{R(s)} = \frac{72(s + 2)}{s^4 + 26.4s^3 + 180.4s^2 + 229s + 144}. \tag{7.170}$$

Comparing Eqs. (7.163) and (7.170), we notice that $G(s)$ has a denominator of only third order while $C(s)/R(s)$ has a denominator of fourth order. Therefore, we must add another pole factor, $s + \alpha$, to the denominator of $G(s)$. In addition, since the numerator of the open-loop transfer function must be the same as that of the closed-loop transfer function, we must also add the factor

$$(s + 2)$$

to the numerator of $G(s)$. The resulting compensating network to be added to $G(s)$ is given by

$$G_c(s) = \frac{s + 2}{s + \alpha}, \tag{7.171}$$

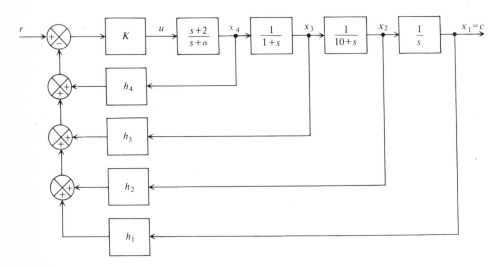

Fig. 7.38   State-variable feedback representation of system.

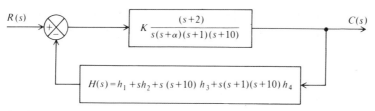

Fig. 7.39 Equivalent block diagram for system illustrated in Fig. 7.38.

where α is a pole of the open-loop transfer function which is to be determined. The resulting linear-state-variable feedback system is illustrated in Fig. 7.38. The problem remaining is to select the values of K, α, and **h**.

An equivalent block diagram of this system is illustrated in Fig. 7.39. The resulting closed-loop transfer function from this equivalent model is given by

$$\frac{C(s)}{R(s)} = \frac{K(s + 2)}{\{(Kh_4 + 1)s^4 + [K(h_3 + 13h_4) + (11 + \alpha)]s^3}$$
$$+ [K(h_2 + 12h_3 + 32h_4) + (10 + 11\alpha)]s^2$$
$$+ [K(h_1 + 2h_2 + 20h_3 + 20h_4) + 10\alpha]s + 2Kh_1\}. \qquad (7.172)$$

Equating the two forms of $C(s)/R(s)$ given by Eqs. (7.170) and (7.172), the following set of equations is obtained:

$$K = 72, \qquad (7.173)$$

$$Kh_4 + 1 = 1, \qquad (7.174)$$

$$K(h_3 + 13h_4) + (11 + \alpha) = 26.4, \qquad (7.175)$$

$$K(h_2 + 12h_3 + 32h_4) + (10 + 11\alpha) = 180.4, \qquad (7.176)$$

$$K(h_1 + 2h_2 + 20h_3 + 20h_4) + 10\alpha = 229, \qquad (7.177)$$

$$2Kh_1 = 144. \qquad (7.178)$$

Notice that we have six simultaneous equations with six unknowns (K, h_1, h_2, h_3, h_4, and α). Solving these equations, we obtain the following expressions:

$$K = 72, \qquad h_1 = 1, \qquad h_2 = 0.0134,$$
$$h_3 = 0.0014, \qquad h_4 = 0, \qquad \alpha = 15.3.$$

From Eq. (7.171), the resulting compensation network, $G_c(s)$, is given by

$$G_c(s) = \frac{s + 2}{s + 15.3},$$

which is a lead network.

It is important to emphasize that α could have turned out to be negative for a different set of specifications. This could be undesirable since it might make the plant open-loop unstable, depending on the value of the gain K. If it turns out to be open-loop unstable, we would put feedback around $G(s)$ in order to modify it to a new open-loop function $G'(s)$ which is open-loop stable, and then solve the problem with the new value of $G'(s)$.

Our results can be evaluated most conveniently on a root-locus plot. It is left as an exercise to the reader to determine the root-locus plot of the compensated system. (See Problem 7.32.)

Notice in this example, as in the first one, that it was not necessary to utilize the matrix relationships previously derived: the block-diagram/transfer-function representation was adequate. The reader is again reminded that the matrix representation should be utilized if there is a large amount of interrelation among the state variables. In addition, if a computer is being used for solution, the matrix representation should be utilized since it is usually easier for a computer to work directly with the matrices.

It is important to emphasize again that the discussion of linear-state-variable feedback in this and the preceding section has assumed that all of the state variables are accessible. This is not always the case. In general, if all of the state variables are not measurable, one could proceed in one of two directions:

1. Measure the state variables that can be measured and use them for control. However, the feedback of the available state variable is not through constant, frequency-insensitive elements, as before. Instead, feedback is through frequency-sensitive transfer functions. For example, if x_2 in Fig. 7.38 were not accessible, we still could feed back x_2 by differentiating state x_1 and then connecting it to h_2.

2. Use estimating filters as a Kalman filter [20–22].

The concepts of measurable and accessible state variables are discussed further in Chapter 9, when the concepts of controllability and observability are discussed.

PROBLEMS

7.1 Determine the circuit structure, the values of resistance and capacitance, the gains of any amplifiers required, and the complex-plane plot for first-order networks having the following characteristics:

a) Phase lead of $60°$ at $\omega = 4$ rad/sec, a minimum input impedance of 50,000 Ω, and an attenuation of 10 db at dc.

b) Phase lag of $60°$ at $\omega = 4$, a minimum input impedance of 50,000 Ω, and a high-frequency attenuation of -10 db.

c) A lag-lead network having an attenuation of 10 db for a frequency range of $\omega = 1$ to $\omega = 10$ rad/sec and an input impedance of 50,000 Ω.

In all cases, limit the maximum values of resistance to 1 MΩ and capacitance to 10 μF. Furthermore, assume that the loads on the networks have essentially infinite impedance.

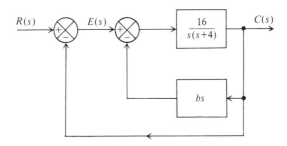

Figure P7.2

⌐7.2 The system illustrated in Fig. P7.2 consists of a unity feedback loop containing a minor rate feedback loop.

a) Without any rate feedback ($b = 0$), determine the damping factor, natural frequency, peak overshoot of the system to a unit step input, and the steady-state error resulting from a unit ramp input.

b) Determine the rate feedback constant b which will increase the equivalent damping factor of the system to 0.8.

c) With rate feedback and a damping factor of 0.8, determine the peak overshoot of the system to a unit step input and the steady-state error resulting from a unit ramp input.

d) Illustrate how the resulting steady-state error of the system with rate feedback to ramp inputs can be reduced to the same level, if rate feedback were not used, and still maintain a damping factor of 0.8.

7.3 Repeat Problem 7.2 for the forward transfer function of the system given by $20/s(1 + s)$.

7.4 Figure P7.4 illustrates the block diagram of a roll control system used to limit the roll rate excursions of a missile by providing sufficient dynamic reaction to disturbing moments [12]. The disturbance moments result from changes in bank angle and steering control deflections. The basic limitation which determines the effectiveness of the roll control system is the response of the aileron servo.

a) Determine the transfer function, $C(s)/R(s)$, of the system illustrated in Fig. P7.4.

b) Since the transient response is governed by a pair of dominant complex-conjugate poles, specify the requirements of the aileron servo parameters in order that the equivalent damping factor of the system is approximately 0.5, and the equivalent natural frequency of the system is approximately 4 rad/sec.

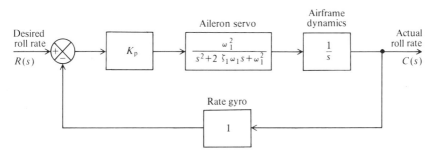

Figure P7.4

7.5 A unity feedback system has a forward transfer function given by

$$G(s) = \frac{28(1 + 0.05s)}{s(1 + s)}.$$

It is desired to compensate this system so that the resulting damping factor is unity (critically damped).

a) Using the classical approach, determine the time constant of a cascaded lead network, containing a zero factor only, that can achieve this.

b) Using the classical approach, determine the rate feedback constant of a minor rate feedback loop which can achieve critical damping.

7.6 Repeat Problem 7.5 for the forward transfer function of the system given by

$$\frac{100(1 + 0.1s)}{s(1 + 10s)}.$$

✓ **7.7** It is desired that the system considered in Problem 6.17 have a phase margin of 65° and a gain margin of 6 db. Cascade compensation is to be employed.

a) Specify the time constant of a lead network (or networks) that can achieve this.

b) Repeat part (a) for a lag network.

7.8 It is desired that the system considered in Problem 6.18 have a phase margin of 65° and a gain margin of 34 db. Cascade compensation is to be employed. Determine the time constant of a lag network (or networks) that can achieve this.

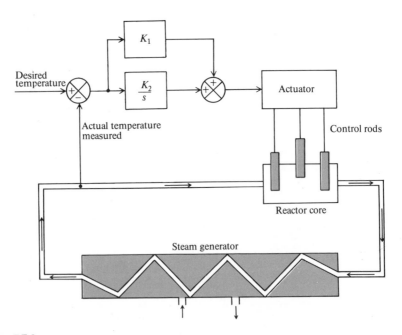

Figure P7.9

7.9 The temperature control loop of a nuclear power plant is illustrated in Fig. P7.9. The transfer function of the nuclear reactor can be adequately represented by

$$G_R(s) = \frac{e^{-0.2s}}{0.4s + 1} .$$

A time delay is included in this transfer function to account for the time required to transport the fluid from the reactor to the measurement point. Using the Bode diagram, determine the values of K_1 and K_2 in order to achieve a phase margin of 30°.

7.10 It is desired that the system considered in Problem 6.19 have a phase margin of 45° at the crossover frequency. Determine the stabilizing element required to achieve this.

7.11 It is desired that the system considered in Problem 6.20 have a phase margin of 45° at the crossover frequency. Determine the stabilizing element required to achieve this.

7.12 The H.S. Denison, shown in Fig. P7.12(a), is the first large hydrofoil seacraft built and operated in the United States [13]. The craft was designed and built by the Grumman Aerospace Corporation for the Maritime Administration of the U.S. Department of Commerce. The 80-ton hydrofoil is capable of operating at speeds of 60 knots in seas containing waves 9 feet high. A simplified schematic of the automatic control system of the H.S. Denison is illustrated in Fig. P7.12(b). It consists of transducers for sensing craft motions and a computer for transmitting commands to the electrohydraulic actuators [13]. Heave rate is fed symmetrically to the forward flaps, roll and roll rate are fed differentially to the forward flaps; pitch rate is fed to the stern foil. The stabilization control system maintains level flight by means of two main surface piercing foils located ahead of the center of gravity and an all-movable submerged foil aft. An equivalent block diagram of the pitch control system is illustrated in Fig. P7.12(c). It is desired that the craft maintain a constant level of travel despite a wave disturbance $U(s)$ whose energy is concentrated at 1 rad/sec. Assume that the specifications require that the pitch loop maintain a gain of 40 db at 1 rad/sec in order to minimize the wave disturbance, and a crossover of 10 rad/sec for adequate response time. In addition, it is desired to have a phase margin of at least 45° at the gain crossover frequency of 10 rad/sec. Select the amplifier gain K_a and compensation network $G_c(s)$ in order to achieve these requirements.

7.13 It is desired that the system considered in Problem 6.22 have a phase margin of 45° at the crossover frequency. In order to achieve this, one or more phase-lead networks are introduced into the controller. Determine the compensation required and the resulting new crossover frequency to meet this specification.

7.14 Space vehicles using wings to maneuver while reentering the earth's atmosphere present an interesting control problem. Figure P7.14(a) illustrates a conceptual design of such a system and Fig. P7.14(b) indicates the block diagram of the pitch-rate control system of such a system [14].

a) Draw the Bode diagram of this system with $K_1 = 1$ and $K_2 = 0$. What are the resulting gain and phase margins?

b) Select the values of K_1 and K_2 which will result in a gain crossover frequency of 1 rad/sec, a phase margin of at least 40° and a gain margin of at least 45 db.

(a)

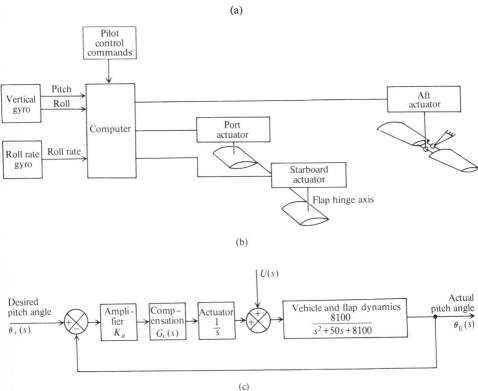

(b)

(c)

Fig. P7.12 (a) Photograph of H. S. Denison. (Courtesy of Grumman Aerospace Corporation) (b) The automatic control system. (c) Block diagram of the pitch control system.

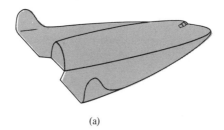

(a)

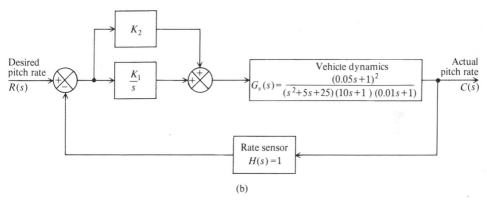

(b)

Figure P7.14

7.15 It is desired that the system considered in Problem 6.23 have a phase margin of $65°$ at the crossover frequency. In order to achieve this, two phase-lead networks are introduced into the controller. This results in the controller having a transfer function given by

$$G_1(s) = K \frac{1 + T_1 s}{1 + T_2 s} \frac{1 + T_3 s}{1 + T_4 s},$$

where K is the value of gain found in part (b) of Problem 6.23. Determine T_1, T_2, T_3, T_4, and the resulting new crossover frequency to meet this specification.

7.16 The design of the Lunar Excursion Module (LEM) shown in Fig. P7.16(a), is an extremely interesting problem [15]. The control, guidance, and navigation for the LEM are provided by an all-digital system from the sensors to the gas-jet propulsion units. For purposes of this analysis, the vehicle dynamics can be approximated by a double integration, as indicated in Fig. P7.16(b), which illustrates one axis of the attitude control system. In addition, the torque $T(s)$ is assumed to be proportional to the control signal $U(s)$. Assume that $J = 0.25$ and

$$T(s) = 2U(s).$$

Utilizing the Bode diagram for solution, determine a lead-compensation network, $G_c(s)$, which will result in a crossover frequency of 6 rad/sec and a phase margin of $60°$.

(a)

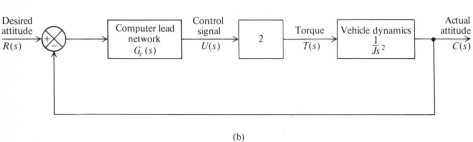

(b)

Fig. P7.16 (a) Apollo 11 Astronauts Neil Armstrong and Edwin Aldrin are inside the lunar module separated from the Apollo command module. (Official NASA photo) (b) One axis of the attitude control system.

7.17 It is desired to add cascade compensation to the system considered in Problem 6.25 in order that the peak overshoot to a step input be approximately 10%.

a) Using the Nichols chart, design a phase-lead network which can achieve this.
b) Repeat part (a) using a phase-lag network.
c) With the compensation networks chosen in parts (a) and (b), determine the closed-loop amplitude and phase-frequency response for each part.
d) What conclusions can you draw from part (c)?

7.18 It is desired to add cascade compensation to the system considered in Problem 6.27 in order that $M_p = 1.2$ while the same steady-state error is maintained.

a) Design a phase-lag network to achieve this.
b) Repeat part (a) for a phase-lead network.
c) With the compensation networks chosen in parts (a) and (b), determine the closed-loop amplitude and phase frequency responses for each part.
d) What conclusions can you draw from part (c)?

7.19 It is desired that the system considered in Problem 7.18 have a peak overshoot of approximately 15% to a step input.

a) Utilizing a minor rate feedback loop, specify the tachometer constant which can achieve this.
b) What will be the resulting system steady-state error to a unit ramp input with the minor rate feedback loop added?
c) Utilizing a simple, high-pass, RC filter in cascade with the tachometer, determine the time constant of the network and the tachometer constant which will result in a 15% overshoot to a step input.
d) What will be the steady-state error to a unit ramp when the high-pass filter is cascaded with the tachometer?

7.20 It is desired that the system considered in Problem 6.30 have a damping factor of 0.75 for the dominant complex roots. Using the root-locus method,

a) Determine the lag network $(s + n\alpha)/(s + \alpha)$ which can achieve this. Assume $K_v = 15$.
b) Determine the lead network $(s + n\alpha)/(s + \alpha)$ which can achieve $K_v = 4.15$.

7.21 The signal-flow graph of the temperature control loop for a xylene chemical process [16] is shown in Fig. P7.21. The temperature of the process, $C(s)$, is related to the heat

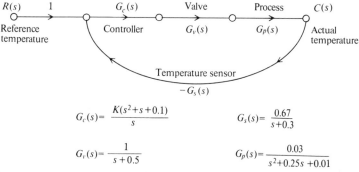

$$G_c(s) = \frac{K(s^2+s+0.1)}{s} \qquad G_s(s) = \frac{0.67}{s+0.3}$$

$$G_v(s) = \frac{1}{s+0.5} \qquad G_p(s) = \frac{0.03}{s^2+0.25s+0.01}$$

Figure P7.21

supplied to the process by the quadratic transfer function $G_p(s)$. Temperature is measured by a sensor having a pole at $s = -0.3$, and the output of the sensor in the form of air pressure is compared with the desired value of temperature as indicated by the reference pressure $R(s)$. The pressure difference (a measure of temperature error) actuates a pneumatic controller which provides as its output a pneumatic actuating signal applied to a steam valve. The valve, in turn, controls the flow of heat to the xylene column in order to minimize the error.

a) Draw the root locus for this system.
b) Determine the required gain K for a damping factor of 0.5.

7.22 It is desired that the system considered in Problem 6.31(b) have a damping factor of 0.75 for the dominant complex roots. Using the root-locus method,

a) Determine the lag network $(s + n\alpha)/(s + \alpha)$ which can achieve this. Assume $K_a = 15$.
b) Determine the lead network $(s + n\alpha)/(s + \alpha)$ which can achieve this. Assume $K_a = 15$.
c) Determine the steady-state error coefficients, ω_p and M_p for parts (a) and (b).

7.23 A turbine speed control system is illustrated in Fig. P7.23. Assume that the transfer function of the control valve, turbine, and speed converter are as follows:

$$G_1(s) = \frac{1}{s + 0.1},$$

$$G_2(s) = \frac{0.5}{s^2 + 3s + 2},$$

$$H(s) = 1.$$

Assuming that the transient response is governed by a pair of dominant complex-conjugate poles, determine the value of the controller gain K in order that the system having a damping factor of 0.5.

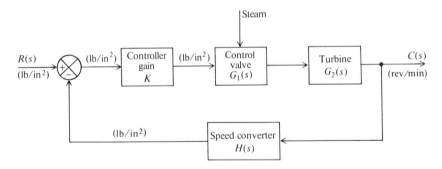

Figure P7.23

7.24 It is desired that the system considered in Problem 6.31(c) have a damping factor of 0.75 for the dominant complex roots. Using the root-locus method,

a) Determine the lag network $(s + n\alpha)/(s + \alpha)$ which can achieve this. Assume $K_v = 15$.
b) Determine the lead network $(s + n\alpha)/(s + \alpha)$ which can achieve this. Assume $K_v = 15$.
c) Determine the steady-state error coefficients, ω_p and M_p for parts (a) and (b).

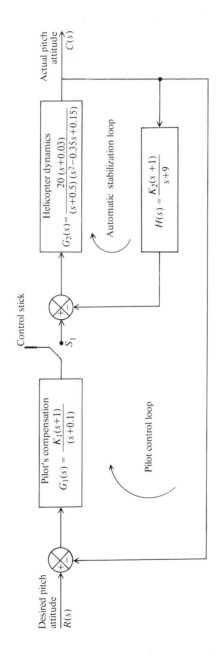

Figure P7.26

7.25 Repeat Problem 7.22 for the system considered in Problem 6.31(d).

7.26 Unlike fixed-wing aircraft which possess a moderate degree of inherent stability, the helicopter is very unstable and requires the use of feedback loops for stabilization. A typical control system involves the use of an inner automatic stabilization loop, and an outer loop which is controlled by the pilot, who inserts commands into it based on attitude errors displayed to the pilot. Figure P7.26 illustrates the pitch control system used on the S-55 helicopter [17]. When the pilot is not utilizing the control stick, the switch S_1 is open which disengages the pilot control loop. The model of the pilot's transfer function, $G_1(s)$, includes a gain factor, an anticipation time constant of 1 sec, and an error-smoothing time constant of 10 sec [18].

a) With the pilot control loop open, plot the root locus for the automatic stabilization loop and determine the gain K_2 which results in a damping factor of 0.5 for the dominant complex roots

b) Draw the root locus of the pilot control loop with K_2 set at the value determined in part (a). Determine the value of the pilot's gain compensation factor K_1 in order that the pilot control loop have a damping factor of 0.5.

7.27 Many modern control systems are designed to be adaptive, in order that they can achieve a desired response in the presence of extreme changes in the system parameters and major external disturbances. Adaptive control systems are usually characterized by devices which automatically measure the dynamics of the controlled system and by other devices which automatically adjust the characteristics of the controlled elements based on a comparison of the measurements with some optimum figure of merit. Figure P7.27 illustrates an adaptive pitch flight control system proposed by the Sperry Gyroscope Company for adaptive flight control [19]. It attempts to measure the exact location of a pair of dominant, variable, servo actuator poles which move in the complex plane, as a function of the flight conditions. The adaptive feature overcomes this problem by adjusting the gain in order to keep the location of these sensitive poles fixed in the complex plane. A test impulse train is injected into the system when the error is small. The performance computer determines the transient response of the system and compares it with an optimum desired response that is set at 2 half-cycles of a transient response over a 3-sec interval of time. The performance computer is designed so that a count less than 2 over a 3-sec period will cause the adaptor motor to increase K, while a count greater than 2 over a 3-sec period will cause the adaptor motor to decrease K. Assume that the poles and zeros of the system are located in the complex plane as follows:

	Poles	Zeros	Gain
Aircraft aerodynamics	$P_A = -4 \pm j\,2$ $P_B = -1 \pm j\,1$	$\omega_1 = -3$ $\omega_2 = -2$	$K_A = 0.5$
Servo actuator	$P_C = -3 \pm j6$	—	—
Autopilot computer	—	$\omega_c = -6$	—

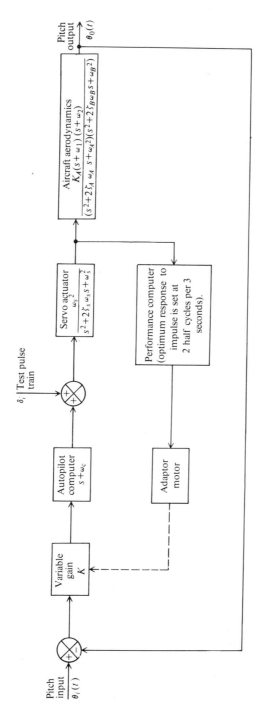

Figure P7.27

a) Draw the root locus of this system.

b) Determine the variable gain K that will result in a damping factor of 0.3, assuming that the transient response is governed by a pair of dominant complex-conjugate poles.

7.28 A unity feedback control system has a forward transfer function given by

$$G(s) = \frac{K(s + 1)}{s(s^2 + 6s + 9)} .$$

It is desired that the system have a velocity constant of 15 and a damping factor of 0.75. Using the root-locus method, determine the lag network $(s + n\alpha)/(s + \alpha)$ which can achieve this, assuming that the transient response is governed by a pair of dominant complex-conjugate poles.

7.29 Synthesize a system utilizing linear-state-variable feedback which has closed loop poles existing at $-9, 0$ and $-16, 0$ and can satisfy the following specifications:

$$K_v = 1, \qquad \zeta = 0.707, \qquad \omega_n = 1.$$

It is assumed that the process to be controlled has a transfer function given by $G(s) = 20/[s(s + 1)(s + 10)]$, and that the transient response is governed by a pair of dominant complex-conjugate poles.

7.30 Repeat Problem 7.29 for the following specifications:

$$\frac{C(s)}{R(s)} = \frac{72(s + 2)}{(s^2 + 1.414s + 1)(s + 9)(s + 16)} ,$$

$$G(s) = \frac{10}{s(1 + 5s)(1 + 0.5s)} .$$

7.31 Repeat Problem 7.29 for the following specifications:

$$\frac{C(s)}{R(s)} = \frac{8}{s^3 + 6s^2 + 10s + 8} ,$$

$$G(s) = \frac{1}{s(s + 1)} .$$

7.32 Draw the root locus and analyze the stability of the resulting system of Fig. 7.38 with the system parameters determined.

7.33 The system shown in Fig. P7.33 contains a proportional plus integral controller.

a) Determine the steady-state error of this system to a unit step input.

b) Determine the steady-state error of this system to a unit ramp input.

c) Determine the range of gain K for which this system is stable.

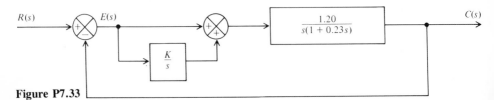

Figure P7.33

7.34 A negative feedback system containing unity feedback has a forward transfer function given by

$$G(s) = \frac{K(s + A)}{s(s + 2)(s + 4)}.$$

The zero factor $(s + A)$ in the numerator of this transfer function is to be used to compensate the system. Utilizing the root locus, analyze the effects on system stability of the following values of A:

a) $A = 1$,   b) $A = 2$,   c) $A = 3$,
d) $A = 4$,   e) $A = 6$.

What conclusions can you draw from your results on the best value of A for compensating this system?

7.35 The transfer functions of a negative feedback system are given by the following:

$$G(s) = \frac{K}{s(s^2 + 6s + 10)},$$

$$H(s) = 1.$$

a) Sketch the root locus.
b) Determine $C(s)/R(s)$, with denominator in factored form, if a damping factor of 0.5 is required for the dominant roots.

7.36 Determine the lag network compensation required to stabilize a unity feedback system whose forward transfer function is given by

$$G(s) = \frac{K}{s(s + 3)(s + 4)}.$$

The requirements for the system damping factor is 0.707, and the velocity constant is 100/sec.

7.37 Repeat Problem 7.28 for

$$G(s) = \frac{K(s + 1)}{s(s + 3)(s + 6)}.$$

REFERENCES

1. W. R. Ahrendt, *Servomechanism Practice*, McGraw-Hill, New York (1954).
2. J. G. Truxal, *Automatic Feedback Control System Synthesis*, McGraw-Hill, New York (1955).
3. A. D. Groner, "AC stabilizing networks," *Control Eng.* **55**, 55–57 (September 1954).
4. G. E. Valley, Jr., and H. Wallman, *Vacuum Tube Amplifiers*, McGraw-Hill, New York (1948).
5. H. Lauer, R. N. Lesnick, and L. E. Matson, *Servomechanism Fundamentals*, McGraw-Hill, New York (1960).
6. J. G. Truxal (Ed.), *Control Engineer's Handbook*, McGraw-Hill, New York (1955).
7. H. W. Bode, *Network Analysis and Feedback Amplifier Design*, Van Nostrand New York (1945).

8. J. L. Melsa and D. G. Schultz, *Linear Control Systems*, McGraw-Hill, New York (1969).

9. L. A. Gould, *Chemical Process Control: Theory and Applications*, Addison-Wesley, Reading, Mass. (1969).

10. H. Chestnut and R. W. Mayer, *Servomechanisms and Regulating System Design* (2nd Edn.), Vol. 1, Wiley, New York (1959).

11. J. L. Bower and P. M. Schultheiss, *Introduction to the Design of Servomechanisms*, Wiley, New York (1958).

12. W. K. Waymeyer and R. W. Sporing, "Closed Loop Adaptation Applied to Missile Control," in *Proceedings of the 1962 Joint Automatic Control Conference*, p. 18-3.

13. R. M. Rose, *The Rough Water Performance of the H. S. Denison*, American Institute of Aeronautics and Astronautics, Paper no. 64–197 (May 1964).

14. R. P. Kotfile and S. S. Oseder, "Stabilization and control of maneuvering reentry vehicle," *Sperry Engineering Review* 18, 2–10 (1965).

15. F. Doennebrink and J. Russel, "LEM stabilization and control system," *AIAA/ION Guidance and Control Conference*, August 16–18, 1965, pp. 430–41.

16. W. A. Lynch and J. G. Truxal, *Principles of Electronic Instrumentation*, McGraw-Hill, New York, (1962), p. 686.

17. L. Kaufman, "Helicopter control stick steering," *Sperry Engineering Review* 11, 41–48 (1958).

18. S. M. Shinners, *Techniques of System Engineering*, McGraw-Hill, New York (1967).

19. F. C. Gregory (Ed.), *Proceedings of the Self-Adaptive Flight Control Systems Symposium*, WADC Technical Report 59-49, ASTIA Document No. AD 209389 (March 1959).

20. R. E. Kalman and R. W. Koepcke, "Optimal synthesis of linear sampling control systems using generalized performance indices," *ASME Trans*. 80, 1820 (1958).

21. R. E. Kalman, "A new approach to linear filtering and prediction problems," *ASME Trans., Series D, J. Basic Eng*. 82, 35 (1960).

22. R. S. Bucy and R. E. Kalman, "New results in linear filtering and prediction problems," *ASME Trans., Series D., J. Basic Eng*. 83, 95 (1961).

8 NONLINEAR FEEDBACK CONTROL-SYSTEM DESIGN

8.1 INTRODUCTION

The feedback control-system design methods presented in previous chapters were restricted to linear constant systems, that is, systems that can be represented by linear differential equations with constant coefficients. In practice, linear systems possess the property of linearity only over a certain range of operation; all physical systems are nonlinear to some degree. Therefore it is important that one acquire a facility for analyzing feedback control systems with varying degrees of nonlinearity.

Any attempt to restrict attention strictly to linear systems can only result in severe complications in system design. To operate linearly over a wide range of variation of signal amplitude and frequency would require components of an extremely high quality; such a system would probably be impractical from the viewpoints of cost, space, and weight. In addition, the restriction of linearity severely limits the system characteristics that can be realized.

In practice, linear operation is required only for small deviations about a quiescent operating point. The saturation of amplifying devices having large deviations about the quiescent operating point is usually acceptable. The presence of nonlinearities in the form of dead zones for small deviations about the quiescent operating point is also usually acceptable. In both cases one attempts to limit the effects of nonlinearities to acceptable tolerances, since it is impractical to eliminate the problem entirely.

It is worth noting that nonlinearities may be intentionally introduced into a system in order to compensate for the effects of other undesirable nonlinearities, or to obtain better performance than could be achieved using linear elements only. A simple example of an intentional nonlinearity is the use of a nonlinear damped system to optimize response in accordance with the magnitude of the error [1]. The on–off contactor (relay) servo, where full torque is applied as soon as the error exceeds a specified value, is another case of an intentionally nonlinear system.

The purpose of this chapter is to examine the broad aspects of nonlinear systems. We first study the characteristics of nonlinearities and then present several methods for analyzing unintentional nonlinear feedback control systems. We follow in Chapter 9 with several illustrations of synthesis of nonlinear systems having intentional nonlinearities utilizing optimal control theory.

We should emphasize here that methods of analyzing nonlinear systems have not progressed as rapidly as have techniques for analyzing linear systems. Comparatively speaking, at the present time we are still in the developmental stage. However, the various methods presented in this chapter will enable one to analyze and synthesize nonlinear feedback control systems quantitatively.

8.2 NONLINEAR DIFFERENTIAL EQUATIONS

A linear differential equation of the nth order, with constant coefficients, is written

$$A_n \frac{d^n y}{dt^n} + A_{n-1} \frac{d^{n-1} y}{dt^{n-1}} + \cdots + A_0 y = x(t), \tag{8.1}$$

where $x(t)$ represents the input to the system, t represents time and is the independent variable, $y(t)$ represents the dependent variable, or the output of the system, and $A_n, A_{n-1}, \ldots, A_0$ are constants.

This equation is of the form derived for several representative mechanical and electrical systems in Chapter 3. For example, Eq. (3.19), which is repeated below, gave the differential equation of motion for a mechanical system which consists of a force $f(t)$ applied to a mass, damper, and spring:

$$M \frac{d^2 y}{dt^2} + B \frac{dy}{dt} + Ky = f(t).$$

The mass of the system is represented by the constant M, the damping factor by the constant B, and the spring constant by K.

Detailed solutions for the class of differential equations having the form shown in Eq. (8.1) are available. They have been studied extensively, and several powerful techniques, such as the Laplace transformation, exist for their solution. All the analytical methods discussed in Chapters 6 and 7 are based on systems which can be represented by simple differential equations having this general form.

If any of the coefficients $A_n, A_{n-1}, \ldots, A_0$ are functions of the independent variable time, then the linear differential equation is said to have variable coefficients. In this case, the differential equation takes the following form:

$$A_n(t) \frac{d^n y}{dt^n} + A_{n-1}(t) \frac{d^{n-1} y}{dt^{n-1}} + \cdots + A_0(t) y = x(t), \tag{8.2}$$

where $A_n, A_{n-1}, \ldots, A_0$ are all functions of time. Except in special cases (such as when the coefficients are polynomials), the solution of linear time-variable equations is quite difficult and one usually has to resort to a computer.

If the coefficients of the differential equation are functions of the dependent variable y, then a nonlinear differential equation results. Its general form is

$$A_n \frac{d^n y}{dt^n} + A_{n-1} \frac{d^{n-1} y}{dt^{n-1}} + \cdots + A_1 \frac{dy}{dt} + A_0 y + \epsilon f\left(y, \frac{dy}{dt}, \ldots, \frac{d^{n-1} y}{dt^{n-1}}\right) = x(t), \tag{8.3}$$

where $x(t)$ represents the input to the system, t represents time and is the independent variable, $y(t)$ represents the dependent variable and the output of a system, $A_n, A_{n-1}, \ldots, A_0$ are constants, ϵ is a constant indicating the degree of nonlinearity present, and $f(y, dy/dt, \ldots, d^{n-1}y/dt^{n-1})$ is a nonlinear function.

Notice that if $\epsilon = 0$, Eq. (8.3) reduces to Eq. (8.1), which represents a linear differential equation having constant coefficients. This leads us to the qualitative rule, that a small amount of nonlinearity in a system means that ϵ is small in comparison with the coefficients $A_n, A_{n-1}, \ldots, A_0$. In addition, a large amount of nonlinearity means that ϵ is large compared with $A_n, A_{n-1}, \ldots, A_0$.

8.3 PROPERTIES OF LINEAR SYSTEMS THAT ARE NOT VALID FOR NONLINEAR SYSTEMS

Several inherent properties of linear systems, which greatly simplify the solution for this class of systems, are not valid for nonlinear systems. The fact that nonlinear systems do not have these properties further complicates their analysis.

Superposition is a fundamental property of linear systems. As a matter of fact, this property can be considered to be a definition of a linear system. The principle of superposition states that if $c_1(t)$ is the response of a system to $r_1(t)$ and $c_2(t)$ is its response to $r_2(t)$, then the system's response to $a_1r_1(t) + a_2r_2(t)$ is $a_1c_1(t) + a_2c_2(t)$. Unfortunately, the superposition principle does not apply to nonlinear systems. Therefore several mathematical procedures used in the design of linear systems cannot be used for nonlinear systems.

Stability of linear systems has been shown (in Chapter 6) to depend only on the system's parameters. The stability of nonlinear systems, however, depends on the initial conditions and the nature of the input signal as well as the system parameters. One cannot expect a nonlinear system that exhibits a stable response to one type of input to have a stable response to other types of input. We shall shortly illustrate nonlinear systems that are stable for very small or very large signals, but not for both.

We normally expect the output of a linear system, excited by a sinusoidal signal, to have the same frequency as the input, although its amplitude and phase may differ. However, the output of nonlinear systems usually contains additional frequency components and may, in fact, not contain the input frequency.

For linear systems, interchanging two elements in cascade does not affect behavior. This is not true if one of the elements is nonlinear.

The question of stability is clearly defined for linear constant systems: A system is either stable or unstable. An unstable linear constant system has an output that grows without bound, either exponentially or in an oscillatory mode with the envelope of the oscillations increasing exponentially. In nonlinear systems, system instability may mean a constant-amplitude output having an arbitrary waveform. It is important to emphasize that an oscillator is stable according to Liapunov. The exponentially decaying system, which we have referred to in this book as being stable, is described by Liapunov as being asymptotically stable.*

Another important difference concerns the amplitude of oscillation in linear and nonlinear systems. A linear oscillator oscillates at an amplitude which is determined by the initial conditions, whereas a nonlinear oscillator exhibits an amplitude which is usually independent of the initial conditions.

8.4 UNUSUAL CHARACTERISTICS
THAT ARE PECULIAR TO NONLINEAR SYSTEMS

This section describes in detail some of the unusual characteristics that are peculiar to nonlinear systems. These phenomena, which do not occur in linear systems, may be desirable or undesirable depending on the application. We discuss specifically the following behavior: limit cycle, soft and hard self-excitation, hysteresis, jump resonance, and subharmonic generation.

Limit cycles are oscillations of fixed amplitude and period that occur in nonlinear systems. Depending on whether the oscillation converges or diverges from the conditions represented, limit cycles can be either stable or unstable. It is possible that conditionally stable systems may contain both a stable and an unstable limit cycle. The occurrence of limit cycles in nonlinear systems makes it necessary to define instability† in terms of acceptable magnitudes of oscillation, since a very small nonlinear oscillation may not be detrimental to the performance of a system.

Self-excited oscillations occurring in systems that are unstable in the presence of very small signals are called *soft self-excitation*. Self-excited oscillations occurring in systems that are unstable in the presence of very large signals are called *hard self-excitation*. Since soft and hard types of oscillation can occur, the control engineer must specify the dynamic range of operation completely when designing a nonlinear system. A feedback control system containing an element having saturation characteristics, such as illustrated in Fig. 8.1(a), could exhibit soft self-excitation. A feedback control system containing an element having dead-zone characteristics, such as illustrated in Fig. 8.1(b), could exhibit hard self-excitation.

Hysteresis is a nonlinear phenomenon which is most usually associated with magnetization curves or backlash of gear trains. A conventional magnetization curve whose path depends on whether the magnetizing force **H** is increasing or decreasing is shown in Fig. 8.1(c).

Jump Resonance [2], another form of hysteresis, is of considerable interest. It exhibits itself in the closed-loop frequency response of certain nonlinear systems, as illustrated in Fig. 8.1(d). As the frequency ω is increased and the input amplitude R is held constant, the response follows the curve AFB. At point B, a small change

---

* Asymptotic stability and nonlinear system stability classified in terms of a regional basis is discussed in Section 8.17 where Liapunov's stability criterion is presented.

† As far as control-system design is concerned, a steady oscillation is treated as being unstable.

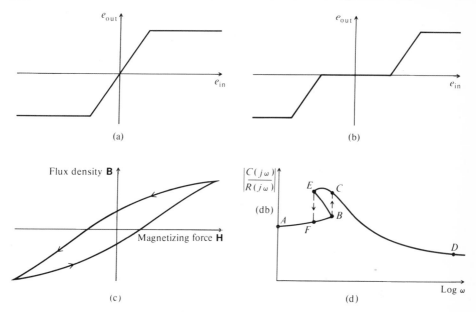

Fig. 8.1 (a) Saturation characteristics. (b) Dead zone characteristics. (c) Conventional hysteresis loop. (d) Closed-loop response of a system with jump resonance.

in frequency results in a discontinous jump to point C. The response then follows the curve to point D upon further increase in frequency. As the frequency is decreased from point D, the response follows the curve to points C and E. At point E, a small change in frequency results in a discontinuous jump to point F. The response follows the curve to point A for further decreases in frequency. Observe from this description that the response never actually follows the segment BE. This portion of the curve represents a condition of unstable equilibrium. The system must be of the second order, or higher, for the phenomenon of jump resonance to occur.

Subharmonic generation [3] refers to nonlinear systems whose output contains subharmonics of the input's sinusoidal excitation frequency. The transition from normal harmonic operation to subharmonic operation is usually quite sudden. Once the subharmonic operation is established, however, it is usually quite stable. In general, if sinusoidal signals f_1 and f_2 are added and their sum is applied to a nonlinear device, the output contains frequency components $af_1 \pm bf_2$, where a and b assume all possible integers including zero.

8.5 METHODS AVAILABLE FOR ANALYZING NONLINEAR SYSTEMS

Several tools are available for the analysis of nonlinear systems. All these techniques depend on the severity of the nonlinearity and/or the order of the system under consideration. We consider most of the useful and popular techniques in this chapter and illustrate their practical application.

The analysis of nonlinear systems is concerned with the existence and effects of limit cycles, soft and hard self-excitation, hysteresis, jump resonance, and subharmonic generation. In addition, the response to specific input functions must be determined. The major difficulty of analyzing nonlinear systems is that no single technique is generally applicable to all problems.

Quasilinear systems, where the deviation from linearity is not too large, permit the use of certain *linearizing approximations* [12]. The *describing-function* approach, which is applicable to nonlinear systems of any order and is concerned with discovering limit cycles, simplifies the problem by assuming that the input to the nonlinear system is sinusoidal and the only significant frequency component of the output is that component having the same frequency as the input [4–7].

Nonlinear systems can often be approximated by several linear regions. The *piecewise-linear* approach permits the segmented linearization of any nonlinearity for any order of system. The *phase-plane method* is a very useful technique for analyzing the response of a second-order nonlinear system [7, 17–19]. *Liapunov's stability methods* are very powerful techniques for determining the steady-state stability of nonlinear systems based on generalizations of energy notions [20–21]. *Popov's method* is very useful for determining the stability of time-variable, nonlinear systems [22–29]. *The generalized circle criterion* is applicable to time-variable, nonlinear systems whose linear portion is not necessarily open-loop stable [25–26].

Systems of very high order having several nonlinearities have hardly been dealt with in general analytical terms. This problem usually requires the use of *numerical methods* utilizing *digital computers* for a solution. It is worth emphasizing at this point that any nonlinear differential equation can be solved by these techniques provided many small increments are used [32, 33, 36–38]. However, the resulting solution is valid only for the specific problem being considered. It is very difficult to extend the result and obtain a general solution which can be used for other problems.

8.6 LINEARIZING APPROXIMATIONS

In quasilinear systems, where the deviation from linearity is not too great, linear approximations may permit the extension of ordinary linear concepts. This approach acknowledges that certain system characteristics change from operating point to operating point, but it assumes linearity in the neighborhood of a specific operating point. The technique of linearizing approximations is universally used by the engineer and may be more familiar to the reader under the names *small-signal theory* and/or *theory of small perturbations.*

Linearizing approximations were utilized when we discussed the two-phase ac servomotor in Section 3.4. For this device, Fig. 3.16 illustrated the quasilinear characteristics relating developed torque and speed. However, by approximating the torque–speed curves with straight lines, the linear differential equation (3.98) was formulated. We then obtained the transfer function of the two-phase ac servomotor, assuming that it was a linear device. It is left as an exercise to the reader in Problem 8.1 to determine the effect of various linearizing approximations.

The effects of a small amount of nonlinearity can be studied analytically by considering small perturbations or changes in the variables about some average value of the variables. This can be represented analytically by

$$A_n \frac{d^n y}{dt^n} + A_{n-1} \frac{d^{n-1} y}{dt^{n-1}} + \cdots + A_1 \frac{dy}{dt} + A_0 y + \epsilon f\left(y, \frac{dy}{dt}, \ldots, \frac{d^{n-1} y}{dt^{n-1}}\right) = x(t).$$

An expansion of the solution to this differential equation, for small nonlinearities, can be written as a power series in ϵ as

$$y(t) = y_{(0)}(t) + \epsilon y_{(1)}(t) + \epsilon^2 y_{(2)}(t) + \epsilon^3 y_{(3)}(t) + \cdots.$$

From this equation, $y(t)$ may be interpreted as composed of a linear component $y_{(0)}(t)$ and several deviation factors: $\epsilon y_{(1)}(t) + \epsilon^2 y_{(2)}(t) + \cdots$. Assuming that ϵ is very small, the nonlinear components will not seriously affect the system's behavior if a linear approximation is assumed. Therefore, within the realm of reasonable engineering approximations, the control engineer may be able to extend linear theory for certain feedback control systems which exhibit a small amount of nonlinearity. It is very interesting that this is just the reason why linear theory has had such good results even though practical systems are never purely linear.

Linearization techniques can also be applied to those problems where it is desired to linearize nonlinear equations by limiting attention to small perturbations about a reference state [12]. This technique is often used in the design of space navigation and control systems where it is desired to maintain a space vehicle along a specified reference trajectory. It will now be shown how the corrective control forces required to keep the vehicle on the desired flight trajectory can be synthesized from a set of linear differential equations, although the basic differential equations describing the reference flight trajectory are nonlinear.

To illustrate this, let us assume that the equation of the system is given by

$$\dot{\mathbf{x}} = \mathbf{f}(\mathbf{x}, \mathbf{u}),$$

where the function $\mathbf{f}$ is nonlinear. Figure 8.2 illustrates the reference trajectory of the space vehicle (solid line) which satisfies the equation

$$\dot{\mathbf{x}}^0 = \mathbf{f}(\mathbf{x}^0, \mathbf{u}^0), \tag{8.4a}$$

where the superscript zero refers to parameters occurring along the reference trajectory.* These reference parameters are related to the parameters of the actual trajectory (dashed line) as follows:

$$\mathbf{x} = \mathbf{x}^0 + \delta\mathbf{x},$$

$$\mathbf{u} = \mathbf{u}^0 + \delta\mathbf{u}.$$

---

* It is important to emphasize that both $\mathbf{x}^0$ and $\mathbf{u}^0$ may be functions of time. Therefore, it would have been more general to use the symbols $\mathbf{x}^0(t)$ and $\mathbf{u}^0(t)$. However, the abbreviated symbols $\mathbf{x}^0$ and $\mathbf{u}^0$ will be used for simplicity.

Figure 8.2 illustrates the reference and actual trajectories, where the actual state $\mathbf{x}$ is perturbed from the reference state $\mathbf{x}^0$ by $\delta \mathbf{x}$. Physically, this means that the actual trajectory of the space vehicle is perturbed, or slightly different, from that of the desired reference trajectory. The vector $\delta \mathbf{u}$ represents the deviation of the control input from the desired reference input $\mathbf{u}^0$ which would result in the desired system response $\mathbf{x}^0$.

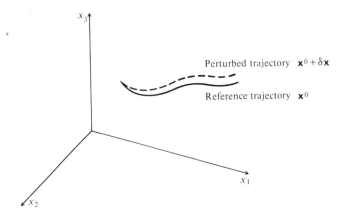

Fig. 8.2   Reference and perturbed trajectories of a space vehicle [12].

What kind of relationship can we derive between $\mathbf{x}^0$, $\delta \mathbf{x}$, $\mathbf{u}^0$, and $\delta \mathbf{u}$? The basic nonlinear equation of the system,

$$\dot{\mathbf{x}} = \mathbf{f}(\mathbf{x}, \mathbf{u})$$

can be expressed as follows:

$$\frac{d}{dt}(\mathbf{x}^0 + \delta \mathbf{x}) = \dot{\mathbf{x}}^0 + \delta \dot{\mathbf{x}} = \mathbf{f}(\mathbf{x}^0 + \delta \mathbf{x}, \mathbf{u}^0 + \delta \mathbf{u}).$$

Since we are assuming that the actual perturbations of the system are small, we can expand the jth component of this equation in a Taylor series about the reference trajectory as follows:

$$\dot{x}_j^0 + \delta \dot{x}_j \approx f_j(\mathbf{x}^0, \mathbf{u}^0) + \frac{\partial f_j}{\partial x_1} \delta x_1 + \cdots + \frac{\partial f_j}{\partial x_m} \delta x_m + \frac{\partial f_j}{\partial u_1} \delta u_1 + \cdots + \frac{\partial f_j}{\partial u_n} \delta u_n.$$

$$(8.4b)$$

Using Eq. (8.4a), we can rewrite Eq. (8.4b) as follows:

$$\delta \dot{x}_j \approx \left(\frac{\partial f_j}{\partial x_1}\right)^0 \delta x_1 + \cdots + \left(\frac{\partial f_j}{\partial x_m}\right)^0 \delta x_m + \left(\frac{\partial f_j}{\partial u_1}\right)^0 \delta u_1 + \cdots + \left(\frac{\partial f_j}{\partial u_n}\right)^0 \delta u_n, \quad (8.5)$$

where $j = 1, 2, 3, \ldots, m$.

Equation (8.5) can be simplified by utilizing the Jacobian matrices* that are defined as follows:

$$
\mathbf{A} = \begin{bmatrix}
\dfrac{\partial f_1}{\partial x_1} & \dfrac{\partial f_1}{\partial x_2} & \cdots & \dfrac{\partial f_1}{\partial x_m} \\[2ex]
\dfrac{\partial f_2}{\partial x_1} & \dfrac{\partial f_2}{\partial x_2} & \cdots & \dfrac{\partial f_2}{\partial x_m} \\[1ex]
\cdot & \cdot & & \cdot \\
\cdot & \cdot & & \cdot \\
\cdot & \cdot & & \cdot \\
\dfrac{\partial f_n}{\partial x_1} & \dfrac{\partial f_n}{\partial x_2} & \cdots & \dfrac{\partial f_n}{\partial x_m}
\end{bmatrix}_{\substack{x = x^0 \\ u = u^0}} , \quad
\mathbf{B} = \begin{bmatrix}
\dfrac{\partial f_1}{\partial u_1} & \dfrac{\partial f_1}{\partial u_2} & \cdots & \dfrac{\partial f_1}{\partial u_n} \\[2ex]
\dfrac{\partial f_2}{\partial u_1} & \dfrac{\partial f_2}{\partial u_2} & \cdots & \dfrac{\partial f_2}{\partial u_n} \\[1ex]
\cdot & \cdot & & \cdot \\
\cdot & \cdot & & \cdot \\
\cdot & \cdot & & \cdot \\
\dfrac{\partial f_n}{\partial u_1} & \dfrac{\partial f_n}{\partial u_2} & \cdots & \dfrac{\partial f_n}{\partial u_n}
\end{bmatrix}_{\substack{x = x^0 \\ u = u^0}} .
$$

It is important to emphasize that all of the partial derivatives in the Jacobian matrices are evaluated along the reference trajectory of the space vehicle. Based on the Jacobian matrices, Eq. (8.5) can be rewritten in the following simplified form:

$$\delta \dot{\mathbf{x}} \approx \mathbf{A}\, \delta\mathbf{x} + \mathbf{B}\, \delta\mathbf{u}.$$

This resulting equation is very important. It states that the differential equation describing the perturbations about the reference trajectory are approximately linear, although the basic system differential equations describing the reference flight trajectory are nonlinear. Therefore, we have succeeded in linearizing the problem.

We can also linearize a system if we can adapt it to behave like a linear system. In order to demonstrate this, let us consider a two-position relay that controls the rotation of a motor in either direction. It is assumed that the control voltage applied by the relay to the motor, e_c, is given by

$$e_c = E \sin \omega t$$

and the resulting motor torque developed, T, would be a square wave due to the switching action. Both e_c and T are illustrated in Fig. 8.3. Observe from this figure that the average, or mean, value of both functions is zero.

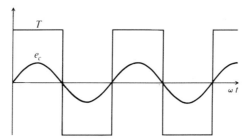

Fig. 8.3 Relay-controlled motor characteristics—Case 1.

---

* Note that when linearizing along a trajectory, the matrices **A** and **B** will be functions of time.

Let us next assume that the control voltage has a finite mean value E_0 where

$$e_c = E_0 + E \sin \omega t.$$

For this case, the torque is a periodic function whose mean value is some nonzero value T_0, since the time intervals where e_c is positive or negative are not equal, as indicated in Fig. 8.4. Note that E_0 is also a function of time, but it is assumed that it is very slowly varying compared with ω. Assuming, in addition, that $E_0 \ll E$, it can easily be shown that the mean value of T_0 is given by the linear relationship

$$T_0 = \frac{2T}{\pi E} E_0.$$

Therefore, the mean value of the torque, T_0, is proportional to the mean value of the controlling voltage.

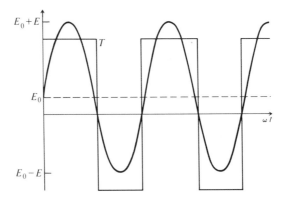

Fig. 8.4   Relay-controlled motor characteristics—Case 2.

This is a very important result. It shows that by means of a nonlinear element like a relay, a linear relationship can be obtained between the mean value of the controlling voltage and the mean value of the motor torque developed. The basic linearization technique utilized has taken the mean value of the applied voltage to the relay as the input, and superimposed on it a sinusoidal function of time whose amplitude and frequency are very high relative to that of the input.

In the following section, we extend our linearization concepts and attempt to apply them to nonlinear systems. Although the notion of a transfer function is inapplicable for nonlinear systems, an equivalent approximate transfer characteristic for a nonlinear device can be derived which can be manipulated as a transfer function under certain circumstances. We define this approximate transfer characteristic as the *describing function*. It is a very useful notion and is frequently employed in practice.

8.7 THE DESCRIBING-FUNCTION CONCEPT

The use of describing functions is an attempt to extend the very powerful transfer function approach of linear systems to nonlinear systems [4, 5, 7]. A describing function is defined as the ratio of the fundamental component of the output of a nonlinear device to the amplitude of a sinusoidal input signal. In general, the describing function depends on the input signal's amplitude and frequency and is complex because phase shift may occur between the input and the fundamental component of the output. We study the describing-function method of analysis and compare it with the transfer-function concept for linear systems.

If the input to a nonlinear element is a sinusoidal signal, the describing-function analysis assumes that the output is a periodic signal having the same fundamental period as that of the input signal. Therefore, the analysis is concerned only with the fundamental component of the output waveform. All harmonics, subharmonics, and any dc component are neglected. This assumption is reasonable since the harmonic terms are often small when compared with the fundamental term. In addition, a feedback control system usually provides additional attenuation of the harmonic terms because of its inherent filtering action. Many nonlinear elements do not generate a dc term since they are symmetrical, nor do they generate any subharmonic terms [3]. Therefore, in many (but not all) situations the fundamental term is the only significant component of the output of the nonlinear element. In addition, it is assumed that there is only one nonlinear element in the feedback control system and that it is not time varying. If a system contains more than one nonlinearity, we must lump all the nonlinearities together and obtain an overall describing function.

An examination of these limitations indicates that the describing function is based on a restricted mathematical foundation. The technique does give, however, reasonable results and does have an advantage that it can be used for systems of any order and is fairly simple to apply. It is recommended that the results always be checked with a computer simulation.

Given its limitations, the describing-function technique is still a very useful tool for analyzing and designing nonlinear systems. The describing function should be thought of as a generalized transfer function for nonlinear systems. Linear criteria of stability, for example, the Nyquist diagram, the Nichols chart, and the root locus, can be used to interpret system stability, with the added constraint that stability must be analyzed as a function of signal level and frequency of the input signal.

In order to derive a mathematical expression for the describing function, let us consider the nonlinear system illustrated in Fig. 8.5. In accordance with the definition of the describing function, let us assume that the input to the nonlinear element N is given by

$$m(\omega t) = M \sin \omega t. \tag{8.6}$$

In general, the steady-state output of the nonlinear device can be represented by the series

$$n(\omega t) = N_1 \sin (\omega t + \phi_1) + N_2 \sin (2\omega t + \phi_2) + N_3 \sin (3\omega t + \phi_3) + \cdots. \tag{8.7}$$

By definition, the describing function is

$$N(M, \omega) = \frac{N_1}{M} e^{j\phi_1}. \tag{8.8}$$

Notice that the describing function depends on the amplitude and frequency of the input signal. The nonlinear element is thus considered to have a gain and phase shift varying with the amplitude and frequency of the input signal.

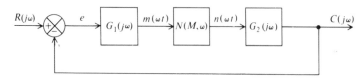

Fig. 8.5   General nonlinear system.

8.8 DERIVATION OF DESCRIBING FUNCTIONS FOR COMMON NONLINEARITIES

The describing functions for several common nonlinearities are derived in this section [5, 6, 7]. The procedure most commonly used is to determine the Fourier series of the output waveshape from the nonlinear device and consider only the fundamental component. Let us consider the nonlinear element N in an overall feedback control system, as shown in Fig. 8.5. Assuming that the input m is given by a sinusoidal signal where

$$m(\omega t) = M \sin \omega t, \tag{8.9}$$

we can represent the output waveshape by a Fourier series given by the expression

$$n(\omega t) = \frac{A_0}{2} + \sum_{K=1}^{K=\infty} A_K \cos K\omega t + \sum_{K=1}^{K=\infty} B_K \sin K\omega t, \tag{8.10}$$

where

$$A_K = \frac{2}{T} \int_{-T/2}^{T/2} n(\omega t) \cos K\omega t \, d(\omega t), \quad K = 0, 1, 2, \dots \tag{8.11}$$

$$B_K = \frac{2}{T} \int_{-T/2}^{T/2} n(\omega t) \sin K\omega t \, d(\omega t), \quad K = 1, 2, 3, \dots. \tag{8.12}$$

In general, if $n(\omega t) = -n(-\omega t)$, then the function is odd and $A_K = 0$. In addition, if $n(\omega t) = n(-\omega t)$, then the function is even and $B_K = 0$.

Since we are only concerned with the fundamental component of the output, it is necessary to determine only A_1 and B_1. For $m(\omega t) = M \sin \omega t$, the describing function can then be obtained from the expression

$$N(M, \omega) = \frac{B_1}{M} + j \frac{A_1}{M} = \left[\left(\frac{B_1}{M} \right)^2 + \left(\frac{A_1}{M} \right)^2 \right]^{1/2} \Big/ \tan^{-1} \frac{A_1}{B_1}. \tag{8.13}$$

The control engineer is usually concerned with the nonlinearities due to dead zones, saturation, backlash, on–off relay control systems, coulomb friction, and stiction. We specifically derive and catalog their describing functions so that a handy reference for some common describing functions will be available. In addition, the procedure illustrated should enable one to develop the facility for calculating the describing function of any nonlinearity encountered.

1. Describing Function of a Dead Zone Figure 8.6 illustrates the dead-zone characteristic. The relationships between input and output of this nonlinearity can be expressed by the equations

$$n(\omega t) = 0 \qquad\qquad\qquad \text{for } -D < m < D, \qquad (8.14)$$

$$n(\omega t) = K_1 M(\sin \omega t - \sin \omega t_1) \qquad \text{for } m > D, \qquad (8.15)$$

$$n(\omega t) = K_1 M(\sin \omega t + \sin \omega t_1) \qquad \text{for } m < -D. \qquad (8.16)$$

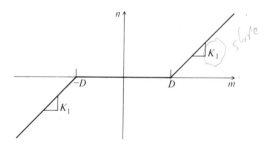

Fig. 8.6 Nonlinear characteristics of a dead zone.

Figure 8.7 illustrates typical input and output waveshapes. Notice that the output is an odd function, and therefore $A_K = 0$. The symmetry over the four quarters of the period allows us to evaluate the expression for the Fourier coefficient, B_1, by taking four times the integral over one quarter of a cycle, as follows:

$$B_1 = \frac{4}{\pi} \int_0^{\pi/2} n(\omega t) \sin \omega t \, d(\omega t). \qquad (8.17)$$

Substituting Eqs. (8.14) and (8.15) into Eq. (8.17), we obtain

$$B_1 = \frac{4}{\pi} \left[\int_0^{\omega t_1} (0) \sin \omega t \, d(\omega t) + \int_{\omega t_1}^{\pi/2} K_1 M(\sin \omega t - \sin \omega t_1) \sin \omega t \, d(\omega t) \right], \quad (8.18)$$

where

$$\omega t_1 = \sin^{-1} (D/M).$$

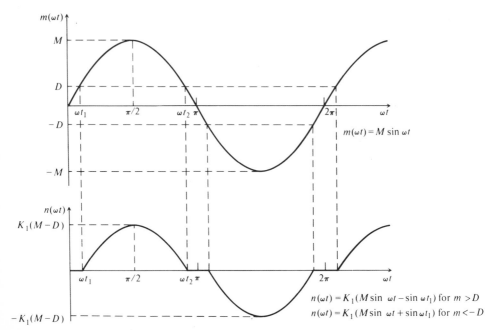

Fig. 8.7   Input and general form of the output waveshape from a nonlinear device having dead-zone characteristics.

Evaluation of Eq. (8.18) results in the expression

$$B_1 = \frac{2K_1M}{\pi}\left(\frac{\pi}{2} - \frac{D}{M}\cos\sin^{-1}\frac{D}{M} - \sin^{-1}\frac{D}{M}\right).\tag{8.19}$$

From Eq. (8.19), since the describing function is the ratio of the amplitude of the fundamental component of the output B_1 to M, it can be expressed as

$$N_{dz}(M) = \frac{B_1}{M} = \frac{2K_1}{\pi}\left(\frac{\pi}{2} - \frac{D}{M}\cos\sin^{-1}\frac{D}{M} - \sin^{-1}\frac{D}{M}\right).\tag{8.20}$$

Notice that the describing function for a dead zone is only a function of the amplitude of the input and not of frequency. Figure 8.8, obtained from Eq. (8.20), is a sketch of the normalized value of the describing function N/K_1 as a function of the ratio D/M. For very small values of D/M the normalized describing function approaches unity. For values of $D/M \geqslant 1$ it equals zero, which implies that the input must be greater than the dead-zone magnitude in order to obtain an output. Notice that the describing function for a dead zone does not have any phase shift.

2. Describing Function of Saturation   Figure 8.9 illustrates the saturation characteristic. The relationship between input and output of this nonlinearity can be expressed

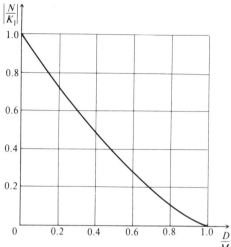

Fig. 8.8 Normalized describing function for a dead zone.

by the following equations

$$n(\omega t) = K_1 M \sin \omega t \qquad \text{for } -S < m < S, \qquad (8.21)$$

$$
\begin{aligned}
n &= K_1 S & \text{for } m > S, \\
n &= -K_1 S & \text{for } m < -S.
\end{aligned}
\qquad (8.22)
$$

Figure 8.10 illustrates typical input and output waveshapes.

Notice that the output waveshape is an odd function for the case of saturation just as it was for the case of a dead zone. Therefore, the expression for the Fourier coefficient B_1 of the output waveshape is

$$B_1 = \frac{2}{T} \int_{-T/2}^{T/2} n(\omega t) \sin \omega t \, d(\omega t). \qquad (8.23)$$

As was true for dead zone, the expression for the Fourier coefficient B_1 can be obtained by taking four times the integral over one quarter of a cycle because of the symmetry

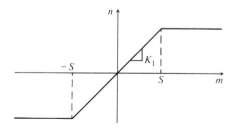

Fig. 8.9 Nonlinear characteristics of saturation.

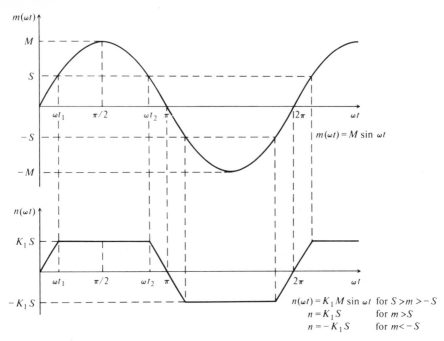

Fig. 8.10 Input and general form of the output waveshape from a nonlinear device having saturation characteristics.

over the four quarters of the period. This results in the expression

$$B_1 = \frac{4}{\pi} \int_0^{\pi/2} n(\omega t) \sin \omega t \, d(\omega t). \tag{8.24}$$

Substituting Eqs. (8.21) and (8.22) into Eq. (8.24), we obtain

$$B_1 = \frac{4}{\pi} \left[\int_0^{\omega t_1} K_1 M \sin \omega t \sin \omega t \, d(\omega t) + \int_{\omega t_1}^{\pi/2} K_1 M \sin \omega t_1 \sin \omega t \, d(\omega t) \right], \tag{8.25}$$

where

$$\omega t_1 = \sin^{-1} \frac{S}{M}.$$

Evaluation of Eq. (8.25) results in the expression

$$B_1 = \frac{2 K_1 M}{\pi} \left(\frac{S}{M} \cos \sin^{-1} \frac{S}{M} + \sin^{-1} \frac{S}{M} \right). \tag{8.26}$$

Since the describing function is defined as the ratio of the amplitude of the fundamental component of the output, B_1, to M, the describing function can be expressed as

$$N_{\text{sat}}(M) = \frac{B_1}{M} = \frac{2 K_1}{\pi} \left(\frac{S}{M} \cos \sin^{-1} \frac{S}{M} + \sin^{-1} \frac{S}{M} \right). \tag{8.27}$$

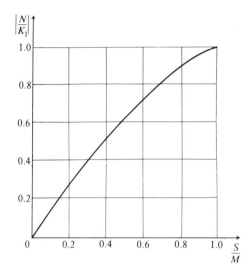

Fig. 8.11 Normalized describing function for saturation.

Notice that the describing function for saturation is only a function of the amplitude of the input and not of frequency. Figure 8.11, obtained from Eq. (8.27), is a sketch of the normalized value of the describing function N/K_1 as a function of the ratio S/M. For very small values of S/M, the normalized describing function approaches zero. For values of $S/M \geqslant 1$ it equals unity, which implies that the output is unaffected by the saturation level if $S \gg M$. Notice that the describing function for saturation does not introduce any phase shift.

It is important to emphasize that the describing functions for dead zone and saturation could have been obtained from one nonlinear characteristic containing

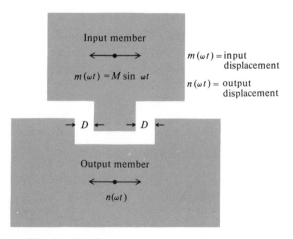

Fig. 8.12 Physical model of backlash.

both types of nonlinearities. Then, the resulting describing function could be reduced to the describing function of dead zone (Eq. 8.20) by letting the saturation level approach infinity, or it could be reduced to the describing function of saturation (Eq. 8.27) by letting the dead-zone width approach zero. It is recommended that the reader derive Eqs. (8.20) and (8.27) using this approach (see Problem 8.2).

3. Describing Function of Backlash Backlash or mechanical hysteresis is due to the difference in motion between an increasing and a decreasing output. Figure 8.12 illustrates a model of backlash, and Fig. 8.13 illustrates its characteristics. The source of backlash that usually receives the most attention is the "looseness" inherent in mechanical gearing. Although attempts have been made to design gears and other mechanical transmission devices so as to fit their mating members very tightly, it is practically impossible to eliminate backlash entirely.

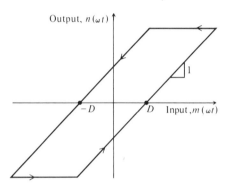

Fig. 8.13   Characteristics of backlash.

From Fig. 8.14, the relationship between input and output for this nonlinearity can be expressed by the following equations:

$$n(\omega t) = -(M - D) \qquad \text{for } 0 \leqslant \omega t < \omega t_1, \tag{8.28}$$

$$n(\omega t) = M \sin \omega t - D \qquad \text{for } \omega t_1 \leqslant \omega t < \frac{\pi}{2}, \tag{8.29}$$

$$n(\omega t) = M - D \qquad \text{for } \frac{\pi}{2} \leqslant \omega t < \omega t_2, \tag{8.30}$$

$$n(\omega t) = M \sin \omega t + D \qquad \text{for } \omega t_2 \leqslant \omega t < \frac{3\pi}{2}, \tag{8.31}$$

$$n(\omega t) = -(M - D) \qquad \text{for } \frac{3\pi}{2} \leqslant \omega t \leqslant 2\pi. \tag{8.32}$$

Notice that the output waveshape is neither an odd nor an even function. This means that the Fourier series of the output waveshape has A_K and B_K components.

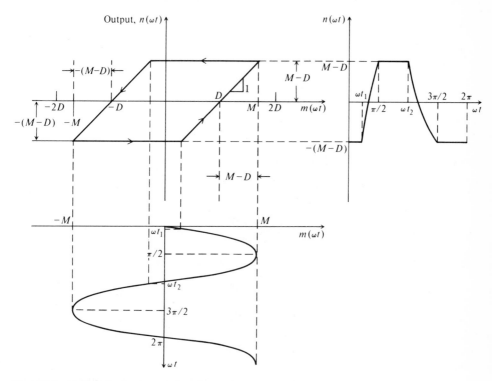

Fig. 8.14 Backlash characteristics where $D < M < 2D$.

Since the describing function is concerned only with the fundamental component of the output, however, we are interested only in A_1 and B_1. From Eqs. (8.28) through (8.32), we can express A_1 as

$$A_1 = \frac{2}{2\pi}\left[\int_0^{\omega t_1} -(M-D)\cos\omega t\,d(\omega t) + \int_{\omega t_1}^{\pi/2}(M\sin\omega t - D)\cos\omega t\,d(\omega t)\right.$$

$$+\int_{\pi/2}^{\omega t_2}(M-D)\cos\omega t\,d(\omega t) + \int_{\omega t_2}^{3\pi/2}(M\sin\omega t + D)\cos\omega t\,d(\omega t)$$

$$\left.+\int_{3\pi/2}^{2\pi}-(M-D)\cos\omega t\,d(\omega t)\right], \tag{8.33}$$

where

$$\omega t_1 = \sin^{-1}\left(\frac{2D}{M} - 1\right) \tag{8.34}$$

and

$$\omega t_2 = \omega t_1 + \pi, \tag{8.35}$$

and B_1 can be expressed as

$$B_1 = \frac{2}{2\pi}\left[\int_0^{\omega t_1} - (M - D)\sin \omega t\, d(\omega t) + \int_{\omega t_1}^{\pi/2} (M \sin \omega t - D)\sin \omega t\, d(\omega t) \right.$$

$$+ \int_{\pi/2}^{\omega t_2}(M - D)\sin \omega t\, d(\omega t) + \int_{\omega t_2}^{3\pi/2}(M \sin \omega t + D)\sin \omega t\, d(\omega t)$$

$$\left. + \int_{3\pi/2}^{2\pi} - (M - D)\sin \omega t\, d(\omega t)\right]. \tag{8.36}$$

Integrating Eqs. (8.33) and (8.36), we obtain the following results:

$$A_1 = \frac{2D}{\pi M}\left(\frac{2D}{M} - 2\right)M, \tag{8.37}$$

$$B_1 = \frac{1}{\pi}\left[\frac{\pi}{2} - \sin^{-1}\left(\frac{2D}{M} - 1\right) - \left(\frac{2D}{M} - 1\right)\cos \sin^{-1}\left(\frac{2D}{M} - 1\right)\right]M. \tag{8.38}$$

Therefore, the describing function for backlash is given by

$$N_{\text{backlash}}(M) = \frac{1}{M}\sqrt{A_1^2 + B_1^2}\,\underline{/\tan^{-1}(A_1/B_1)}. \tag{8.39}$$

This expression is valid only when the positive slope of the backlash characteristic as shown in Fig. 8.14 is unity. If it is any other value, such as K_1, then Eq. (8.39) is modified, since the right-hand sides of the defining equations (8.29) and (8.31) would have to be multiplied by K_1.

Notice that the describing function for backlash is only a function of the amplitude of the input and not of the frequency. Figures 8.15 and 8.16, which have been obtained from Eqs. (8.37) through (8.39), are sketches of the amplitude and phase

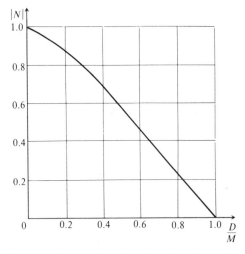

Fig. 8.15 Amplitude characteristics for the describing function of backlash.

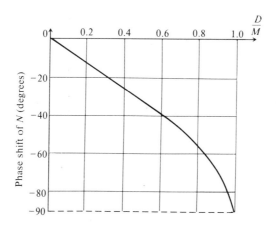

Fig. 8.16 Phase characteristics for the describing function of backlash.

characteristics, respectively, of the describing function as a function of the ratio D/M. Notice that a phase lag occurs at low amplitudes. This phase lag may introduce problems of stability in a feedback system.

4. Describing Function of an On–Off Element Having Hysteresis A class of systems of great practical importance is that of "on–off" control systems. In these systems, as soon as the error signal exceeds a certain level, a relay switches on full corrective torque having proper polarity. When the error falls below a certain level, all the corrective torque is removed. These simple and relatively inexpensive devices find many practical uses in thermostatic control of heat, in automobile voltage regulators, in aircraft and space-vehicle control applications where space and weight limitations are very critical, and so on.

The heart of an on–off control system, the relay or contactor, has a variety of characteristics. For the purpose of deriving a describing function, we consider a three-position contactor exhibiting hysteresis characteristics; this includes the two-position contactor as a limiting case. Figure 8.17 illustrates the input–output characteristic and the waveshapes for such a device. The hysteresis effect occurs because of the different values of control signal required for corrective torque application and its removal. Torque is applied when the control signal reaches $\pm(D + h)$, but it is not removed until the control signal equals $\pm D$. The relationship between input and output for this nonlinearity can be expressed by the following equations:

$$n(\omega t) = 0 \qquad \text{for } \omega t_4 \leqslant \omega t < \omega t_1, \qquad (8.40)$$

$$n(\omega t) = K_1 \qquad \text{for } \omega t_1 \leqslant \omega t < \omega t_2, \qquad (8.41)$$

$$n(\omega t) = 0 \qquad \text{for } \omega t_2 \leqslant \omega t < \omega t_3, \qquad (8.42)$$

$$n(\omega t) = -K_1 \qquad \text{for } \omega t_3 \leqslant \omega t < \omega t_4. \qquad (8.43)$$

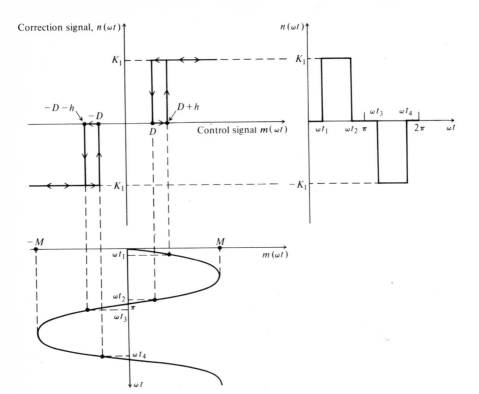

Fig. 8.17 On–off characteristics for a three-position contactor having hysteresis.

Notice that the output waveshape is neither an odd nor an even function. The Fourier series of the output waveshape, therefore, contains A_K and B_K components. From Eqs. (8.40) through (8.43), we can express A_1 as

$$A_1 = \frac{2}{2\pi}\left[\int_{\omega t_4}^{\omega t_1}(0)\cos\omega t\,d(\omega t) + \int_{\omega t_1}^{\omega t_2}(K_1)\cos\omega t\,d(\omega t)\right.$$

$$\left. + \int_{\omega t_2}^{\omega t_3}(0)\cos\omega t\,d(\omega t) + \int_{\omega t_3}^{\omega t_4}(-K_1)\cos\omega t\,d(\omega t)\right], \tag{8.44}$$

where

$$\omega t_1 = \sin^{-1}\frac{D+h}{M}, \tag{8.45}$$

$$\omega t_2 = \pi - \sin^{-1}\frac{D}{M}, \tag{8.46}$$

$$\omega t_3 = \omega t_1 + \pi, \tag{8.47}$$

$$\omega t_4 = \omega t_2 + \pi, \tag{8.48}$$

and B_1 as

$$B_1 = \frac{2}{2\pi}\left[\int_{\omega t_4}^{\omega t_1} (0) \sin \omega t \, d(\omega t) + \int_{\omega t_1}^{\omega t_2} (K_1) \sin \omega t \, d(\omega t)\right.$$

$$\left. + \int_{\omega t_2}^{\omega t_3} (0) \sin \omega t \, d(\omega t) + \int_{\omega t_3}^{\omega t_4} (-K_1) \sin \omega t \, d(\omega t)\right]. \tag{8.49}$$

Integrating Eqs. (8.44) and (8.49), we obtain the expressions

$$A_1 = -\frac{2K_1}{\pi}\left(\frac{h}{M}\right), \tag{8.50}$$

$$B_1 = \frac{2K_1}{\pi}\left[\cos \sin^{-1}\frac{D+h}{M} - \cos\left(\pi - \sin^{-1}\frac{D}{M}\right)\right]. \tag{8.51}$$

The describing function is given by the expression

$$DN_{\text{on-off}}(M) = \frac{D}{M}\sqrt{A_1^2 + B_1^2}\bigg/ \tan^{-1}\frac{A_1}{B_1}. \tag{8.52}$$

Notice that the describing function for this device is a function only of the amplitude of the input and not of frequency. Figures 8.18 and 8.19, which have been obtained from Eqs. (8.50) and (8.51), are sketches of the normalized amplitude and phase characteristics of the describing function as a function of the ratio M/D, respectively. Notice that the phase lag is zero when hysteresis is not present and grows progressively worse as the hysteresis increases.

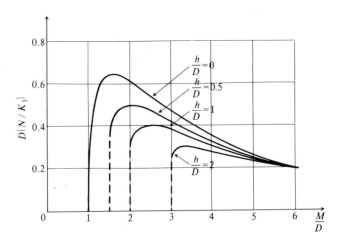

Fig. 8.18 Amplitude characteristics for the describing function of an on–off device having hysteresis.

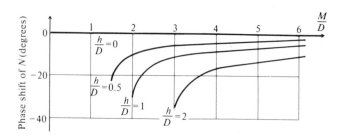

Fig. 8.19 Phase characteristics for the describing function of an on–off device having hysteresis.

5. Describing Function of Coulomb Friction and Stiction In Chapter 3 we considered the effect of damping in linear systems. Damping, a form of friction, is known as *viscous friction*. Its characteristic is that its magnitude is always proportional to velocity, as illustrated in Fig. 8.20(a). The damping factor B is the slope of this characteristic. Another type of frictional force commonly found in control systems is known as *coulomb friction*. Unlike viscous friction, it is not proportional to velocity but is a constant force that always opposes the velocity. This nonlinear phenomenon is illustrated in Fig. 8.20(b), where the coulomb friction force is denoted by $\pm F_c$. Another nonlinear form of frictional force, known as static friction or *stiction*, is the value of the frictional force at zero velocity. It is usually denoted by $\pm F_s$. Figure 8.20(c) illustrates the composite frictional-force characteristics generally encountered when controlling some load.

To determine the describing function of coulomb friction, we can express the relationship between the input and output as

$$n(\omega t) = m(\omega t) \pm F_c$$

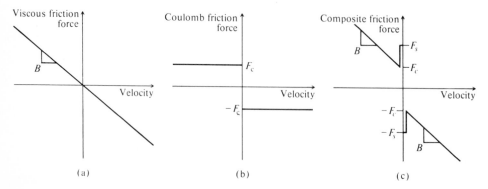

Fig. 8.20 (a) Viscous friction characteristics. (b) Coulomb friction characteristics. (c) Composite friction characteristics illustrating viscous, coulomb, and static friction (stiction).

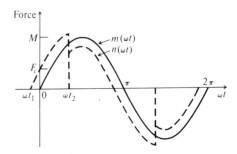

Fig. 8.21 Coulomb friction characteristics.

where

$$m(\omega t) = \text{applied force,}$$
$$F_c = \text{force necessary to overcome coulomb friction,}$$
$$n(\omega t) = \text{output force.}$$

The corresponding steady-state waveforms are given by Fig. 8.21. It should be noted that the discontinuities of the output waveform correspond to zero velocity since the force required to overcome coulomb friction changes sign at these instants. The relationship between the input and output forces is

$$n(\omega t) = M \sin \omega t + F_c \qquad \text{for } -\omega t_1 < \omega t < \omega t_2, \qquad (8.53)$$

$$n(\omega t) = M \sin \omega t - F_c \qquad \text{for } \omega t_2 < \omega t < \pi, \qquad (8.54)$$

where

$$\omega t_1 = \sin^{-1} \gamma, \qquad \omega t_2 = \cos^{-1}(\pi \gamma / 2), \qquad \gamma = F_c / M.$$

The fundamental components of the output A_1 and B_1 are

$$A_1 = \frac{2}{\pi} \int_{-\omega t_1}^{\omega t_2} M(\sin \omega t + \gamma) \cos \omega t \, d(\omega t)$$
$$+ \frac{2}{\pi} \int_{\omega t_2}^{\pi - \omega t_1} M(\sin \omega t - \gamma) \cos \omega t \, d(\omega t), \qquad (8.55)$$

$$B_1 = \frac{2}{\pi} \int_{-\omega t_1}^{\omega t_2} M(\sin \omega t + \gamma) \sin \omega t \, d(\omega t)$$
$$+ \frac{2}{\pi} \int_{\omega t_2}^{\pi - \omega t_1} M(\sin \omega t - \gamma) \sin \omega t \, d(\omega t). \qquad (8.56)$$

Integrating Eqs. (8.55) and (8.56), we obtain the following expressions:

$$A_1 = 2M\gamma \left(\frac{4}{\pi^2} - \gamma^2\right)^{1/2}, \qquad (8.57)$$

$$B_1 = M[1 - 2\gamma^2]. \qquad (8.58)$$

The resulting expression for coulomb friction is

$$N_c(\gamma) = \left[1 - 4\left(1 - \frac{4}{\pi^2}\right)\gamma^2\right]^{1/2} \bigg/ \tan^{-1}\frac{2\gamma(4/\pi^2 - \gamma^2)^{1/2}}{1 - 2\gamma^2} \qquad \text{for} \quad \gamma \leqslant 0.536 \quad (8.59)$$

$$N_c(\gamma) = \frac{1}{\pi}\{[\pi - (\omega t_1 - \omega t_2) - \sin \omega t_1(\cos \omega t_1 + \cos \omega t_2)$$

$$- \cos \omega t_2(\sin \omega t_1 + \sin \omega t_2)]^2 + [\sin \omega t_1 + \sin \omega t_2]\}^{1/2}$$

$$\times \bigg/ \tan^{-1}\frac{\sin \omega t_1 + \sin \omega t_2}{\{\pi - (\omega t_1 - \omega t_2) - \sin \omega t_1(\cos \omega t_1 + \cos \omega t_2)}$$

$$- \cos \omega t_2(\sin \omega t_1 + \sin \omega t_2)\}$$

$$\text{for} \quad \gamma > 0.536. \quad (8.60)$$

Observe that the describing function for coulomb friction depends only on the amplitude of the input and not on its frequency. The gain–phase relationship of the describing function for coulomb friction is illustrated in Fig. 8.22.

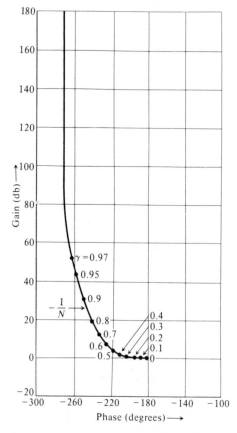

Fig. 8.22 Gain–phase characteristics for the describing function of coulomb friction.

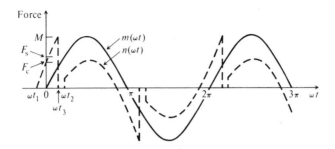

Fig. 8.23   Coulomb friction and stiction characteristics.

The describing function for the simultaneous occurrence of both coulomb friction and stiction is considered next. Waveform relationships between the applied force $m(\omega t)$, the output force $n(\omega t)$, and the forces necessary to overcome coulomb friction and stiction F_c and F_s are given in Fig. 8.23. The expressions for the fundamental components of the Fourier coefficients are

$$A_1 = \frac{2}{\pi} \int_{\omega t_2}^{\pi + \omega t_3} M(\sin \omega t - \gamma) \cos \omega t \, d(\omega t), \tag{8.61}$$

$$B_1 = \frac{2}{\pi} \int_{\omega t_2}^{\pi + \omega t_3} M(\sin \omega t - \gamma) \sin \omega t \, d(\omega t), \tag{8.62}$$

where

$$\omega t_1 = \sin^{-1} \gamma,$$
$$\omega t_2 = \sin^{-1} (F_s/M),$$
$$\omega t_3 = \cos^{-1} (\pi\gamma/2),$$
$$\gamma = F_c/M.$$

The describing function for the combined case of coulomb friction and stiction is obtained by integrating Eqs. (8.61) and (8.62). The resultant expression is

$$N(\gamma) = \frac{1}{\pi} \{[\pi - (\omega t_2 - \omega t_3) - \cos \omega t_2(2 \sin \omega t_1 - \sin \omega t_2)$$

$$- \cos \omega t_3(2 \sin \omega t_1 + \sin \omega t_3)]^2$$

$$+ [\sin \omega t_3(2 \sin \omega t_1 + \sin \omega t_3) + \sin \omega t_2(2 \sin \omega t_1 - \sin \omega t_2)]^2\}^{1/2}$$

$$\times \, \Big/ \tan^{-1} \frac{\sin \omega t_3(2 \sin \omega t_1 + \sin \omega t_3) + \sin \omega t_2(2 \sin \omega t_1 - \sin \omega t_2)}{\pi - (\omega t_2 - \omega t_3) - \cos \omega t_2(2 \sin \omega t_1 - \sin \omega t_2)}$$
$$- \cos \omega t_3(2 \sin \omega t_1 + \sin \omega t_3) \, .$$

$$\tag{8.63}$$

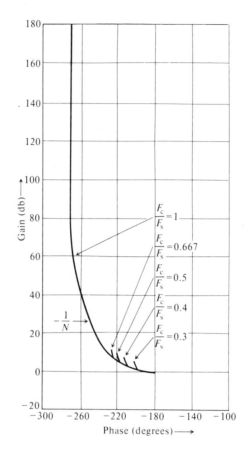

Fig. 8.24 Gain–phase characteristics for the describing function of coulomb friction and stiction.

Observe from this expression that the describing function for the combined case of coulomb friction and stiction is a function of the amplitude of the input and the relative magnitudes of friction but not of frequency. Figure 8.24 illustrates the gain–phase relationship of the describing function for the combined case of coulomb friction and stiction.

8.9 USE OF THE DESCRIBING FUNCTION TO PREDICT OSCILLATIONS

The describing function of a nonlinear element can be utilized to determine the existence of a limit cycle in a nonlinear feedback control system in an approximating manner [8]. Let us consider the system illustrated in Fig. 8.5. If we assume that the describing function of the nonlinearity is $N(M, \omega)$ and $R(j\omega) \equiv 0$, let us determine the conditions under which an assumed oscillation $m(\omega t)$ can be sustained, where

$$m(\omega t) = M \cos \omega t, \tag{8.64}$$

The fundamental component of $n(\omega t)$ is given by

$$n_1(\omega t) = |N_1(M, \omega)| \, M \cos [\omega t + \phi_1(M, \omega)]. \tag{8.65}$$

Expressing Eqs. (8.64) and (8.65) as the real part of a complex exponential, we obtain the following expressions:

$$m(\omega t) = \text{Re} \, [M e^{j\omega t}],$$
$$n_1(\omega t) = \text{Re} \, [|N_1(M, \omega)| \, M e^{j(\omega t + \phi_1)}].$$

The output of the linear elements is given by

$$\text{Re} \, [|N_1(M, \omega)| \, M \, |G(j\omega)| \, e^{j(\omega t + \phi_1 + \alpha)}], \tag{8.66}$$

where

$$G(j\omega) = G_1(j\omega)G_2(j\omega),$$
$$\alpha = \text{phase shift due to } G(j\omega).$$

Equation (8.66) must equal $-m(\omega t)$ if the initial assumption is to hold. Thus, dropping the real-part notation, we obtain the following expression:

$$-M e^{j\omega t} = |N_1(M, \omega)| \, M \, |G(j\omega)| \, e^{j(\omega t + \phi_1 + \alpha)}.$$

This can be rewritten as

$$[|N_1(M, \omega)| \, |G(j\omega)| \, e^{j(\phi_1 + \alpha)} + 1]M e^{j\omega t} = 0. \tag{8.67}$$

For a sustained oscillation, the bracketed term in Eq. (8.67) must vanish, since $M \neq 0$. Therefore,

$$1 + N(M, \omega)G(j\omega) = 0 \tag{8.68}$$

and any combination of values of input amplitude M and frequency ω which satisfy the equation

$$G(j\omega) = -\frac{1}{N(M, \omega)} \tag{8.69}$$

are capable of providing a limit cycle. If a combination of amplitude and frequency can be found which satisfies Eq. (8.69), the feedback control system can have a sustained oscillation.

From experience, I have found the Nyquist diagram and the Nichols chart (gain–phase plot) to be the most revealing techniques for stability analysis when utilizing the describing-function method. Two separate sets of loci, corresponding to $G(j\omega)$ and $-1/N(M, \omega)$ of Eq. (8.69), are sketched on the same graph for either of these methods. Generally, the sketch of $-1/N(M, \omega)$ will be a family of curves for

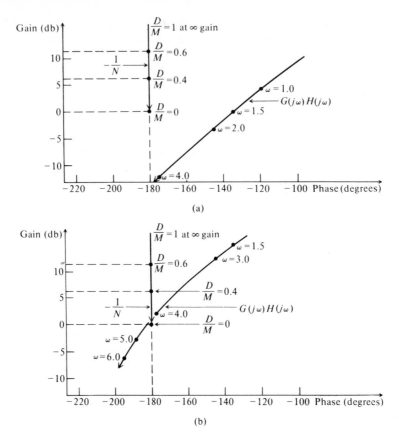

Fig. 8.25 (a) Gain–phase diagram stability analysis of nonlinear system containing a dead zone where

$$G(j\omega)H(j\omega) = \frac{2}{j\omega(1 + 0.5j\omega)(1 + 0.1j\omega)}.$$

(b) Gain–phase diagram stability analysis of nonlinear system containing a dead zone where

$$G(j\omega)H(j\omega) = \frac{17}{j\omega(1 + 0.5j\omega)(1 + 0.1j\omega)}.$$

different input magnitudes M and frequencies ω. Intersections of the two loci indicate possible solutions to Eq. (8.69) and yield information as to the magnitude and frequency ω of sustained oscillation. If no intersections result, an oscillation is unlikely. The distance to a possible intersection can be used as a criterion of closeness to oscillation. This is next illustrated on the gain–phase diagram for several representative systems.

Figure 8.25(a) illustrates an analysis via the gain–phase diagram for a nonlinear system containing a dead zone where $K_1 = 1$. The figure illustrates a stable system where a dead zone is present and

$$G(j\omega)H(j\omega) = \frac{2}{j\omega(1 + 0.5j\omega)(1 + 0.1j\omega)} . \tag{8.70}$$

Figure 8.25(b) illustrates a system where a dead zone is present and

$$G(j\omega)H(j\omega) = \frac{17}{j\omega(1 + 0.5j\omega)(1 + 0.1j\omega)} . \tag{8.71}$$

An intersection occurs at a frequency of approximately 4.4 and a value of D/M of 0.09. This is to be interpreted as the frequency and amplitude which satisfy Eq. (8.69) and which result in a limit cycle. Notice that the system is unstable from a linear viewpoint since the $-1,0$ point is enclosed.

The interpretation of this situation is quite illuminating. Since the normalized describing function of a dead zone is less than one, its effect is to reduce the overall system gain. When multiplied by the transfer function given by Eq. (8.70), which would produce a stable feedback system by itself, its effect is to make the system even more stable. Therefore, in order to illustrate a limit cycle in this nonlinear system containing a dead zone, it is necessary to illustrate a system that produces an unstable feedback system from a linear viewpoint as well. The frequency function of Eq. (8.71) satisfies this requirement, as indicated in Fig. 8.25(b).

Figure 8.26 illustrates the gain–phase diagram analysis for a nonlinear system containing backlash, where

$$G(j\omega)H(j\omega) = \frac{1.5}{j\omega(1 + j\omega)^2} . \tag{8.72}$$

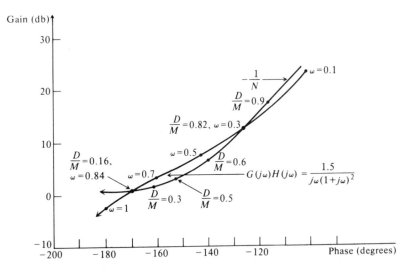

Fig. 8.26   Gain–phase diagram analysis of nonlinear system containing backlash.

Notice that the system has two points of intersection corresponding to a pair of limit cycles. They occur at $\omega = 0.84$, $D/M = 0.16$ and $\omega = 0.3$, $D/M = 0.82$. These two limit cycles must now be examined to determine whether they are stable or unstable.

A generalized rule can be established for determining whether a limit cycle is stable or unstable [39]. If the two loci are assigned a sense of direction so that the linear locus $G(j\omega)$ is pointing in the direction of increasing frequency and the nonlinear locus $-1/N$ is pointing in the direction of increasing amplitude M (decreasing D/M in our example), then a stable limit cycle occurs when the nonlinear locus appears to an observer, stationed on the linear locus and facing in the direction of increasing frequency, to cross from left to right in the direction of increasing amplitude M. If the opposite occurs, then the limit cycle is unstable and the state of the system is divergent. As an example of applying this rule, consider the gain–phase plot of Fig. 8.27. Here, we see that there are two unstable limit cycles (divergent states) and three stable limit cycles (convergent states). The unstable limit cycles cannot maintain themselves in the presence of minute disturbances; the stable limit cycles will always maintain themselves in the presence of disturbances.

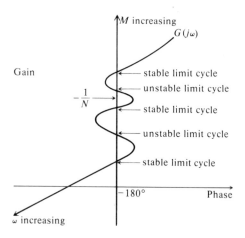

Fig. 8.27   Illustration of stable and unstable limit cycles.

Applying this generalized rule for determining whether the limit cycles in Fig. 8.26 are stable or unstable, we find that $\omega = 0.84$, $D/M = 0.16$ corresponds to a stable limit cycle and the other intersection corresponds to an unstable limit cycle. The stable limit cycle corresponds to a convergent point because disturbances at either side tend to converge to these conditions. This is contrasted with the unstable limit cycle at $\omega = 0.3$, $D/M = 0.82$, which is a divergent point, since disturbances which are not large enough to reach this intersection will decay, and disturbances which are larger will result in oscillations which tend to increase in amplitude until the stable limit cycle is reached.

8.10 DESIGN OF NONLINEAR FEEDBACK CONTROL SYSTEMS USING DESCRIBING FUNCTIONS

The purpose of this section is to illustrate the procedure for compensating for undesirable nonlinearities in a system. As an example, we consider the nonlinearity to be backlash. The gain–phase plot will be used for analysis, although we could use the Nyquist diagram just as well.

Oscillating input signals, having frequencies very much greater in magnitude than the system bandwidth, can be used to maintain the output of a system containing backlash at its correct average value. This technique is known as *dither*. It is effective in reducing the influence of very small amplitude nonlinearities. However, the resultant increased wear on the system is a serious disadvantage of this simple approach to the problem. The effect of dither on the describing function is analyzed in References 9 through 14 where the "dual input describing function" is discussed.* For systems which are to operate continuously for long periods of time, it is necessary to utilize other approaches which will eliminate backlash. Reducing the system gain, addition of phase-lead networks, and introducing rate feedback are all relatively simple methods which can be used to minimize the effects of backlash. We next demonstrate the theoretical effects of these techniques on backlash [1, 8].

Let us reconsider the nonlinear system analysis of Fig. 8.26. This sketch illustrated that a nonlinear system having the form shown in Fig. 8.5, where

$$G(j\omega) = G_1(j\omega)G_2(j\omega) = \frac{1.5}{j\omega(1 + j\omega)^2},$$

and having a backlash element, was indeed oscillatory. We demonstrate how this system can be stabilized by each of the following electrical methods:

a) Reducing the system gain.

b) Adding a phase-lead network.

c) Introducing rate feedback.

1. Reducing the System Gain   In order to consider the effects of gain changes, let us rewrite $G(j\omega)$ of Eq. (8.68) as

$$G(j\omega) = KG'(j\omega),$$

where K represents the system gain and $G'(j\omega)$ represents only the poles and zeros of the linear part of the system. Therefore, Eq. (8.69) can be rewritten as

$$G'(j\omega) = -\frac{1}{KN(M, \omega)}.$$

* The dual-input describing function is a modified describing function which is dependent on two frequency components: the intelligence signal and the dither signal.

By reducing the gain K, the limit cycle is eliminated because the curve of $-1/KN$ is moved upward. Figure 8.28 illustrates how the oscillation can be eliminated by reducing the system gain from 1.5 to unity. At a gain setting of approximately 1.3, the curves of $G'(j\omega)$ and $-1/KN$ just clear each other. A gain setting of unity was chosen in order to maintain some margin of safety.

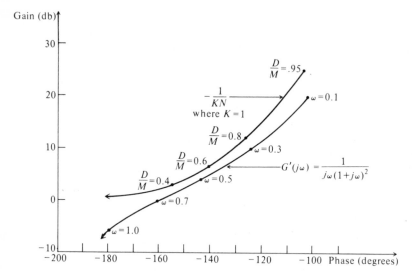

Fig. 8.28 Compensation of a nonlinear system containing backlash, by reduction of the system gain.

2. Addition of a Phase-Lead Network A passive phase-lead network can also be used to eliminate the oscillation. The transfer function of this network is given by

$$G(j\omega)_{\text{lead}} = \frac{1 + \alpha T(j\omega)}{1 + T(j\omega)} \qquad (8.73)$$

where $\alpha T > T$. (The attenuation $1/\alpha$ is nullified by increasing the system gain by α.) The compensated value of $G(j\omega)$, $G(j\omega)_{\text{comp}}$, is given by

$$G(j\omega)_{\text{comp}} = \frac{1.5}{j\omega(1 + j\omega)^2} \times \frac{1 + \alpha T(j\omega)}{1 + T(j\omega)}. \qquad (8.74)$$

For the system under consideration, a value of $\alpha T = 0.8$ and $T = 0.4$ will eliminate the limit cycle, as is illustrated in Fig. 8.29. This is not the only lead network which could be used for compensation; it is one of many possible solutions, as can be noted from studying the gain–phase diagram.

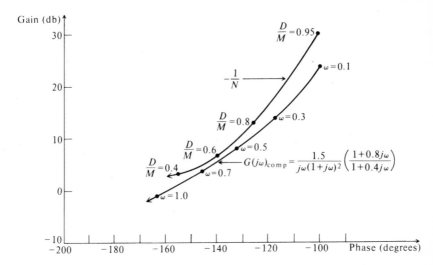

Fig. 8.29 Compensation of a nonlinear system containing backlash, by addition of a phase-lead network.

3. Introduction of Rate Feedback Addition of rate feedback can also be a very effective method for eliminating the oscillation. For this configuration, which is illustrated in Fig. 8.30, the value of the system feedback element $H(j\omega)$ is $1 + bj\omega$. Therefore, the oscillation criterion for this configuration must be modified from that given by Eq. (8.69) to the following expression:

$$G(j\omega)H(j\omega) = - \frac{1}{N(M, \omega)}. \tag{8.75}$$

With rate feedback, the value of $G(j\omega)H(j\omega)$ for the system being considered is given by

$$G(j\omega)H(j\omega) = \frac{1.5}{\omega(1 + j\omega)^2}(1 + bj\omega). \tag{8.76}$$

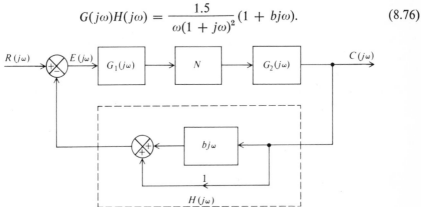

Fig. 8.30 Nonlinear system containing rate and position feedback.

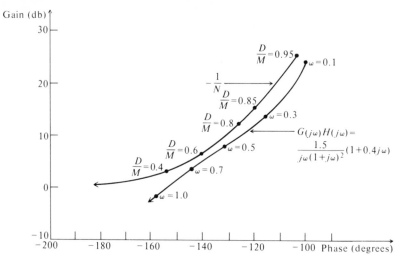

Fig. 8.31 Compensation of a nonlinear system containing backlash, by the introduction of rate feedback.

A value of $b = 0.4$ will eliminate the limit cycle, as is illustrated in Fig. 8.31, and a relatively safe margin is achieved. At a value of b approximately equal to 0.25, the curves of $G(j\omega)H(j\omega)$ and $-1/N$ just clear each other.

Since the describing-function analysis is an amplitude-sensitive method applicable to nonlinear systems, and since superposition of signals is not valid for these systems, we cannot extend this method and obtain data on low-amplitude transient responses. Our main concern has been the elimination of limit cycles.

8.11 DIGITAL COMPUTER COMPUTATION OF THE DESCRIBING FUNCTION

The digital computer is very useful for the construction of the describing function [40, 42]. The computer can compute $-1/N$ and, as indicated in Chapter 6, $G(j\omega)$. This section illustrates the procedure used to analyze and compensate a practical non-linear system containing backlash with the aid of a digital computer. It is based on the analytic method illustrated in Sections 8.7 through 8.10. The BASIC language will be used for the program [34, 35].

Let us consider the system illustrated in Fig. 8.5, where

$$G(j\omega) = \frac{4(1 + 3j\omega)}{j\omega(1 + 2j\omega)^2} \tag{8.77}$$

and N corresponds to backlash. We know from Eq. (8.39) that

$$N_{\text{backlash}}(M) = \frac{1}{M} \sqrt{A^2 + B^2} \,\underline{/\tan^{-1}(A/B)}, \tag{8.78}$$

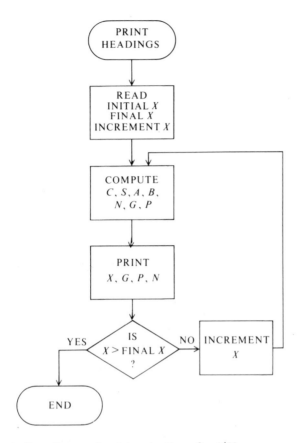

Fig. 8.32 (a) Logic flow diagram for determination of $-1/N$.

where

$$A = \frac{2D}{\pi M}\left(\frac{2D}{M} - 2\right)M,$$

$$B = \frac{1}{\pi}\left[\frac{\pi}{2} - \sin^{-1}\left(\frac{2D}{M} - 1\right) - \left(\frac{2D}{M} - 1\right)\cos\sin^{-1}\left(\frac{2D}{M} - 1\right)\right]M.$$

Note that the subscripts have been dropped from the A- and B-terms for simplicity. The coding symbols used are as follows:

$$X = D/M, \qquad S = \sin^{-1} C = \tan^{-1}\left(\frac{C}{\sqrt{1 - C^2}}\right), \qquad P = -\pi - \tan^{-1}(A/B),$$

$$C = 2X - 1, \qquad G = 20\log_{10}\frac{1}{N}.$$

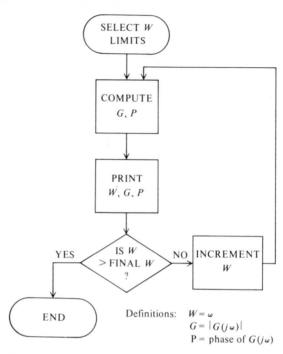

Fig. 8.32 (b) Logic flow diagram for determination of $G(j\omega)$.

Figure 8.32(a) illustrates the logic flow diagram for developing the program for computing $-1/N$. Table 8.1 illustrates the actual program for computing $-1/N$. The reader should compare these two in order fully to understand the digital computer's program. Table 8.2 illustrates the computer run for calculating $-1/N$. Notice that it was necessary to compute additional values of $0.9 < D/M < 0.99$ at the end of the computer run since it was found that a limit cycle existed in this region. Figure 8.32(b) illustrates the logic flow diagram for determining $G(j\omega)$. In the coding, G represents the gain of $G(j\omega)$, P represents its phase and W represents ω. Table 8.3 illustrates the actual program for computing $G(j\omega)$ and Table 8.4 illustrates the computer run for calculating $G(j\omega)$.

The values for $-1/N$ and $G(j\omega)$ are illustrated on the gain–phase diagram in Fig. 8.33. It indicates limit cycles at $\omega = 1.36$, $D/M = 0.41$ (stable) and $\omega = 0.27$, $D/M = 0.945$ (unstable). As illustrated previously in Section 8.10, this nonlinear system can be compensated by lowering the gain, adding a phase–lead network or rate feedback. A similar analysis indicates that at $K = 1.65$, a lead network given by

$$G(j\omega) = \frac{1 + 1.2j\omega}{1 + 0.4j\omega}$$

Table 8.1 Computer program for computing $-1/N$ (BASIC program).

```
  1   REM DESCRIBING FUNCTION FOR BACKLASH
 10   PRINT "D/M","GAIN(DB)","PHASE(DEGREES)","N"
 20   READ X1,X2,D
 30   FOR X=X1 TO X2 STEP D
 40   LET C=2*X−1
 50   LET S=ATN(C/SQR(1−C↑2))
 60   LET A=1.27324*X*(X−1)
 70   LET B=0.31831*(1.570796−S−C*COS(S))
 80   LET N=SQR(A↑2+B↑2)
 90   LET G=20*0.43429448*LOG(1/N)
100   LET P=−180−57.29578*ATN(A/B)
110   PRINT X,G,P,N
120   NEXT X
130   DATA 0.05, 0.95, 0.05
140   END
RUN
```

Table 8.2 Results of computer analysis for $-1/N$

Balash D/M	Gain(DB)($-1/N$)	Phase(Degrees)	N
0.05	0.147435	-176.473	0.983169
0.1	0.401232	-173.107	0.954857
0.15	0.720751	-169.841	0.92037
0.2	1.0957	-166.638	0.881485
0.25	1.52297	-163.472	0.839173
0.3	2.00298	-160.322	0.794056
0.35	2.53864	-157.17	0.746565
0.4	3.13505	-153.998	0.697024
0.45	3.79968	-150.787	0.645678
0.5	4.54295	-147.518	0.592724
0.55	5.37919	-144.17	0.53832
0.6	6.32826	-140.714	0.4826
0.65	7.41848	-137.119	0.425673
0.7	8.69166	-133.34	0.367635
0.75	10.213	-129.314	0.308567
0.8	12.092	-124.95	0.248541
0.85	14.5345	-120.088	0.187619
0.9	18.0025	-114.426	0.125856
0.95	23.9717	-107.175	6.33018 E-2
Time: 1 Secs			

KEY
READY.
130 DATA 0.9, 0.99, 0.01 (Modification to Address 130 of Computer Program shown in Table 8.1)

RUN

Balash D/M	Gain(DB)($-1/N$)	Phase(Degrees)	N
0.9	18.0025	-114.426	0.125856
0.91	18.9072	-113.146	0.113407
0.92	19.9199	-111.798	0.100927
0.93	21.0694	-110.367	8.84157 E-2
0.94	22.3982	-108.836	7.58739 E-2
0.95	23.9717	-107.175	6.33018 E-2
0.96	25.8999	-105.345	5.06999 E-2
0.97	28.3887	-103.275	3.80685 E-2
0.98	31.9007	-100.827	2.54078 E-2
0.99	37.9115	-97.6472	1.27182 E-2
Time: 1 Secs			

Table 8.3 Computer program for computing $G(j\omega)$ (BASIC program).

RESPG

```
10   PRINT "OMEGA", "GAIN(DB)", "PHASE(DEG)"
20   READ K
30   READ W1, W2, D
40   FOR W = W1 TO W2 STEP D
50   LET G = 4.3429448*LOG(K*K*(1 + 9*W*W)/(W*W*(1 + 4*W*W)↑2))
60   LET P = −90 + 57.29578*(ATN(3*W) − 2*ATN(2*W))
70   PRINT W, G, P
80   NEXT W
90   DATA 4, 0.1, 2.0, 0.05
200  END
```

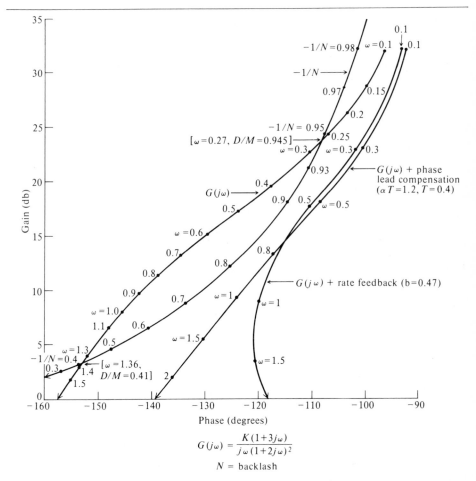

$$G(j\omega) = \frac{K(1+3j\omega)}{j\omega(1+2j\omega)^2}$$

N = backlash

Fig. 8.33 Describing-function analysis based on computer results.

Table 8.4 Results of computer analysis for $G(j\omega)$

RESPG Omega	Gain(DB)	Phase(Deg)
0.1	32.0748	−95.9206
0.15	28.5717	−99.1707
0.2	26.0668	−102.639
0.25	24.0824	−106.26
0.3	22.4048	−109.94
0.35	20.9235	−113.587
0.4	19.577	−117.125
0.45	18.3297	−120.503
0.5	17.16	−123.69
0.55	16.0544	−126.671
0.6	15.004	−129.443
0.65	14.0026	−132.012
0.7	13.0454	−134.388
0.75	12.1288	−136.582
0.8	11.2499	−138.609
0.85	10.4059	−140.482
0.9	9.59458	−142.214
0.95	8.81385	−143.818
1.	8.0618	−145.305
1.05	7.33669	−146.686
1.1	6.63691	−147.97
1.15	5.96097	−149.167
1.2	5.3075	−150.284
1.25	4.67521	−151.329
1.3	4.06293	−152.306
1.35	3.46955	−153.223
1.4	2.89405	−154.085
1.45	2.33549	−154.895
1.5	1.79296	−155.659
1.55	1.26566	−156.379
1.6	0.752819	−157.06
1.65	0.253712	−157.704
1.7	−0.232325	−158.315
1.75	−0.705916	−158.894
1.8	−1.16765	−159.443
1.85	−1.61806	−159.966
1.9	−2.05767	−160.464
1.95	−2.48695	−160.938
2.	−2.90636	−161.39
Time: 2 Secs.		

and rate feedback having a constant of 0.47 result in a 3 db margin of safety. The system compensated with a lead network and rate feedback is illustrated in Fig. 8.33.

8.12 PIECEWISE-LINEAR APPROXIMATIONS

Approximating any nonlinearity by means of piecewise-linear segmentation is a very useful tool for analysis. Each segment leads to a relatively simple linear differential equation which can be solved by conventional linear techniques. This method, which is not limited to quasilinear systems, has the advantage of yielding an exact solution for nonlinearities of any order, if the nonlinearity is itself piecewise linear or can be approximated by piecewise-linear segments. We illustrate its application by means of an example.

Let us consider saturation. Figure 8.34 illustrates a simple feedback control system containing an integrator and an amplifier which saturates. The amplifier gain is 5 over an input voltage range of ± 1 volt. For input voltages greater than this, the amplifier saturates. It is quite evident that two distinct linear operating regions for the amplifier exist. Each of these linear regions can be considered separately in a piecewise-linear manner in order to obtain the composite response of the system.

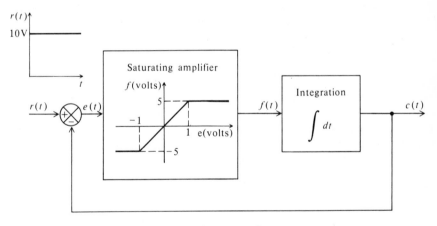

Fig. 8.34 Feedback control system containing saturation.

For the unsaturated region, the relationships depicting the system operation are

$$e(t) = r(t) - c(t), \tag{8.79}$$

$$f(t) = 5e(t), \tag{8.80}$$

$$c(t) = \int f(t)\, dt. \tag{8.81}$$

During saturation, Eqs. (8.79) and (8.81) are still valid. However, (8.80) changes to

$$f(t) = 5 \qquad \text{for} \qquad e(t) > 1 \tag{8.82}$$
$$f(t) = -5 \qquad \text{for} \qquad e(t) < -1. \tag{8.83}$$

Assuming zero initial conditions and a step input of 10 V, the expression for the output during the saturated region of operation $c_{\text{sat}}(t)$ is given by

$$c_{\text{sat}}(t) = \int_0^t 5 \, dt = 5t. \tag{8.84}$$

The expression for the output during the unsaturated region is given by

$$c_{\text{us}}(t) = 5 \int_{t_1}^t (10 - c) \, dt, \qquad \text{or} \qquad \frac{dc_{\text{us}}}{dt} + 5c = 50. \tag{8.85}$$

The time t_1 is the time at which the amplifier becomes unsaturated. When $c = 9$, $e = 1$, and t_1 is 1.8 sec. Using conventional techniques, the solution to Eq. (8.85) is

$$c_{\text{us}}(t) = 10 - e^{-5(t-1.8)}. \tag{8.86}$$

The initial value for this region, $c_{\text{us}}(0)$, is the same as the final value of the saturated region, $c_{\text{sat}}(1.8) = 9$; this continuity of the output is imposed by the integrator. Therefore, the composite solution for this problem, obtained by a piecewise-linear analysis, is

$$c_{\text{sat}}(t) = 5t \qquad\qquad \text{for } 0 \leqslant t \leqslant 1.8, \tag{8.87}$$
$$c_{\text{us}}(t) = 10 - e^{-5(t-1.8)} \qquad \text{for } t > 1.8. \tag{8.88}$$

The response of this system to a step input of 10 V is sketched in Fig. 8.35.

The piecewise-linear approach illustrated in the preceding problem can be extended to very complex nonlinearities. It is important to emphasize that the boundary conditions between linear regions are continuous whenever the transfer

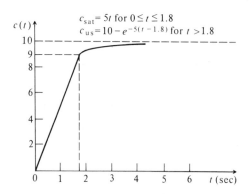

Fig. 8.35   Step response of saturating system obtained from piecewise-linear analysis.

function following the nonlinearity is a proper rational function. The resulting differential equation for each segmented region is linear and can be easily solved by conventional linear techniques.

8.13 STATE-SPACE ANALYSIS: THE PHASE PLANE

A useful technique of applying the state-space approach to nonlinear systems is the phase-plane method [7, 17–19]. It is a technique for analyzing the transient response of a nonlinear feedback control system to a step input or for solving an initial-condition problem. The phase-plane method is limited to second-order systems. The variation of the displacement is plotted against velocity on a graph known as the *phase plane*. A curve for a specific step input is known as a *trajectory*. A set of curves of displacement versus velocity of a specific system, which are repeated for several step amplitudes and are plotted on the same phase plane, is called a *phase portrait*.

The starting point of a trajectory is the initial displacement, $x(0)$, and velocity, $(dx/dt)(0)$. The future path of the trajectory after it leaves its initial starting point represents the behavior of the system for some step-input excitation. If the trajectory approaches infinity on the phase plane, the system is unstable. If the phase trajectory approaches the vicinity of the origin, however, the system is stable. If the trajectory circles the origin continuously in a closed curve after excitation, a sustained oscillation known as a *limit cycle* exists.

The phase-plane method is specifically concerned with the solution of second-order, nonlinear differential equations having the following general form:

$$\frac{d^2x}{dt^2} + A_1\left(x, \frac{dx}{dt}\right)\frac{dx}{dt} + A_2\left(x, \frac{dx}{dt}\right)x = 0. \tag{8.89}$$

The initial conditions, $x(0)$ and $(dx/dt)(0)$, represents the input to the system. Equation (8.89) immediately emphasizes some serious limitations of the phase-plane method. It is only useful for analyzing second-order systems. Since the right-hand side of Eq. (8.89) is zero, the analysis is further limited to initial value problems, or to step inputs where appropriate transformations can be made to produce a zero right-hand side. It is interesting to compare the limitations of the phase-plane method with those of the describing-function analysis, where the response to sinusoidal inputs of systems having any order could be determined. The phase-plane method could be extended to higher-order systems, but it is too impractical.*

The following section discusses the techniques which can be used for constructing the phase portrait from the differential equation of a system. Then we examine the

---

* A system described by an nth-order differential equation requires an n-dimensional phase space with a knowledge of n initial conditions. It is, indeed, an arduous task to visualize this for third- and higher-order systems, and is rarely used.

properties and interpretation of the phase portrait. The procedure to be followed for applying the phase-plane approach to some representative design problems is then presented.

8.14 CONSTRUCTION OF THE PHASE PORTRAIT

Three procedures that can be used to construct the phase portrait of a system are

1. direct solution of the differential equation,

2. transformation of the second-order differential equation to a first-order equation,

3. the method of isoclines.

The first two methods are analytical techniques; the last is graphical. We shall describe these methods next, together with some simple illustrative examples.

1. Direct Solution of The Differential Equation This is the most straightforward method for obtaining the phase portrait. It is usually the most useless method from a practical viewpoint since we do not have to resort to the phase-plane representation for a solution if the differential equation is integrable. However, this method does give a physical "feel" for the situation.

The procedure is to solve the differential equation for the dependent variable x. The solution for x is then differentiated in order to obtain the derivative of the dependent variable, dx/dt. The independent variable, time, is then eliminated between the two resulting equations. A single equation that relates x and dx/dt results, and can be used to plot the phase portrait directly. However, the approach has the disadvantage that it requires the solution of a nonlinear, second-order, differential equation.

Let us illustrate the procedure in detail by first considering the simple linear system illustrated in Fig. 3.7. That configuration illustrated a system where a torque $T(t)$ was applied to a body having a moment of inertia J, a twisting shaft having a stiffness factor K, and a damper having a damping factor B. Applying Newton's second law of motion to this system resulted in the following relationship:

$$J \frac{d^2\theta(t)}{dt^2} + B \frac{d\theta(t)}{dt} + K\theta(t) = T(t), \qquad (8.90)$$

where $\theta(t)$ is the displacement. The differential equation is second order, and the phase-plane method is certainly applicable. Assuming that the system is unexcited, the resulting differential equation is given by

$$\frac{d^2\theta(t)}{dt^2} + \frac{B}{J} \frac{d\theta(t)}{dt} + K \frac{\theta(t)}{J} = 0. \qquad (8.91)$$

Equation (8.91) can be written in terms of damping factor ζ and natural frequency ω_n, as follows.

$$\frac{d^2\theta(t)}{dt^2} + 2\zeta\omega_n \frac{d\theta(t)}{dt} + \omega_n^2\theta(t) = 0, \tag{8.92}$$

where

$$B/J = 2\zeta\omega_n,$$

$$K/J = \omega_n^2.$$

Before sketching the phase portrait, let us consider the simpler case of an undamped system where $\zeta = 0$. For this situation, Eq. (8.92) reduces to

$$\frac{d^2\theta(t)}{dt^2} + \omega_n^2\theta(t) = 0. \tag{8.93}$$

From elementary calculus, the solution to Eq. (8.93) is that of simple harmonic motion:

$$\theta(t) = R\sin(\omega_n t + \phi). \tag{8.94}$$

In order to obtain a relationship between $\theta(t)$ and $d\theta(t)/dt$, we differentiate Eq. (8.94) and then eliminate time between the resulting equations, as follows:

$$\frac{d\theta(t)}{dt} = R\omega_n \cos(\omega_n t + \phi). \tag{8.95}$$

Eliminating time between Eqs. (8.94) and (8.95), we obtain the expression

$$\left[\frac{1}{\omega_n}\frac{d\theta(t)}{dt}\right]^2 + [\theta(t)]^2 = R^2. \tag{8.96}$$

The phase portrait for this system can be drawn directly from Eq. (8.96). Observe that Eq. (8.96) describes a family of concentric circles in the $(1/\omega_n)(d\theta(t)/dt)$ versus $\theta(t)$ plane having a radius of R. Therefore, the phase portrait for this system (shown in Fig. 8.36) is a family of concentric circles if a normalized ordinate axis is used. If the ordinate axis is not normalized, a family of ellipses results. Any set of initial conditions, such as points R_1, R_2, R_3, and R_4, specifies a particular circle. For $t > 0$, the motion is in the indicated direction.

Let us next sketch the phase portrait for the case where the damping factor of this system is finite. The general solution for Eq. (8.92), when it is excited by a step input and has a set of initial conditions, was derived in Chapter 4 (see Eq. 4.45). The form of the solution for this system when it is unexcited by a step input and only has a set of initial conditions, $\theta(t_0)$ and $\dot{\theta}(t_0)$, present is given by

$$\theta(t) = \frac{\omega_n^2}{\sqrt{1-\zeta^2}}e^{-\zeta\omega_n(t-t_0)}\sin[\omega_n\sqrt{1-\zeta^2}(t-t_0) + \phi_1]\theta(t_0)$$

$$+ \frac{\omega_n}{\sqrt{1-\zeta^2}}e^{-\zeta\omega_n(t-t_0)}\sin[\omega_n\sqrt{1-\zeta^2}(t-t_0)]\dot{\theta}(t_0). \tag{8.97}$$

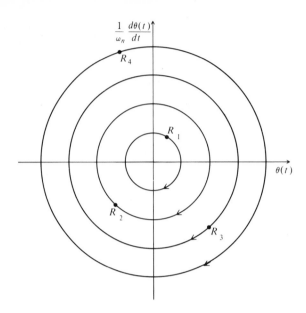

Fig. 8.36   Phase portrait for a second-order linear system having no damping.

Proceeding as in the previous example, the derivative of Eq. (8.97) is obtained:

$$
\dot{\theta}(t) = \omega_n^3 e^{-\zeta\omega_n(t-t_0)} \cos\left[\omega_n\sqrt{1-\zeta^2}\,(t-t_0) + \phi_1\right]\theta(t_0)
$$

$$
-\frac{\zeta\omega_n^3}{\sqrt{1-\zeta^2}} e^{-\zeta\omega_n(t-t_0)} \sin\left[\omega_n\sqrt{1-\zeta^2}\,(t-t_0) + \phi_1\right]\theta(t_0)
$$

$$
+ \omega_n^2 e^{-\zeta\omega_n(t-t_0)} \cos\left[\omega_n\sqrt{1-\zeta^2}\,(t-t_0)\right]\dot{\theta}(t_0)
$$

$$
-\frac{\zeta\omega_n^2}{\sqrt{1-\zeta^2}} e^{-\zeta\omega_n(t-t_0)} \sin\left[\omega_n\sqrt{1-\zeta^2}\,(t-t_0)\right]\dot{\theta}(t_0). \qquad (8.98)
$$

The complexity of Eqs. (8.97) and (8.98) makes it quite difficult to eliminate time between them. Therefore we use an alternative approach. Equations (8.97) and (8.98) will be evaluated for several values of time to obtain the corresponding coordinates in a normalized phase plane. The result is plotted in Fig. 8.37 for a damping factor of 0.7. Notice that the phase portrait is a collection of noncrossing paths describing system behavior for all possible initial conditions. Since all the trajectories approach the origin, the system is stable as we would expect it to be.

The projection of a specific trajectory onto the abscissa gives the variation of $\theta(t)$ with time, and its projection onto the ordinate gives the variation of $d\theta(t)/dt$ with time. Time appears in the portrait implicitly as a parameter along all phase trajectories. We discuss its computation later in the chapter.

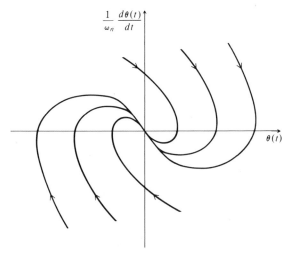

Fig. 8.37   Phase portrait of a stable node.

Let us next illustrate the application of this method to a relatively simple nonlinear differential equation. We consider the problem of nonlinear friction which was discussed previously in Section 8.8. We first determine the phase portrait for coulomb friction and then extend it to the case of stiction. Consider a system containing coulomb friction $\pm F_c$, the moment of inertia of the system being J and the spring constant K. The differential equation of the motion of the system is given by

$$J \frac{d^2c(t)}{dt^2} + Kc(t) \pm F_c = 0. \tag{8.99}$$

Defining

$$\omega = (K/J)^{1/2}, \tag{8.100}$$

$$\gamma = \frac{F_c}{J\omega^2}, \tag{8.101}$$

$$\alpha = \begin{cases} c - \gamma & \text{when} \quad \dot{c} < 0, \\ c + \gamma & \text{when} \quad \dot{c} > 0, \end{cases} \tag{8.102}$$

the normalized equation of motion is

$$\ddot{\alpha} + \omega^2\alpha = 0. \tag{8.103}$$

The phase portrait for Eq. (8.103) in the α versus $\dot{\alpha}$ plane is the same as that for the linear, second-order system considered previously (see Fig. 8.36). However, in the $\dot{c}/\omega$ versus c plane, the phase portrait for Eq. (8.103) is quite different, and can be obtained by considering Eq. (8.103) in two parts: one for the upper half-plane where $\dot{c}/\omega > 0$, and the other for the lower half-plane where $\dot{c}/\omega < 0$. A little thought shows that the phase portrait in the upper half-plane is a family of semicircles, centered about $c = -\gamma$, since c is related to α merely by a simple translation. In a similar manner, the phase portrait for the lower half-plane is a family of semicircles centered about $c = \gamma$. Figure 8.38 illustrates the phase portrait.

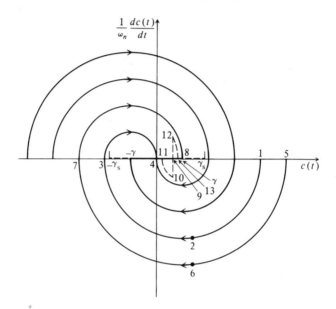

Fig. 8.38   Phase portrait for a system containing coulomb friction.

It is interesting to observe from Fig. 8.38 that as soon as the displacement has a value within the interval on the $c(t)$-axis given by

$$-\gamma \leqslant c \leqslant \gamma, \tag{8.104}$$

all motion stops. This gives rise to the possibility of large, steady-state errors. For example, if the initial conditions are at point 1, the trajectory will be 1–2–3–4 and the system will not have any steady-state error. If the initial conditions are at point 5, however, the trajectory followed will be 5–6–7–8 and the system will have a steady-state error equal to γ.

We have already seen that dither is useful for eliminating the steady-state error due to coulomb friction. Its effect on the steady-state error can easily be understood by studying the phase plane. For example, let us assume that a finite steady-state error exists corresponding to point 9 on Fig. 8.38. With dither, the effect of a negative disturbance, segment 9–10, results in the system returning to point 11 on the $c(t)$-axis, while a positive disturbance, segment 9–12, results in the system returning to point 13 on the $c(t)$-axis. Since the projection 11–9 is greater than the projection 9–13, the system will tend to move towards the origin.

Before leaving this system, let us consider the required modification to the phase portrait in Fig. 8.38 for stiction. Since stiction occurs only for zero velocity and is greater than coulomb friction, its effect is to extend the termination line: $\pm \gamma$ to $\pm \gamma_s$. These extended terminations are illustrated in Fig. 8.38.

2. Transformation of the Second-Order Differential Equation to a First-Order Equation

If the differential equation of the system cannot be easily integrated, a new differential equation in terms of the phase variables may be formed, from which the phase portrait can be obtained directly. This method can best be illustrated by considering the linear, second-order, undamped system discussed previously, namely,

$$\frac{d^2\theta(t)}{dt^2} + \omega_n^2\theta(t) = 0. \tag{8.105}$$

Using the dot notation, we have

$$\ddot{\theta} + \omega_n^2\theta = 0. \tag{8.106}$$

Defining the state variables as

$$x_1 = \theta, \tag{8.107}$$

$$x_2 = \dot{\theta}, \tag{8.108}$$

we have

$$\dot{x}_1 = x_2, \tag{8.109}$$

$$\dot{x}_2 = -\omega_n^2 x_1. \tag{8.110}$$

Dividing Eq. (8.110) by Eq. (8.109), the following is obtained:

$$\frac{dx_2}{dx_1} = -\frac{\omega_n^2 x_1}{x_2}. \tag{8.111}$$

Equation (8.111) can be rewritten as

$$x_2 \, dx_2 + \omega_n^2 x_1 \, dx_1 = 0.$$

Integrating, we obtain

$$x_2^2 + \omega_n^2 x_1^2 = C,$$

where C is an arbitrary constant easily determined from the initial conditions. Solving for x_2, we obtain

$$x_2 = \pm\sqrt{C - \omega_n^2 x_1^2}.$$

In terms of the phase variable, this equation can be written as

$$\dot{\theta}(t) = \pm\sqrt{C - \omega_n^2\theta^2(t)}. \tag{8.112}$$

The result of sketching various phase trajectories from Eq. (8.112) will be the phase portrait illustrated in Fig. 8.36, where a normalized ordinate axis is used. The initial conditions determine the particular trajectory followed.

This technique can be easily extended to the systems whose phase portraits are illustrated in Figs. 8.37 and 8.38. This method is by far the most useful analytic technique for obtaining the phase portrait of a system.

3. Method of Isoclines The method of isoclines is a graphical procedure for determining the phase portrait. It can be used even if the differential equation cannot be
solved analytically. In practice, it is a very powerful method to use.

Isoclines are lines in the phase plane corresponding to constant slopes of the
phase portrait. One starts with the differential equation in the form shown by Eq.
(8.111). Here, dx_2/dx_1, or $d\dot\theta/d\theta$, corresponds to the slope of the trajectories that form
the phase portrait. Numerical values are assigned for the slopes $d\dot\theta/d\theta$, and Eq.
(8.111) is used to find the corresponding points in the phase plane having those slopes.
Once a set of isoclines is drawn, a trajectory may be drawn by starting at some point
on an isocline and then proceeding to the next isocline along a straight line whose
slope is the average of the slopes corresponding to the two isoclines. Since the procedure is a numerical approximation, closer spacing of the isoclines increases the
accuracy of the resulting trajectory.

Let us illustrate the application of the isocline method to the linear, second-order,
undamped and damped systems whose portraits are given in Figs. 8.36 and 8.37,
respectively. For the undamped case, the family of isoclines can be drawn from Eq.
(8.111). However, in order to plot the phase portrait on a normalized plane
$[(1/\omega_n)\dot\theta(t)$ versus $\theta(t)]$, let

$$\dot\Theta(t) = \dot\theta(t)/\omega_n. \tag{8.113}$$

So we have

$$\frac{d\dot\Theta(t)}{d\theta(t)} = -\frac{\theta(t)}{\dot\Theta(t)} = m, \tag{8.114}$$

where m represents the slope of the trajectory.

Isoclines associated with slopes corresponding to Eq. (8.114) constitute a family
of straight lines passing through the origin and are illustrated in Fig. 8.39. Also
shown is the construction of the phase trajectory starting with a point that lies on the
isocline corresponding to a slope of -1.5. The motion of the trajectory drawn from
point A on the -1.5 isocline, to the isocline whose slope is -2, has a slope in the
phase plane that is the average of these two isoclines, or -1.75. This is indicated on
Fig. 8.39 as line segment AB. In addition, the following line segment, BC, whose
slope is -2.5, is illustrated. The complete trajectory is shown dotted. It is obvious
from this simple example that the accuracy of the isocline method depends on the
number of isoclines drawn.

Let us next construct the phase portrait for the linear, second-order, damped
system considered previously. From Eq. (8.92), the differential equation of the system
is

$$\frac{d^2\theta(t)}{dt^2} + 2\zeta\omega_n\frac{d\theta(t)}{dt} + \omega_n^2\theta(t) = 0. \tag{8.115}$$

Setting $\zeta = 0.5$ and using the dot notation, we have

$$\ddot\theta + \omega_n\dot\theta + \omega_n^2\theta = 0. \tag{8.116}$$

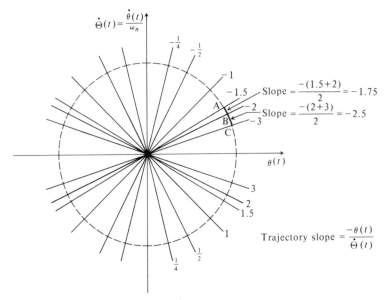

Fig. 8.39 Construction of a phase trajectory for a linear, second-order, undamped system using the method of isoclines.

Defining the state variables as

$$x_1 = \theta, \tag{8.117}$$

$$x_2 = \dot{\theta}, \tag{8.118}$$

then we have

$$\dot{x}_1 = x_2, \tag{8.119}$$

$$\dot{x}_2 = -\omega_n^2 x_1 - \omega_n x_2. \tag{8.120}$$

Dividing Eq. (8.120) by Eq. (8.119), we obtain the following:

$$\frac{dx_2}{dx_1} = -\omega_n^2 \frac{x_1}{x_2} - \omega_n. \tag{8.121}$$

Defining a normalized state variable

$$x_3 = \frac{x_2}{\omega_n} = \frac{\dot{\theta}}{\omega_n} = \dot{\Theta}(t),$$

we can write Eq. (8.121) as follows:

$$\frac{dx_3}{dx_1} = -1 - \frac{x_1}{x_3},$$

or

$$\frac{d\dot{\Theta}(t)}{d\theta(t)} = -1 - \frac{\theta(t)}{\dot{\Theta}(t)}.$$

From this equation, the slope of the trajectories in the $\dot{\Theta}(t)$ versus $\theta(t)$ plane is given by

$$\text{trajectory slope} = -1 - \frac{\theta(t)}{\dot{\Theta}(t)}. \tag{8.122}$$

Isoclines associated with slopes corresponding to Eq. (8.122) constitute a family of straight lines passing through the origin in the phase plane. This is illustrated in Fig. 8.40. The construction of the trajectory starting with a point which lies on the isocline corresponding to a slope of 2 is illustrated. The segment of the trajectory drawn from point A on the isocline whose slope is 2 to that whose slope is 1 would be a straight line whose slope is the average of 2 and 1, or 1.5. This is indicated in Fig. 8.40 as line segment AB. In addition, the following line segment BC, whose slope is 0.5, is illustrated. The remaining trajectory is shown dashed. As for the undamped system, the accuracy depends greatly on the number of isoclines drawn.

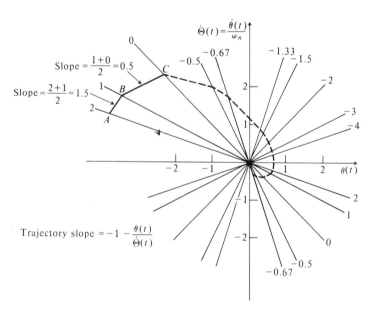

Fig. 8.40 Construction of a phase trajectory for a linear, second-order, damped system using the method of isoclines.

Before leaving this discussion, it should be mentioned that there are several other methods available for obtaining a phase portrait. Most notable of these other methods are the method of tangents and Lienard's construction. The former technique is very similar to the method of isoclines. It consists of deriving the tangent to a trajectory at many points in the phase plane. The phase trajectory is then drawn tangent to these lines. This technique is usually a little more tedious than the method of iso-clines. The reader should consult Ref. 17 for a further discussion of the method of tangents. The latter technique, Lienard's construction, is useful if the second-order differential equation has the following special form:

$$\frac{d^2x}{dt^2} + f\left(\frac{dx}{dt}\right) + Kx = 0 \tag{8.123}$$

where $f(dx/dt)$ represents nonlinear damping. The reader should consult Ref. 18 for a further discussion of this specialized construction.

8.15 CHARACTERISTICS OF THE PHASE PORTRAIT

Several properties of the phase portrait need to be singled out; their correct interpre-tation is important for the intelligent analysis of feedback control systems. We begin by defining and illustrating the notion of *singular points*. Then we illustrate limit cycles and follow this by showing how to determine time from the phase portrait. The section concludes with examples of several interesting and representative phase portraits.

1. Singular Points Singular points are points of the phase plane where the system is in a state of equilibrium. At these points both velocity and acceleration of the system are zero. The origin is the only singular point in the phase portrait of the linear system illustrated in Fig. 8.37.

Consider the general second-order, nonlinear differential equation (8.89)

$$\frac{d^2x}{dt^2} + A_1\left(x, \frac{dx}{dt}\right)\frac{dx}{dt} + A_2\left(x, \frac{dx}{dt}\right)x = 0.$$

If this equation is written in the state-space form, the singular points are defined as the points that make the quantities in Eqs. (8.124) and (8.125) equal to zero:

$$\frac{dx}{dt} = P(x, y), \tag{8.124}$$

$$\frac{dy}{dt} = Q(x, y). \tag{8.125}$$

Here x and y are the state variables. The characteristics of singular points may vary greatly depending on the variations of the coefficients of the first-order differential equations given by Eqs. (8.124) and (8.125). The type of singular point may be found by means of a Taylor-series expansion of Eqs. (8.124) and (8.125) in the vicinity of the singular point. Assuming that the singularity occurs at $x = A$ and $y = B$, the result of expanding $P(x, y)$ and $Q(x, y)$ about these points is

$$\frac{dx}{dt} = A_1(x - A) + A_2(y - B) + A_3(x - A)^2$$

$$+ A_4(x - A)(y - B) + A_5(y - B)^2 + \cdots . \quad (8.126)$$

$$\frac{dy}{dt} = B_1(x - A) + B_2(y - B) + B_3(x - A)^2$$

$$+ B_4(x - A)(y - B) + B_5(y - B)^2 + \cdots . \quad (8.127)$$

We assume that the character of the singular points is determined entirely by the coefficients of the linear terms only: A_1, A_2, B_1, and B_2. This is certainly reasonable if a sufficiently small region is chosen in the vicinity of the singularity. What we have done is to characterize the system by its linear part in the vicinity of the singular point. Therefore, Eqs. (8.126) and (8.127) reduce to

$$\frac{dx}{dt} = A_1(x - A) + A_2(y - B), \quad (8.128)$$

$$\frac{dy}{dt} = B_1(x - A) + B_2(y - B). \quad (8.129)$$

We further simplify the problem by assuming that the singularity occurs at the origin. Therefore A and B are both zero. Equations (8.128) and (8.129) reduce to the form

$$\frac{dx}{dt} = A_1x + A_2y, \quad (8.130)$$

$$\frac{dy}{dt} = B_1x + B_2y. \quad (8.131)$$

We have not lost any generality in using this assumption, since the same result can always be obtained merely by changing the variables as follows:

$$x - A = p, \qquad y - B = q. \quad (8.132)$$

The characteristics of the singular point can be determined by eliminating one of the two variables of Eqs. (8.130) and (8.131) and studying the resulting characteristic equation. The result is

$$X(s) = \frac{(s - B_2)x(0) + A_2y(0)}{s^2 - (A_1 + B_2)s + (A_1B_2 - A_2B_1)}, \quad (8.133)$$

and the characteristic equation is given by

$$s^2 - (A_1 + B_2)s + (A_1B_2 - A_2B_1) = 0. \quad (8.134)$$

Assuming real coefficients, six different characteristic sets of roots of Eq. (8.134) are possible. These sets of roots result in singular points which can be classified as belonging to one of four types.

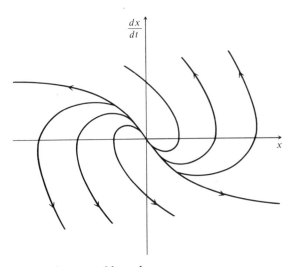

Fig. 8.41   Phase portrait of an unstable node.

a) Node. A node is a point in the phase plane consisting of a family of trajectories which directly converge and approach it (stable node) or radiate from it (unstable node). A stable node occurs when the roots are both real and both in the left half of the *s*-plane. An unstable node occurs when the roots are both real and both lie in the right half of the *s*-plane. Figure 8.37 illustrates a stable node and Fig. 8.41 an unstable node.

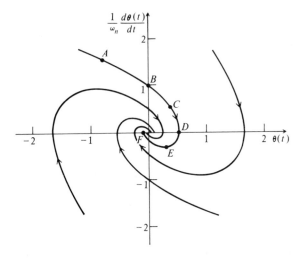

Fig. 8.42   Phase portrait of a stable focus.

b) Focus. A focus is a point in the phase plane consisting of a family of spiral trajectories which either converge on the point (stable focus) or diverge from it (unstable focus). A stable focus occurs when the roots are complex conjugate and lie in the left half of the *s*-plane. An unstable focus occurs when the roots are complex conjugate and lie in the right half-plane. The origin of the phase portrait in Fig. 8.42 is a stable focus. Figure 8.43 illustrates an unstable focus.

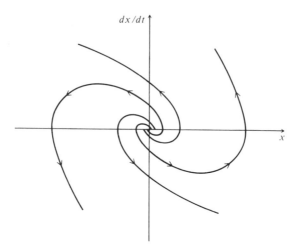

Fig. 8.43   Phase portrait of an unstable focus.

c) Center. A center is a point in the phase plane consisting of a family of closed curves encircling it. This occurs when the roots are complex conjugate and lie on the *jω*-axis. The origin of the phase portrait in Fig. 8.36 is a center.

d) Saddle point. A saddle point is a point in the phase plane that is characterized by the phase portrait illustrated in Fig. 8.44. This occurs when the roots are real, with one in the right half-plane and the other in the left half-plane. Except for the case where the numerator of Eq. (8.133) contains a zero which exactly cancels the zero of the denominator lying in the right half-plane, this type of singularity always represents an unstable situation.

2. Limit Cycles   Limit cycles are defined as isolated closed paths of the phase portrait. The location and determination of the type of limit cycle, together with the singular points which exist in the phase plane, offer a complete description of the behavior of a nonlinear second-order system.

A limit cycle can be stable or unstable, depending on whether the paths in the neighborhood of the limit cycle converge toward it or diverge away from it. They can result from either soft or hard self-excitation. These situations can be portrayed on the phase plane. Figure 8.45 illustrates a system with a soft self-excitation. Physically, this phase portrait may represent a system which has excessive gain for small signals

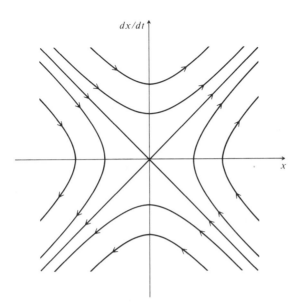

Fig. 8.44   Phase portrait of a saddle point.

and the output builds up in an unstable manner. With large signals the output also approaches the stable limit cycle from the outside as shown in Fig. 8.45. A stable limit cycle exists between these two conditions and a sustained oscillation occurs. Figure 8.46 illustrates a system where a hard self-excitation exists. The generation of an oscillation depends on the initial conditions. For example, let us assume that the

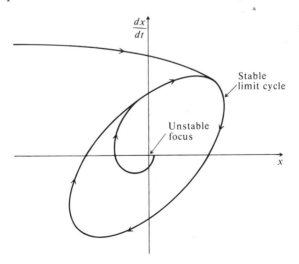

Fig. 8.45   Phase portrait of soft self-excitation.

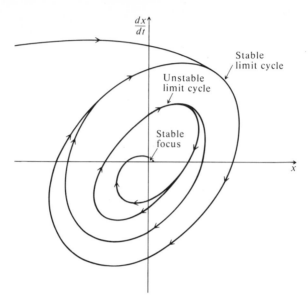

Fig. 8.46   Phase portrait of hard self-excitation.

initial state is at the stable focus. For hard self-excitation to occur, a very large disturbance is required to change the state of the system to a region outside the unstable limit cycle. If this disturbance is sufficient for the operating point of the system to reach the stable limit cycle, a steady oscillation occurs.

3. Determination of Time from the Phase Plane   Although we cannot solve for x and dx/dt as functions of time directly, it is possible to obtain time from the phase portrait. The variation of time can be easily found from the equation

$$y = dx/dt. \tag{8.135}$$

Solving for t, we obtain the relationship

$$t = \int_A^B \frac{1}{y}\, dx. \tag{8.136}$$

Equation (8.136) shows that if the phase portrait is replotted with $1/y$ or $(dx/dt)^{-1}$ as the ordinate and x as the abscissa, the area under the resulting curve represents time. This area can be evaluated with a planimeter or by approximation using a series of rectangles. Let us use the phase trajectory labeled $ABCDEF$ on the phase portrait of Fig. 8.42 to illustrate the determination of time. We assume that $\omega_n = 1$. Our first problem is to determine the time it takes for the motion to go from A to C. Figure 8.47 represents a plot of $[d\theta(t)/dt]^{-1}$ versus $\theta(t)$ for the interval ABC. Using a series of rectangles for area summation, we find that the time is approximately equal to 1.12 sec. If we attempt to find the time it takes for the trajectory to pass from C to E, we run into a problem since $d\theta(t)/dt$ equals zero at point D and the reciprocal

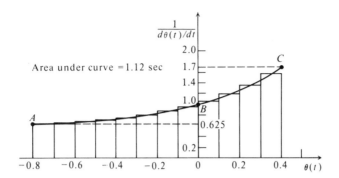

Fig. 8.47 Finding time from a reciprocal phase-plane plot.

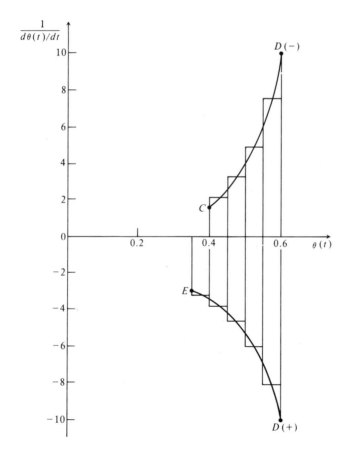

Fig. 8.48 Approximating time from a reciprocal phase-plane plot when velocity passes through zero.

phase-plane plot goes to infinity. This certainly is not the true situation, and we must resort to an approximation in the vicinity of point D. The most practical approximation is to find the time it takes to go from point C to a small finite distance on the phase trajectory before point D, $D(-)$, and then the time it takes to go from a small finite distance past point D, $D(+)$, to point E. This technique is illustrated in Fig. 8.48. Using the rectangular area summation technique, we find that it takes approximately 2.18 sec to go from point C to point E. Therefore, the total time it takes to traverse the segment $ABCDE$ is approximately 3.3 sec.

Ordinarily, the reciprocal phase-plane plot is sufficient for obtaining time. For academic completeness, several other methods also exist for obtaining time from the phase plane. References 7 and 18 deal with these approaches thoroughly.

4. Representative Phase Portraits In order to develop a greater facility for intelligently interpreting phase portraits, we next present additional representative phase portraits. These, with the portraits already discussed, should provide a good facility for interpreting phase portraits. Specifically, we consider the phase portraits of undamped, second-order control systems that have the nonlinear characteristics of rate and position limiting. We compare the results with the portrait illustrated in Fig. 8.39.

a) Rate Limiting of an Undamped, Second-Order System. Consider the linear, undamped, second-order system whose phase trajectory was illustrated in Fig. 8.39. We shall change this linear system to a nonlinear one by adding a governor to the servomotor driving the load so that the maximum rate is limited. The resulting phase portrait is illustrated in Fig. 8.49. If the initial conditions are such that the resulting phase trajectory lies anywhere within the dashed trajectory, the system will oscillate as the undamped linear system did, as shown in Fig. 8.39. However, if the initial

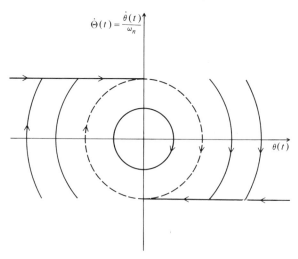

Fig. 8.49 Phase portrait of a second-order, undamped system having rate limiting.

conditions are such that they lie outside the dashed trajectory, the output rate is limited to the maximum value allowed by the governor, as shown in Fig. 8.49 by the dashed trajectory. Thereafter the system will oscillate indefinitely following the dashed trajectory.

b) *Position Limiting of an Undamped, Second-Order System.* Next consider the linear, undamped, second-order system that is made nonlinear by adding limit stops to the output shaft, so that the maximum and minimum positions are limited. The resulting phase portrait is given in Fig. 8.50. The interpretation of this phase portrait is analogous to that of Fig. 8.49, except that it is shifted by 90°.

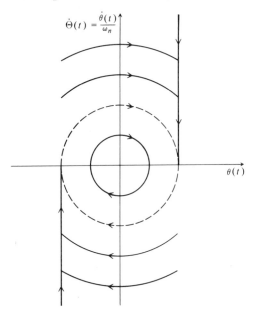

Fig. 8.50   Phase portrait of a second-order, undamped system having position limiting.

8.16 DESIGN OF NONLINEAR FEEDBACK CONTROL SYSTEMS USING THE PHASE-PLANE METHOD

This section illustrates the design of a nonlinear feedback control system using the phase-plane method [19]. We use the analytic tools developed in Sections 8.13 to 8.15 to demonstrate the procedure to follow when designing a nonlinear system. Specifically, we consider an on–off control system. The primary function of this example is to use the transient response of the system as a guide for choosing the parameters. One of our main objectives will be to determine the existence of limit cycles as the parameters are varied. We will not, however, be able to obtain the margin of stability for the system from the phase plane. The following analysis should be compared with that obtained from the describing function in Section 8.10.

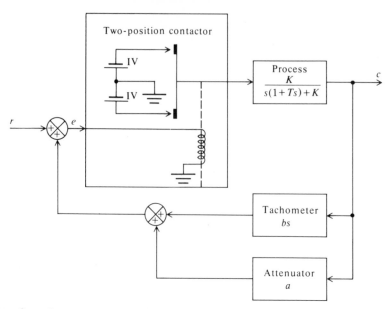

Fig. 8.51 On–off control system having variable position and velocity feedback.

Let us consider the on–off control system illustrated in Fig. 8.51. It has a two-position contactor which applies a corrective signal having the proper phase for control action. The problem is to determine a good polarity combination of the variable position-feedback constant, a, and the variable velocity feedback constant, b, for acceptable transient performance. We assume that the input r is zero and that the forcing function generated by the contactor is unity; therefore the general form of the differential equation for this system is given by

$$\frac{d^2c}{dt^2} + 2\zeta\omega_n \frac{dc}{dt} + \omega_n^2 c = \text{sgn}\left(ac + b\frac{dc}{dt}\right), \tag{8.137}$$

where $0 < \zeta < 1$. The sign of the unit forcing function is given by the sign of $ac + b(dc/dt)$. Assuming that $\zeta = 0.5$ and $\omega_n = 1$, the equation reduces to

$$\frac{d^2c}{dt^2} + \frac{dc}{dt} + c = \text{sgn}\left(ac + b\frac{dc}{dt}\right). \tag{8.138}$$

Specifically, we shall determine the transient response of this system for the following polarity combinations of a and b:

$$
\begin{array}{lll}
\text{(A)} & a > 0, & b > 0, \\
\text{(B)} & a > 0, & b < 0, \\
\text{(C)} & a < 0, & b > 0, \\
\text{(D)} & a < 0, & b < 0.
\end{array}
\tag{8.139}
$$

In general the phase portrait for any of these cases may be obtained in a similar manner. We draw heavily on the results of our studies of singular points in drawing the various phase portraits. The basic relation analyzed in our discussion of singular points was Eq. (8.89):

$$\frac{d^2c}{dt^2} + A_1\left(c, \frac{dc}{dt}\right)\frac{dc}{dt} + A_2\left(c, \frac{dc}{dt}\right)c = 0. \tag{8.140}$$

If we are considering an underdamped, linear second-order system, we know that the differential equation (8.92) results in a phase portrait similar to that in Fig. 8.37, where the origin acts as a stable node. In the case of an on–off system, however, we find that it is represented by a differential equation which is given by Eq. (8.138). The right-hand side of this equation can be thought of as either a unit positive or negative forcing function. The phase portrait of a system having a unit positive forcing function will have spirals that converge towards a stable focus at $c = 1$, $dc/dt = 0$. The phase portrait of a system having a unit negative forcing function will have spirals that converge towards a stable focus at $c = -1$, $dc/dt = 0$. The on–off system we are considering actually has two stable foci because of the action of the two-position contactor. A switching line, defined by

$$ac + b\frac{dc}{dt} = 0 \tag{8.141}$$

separates the regions where the phase trajectory spirals towards $c = 1$, $dc/dt = 0$ or $c = -1$, $dc/dt = 0$. Using this general approach, which is valid for all four cases, the transient response can be readily determined.

Case A. $a > 0$, $b > 0$. The switching line is a straight line given by the following relationship:

$$\frac{dc}{dt} = -\frac{a}{b}c. \tag{8.142}$$

It passes through the origin and lies in the second and fourth quadrants. The sign of $ac + b(dc/dt)$ is positive in the region to the right of it and negative to the left of it. The phase portrait for this system, which is given in Fig. 8.52, can be constructed using the method of isoclines or by transforming the second-order differential equation to a first-order equation. Observe that three trajectories are possible, depending on the initial conditions. Two represent convergent motion where the phase portrait terminates on the foci at -1 or 1, depending on the final switching input. The third possible trajectory represents a limit cycle.

Any initial condition occurring beyond the limit cycle or very close to it in the enclosed area will eventually result in the trajectory reaching some part of the limit cycle, $ABCDA$. For a limit cycle to occur, the distance DE from the switching line to the corresponding focus must be greater than the distance BE, which represents the distance from the subsequent crossing of the line to the focus. This must be true with respect to the other focus as well.

Trajectories that start sufficiently inside the limit cycle will spiral into one of the foci. However, there is no chance of these trajectories ultimately reaching the origin.

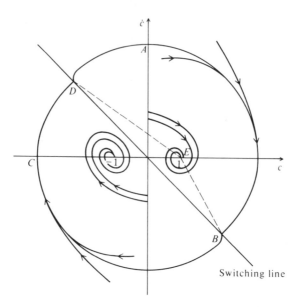

Fig. 8.52   Phase portrait for an on–off control system where $a > 0$ and $b > 0$.

Case B. $a > 0, b < 0$. The switching line is a straight line given by

$$\frac{dc}{dt} = \frac{a}{b}c. \tag{8.143}$$

It passes through the origin and lies in the first and third quadrants. In a manner similar to the first case, we can show the phase portrait to be as shown in Fig. 8.53. Observe that any set of initial conditions which results in a trajectory intersecting the switching line inside the interval AB results in a motion spiraling towards one of the two stable foci. However, if the intersection occurs outside the interval AB, the trajectory theoretically ends on the switching line. Figure 8.53 indicates these points by C and D. Points E and F represent points of tangency with the switching lines for limiting trajectories. In reality, however, the system cannot just end at these points. This inconsistency is resolved by the fact that switching action of a contactor always has a certain time lag due to its dynamics. Therefore, when a solution reaches the switching line it actually proceeds for some small distance past it, before there is a change of sign in the forcing function. This results in a zigzag action along the switching line. Eventually, the trajectory spirals into one of the foci as shown in Fig. 8.53. Physically, this is audible as a "chattering" of the contactor.

Case C. $a < 0, b > 0$. The switching line has the same form as that of case B. The phase portrait is given in Fig. 8.54. For this case a stable limit cycle always prevails. Regardless of the initial conditions, the final solution always winds up as a stable limit cycle.

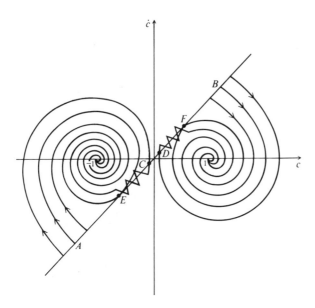

Fig. 8.53   Phase portrait for an on–off control system where $a > 0$, $b < 0$.

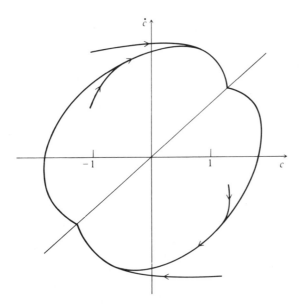

Fig. 8.54   Phase portrait for an on–off control system where $a < 0$, $b > 0$.

Case D. $a < 0, b < 0$. The switching line has the same form as that of case A. The phase portrait, given in Fig. 8.55, shows that no periodic solution exists. The solutions tend to end on the switching line. Because of the time lag of the relay, however, the trajectory zigzags towards the origin of the phase plane. The system finally oscillates at very high frequency and small amplitude around the origin.

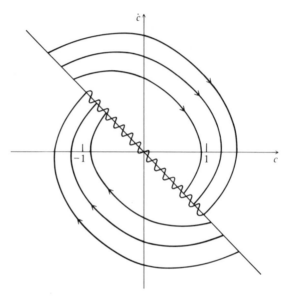

Fig. 8.55   Phase portrait for an on–off control system where $a < 0$, $b < 0$.

It is interesting to observe that of the four cases considered, the control-system engineer would prefer the phase portrait of case **D**, the only configuration which resulted in a stable equilibrium state occurring around the origin. However, we would have to tolerate some chattering around the origin with this linear switching system. To eliminate the chattering, one would have to use nonlinear switching techniques [19].

An analog computer is very useful for studying the phase portraits of nonlinear systems as certain parameters are varied.* Figure 8.56 illustrates the analog computer circuitry required for simulating the system whose block diagram was shown in Fig. 8.51. Switch S_1 permits choices of $a > 0$ or $a < 0$, and switch S_2 permits choices of $b > 0$, or $b < 0$. By connecting the points x and y to the horizontal and vertical deflection plates of an oscilloscope, respectively, the phase portraits shown in Figs. 8.52 through 8.55 can be obtained as switches S_1 and S_2 are varied. The reader should try this useful exercise.

---

* The reader should consult References 41 and 42 for a discussion of analog computers.

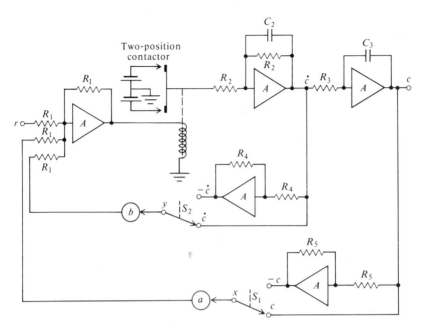

Fig. 8.56 Analog computer circuit for simulating system shown in Fig. 8.51.

8.17 LIAPUNOV'S STABILITY CRITERION

A. M. Liapunov [20] developed a fundamental method of determining the stability of a dynamic system based on generalization of energy considerations. This section presents the first and second methods of Liapunov and illustrates their application to nonlinear feedback control systems [21].

A. Liapunov's First Method Liapunov divided the general problem of analyzing the stability of nonlinear systems into two classes. The first class consists of all those methods in which the differential equation of the system can be solved. System stability or instability is determined from this solution. This approach, which is known as *Liapunov's first method*, does not say anything of particular importance concerning the solution of the nonlinear differential equations. However, Liapunov did point out in his first method that the solution may be obtained in the form of a series from which stability can be determined using his second method. In addition, he proved that approximate solutions of nonlinear differential equations often yield useful stability information.

In order to illustrate Liapunov's first method, let us assume that the nonlinearity is single valued (has no hysteresis present) and has derivatives of every order in the vicinity of a point A. Then the nonlinear function, $y = f(x)$, can be expanded into a Taylor series as follows:

$$y = f(x) = y(A) + (x - A)\left(\frac{dy}{dx}\right)_A + \frac{1}{2!}(x - A)^2\left(\frac{d^2y}{dx^2}\right)_A$$

$$+ \frac{1}{3!}(x - A)^3\left(\frac{d^3y}{dx^3}\right)_A + \cdots + \frac{1}{n!}(x - A)^n\left(\frac{d^ny}{dx^n}\right)_A + \cdots. \quad (8.144)$$

Note that the first two terms of the series represent the linear approximation about the operating point of the actual nonlinearity. Liapunov proved that if the real parts of the roots of the characteristic equation corresponding to the differential equation of the linear approximation are different from zero, the equations of the linear approximation always give a correct answer to the question of stability of a nonlinear system [21]. This theorem means that we can use a linear approximation of the nonlinear equation and determine stability from it. If the roots of the linearized characteristic equation are negative, the motion is stable about the point in question. However, if any of the real parts of the roots of the characteristic equation of the linear approximation are positive, the motion is unstable about the operating point. In the special case of roots of the linearized characteristic equation having zero real parts, no conclusion may be drawn.

To illustrate the application of Liapunov's first method, consider the Van der Pol equation which describes the voltage build-up of an oscillator,

$$\ddot{v} + u(v^2 - 1)\dot{v} + Kv = Q, \quad (8.145)$$

where

$$u < 0, K > 0.$$

The equilibrium point of this system is determined when the velocity and acceleration are zero. Then,

$$Kv = Q, \quad (8.146)$$

or

$$v = \frac{Q}{K} = V = \text{equilibrium point.} \quad (8.147)$$

In order to determine behavior in the area of the equilibrium point, V, let

$$v = V + v_i, \quad (8.148)$$

where

$$\ddot{V} = \dot{V} = 0.$$

Substituting Eq. (8.148) into Eq. (8.145) we obtain the following expression:

$$\ddot{v}_i + u[(V^2 + 2Vv_i + v_i^2) - 1]\dot{v}_i + K(V + v_i) = Q. \quad (8.149)$$

The linear approximation to this equation is given by

$$\ddot{v}_i + u[V^2 - 1]\dot{v}_i + Kv_i = 0. \quad (8.150)$$

The characteristic equation is given by

$$s^2 + u[V^2 - 1]s + K = 0. \tag{8.151}$$

Applying the Routh-Hurwitz stability criterion, we find that, for stability,

$$u[V^2 - 1] > 0. \tag{8.152}$$

Equation (8.152) states that when $u < 0$, the system is stable for all $V < 1$. Liapunov's first method is not applicable when $V = 1$, since this condition results in zero real parts for the roots of the characteristic equation.

It is important to emphasize that Liapunov's first method determines stability in the immediate vicinity of the equilibrium point.

B. Liapunov's Second Method This determines stability without actually having to solve the differential equation. In this method a function of the state variables having special properties is formed, that can be compared to the sum of the kinetic and potential energy, and the derivative of the function with respect to time is taken. If this derivative is negative along the trajectories of the system, it can be shown that the system is asymptotically stable. The remainder of this section is devoted to the details and application of the method.

We introduce Liapunov's second method by first considering a linear system. Reconsider the simple mass–spring–damper mechanical system illustrated in Fig. 3.3. It was shown there that this system can be represented by the following differential equation:

$$M \frac{d^2 y(t)}{dt^2} + B \frac{dy(t)}{dt} + Ky(t) = f(t). \tag{8.153}$$

Assume that $M = B = K = 1$ and that $f(t) = 0$. Then we have

$$\ddot{y} + \dot{y} + y = 0. \tag{8.154}$$

Defining the state variables as

$$x_1 = y, \tag{8.155}$$

$$x_2 = \dot{y}, \tag{8.156}$$

the system can be described by the following two first-order differential equations:

$$\dot{x}_1 = x_2 = \dot{y}, \tag{8.157}$$

$$\dot{x}_2 = -x_1 - x_2. \tag{8.158}$$

This simple linear system can easily be solved. Assuming the initial conditions are

$$x_1(0) = 1, \tag{8.159}$$

$$x_2(0) = 0, \tag{8.160}$$

then we obtain the following solutions:

$$x_1 = 1.15e^{-t/2} \sin (0.866t + \pi/3), \tag{8.161}$$

$$x_2 = -1.15e^{-t/2} \sin (0.866t). \tag{8.162}$$

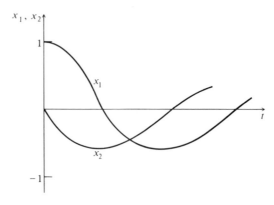

Fig. 8.57 Time-domain response of a simple mechanical system.

Equations (8.161) and (8.162) are plotted in the time domain in Fig. 8.57, and in the phase plane in Fig. 8.58. These two figures completely determine the dynamics and stability of this simple mechanical system. The system is stable and the states x_1 and x_2 behave as indicated.

Now, let us look at this simple system from the viewpoint of energy. The total stored energy is given by

$$V = \tfrac{1}{2}Kx_1^2 + \tfrac{1}{2}Mx_2^2. \tag{8.163}$$

Since $K = M = 1$ in our simple example,

$$V = \tfrac{1}{2}x_1^2 + \tfrac{1}{2}x_2^2. \tag{8.164}$$

This total energy is dissipated as heat in the damper at the rate of

$$\dot{V} = -B\dot{x}_1 x_2 = -Bx_2^2. \tag{8.165}$$

Since $B = 1$, we obtain

$$\dot{V} = -x_2^2. \tag{8.166}$$

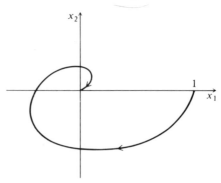

Fig. 8.58 Phase plane of a simple mechanical system.

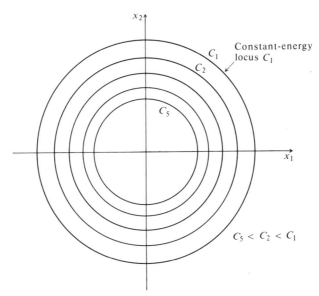

Fig. 8.59 Constant-energy loci on the phase plane illustrating a decrease of energy with time.

Equation (8.164) determines the loci of constant stored energy in the $x_1 x_2$-plane. Clearly they are circular for this simple example. Another important observation from Eq. (8.166) is that the energy rate is always negative and, therefore, these circles must get smaller and smaller with time. Figure 8.59 illustrates this characteristic in the phase plane. For this simple example, we can determine the time variation of V and $\dot{V}$ explicitly by substituting Eqs. (8.161) and (8.162) into Eqs. (8.164) and (8.165). The results are as follows:

$$V = 0.667e^{-t}[\sin^2(0.866t) + \sin^2(0.866t + \pi/3)], \qquad (8.167)$$

$$\dot{V} = -1.333e^{-t}\sin^2(0.866t). \qquad (8.168)$$

Figure 8.60 illustrates the time variation of V and $\dot{V}$. Comparing Figs. 8.59 and 8.60, we conclude that the total stored energy approaches zero as time approaches infinity. This implies that the system is asymptotically stable. By this is meant that the state will return to the origin from any point **x** within a region R enclosing the origin. Asymptotic stability is the type of stability preferred by control engineers.

The stability of nonlinear systems depends on the particular state space in which the state vector ranges in addition to the type and magnitude of the input. Therefore, the stability of nonlinear systems can also be classified on a regional basis as follows [12, 24]:

a) local stability, or stability in the small,

b) finite stability,

c) global stability, or stability in the large.

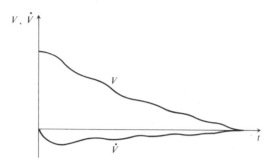

Fig. 8.60 Variation of energy and energy rate with time.

A nonlinear control system is denoted as being *locally stable* if it remains within an infinitesimal region about a singular point when subjected to a small perturbation. *Finite stability* refers to a system which returns to a singular point from any point **x** within a region R of finite dimensions surrounding it. The system is said to be *globally stable* if the region R includes the entire finite state space. Stability of either the local, finite, or global variety does not exclude limit cycles, but rather only excludes the possibility of the state point tending to travel to infinity. If the state point approaches the singularity as time approaches infinity, for any initial conditions within the region under consideration, then the system is described as being *asymptotically stable*. Asymptotic stability excludes a stable limit cycle as a possible dynamic equilibrium condition. The strongest possible condition that can be placed on a nonlinear control system, with parameters that do not vary with time, is global asymptotic stability.*

A major factor in this analysis has been the choice of the energy function V,

$$V = \tfrac{1}{2}x_1^2 + \tfrac{1}{2}x_2^2. \tag{8.169}$$

This function has two very interesting properties. First it is positive for all nonzero values of x_1 and x_2. Secondly, it equals zero when $x_1 = x_2 = 0$. A scalar function having these properties is called a *positive-definite* function. By adding V as a third dimension to the x_1x_2-plane, the positive-definite function $V(x_1, x_2)$ appears as a cup-shaped three-dimensional surface as illustrated in Fig. 8.61.

Liapunov's stability theorem can now be summarized for n-dimensional state space. A dynamic system of nth order is asymptotically stable if a positive-definite function V can be found whose derivative with respect to time is negative along the trajectories of the system. In practice, it is fairly easy to find a function which is positive definite, but usually it is very hard to find a function where, in addition, $\dot{V} < 0$ along the trajectories.

A justification of Liapunov's second method can best be presented by considering the phase plane of Fig. 8.62. Contours of constant V are shown by the curves of C_1,

---

* There are about 30 different classes of stability currently in use. However, the types defined in this book are the important ones most frequently used by the control-system engineer.

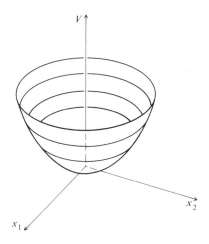

Fig. 8.61 A positive-definite function.

C_2, and C_3. Assume that the phase trajectory of this second-order system, whose initial state is the point p, is described by the following state equations:

$$\dot{x}_1 = F_1(x_1, x_2), \tag{8.170}$$

$$\dot{x}_2 = F_2(x_1, x_2). \tag{8.171}$$

The positive-definite function, V, is assumed to be given by

$$V = a^2 x_1^2 + b^2 x_2^2, \tag{8.172}$$

where a and b are unknown coefficients. The quantity V is permitted to take on successively larger constant values,

$$V = 0, C_1, C_2, C_3, \ldots, \qquad \text{where} \qquad 0 < C_1 < C_2 < C_3 < \cdots. \tag{8.173}$$

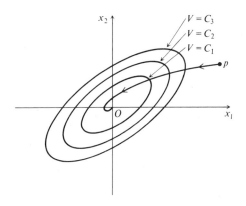

Fig. 8.62 Justification of Liapunov's stability criterion.

Therefore, Eq. (8.172) results in a set of equations:

$$a^2x_1^2 + b^2x_2^2 = 0$$
$$a^2x_1^2 + b^2x_2^2 = C_1 \qquad (8.174)$$
$$a^2x_1^2 + b^2x_2^2 = C_2$$
$$\cdots$$

When $V = 0$, Eq. (8.174) describes the origin of the state space; for other values, the resulting equations describe ellipses in the state space. As shown in Fig. 8.62, each succeeding ellipse contains within itself all of the preceding ellipses. The time derivative of the V function is given by

$$\frac{dV}{dt} = \frac{\partial V}{\partial x_1}\frac{dx_1}{dt} + \frac{\partial V}{\partial x_2}\frac{dx_2}{dt}, \qquad (8.175)$$

or

$$\frac{dV}{dt} = \frac{\partial V}{\partial x_1}\dot{x}_1 + \frac{\partial V}{\partial x_2}\dot{x}_2. \qquad (8.176)$$

Substituting Eqs. (8.170) and (8.171) into Eq. (8.176), we obtain

$$\frac{dV}{dt} = \frac{\partial V}{\partial x_1}F_1(x_1, x_2) + \frac{\partial V}{\partial x_2}F_2(x_1, x_2). \qquad (8.177)$$

Taking the partial derivatives of Eq. (8.172) as indicated, Eq. (8.177) can be rewritten as

$$\frac{dV}{dt} = 2a^2x_1F_1(x_1, x_2) + 2b^2x_2F_2(x_1, x_2). \qquad (8.178)$$

If dV/dt is negative, then the state must move, from its initial state point p, in the direction of smaller values of V and toward the origin. This system would then be asymptotically stable. This illustration can be extended, analogously, to higher-order systems.

To illustrate the application of Liapunov's second method we consider the second-order differential equation

$$\frac{dx}{dt} = y, \qquad (8.179)$$

$$\frac{dy}{dt} = -K_1y - K_2y^3 - x, \qquad (8.180)$$

where K_1 and K_2 are not both zero. Depending on the values of K_1 and K_2, stability or instability can result. For example, if $K_1 > 0$ and $K_2 > 0$, and V is given by

$$V = x^2 + y^2, \qquad (8.181)$$

this system is stable, since the derivative of V with respect to time t, dV/dt, is negative, since

$$\frac{dV}{dt} = 2\left(x\frac{dx}{dt} + y\frac{dy}{dt}\right) = -2(K_1 y^2 + K_2 y^4) < 0. \tag{8.182}$$

Equilibrium occurs at the singularity located at the origin, and the equilibrium point is asymptotically stable.

It is left as an exercise for the reader to prove that the condition $K_1 < 0$ and $K_2 < 0$ corresponds to an unstable equilibrium; the condition $K_1 > 0$ and $K_2 < 0$ corresponds to a stable equilibrium only if $0 < y^2 < K_1/K_2$.

As an additional example, let us consider the following nonlinear problem:

$$\dot{x}_1 = x_2 \tag{8.183}$$

$$\dot{x}_2 = -ax_2 - bx_2^3 - x_1, \tag{8.184}$$

where $a \geqslant 0$ and $b \geqslant 0$ and are not both zero. Let us try the following energy function:

$$V = x_1^2 + x_2^2. \tag{8.185}$$

This function satisfies the Liapunov criterion since it is positive for all nonzero values of x_1 and x_2, and $V(0, 0) = 0$. Now let us examine $\dot{V}$. Differentiating the V-function, we obtain

$$\dot{V} = 2\left(x_1 \frac{dx_1}{dt} + x_2 \frac{dx_2}{dt}\right). \tag{8.186}$$

Substituting Eqs. (8.183) and (8.184) into Eq. (8.186), we obtain

$$\dot{V} = 2[x_1 x_2 + x_2(-ax_2 - bx_2^3 - x_1)], \tag{8.187}$$

$$\dot{V} = -2[ax_2^2 + bx_2^4]. \tag{8.188}$$

Therefore, $\dot{V}$ is negative everywhere except at the origin, and the equilibrium is asymptotically stable. Notice that if $a < 0$ and $b < 0$, then we have a case of an unstable equilibrium as all trajectories move away from the origin. In the case where $a > 0$ and $b < 0$, V is negative only if $0 < x_2^2 < a/b$. This means that stability occurs only for initial disturbances that do not cause the (x_1, x_2)-point to lie outside a horizontal strip in the phase plane.

It is important to recognize that the stability conditions obtained from a particular V-function are usually sufficient, but not necessary. In addition, a Liapunov function for any particular system is not unique. Therefore, if a particular V-function should fail to demonstrate whether a particular system is stable or not, there is no assurance that another function could not be found that does determine stability or instability. There is also no assurance that exceeding the limits based on a particular V-function will actually cause the equilibrium to be unstable. The Liapunov stability criterion is a conservative one.

As can be seen from this presentation, the primary use of Liapunov's second method is not the determination of stability where the answer may be found by other means, but is rather in the study of problems of stability which are not readily determined by other methods. For example, the determination of stability of high-order nonlinear control systems is such a problem.

It is important to emphasize that we have not specified any guidelines for selecting the V-functions. Actually, the choice of the V-function is the main problem in wider application of Liapunov's method in practice. Methods for constructing the V-function is of current research interest [14, 21].

8.18 POPOV'S METHOD

An interesting and very powerful stability criterion for nonlinear systems that are time invariant was introduced in 1959 by V. M. Popov who obtained a frequency-domain criterion as a sufficient condition for asymptotic stability of single-loop control systems [22–29]. The method, as originally developed by Popov, is applicable to single-loop feedback systems containing time-invariant linear elements and time-invariant nonlinearities. An important feature of Popov's approach is that it is applicable to systems of high order. Once the frequency response of the linear element is known, very little additional calculation is required for determining stability of the nonlinear control system.

This section presents Popov's stability criterion in terms of inequality constraints on the nonlinear element in conjunction with a modified frequency plot of the linear element. It will be shown that the most important and appealing feature of the Popov criterion is that it shares all of the desirable characteristics of the Nyquist method.

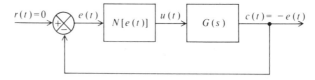

Fig. 8.63 Nonlinear feedback control system considered by Popov [22].

In order to introduce Popov's method, let us consider the nonlinear system illustrated in Fig. 8.63. It is composed of a linear, time-invariant process $G(s)$ and a nonlinear, time-invariant element $N[e(t)]$. The reference input $r(t)$ is assumed to be zero. Therefore, the response of this system can be expressed as

$$e(t) = e_0(t) - \int_0^t g(t - \tau)u(\tau) \, d\tau, \tag{8.189}$$

where

$$g(t) = \mathscr{L}^{-1}[G(s)] = \text{unit impulse response}$$
$$-e_0(t) = \text{initial condition response.}$$

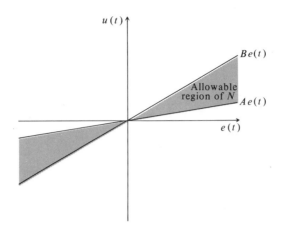

Fig. 8.64   Restricted region of nonlinearity.

In this analysis, special restrictions are placed on the nonlinear and linear elements. For the nonlinear element $N[e(t)]$, it is assumed that the input–output relationship is restricted to lie within a sector bounded by two straight lines that pass through the origin as illustrated in Fig. 8.64. Thus

$$A \leqslant N[e(t)] \leqslant B, \tag{8.190}$$

where

$$u(t) = N[e(t)]e(t). \tag{8.191}$$

Furthermore, it is assumed that for all t, and for every finite value e_m, there is a finite value u_m such that

$$|u(t)| \leqslant u_m < \infty \qquad \text{if} \qquad |e(t)| \leqslant e_m. \tag{8.192}$$

The only assumption concerning the linear element $G(s)$ is that it is output stable of degree n for some value of n. By this we mean that if $n < 0$, the output response to an initial condition or an impulse may diverge, but when the output is multiplied by e^{nt} it will converge towards zero. For the case of $n > 0$, output stable means that the output response to an initial condition or an impulse will converge towards zero faster than the function e^{-nt}. In general, a linear element will be output stable of degree n if its transfer function $G(s)$ and initial condition response function $E_0(s)$ are rational functions of s and its poles all satisfy

$$\text{Re } s < -n.$$

Therefore, n actually represents the settling rate of the linear element.

Popov's method is concerned with the asymptotic behavior of the control signal $u(t)$ and output $-e(t)$ of the linear element. Therefore, in addition to the definitions of asymptotic stability, local stability, finite stability, and global stability that were introduced in Section 8.17 in connection with the Liapunov stability criterion, we are concerned here with control asymptoticity and output asymptoticity. *Control asymptoticity of degree n* exists if a real value n can be found such that for every set of initial conditions

$$\int_0^\infty [e^{nt}u(t)]^2 \, dt < \infty. \tag{8.193}$$

Output asymptoticity of degree n exists if a real value n can be found such that for every set of initial conditions

$$\int_0^\infty [e^{nt}e(t)]^2 \, dt < \infty. \tag{8.194}$$

These stability definitions can be clarified by considering the following lemma.

If the linear element $G(s)$ of Fig. 8.63 is output stable of degree n, the input and output of the nonlinear element are bounded and satisfy Eq. (8.192), and the feedback system is control asymptotic of degree n, then

$$\lim_{t \to \infty} e^{nt}e(t) = 0. \tag{8.195}$$

Therefore, if this lemma is satisfied, $e(t)$ converges towards zero faster than e^{-nt} for $n > 0$.

How can we relate control asymptoticity and output asymptoticity? Obviously, there should be some relationship based on the properties of the linear element $G(s)$. Let us assume that the linear element is output stable of degree n. It can be shown that if the linear element is control asymptotic of degree n, then it is also output asymptotic of degree n. In addition, if for each set of initial conditions a number Q_0 exists such that

$$|e_0(t)| \leqslant Q_0 e^{-nt},$$

then there exists a number Q that is dependent on Q_0 such that

$$|e(t)| \leqslant Q e^{-nt}$$

for all values of t. This appears reasonable since a decaying control signal $u(t)$, that satisfies Eq. (8.193), when it is fed into the linear element whose unit–impulse response decays like $u(t)$, will produce an output $-e(t)$ that also decays in a similar manner.

Popov's fundamental theorem is based on the basic feedback control system illustrated in Fig. 8.63. It is assumed that the linear system is output stable. The theorem states that for the feedback system to be absolutely control and output asymptotic for

$$0 \leqslant N[e(t)] \leqslant K,$$

it is sufficient that a real number q exists such that for all real $\omega \geqslant 0$ and an arbitrarily small $\delta > 0$, the following condition is satisfied:

$$\mathrm{Re} \, [(1 + j\omega q)G(j\omega)] + 1/K \geqslant \delta > 0. \tag{8.196}$$

The relation (8.196) is the Popov criterion. Depending on the type of nonlinearity

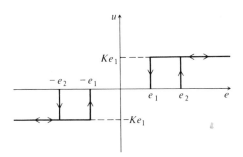

Fig. 8.65 Passive hysteresis nonlinear characteristics.

present, the following restrictions on q and K are imposed:

A. For a single-valued, time-invariant, nonlinearity:

$$-\infty < q < \infty \quad \text{if} \quad 0 < K < \infty,$$
$$0 \leqslant q < \infty \quad \text{if} \quad K = \infty.$$

B. For a nonlinearity having passive hysteresis (see Fig. 8.65):

$$-\infty < q \leqslant 0 \quad \text{and} \quad 0 < K < \infty.$$

C. For a nonlinearity having active hysteresis (see Fig. 8.66):

$$0 \leqslant q < \infty \quad \text{and} \quad 0 < K \leqslant \infty.$$

Examination of these three possible types of nonlinearities shows that the theorem allows for a trade-off between the requirements on the nonlinear and linear elements.

Let us rewrite (8.196) as follows:

$$\text{Re } G(j\omega) > -\frac{1}{K} + \omega q \text{ Im } G(j\omega). \tag{8.197}$$

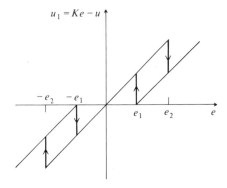

Fig. 8.66 Active hysteresis nonlinear characteristics.

Relation (8.197) states that for each frequency ω, the Nyquist plot of $G(j\omega)$ must lie to the right of a straight line given by

$$\text{Re } G(j\omega) = -\frac{1}{K} + \omega q \text{ Im } G(j\omega). \tag{8.198}$$

This line is called the Popov line and is illustrated in Fig. 8.67. The angles α and β are

$$\alpha = \tan^{-1} \omega q, \tag{8.199}$$

$$\beta = \tan^{-1} \frac{1}{\omega q}, \tag{8.200}$$

and it is clear that the slope of this line depends on the product ωq.

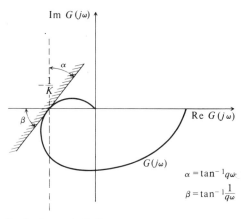

Fig. 8.67   Popov's method when q is finite.

Stability depends on choosing a value of q such that, for each frequency ω, $G(j\omega)$ lies to the right of the Popov line. It is important to recognize from Eqs. (8.199) and (8.200) that the slope of this line is frequency dependent. A Popov line whose slope is not frequency dependent can be found in a modified frequency plane. In order to find the particular frequency-insensitive Popov line, a simple transformation is used. The modified frequency response function $G^*(s)$ is defined as

$$G^*(j\omega) \triangleq \text{Re } G(j\omega) + j\omega \text{ Im } G(j\omega). \tag{8.201}$$

Therefore, Eq. (8.197) can be rewritten as

$$\text{Re } G^*(j\omega) > -\frac{1}{K} + q \text{ Im } G^*(j\omega). \tag{8.202}$$

In the $G^*(j\omega)$-plane, the Popov line is defined by

$$\text{Re } G^*(j\omega) = -\frac{1}{K} + q \text{ Im } G^*(j\omega),$$

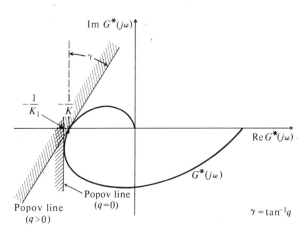

Fig. 8.68   Popov line in the $G^*(j\omega)$-plane for the case where $q \geqslant 0$.

and is frequency insensitive. The Popov line in the $G^*(j\omega)$-plane is illustrated in Figs. 8.68 and 8.69. The angle γ is defined as

$$\gamma = \tan^{-1} q.$$

Notice from Figs. 8.68 and 8.69 that the $\dot{G}^*(j\omega)$-locus passes to the right of a tangent to the locus at the point where $G^*(j\omega)$ intersects the negative real axis. These points are labeled $-1/K$. Therefore, K represents the maximum permissible gain for the system. For the case where $q = 0$, the Popov line expression reduces to

$$\text{Re } G^*(j\omega) = -\frac{1}{K}$$

and the system is stable if it lies to the right of a vertical line passing through the point $-1/K_1$ as is illustrated in Fig. 8.68.

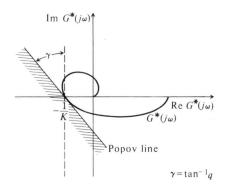

Fig. 8.69   Popov line in the $G^*(j\omega)$-plane for the case where $q < 0$.

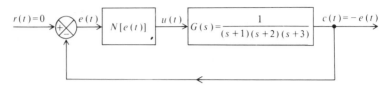

Fig. 8.70   Nonlinear control system example.

As an example of applying Popov's method, consider the system illustrated in Fig. 8.70. For the linear element, the initial condition response $e_0(t)$ is given by

$$e_0(t) = e_{10}e^{-t} + e_{20}e^{-2t} + e_{30}e^{-3t},$$

where e_{10}, e_{20}, and e_{30} depend on the initial conditions. The unit impulse response $g(t)$ is given by

$$g(t) = [0.5e^{-t} - e^{-2t} + 0.5e^{-3t}]U(t) \qquad (8.203)$$

where $U(t)$ is a unit step function. Equation (8.203) indicates that the linear element is output stable and satisfies one of the necessary constraints in order to use Popov's method. The corresponding $G^*(j\omega)$-locus is illustrated in Fig. 8.71. From this diagram, we can conclude that if the nonlinear element corresponds to a single-valued nonlinear element, and if $q = 0.5$, the Popov condition is satisfied when $0 < K \leqslant 60$.

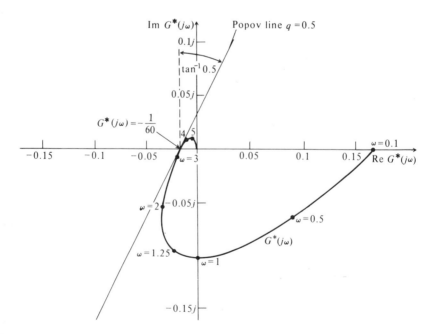

Fig. 8.71   The $G^*(j\omega)$-locus for system illustrated in Fig. 8.70.

In conclusion, we see that the Popov method gives an exact and sufficient condition for determining the absolute stability of feedback systems having the configuration illustrated in Fig. 8.63, with certain restrictions imposed. The inequality (8.196), which was given in terms of $G(j\omega)$ and a real constant q, is the key element of the technique. The importance of Popov's method is due to the fact that Popov's condition is given in terms of $G(j\omega)$ which makes the technique easily applicable to systems having high-order processes which are to be controlled. In addition, the method shares all of the desirable characteristics of the Nyquist method. In the following section, Popov's method is extended to other types of systems which are not necessarily restricted to systems in which the linear portion is output stable, and the nonlinearity is time invariant.

8.19 THE GENERALIZED CIRCLE CRITERION

The generalized circle criterion [25, 26] enables one to investigate the asymptotic behavior of a much wider class of systems than that for which Popov's theorem was originally intended. For example, this technique can be applied to systems having unstable or nonasymptotically stable plants, and time-variable nonlinearities. The generalized circle criterion presented in this section consists of modifying Popov's basic theorem in such a manner that the Popov condition can be applied directly to the original transfer function.

Let us reconsider the basic nonlinear feedback control system illustrated in Fig. 8.63, and allow the nonlinearity to be time variable. It is assumed that the linear element of this system is obtained by applying negative feedback through a constant feedback gain A to the original linear element, and is output stable. The generalized circle criterion is as follows: for the system of Fig. 8.63 to be absolutely control and output asymptotic for

$$A < N[e(t), t] < B$$

then it is sufficient that a real number q exists such that for all real ω, the following conditions are satisfied:

$$\left| G(j\omega) + \frac{(B + A) - j\omega q(B - A)}{2AB} \right|^2 - \left(\frac{B - A}{2AB} \right)^2 (1 + q^2\omega^2) \geqslant \delta > 0,$$
$$\text{for} \quad 1/A > 1/B \quad (8.204)$$

$$\left| G(j\omega) + \frac{(B + A) - j\omega q(B - A)}{2AB} \right|^2 - \left(\frac{B - A}{2AB} \right)^2 (1 + q^2\omega^2) \leqslant -\delta < 0,$$
$$\text{for} \quad 1/A < 1/B. \quad (8.205)$$

The quantities $B - A$ and q are restricted as K and q were in Popov's method, discussed previously in Section 8.18.

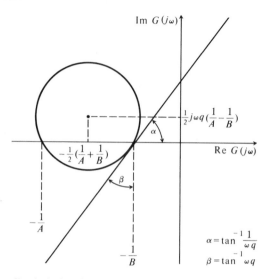

Fig. 8.72 The generalized circle criterion for the case $1/A > 1/B$.

Figure 8.72 illustrates the physical interpretation of relation (8.204) where it is assumed that $1/A > 1/B$. For each value of $\omega \geqslant 0$, the Nyquist plot of $G(j\omega)$ must lie outside the circle centered at*

$$-\frac{1}{2}\left(\frac{1}{A}+\frac{1}{B}\right)+\frac{1}{2}j\omega q\left(\frac{1}{A}-\frac{1}{B}\right), \qquad (8.206)$$

which crosses the real axis at the points $-1/A$ and $-1/B$. It is interesting to note that if $1/B > 1/A$, then the Nyquist plot must lie inside the circle that is centered at the point given by Eq. (8.206).

Analysis of the generalized circle criterion is quite interesting. For example, if we let $A \to 0$ and $B \to K$, then relations (8.204) and (8.205) reduce to (8.197), which corresponds to the Popov condition (8.196). On the other hand, if we let $A \to C$ and $B \to C$, then we have a linear time-invariant system with gain C. For this case, the critical circle reduces to a point $-(1/C)$ in the $G(j\omega)$-plane, which is of course the critical point of the Nyquist diagram.

The generalized critical circle illustrated in Fig. 8.72 and defined by (8.204) and (8.205) is a function of frequency. Although all of the circles pass through the points $-1/A$ and $-1/B$, their centers move up with increasing values of $q\omega$. However, for the general nonlinearity case where the nonlinearity may contain hysteresis and is time variable, $q = 0$ and a set of circles result that are symmetrical about the real axis. Notice also from Fig. 8.72 that trade-offs can be made between the requirements

---

* The circle criterion was originally developed only for the case of $q = 0$ [26]. However, the generalized circle criterion presented here is an extension which is valid for all values of q including zero [25].

on the linear and nonlinear elements. For example, by narrowing the sector

$$A < N[e(t), t] < B \qquad (8.207)$$

the critical circles will be reduced and this will increase the permissible range of $G(j\omega)$.

Let us consider the application of the generalized circle criterion. Unlike the situation for the Popov line of Section 8.18, it is not advantageous to transform the critical circles from the $G(j\omega)$-plane into the $G^*(j\omega)$-plane, since this will only result in a family of curves that are not circles and whose shapes depend on both ω and q. However, it can easily be shown that if a tangent is drawn on the critical circle at the point $-1/B$, its angle α is given by

$$\alpha = \tan^{-1} \frac{1}{\omega q}. \qquad (8.208)$$

Figure 8.72 illustrates this angle and also the angle which is given by

$$\beta = \tan^{-1} \omega q. \qquad (8.209)$$

Comparing Figs. 8.67 and 8.72, we observe that the tangent line on the critical circle has the same slope as the Popov line when $A = 0$. In addition, if

$$B = K$$

and

$$0 \leqslant A < B,$$

then the tangent line on the critical circle becomes identical to the Popov line shown in Fig. 8.67. We also showed in Section 8.18 that the Popov line could be transformed into a frequency-independent line on the $G^*(j\omega)$-plane as was shown in Figs. 8.68 and 8.69. Therefore, if $G^*(j\omega)$ lies to the right of the Popov line of Figs. 8.68 and 8.69, then $G(j\omega)$ lies outside the critical circle illustrated in Fig. 8.72 for all ω.

As an example of the generalized circle criterion, let us consider a nonlinear system with the configuration illustrated in Fig. 8.63, where the transfer function of the linear element is given by

$$G(s) = \frac{1}{(s - 1)(s + 2)(s + 3)}, \qquad (8.210)$$

and the initial-condition response is

$$e_0(t) = e_{10}e^t + e_{20}e^{-2t} + e_{30}e^{-3t}, \qquad (8.211)$$

where e_{10}, e_{20}, and e_{30} are related to the initial conditions for a particular set of state variables. It is important to recognize that we are dealing with a nonlinear element that is time variable and a linear element that is not output stable; a problem which could not be solved using Popov's basic method. It is assumed that the nonlinear element corresponds to the general nonlinearity $N[e(t), t]$, and therefore q must be chosen equal to zero. The solution consists of plotting the frequency locus $G(j\omega)$ as illustrated in Fig. 8.73. The generalized circle criterion results in Popov sectors

$$A < N[e(t), t] < B \tag{8.212}$$

such that for each of these Popov sectors the $G(j\omega)$-locus lies on or outside a circle symmetrical about the real axis and passing through the points $-1/A$ and $-1/B$. For this particular example, the possible Popov sectors (illustrated in Fig. 8.73) which result in stable systems are given by the following conditions:

$$
\begin{aligned}
12.6 &< N[e(t), t] < 15.2, \\
19.6 &< N[e(t), t] < 34.5, \\
41.7 &< N[e(t), t] < 167.0.
\end{aligned}
\tag{8.213}
$$

Notice from Fig. 8.73 that the size of the Popov sector, $B - A$, i.e., the difference between the lower bound A and upper bound B, depends on the size of the critical circle.

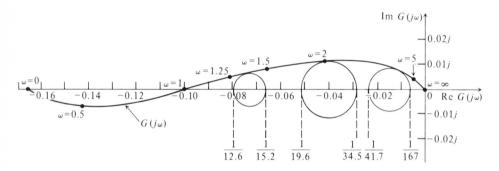

Fig. 8.73   Generalized circle criterion example where

$$G(s) = \frac{1}{(s - 1)(s + 2)(s + 3)}$$

and the nonlinearity corresponds to the general nonlinearity case where $q = 0$.

Therefore, we see that Popov's method can be extended to systems with unstable or nonasymptotically stable plants and time-varying nonlinearities by utilizing the generalized circle criterion. As shown in this section, the method permits the design of nonlinear systems whose linear elements are not output stable and whose nonlinearity can be time variable and correspond to any general nonlinearity. There are many practical situations where this is the very case involved, and the generalized circle criterion provides a very powerful method for solving this class of problems.

PROBLEMS

8.1 The torque–speed characteristics of a two-phase, ac instrument servomotor are illustrated in Fig. P8.1. Assume that the inertia of the rotor is 0.1 oz in² and that the load inertia and coefficient of viscous friction are negligible.

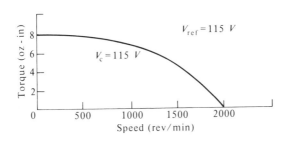

Figure P8.1

a) Derive the transfer function of the motor, relating output position to control voltage, using linearizing approximations. Approximate the characteristics by one straight line that is tangent to the exact characteristics at 1000 rev/min, and by one straight line that goes through the two endpoints. Compare your results.
b) How do the time constants derived in (a) change if the torque–speed characteristics are approximated by two other straight lines: one at low speed that is tangent to the exact characteristics at 250 rev/min, and one at high speed that is tangent to the exact characteristics at 1750 rev/min?
c) Comparing your answers to (a) and (b), what conclusions can you reach?

8.2 Derive the describing function corresponding to the combined nonlinear characteristics of saturation S and dead zone D.

8.3 Derive the describing function of a 2-position contactor which does not exhibit any hysteresis affect.

8.4 An amplifying device in a feedback loop has the following nonlinear characteristics:

1. No output signal for all inputs whose magnitude is less than E_1 V.

$$e_0(t) = 0 \qquad \text{when} \qquad |e_i(t)| < |E_1|.$$

2. Input signals whose magnitude is greater than E_1 V but less than E_2 V are amplified according to the relation

$$e_g(t) = K[e_i(t) - E_1],$$

where

$$|E_1| < |e_i(t)| < |E_2|.$$

3. For all input signals whose magnitude is greater than E_2 V, the output is given by

$$e_0(t) = KE_2,$$

when

$$|e_i(t)| > |E|_2.$$

Sketch the input–output characteristics and derive the describing function.

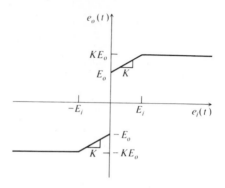

Figure P8.5

8.5 Derive the describing function for an amplifying device which has the nonlinear characteristics illustrated in Fig. P8.5.

8.6 A unity feedback system consists of cascaded elements which include a relay, pure integration, an amplifier which saturates, a two-phase, ac servomotor, and a gear train containing backlash.
a) Draw the block diagram and show the linear transfer functions and the describing functions for the nonlinear elements, symbolically.
b) Qualitatively illustrate how you would predict the presence of an oscillation on a gain–phase diagram.

8.7 A unity feedback instrument servo with a spring-loaded shaft has a dead zone of 2°. Assume that K_1 in Fig. 8.6 equals 1. The transfer function for the linear portion of the system is given by

$$G(j\omega) = \frac{20}{j\omega(j\omega + 0.05)(j\omega + 0.1)}.$$

Utilizing the describing-function method on a gain–phase diagram, determine the conditions necessary for the existence of a limit cycle.

8.8 Repeat Problem 8.7 with the transfer function for the linear portion of the system given by

$$G(j\omega) = \frac{2}{j\omega(j\omega + 0.1)}.$$

8.9 A unity feedback instrument servo is driven by an amplifier which saturates at 70% of rated voltage of the motor. Assume that the gain of the unsaturated amplifier is 40. The transfer function of the linear portion of the system, excluding the amplifier, is given by

$$G(j\omega) = \frac{0.25}{j\omega(j\omega + 2)}.$$

(a)

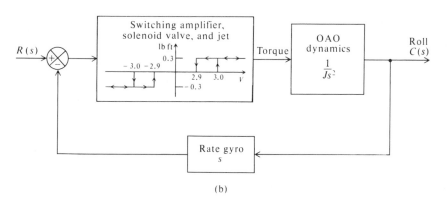

(b)

Fig. P8.12 (a) An artist's conception of NASA's OAO. (Official NASA photo) (b) Equivalent block diagram of the roll coarse solar orientation control loop.

Utilizing the describing function method on a gain–phase diagram, determine whether a limit cycle exists.

8.10   An instrument servo contains a three-position contactor which has a pull-in point and a drop-out point at errors of 0.03 rad and 0.01 rad respectively. The transfer function of the linear portion is given by

$$G(j\omega) = \frac{10}{j\omega(j\omega + 4)}.$$

Determine whether a limit cycle exists utilizing the describing-function analysis on a gain–phase diagram. Assume K_1 in Fig. 8.17 equals 1.

8.11   Repeat Problem 8.10 with the transfer function of the linear portion given by

$$G(j\omega) = \frac{2.5}{j\omega(j\omega + 4)}.$$

8.12   The Orbiting Astronomical Observatory (OAO), shown in Fig. P8.12(a), is designed to provide astronomers with a standardized stable platform in space readily adaptable to a variety of experiments [15]. Orbiting at 500 miles altitude, the OAO permits scientists to examine any point in the heavens, unimpeded by the earth's atmosphere. In addition, it has a precision and stability unsurpassed even by a stationary observatory. After injection into orbit and separation of the satellite from the second booster stage, the high-thrust gas jets stop any tumbling or rolling sensed by rate gyros and align the OAO's optical axis with the earth–sun line to $\pm\frac{1}{4}°$. The OAO utilizes coarse and fine solar orientation control systems. The coarse loop depends on gas jets firing and is a nonlinear control system. The fine loop depends on a momentum-exchange wheel and is a linear control system. Let us consider the nonlinear characteristics of the coarse solar orientation control system in this problem. Figure P8.12(b) illustrates an equivalent block diagram of the roll coarse solar orientation control loop. Basically, it consists of a switching amplifier which has very similar characteristics to the nonlinear on–off element having hysteresis that has been analyzed previously and which controls a solenoid valve and jet. As indicated in Fig. P8.12(b), the jet fires when the error reaches ±3 V and stops firing when the error is reduced to ±2.9 V. Assume that the resultant corrective torque produced by the jet is ±0.3 lb ft, and the inertia in the roll axis is 1000 lb ft sec². The rate gyro, which closes this rate loop, has a sensitivity of 10 V/(degree/sec). Using the describing function analysis on a gain–phase diagram, determine the existence of any limit cycles.

8.13   An instrument servo system used for positioning a load may be adequately represented by the block diagram shown in Fig. P8.13.

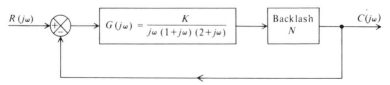

$R(j\omega)$ $\oplus$ $G(j\omega) = \dfrac{K}{j\omega(1+j\omega)(2+j\omega)}$ Backlash N $C(j\omega)$

Figure P8.13

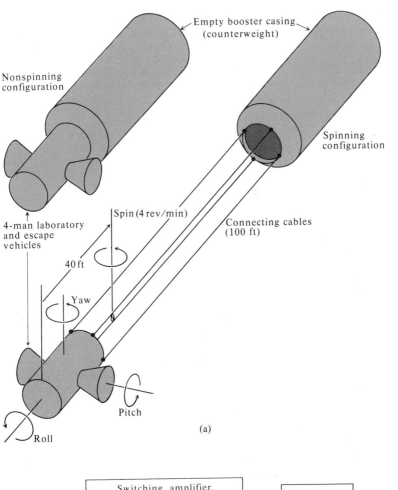

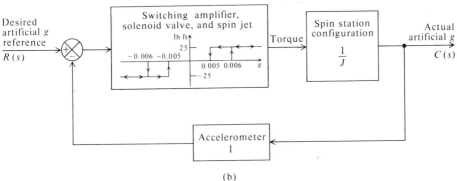

Fig. P8.15 (a) Conceptual design of spinning space station. (Courtesy of the Sperry Rand Corporation) (b) Equivalent block diagram of the spin speed control system.

a) Using the gain–phase diagram, prove that if $K = 4$ the system exhibits a limit cycle(s). Determine the range of frequencies and D/M ratios for which the limit cycle(s) exist. Are they stable or unstable?

b) It is desired to stabilize this configuration by reducing the system gain. Determine the maximum value of gain K which will allow a minimum stability margin of 3 db.

c) It is desired to stabilize this configuration by adding a lead network in cascade with $G(j\omega)$. Determine the time constants of the lead network

$$\frac{1 + j\omega\alpha T}{1 + j\omega T}$$

which will permit a minimum stability margin of 3 db.

d) It is desired to stabilize this configuration by the introduction of rate feedback as shown in Fig. 8.30. Determine the rate feedback constant b which will permit a minimum stability margin of 3 db.

8.14   Repeat Problem 8.13 for

$$G(j\omega) = \frac{K(1 + 3j\omega)}{j\omega(1 + 2.1j\omega)^2}.$$

8.15   It is questionable whether man can tolerate prolonged periods of weightlessness and it may, therefore, be necessary to provide an artificial gravity environment in future space stations. This can be accomplished by spinning the space station to obtain a centripetal acceleration equivalent to 0.2 to 0.5 of the earth's surface gravity. Various studies have been performed on this problem [16]. Figure P8.15(a) illustrates a conceptual design of what such a space configuration might look like. Basically, it consists of a small laboratory that is connected to a counterweight with a cable, and the entire configuration is spun about its composite mass center. From a practical viewpoint, this configuration is launched as a retracted, unmanned vehicle. Crew members arrive later in escape vehicles and dock with the orbiting, nonspinning space station. After entering the laboratory, a transition maneuver is performed to deploy the stations. After this maneuver, the station is spun at 4 rev/min in order to obtain an acceleration of 0.25 g in the laboratory portion. This configuration requires many control systems which include pitch, roll, yaw, and spin control. Let us focus our attention in this problem on the spin control system. The spin speed must be held fairly constant in order to maintain a fixed artificial gravity. Since the space station will be operating at altitudes between 100 to 150 miles, however, residual atmosphere can cause significant reductions in the spin speed. Figure P8.15(b) illustrates a proposed spin speed control system [30]. An accelerometer gives a direct measure of the gravity level and spin jets are activated through the switching amplifier and solenoid valve to correct the errors. As indicated in Fig. P8.15(b), a torque of ± 25 lb ft is produced when errors exceeding $\pm 0.006\,g$ are encountered. This torque remains until the error is reduced to $\pm 0.0005\,g$. Assume the inertia of the configuration is 50,000 lb ft sec^2. Using the describing function analysis on a gain–phase diagram, determine the existence of any limit cycles.

8.16   A positioning system consists of a major feedback path and three minor feedback paths as shown in Fig. P8.16. Utilizing the signal-flow diagram approach, determine whether any limit cycles exist if the nonlinearity corresponds to backlash and the system parameters

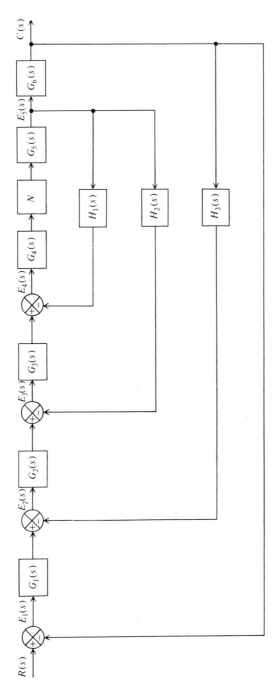

Fig. P8.16

correspond to the following values:

$$G_1(s) = \frac{2(1 + 10s)}{s(1 + s)}, \qquad\qquad H_1(s) = 1.2,$$

$$G_2(s) = \frac{100(0.1s + 1)(0.2s + 1)}{(s + 1)(0.5s + 1)}, \qquad H_2(s) = 0.0282s,$$

$$G_3(s) = \frac{200(0.04s + 1)}{s(0.02s + 1)}, \qquad\qquad H_3(s) = 10s.$$

$$G_4(s) = \frac{4}{(0.1s + 1)},$$

$$G_5(s) = \frac{5}{s},$$

$$G_6(s) = \frac{0.001}{s},$$

8.17 Repeat Problem 8.16 with the nonlinearity corresponding to coulomb friction.

8.18 In 1965 the Ranger unmanned space vehicle, shown in Fig. P8.18(a), investigated the surface of the moon by means of a TV camera and instruments. From a control-system viewpoint, the Ranger attitude control system must stabilize the vehicle from second-stage separation until lunar encounter [31]. The accuracy requirements are especially high since the Ranger vehicle is an unmanned vehicle and its solar panels must point accurately at the sun in order to obtain energy. In addition, the Ranger vehicle transmits data back to earth by means of a narrow-beam antenna which requires very accurate pointing. The equivalent block diagram of the Ranger attitude control system for the pitch axis is illustrated in Fig. P8.18(b). Error signals in position, generated by a sun sensor, are added to velocity error signals generated by a rate gyro. A switching amplifier, which has very similar characteristics to the nonlinear on-off element having hysteresis that has been analyzed previously, controls a solenoid valve and jet. As indicated in Fig. P8.18(b), the jet fires when the dead zone error is equivalent to ± 1 mrad and stops firing when the error decreases to ± 0.96 mrad. Assume that the corrective torque produced by the jet is ± 0.02 lb ft, and the inertia of Ranger in the pitch axis is 110 lb ft sec^2. Utilizing the describing function method, determine the sensitivity of the rate gyro, b, in order to obtain a stability margin of at least $40°$ for $M/D < 6$.

8.19 The control system of Fig. P8.19 illustrates an undamped servo, where the motor torque is proportional to the error e and is not affected by velocity. Consider the composite gain of the forward part of the loop to be 800 dyne cm per radian of error.

a) Draw the phase trajectory for the following set of initial conditions:

$$\theta_0 = \text{initial output displacement (in rad)} = 1 \text{ rad}$$
$$\dot\theta_0 = \text{initial output velocity (in rad/sec)} = 0.$$

b) What conclusions can you reach from the phase trajectory?

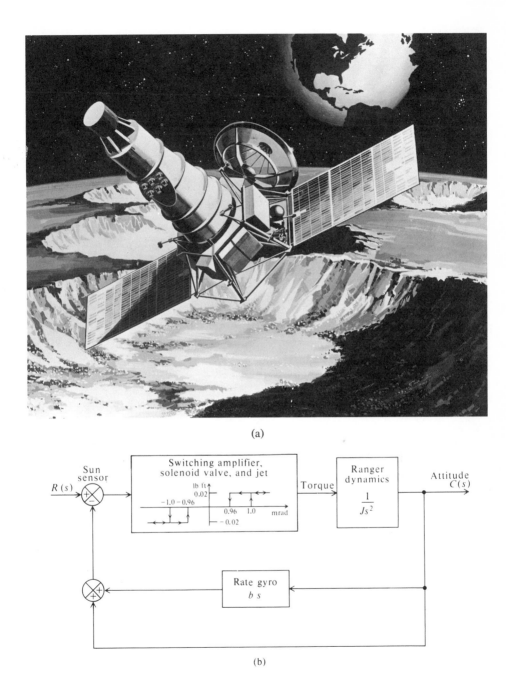

(a)

(b)

Fig. P8.18   (a) An artist's conception of a Ranger spacecraft photographing the moon before impact. (Official NASA photo.) (b) Equivalent block diagram of the Ranger attitude control system for the pitch axis.

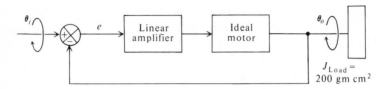

Figure P8.19

8.20 The control system shown in Fig. P8.20 illustrates a positioning servo system.

a) Draw the phase trajectory for the following set of initial conditions:

$$\theta_0 = 2.5 \text{ rad}, \qquad \dot{\theta}_0 = 0.$$

b) What conclusions can you reach from the phase portrait?

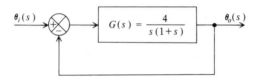

Figure P8.20

8.21 Draw the phase portrait of a linear, unity, feedback control system whose transfer function is given by

$$G(s) = \frac{10}{s(1 + 0.1s)}.$$

8.22 Repeat Problem 8.20 with rate limiters at 4 rad/sec added to the system.

8.23 Repeat Problem 8.20 with position limiters at 1 rad added to the system.

8.24 Repeat Problem 8.21 with the rate limiters at 1 rad/sec added to the system.

8.25 Repeat Problem 8.21 with position limiters at 1 rad added to the system.

8.26 Determine the time it takes for the phase trajectory illustrated in Fig. P8.26 to traverse the following segments:

a) AB,   b) BC,   c) CD,

d) DE,   e) AE.

8.27 Manipulate the following equations into the form

$$\frac{d\dot{\theta}(t)}{d\theta} = f(\theta, t):$$

a) $A\dfrac{d^2\theta(t)}{dt^2} + B\dfrac{d\theta(t)}{dt} + C\theta(t) = D,$

b) $A\dfrac{d^2\theta(t)}{dt^2} + B\dfrac{d\theta(t)}{dt}\left|\dfrac{d\theta(t)}{dt}\right| + C\theta(t) = D,$

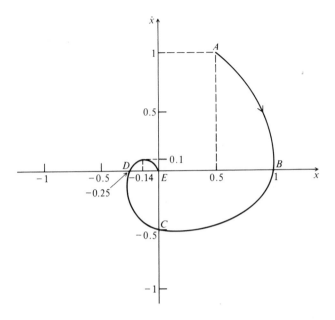

Figure P8.26

c) $A\dfrac{d^2\theta(t)}{dt^2} + B\dfrac{d\theta(t)}{dt}|\theta(t)| + C\theta(t) = D$,

d) $A\dfrac{d^2\theta(t)}{dt^2} + B\dfrac{d\theta(t)}{dt} + C\theta(t)|\theta(t)| = D$.

8.28 Derive the isocline equation for each of the following differential equations and draw the isoclines together with their slope markers:

a) $\dfrac{d^2\theta(t)}{dt^2} + 0.8\dfrac{d\theta(t)}{dt} + 0.4\theta(t) = 0$,

b) $\dfrac{d^2\theta(t)}{dt^2} + 0.8\dfrac{d\theta(t)}{dt} + 0.4\theta(t) = 0.6$,

c) $\dfrac{d^2\theta(t)}{dt^2} + 0.8\dfrac{d\theta(t)}{dt}\left|\dfrac{d\theta(t)}{dt}\right| + 0.4\theta(t) = 0$,

d) $\dfrac{d^2\theta(t)}{dt^2} + 0.8\dfrac{d\theta(t)}{dt}\left|\dfrac{d\theta(t)}{dt}\right| + 0.4\theta(t) = 0.6$.

8.29 Construct the phase portraits for each of the differential equations of Problem 8.28 using the method of isoclines.

8.30 The differential equation of a typical second-order system containing both viscous and coulomb friction is given by

$$\frac{d^2y}{dx^2} + 0.8\frac{dy}{dx} + 0.4y + A\sin\left(\frac{dy}{dx}\right) = 0.$$

Determine and construct the phase portraits for $A = 0.2, 0.4$, and 0.8.

8.31 Determine and construct the phase trajectory for a second-order system containing rate limiting where $\theta(0) = 0.7$. The differential equation for the system is given by

$$\frac{d^2\theta(t)}{dt^2} + 0.8\frac{d\theta(t)}{dt} + 0.4\theta(t) = 0$$

and

$$\left|\frac{d\theta(t)}{dt}\right|_{max} = 0.3.$$

8.32 Determine and construct the phase trajectory of a nonlinear system for which the describing equations are given by

$$\frac{d^2\theta(t)}{dt^2} + 0.8\frac{d\theta(t)}{dt} + \theta(t) = 0.5 \qquad \text{where} \qquad \theta(t) > 0$$

and

$$\frac{d^2\theta(t)}{dt^2} + 0.8\frac{d\theta(t)}{dt} + \theta(t) = -0.5 \qquad \text{where} \qquad \theta(t) < 0.$$

Assume that the initial conditions are $\theta(0) = 0.5$ and $(d\theta/dt)(0) = 0$.

8.33 Utilizing Liapunov's second method, determine the stability of the nonlinear differential equation

$$\ddot{x} + K_1\dot{x} + K_2(\dot{x})^5 + x = 0$$

if

a) $K_1 > 0, \quad K_2 > 0$;   b) $K_1 < 0, \quad K_2 < 0$;

c) $K_1 > 0, \quad K_2 < 0$.

8.34 Repeat Problem 8.33 for the nonlinear differential equation

$$\ddot{x} + [K_1 + K_2(x)^4]\dot{x} + x = 0.$$

8.35 A unity-feedback control system contains a linear and nonlinear element. The transfer function of the linear element is given by

$$G(s) = \frac{(s + 3)}{s(s + 2)^2(s + 1.5)}.$$

The initial-condition response of the linear element is given by

$$e_n(t) = e_{10} + e_{20}e^{-2t} + e_{30}te^{-2t} + e_{40}e^{-1.5t},$$

where e_{10}, e_{20}, e_{30}, and e_{40} are related to the initial conditions and the unit-impulse response is given by

$$g(t) = [0.5 + 3.5e^{-2t} + te^{-2t} - 4e^{-1.5t}]U(t)$$

where $U(t)$ is the unit step input. Using Popov's method, determine the values of K which will result in a stable system, assuming that the nonlinear element is single-valued with $q = 1.0$.

8.36 Repeat Problem 8.35 with $q = 0.75$ and the transfer function of the linear element given by

$$G(s) = \frac{(s + 1)}{s(s + 1.5)(s + 2)^2}.$$

8.37 Repeat Problem 8.35 with $q = 1.11$ and the transfer function of the linear element given by

$$G(s) = \frac{(s + 4)}{s(s + 2)^2(s + 1.5)}.$$

8.38 A unity-feedback control system contains a linear and nonlinear element. The transfer function of the linear element is given by

$$G(s) = \frac{(s + 2)}{(s - 1)(s + 3)(s + 4)}.$$

The initial condition response is

$$e_0(t) = e_{10}e^t + e_{20}e^{-3t} + e_{30}e^{-4t},$$

where e_{10}, e_{20}, and e_{30} arise from the initial conditions. Using the generalized circle criterion, determine possible values of $N[e(t), t]$ which will result in a stable system if the nonlinear element corresponds to the general nonlinearity case ($q = 0$).

8.39 Repeat Problem 8.38 with the transfer function of the linear element given by

$$G(s) = \frac{1}{(s - 1)(s + 3)(s + 4)}.$$

8.40 Repeat Problem 8.38 with the transfer function of the linear element given by

$$G(s) = \frac{(s + 6)}{(s - 1)(s + 3)(s + 4)}.$$

8.41 An instrument servo system used for positioning a load may be adequately represented by the block diagram shown in Fig. P8.41, which indicates that the system contains backlash.

a) Using the gain–phase diagram, determine the existence of any limit cycles if $K = 25$.
b) What is the maximum gain that can be tolerated in $G(j\omega)$ before a limit cycle occurs?

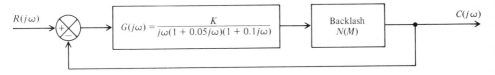

Figure P8.41

8.42 In contrast to the on–off element having dead-zone and hysteresis characteristics present, as analyzed in Fig. 8.17, an ideal relay has no dead-zone or hysteresis characteristics.

a) For the ideal relay characteristics shown in Fig. P8.42, determine its describing function analytically, assuming that the input to the nonlinear element is a sinusoid of $M \sin \omega t$.
b) Check your answer by reducing Eq. 8.51 for the conditions of zero dead zone and hysteresis (A_1 of Eq. 8.50 now becomes zero).

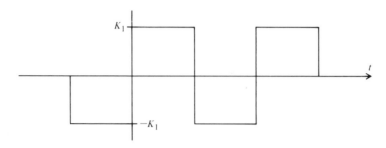

Figure P8.42

8.43 Utilizing the result of your derivation of the describing function for the ideal relay characteristics in Problem 8.42, consider the stability of the control system shown in Fig. P8.43. For the value of $G(j\omega)$ indicated, determine the existence of any limit cycles and whether they are stable or unstable.

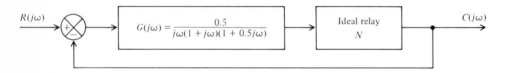

Figure P8.43

8.44 Assume that the block diagram of a load-positioning control system, which contains backlash, is represented as in Fig. P8.41 and that the linear portion's transfer function is given by

$$G(j\omega) = \frac{20}{j\omega(1 + 0.5j\omega)^2}.$$

a) Using the gain–phase diagram, determine the existence of any limit cycles.
b) Determine the gain which can permit a stability margin of 3 db.

8.45 Repeat Problem 8.44 for

$$G(j\omega) = \frac{10}{j\omega(1 + 0.3j\omega)(1 + 0.1j\omega)}.$$

REFERENCES

1. P. M. Lowitt and S. M. Shinners, "Type N—Integral space tracking configuration," *IEEE Trans. Military Electronics* **MIL-9,** 88–98 (1965).
2. E. Levinson, "Some saturation phenomena in servomechanisms," *Trans. AIEE* **72,** 1 (1953).
3. C. A. Ludeke, "The generation and extinction of subharmonics," in *Proceedings of the Symposium on Nonlinear Circuit Analysis,* Polytechnic Institute of Brooklyn, New York (April 1953).
4. R. J. Kochenburger, "A frequency response method for analyzing and synthesizing contactor servo-mechanisms," *Trans. AIEE* **69,** 270 (1950).
5. E. C. Johnson, "Sinusoidal analysis of feedback-control systems containing nonlinear elements," *Trans. AIEE* **71,** 169 (1952).
6. H. D. Grief, "Describing function method of servomechanism analysis applied to most commonly encountered nonlinearities," *Trans. AIEE* **72,** 253 (1953).
7. J. G. Truxal, *Automatic Feedback Control System Synthesis,* McGraw-Hill, New York (1955).
8. P. M. Lowitt and S. M. Shinners," Integrated optimal synthesis for a radar tracker," in *Proceedings of the Seventh National Military Electronics Convention,* Washington, D.C. (September 1963).
9. R. Oldenburger and R. C. Boyer, "Effects of extra sinusoidal inputs to nonlinear systems," in *Proceedings of the ASME Winter Annual Meeting,* New York (1961).
10. S. M. Shinners, "Dual-input describing function," *Control Eng.* **18,** 53–55 (February 1971).
11. O. I. Elgerd, "Continuous control by high frequency signal injection," *Instrum. Control Systems* **37,** 12 (1964).
12. O. I. Elgerd, *Control Systems Theory,* McGraw-Hill, New York (1967).
13. R. C. Boyer, *Sinusoidal Signal Stabilization,* M.S. Thesis, Purdue University, Lafayette, Ind. (1960).
14. J. E. Gibson, *Nonlinear Automatic Control,* McGraw-Hill, New York (1963).
15. O. Romaine, "OAO: NASA's biggest satellite yet," *Space/Aeronautics* **40,** 54–58 (1962).
16. I. N. Hutchinson, and J. L. Keller, "A flexibly coupled spinning space station—Its stabilization and control," *Sperry Engineering Review,* **18,** 23–32 (1965).
17. I. Ritow, "Designing servos by the phase-plane method," *Elec. Mfg.* **62,** 98 (1956).
18. E. Levinson, "Phase-plane analysis," *Electro-Technology* **69,** 118 (1962)
19. H. S. Tsien, *Engineering Cybernetics,* McGraw-Hill, New York (1954).
20. A. M. Liapunov, *On the General Problem of Stability of Motion,* Ph.D. Thesis, Kharkov 1892; reprinted (in French) in *Annals of Mathematics Studies,* Vol. 17, Princeton University Press, Princeton, N.J. (1949).
21. N. Minorsky, *Introduction to Non-linear Mechanics,* J. W. Edwards, Ann Arbor, Michigan, p. 52 (1967).
22. V. M. Popov, "Absolute stability of nonlinear systems of automatic control," *Automation Remote Control* **22,** 857–75 (1961).
23. V. A. Jakubovic, "Frequency conditions for the absolute stability and dissipativity of control systems with a single differentiable nonlinearity," *Soviet Math.* **6,** 98–101 (1965).
24. S. Lefschetz, *Stability of Nonlinear Control Systems,* Academic, New York (1965).

25. J. C. Hsu and A. U. Meyer, *Modern Control Principles and Applications*, McGraw-Hill, New York (1968).
26. G. Zames, "On the input-output stability of time-varying nonlinear feedback systems— Part II: Condition involving circles in the frequency plane and sector nonlinearities," *IEEE Trans. Automatic Control* **AC-11**, 465–76 (1966).
27. C. A. Desoer, "A generalization of the Popov criterion," *IEEE Trans. Automatic Control* **AC–10**, 182–85 (1965).
28. V. M. Popov and A. Halanay, "On the stability of nonlinear automatic control systems with lagging argument," *Automation Remote Control* **23**, 783–86 (1963).
29. A. V. Michailov, "Harmonic analysis in the theory of automatic control," *A.T. Moscow*, No. 3, p. 27 (1938).
30. I. N. Hutchinson, R. F. Morrison, and J. L. Keller, "Stabilization and control of a cable-connected spinning space station," in *Proceedings of the 1965 Joint Automatic Control Conference*, pp. 520–30.
31. W. Turk, *Ranger Block III attitude control system*, Jet Propulsion Laboratory Technical Report, No. 32–663, Pasadena, California (November 1964).
32. B. O. Watkins, *Introduction to Control Systems*, Macmillan, New York (1969).
33. E. M. Grabbe, S. Ramo, and D. E. Wooldridge, *Handbook of Automation, Computation, and Control*, Vol. 1, Wiley, New York (1958).
34. *"BASIC" Language, Reference Manual*, General Electric Co., Information Systems Division, Phoenix, Arizona (June 1965; rev. January 1967).
35. S. A. Hovanessian and L. A. Pipes, *Digital Computer Methods in Engineering*, McGraw-Hill, New York (1969).
36. H. H. Rosenbrick and C. Storey, *Computational Techniques for Chemical Engineers*, Pergamon, Oxford (1966).
37. D. D. McCracken and W. S. Dorn, *Numerical Methods and FORTRAN Programming*, Wiley, New York (1964).
38. R. W. Hamming, *Numerical Methods for Scientists and Engineers*, McGraw-Hill, New York (1962).
39. D. Graham and D. McRuer, *Analysis of Nonlinear Control Systems*, Wiley, New York (1961).
40. T. C. Bartee, *Digital Computer Fundamentals* (2nd Edn.), McGraw-Hill, New York (1966).
41. G. A. Korn and T. M. Korn, *Electronic Analog and Hybrid Computers*, McGraw-Hill, New York (1964).
42. S. M. Shinners, "Which computer—Analog, digital or hybrid?" *Machine Design* **43**, 104–111 (January 21, 1971).

9

OPTIMAL CONTROL THEORY AND APPLICATIONS

9.1 INTRODUCTION

From the presentation of linear and nonlinear systems in Chapters 7 and 8, we recognize that the conventional design of feedback control systems has several disadvantages. The most serious disadvantage is that the design techniques presented depended heavily on trial-and-error procedures. In contrast, optimal control theory is concerned with obtaining a system which is the best possible with respect to a *standard* against which we can measure real performance. We denote this standard as the performance criterion (see Chapter 5 for a discussion of performance criteria). The task of designing control systems which are optimal, in some sense, is one of the most important and complex problems facing control engineers today.

Wiener [30], in the late 1940's, developed the concept of optimum design that was based on optimizing a performance criterion. McDonald [1] first applied the concept of optimization to control systems in 1950. His objective was to minimize the transient response time of a relay-type feedback control system to step inputs. In 1957, Draper and Li [2] wrote a booklet discussing the theoretical concepts of optimal control for an internal combustion engine. Their system attempted to minimize (or optimize) the consumption of fuel. Since that time many papers have been written on optimal control systems both in this country and abroad. Most important is the work of Bellman [12, 13], who developed the concept of dynamic programming, and Pontryagin, Boltyanskii, and Gamkrelidge [15, 16], who developed the maximum principle. The purpose of this chapter is to present the principles of this theory and some examples of its application.

9.2 CHARACTERISTICS OF THE OPTIMAL CONTROL PROBLEM

In this section, we consider the basic problem of optimal control theory [8]. It consists of choosing the input $\mathbf{u}$ to a control system so that the performance of the system is optimum with respect to some performance criterion. The basic goal of optimal control theory is to design control elements which meet a wide variety of requirements in the best possible manner. The structure of the optimal control problem can be described in the following way. One has

1. A controlled process whose dynamics have the following form:

$$\dot{\mathbf{x}} = f(\mathbf{x}, \mathbf{u}, t). \tag{9.1}$$

2. Restrictions on the input **u** and/or the plant state **x**.

3. A reference signal **r** which signifies the desired output response.

4. A performance criterion having the following general form:

$$S = \int_{t_0}^{T} G\big(\mathbf{c}(t), \mathbf{u}(t), \mathbf{r}(t), t\big) dt. \tag{9.2}$$

In Eq. (9.2), the integrand G is referred to as the loss function and represents a measure of instantaneous change from ideal performance [3]. Therefore, the performance criterion is interpreted as the cumulative loss. The optimal control problem consists of the determination of the control input **u** that minimizes the performance criterion S subject to certain constraints on **u** and **x**.

The problem of a unity feedback servo positioning system can be readily formulated using optimal control theory [4]. The objective is to determine the input **u** that causes the output **c** to follow a given reference signal **r** within a certain accuracy. Typically, the loss function could be the squared norm of the magnitude of the error $e = |(\mathbf{c} - \mathbf{r})|$:

$$G = |(\mathbf{c} - \mathbf{r})|^2. \tag{9.3}$$

Thus from Eq. (9.2), the performance is given by

$$S = \int_{t_0}^{T} |(\mathbf{c} - \mathbf{r})|^2 \, dt. \tag{9.4}$$

The limits of the optimization interval are usually specified and can be finite or infinite. (See Problem 9.21 for an application of this performance criterion.)

Generally, it is desirable to revise the performance criterion, as defined by Eq. (9.4), in order to include the effects of excessive control effort. By revising Eq. (9.4) to

$$S = \int_{t_0}^{T} [|(\mathbf{c} - \mathbf{r})|^2 + \lambda(\mathbf{u})^2] \, dt, \tag{9.5}$$

a compromise between small and large error and control amplitudes can be obtained by a proper selection of λ.

The choice of an input **u**, over the operating interval from $t_0 = 0$ to $t = T$ sec, is denoted as the control law [5]. If the input **u** minimizes the performance criterion S, then it is optimal. Under these conditions, the optimal control is denoted by $\mathbf{u}^0$ and the optimal performance is denoted by S^0.

The optimal policy can be either an open-loop policy or a feedback (closed-loop) policy. The designation *open-loop* is used if the controlled input is specified as a function of time. The optimal policy is designated *feedback* if the controlled input is specified as a function of the current state of the plant. The nature of the policy is extremely important since it indicates how $\mathbf{u}^0$ is to be generated from **r** and **c**. The representation of optimal control systems having open-loop and feedback policies is illustrated in Fig. 9.1.

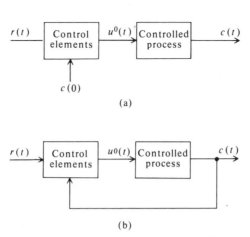

Fig. 9.1 Open-loop and feedback optimal control systems: (a) open-loop optimal control system, (b) feedback optimal control system.

As the preceding chapters have shown, feedback has a great many advantages and, therefore, a feedback policy is preferred for optimal control. For example, feedback operation, in general, tends to make the system less sensitive to variations. In addition, feedback operation makes use of the most recent information on the state of the plant. As a result, if a disturbance within the system occurs in the feedback case, the system operates optimally on the latest measurement of **c**. In an open-loop system, however, the entire input is preprogrammed only on the basis of the initial-state value. In this case, any disturbances within the system destroy the optimality of operation.

9.3 CONTROLLABILITY

The concepts of controllability [6, 7, 9] and observability play a very important role in optimal control theory. Before we can design a system to be optimal, we must first determine whether it is controllable and its states are observable, since the conditions on controllability and observability often govern the existence of a solution to an optimal control system. Kalman [6, 7] first introduced the concepts of controllability and observability in 1960. These concepts are basic in modern optimal control theory. This section develops the controllability concept and the following section presents observability.

In order to introduce the concept of controllability, let us consider the simple open-loop system illustrated in Fig. 9.2. A system is completely controllable if there exists a control which transfers every initial state at $t = t_0$ to any final state at $t = T$ for all t_0 and T. Qualitatively, this means that the system G is controllable if every state variable of G can be affected by the input signal **u**. However, if one (or several) of

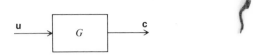

Fig. 9.2   Open-loop system containing several inputs and outputs.

the state variables is (or are) not affected by **u**, then this (or these) state variable(s) cannot be controlled in a finite amount of time by **u** and the system is not completely controllable.

As an example of a system which is not completely controllable, let us consider the signal-flow diagram illustrated in Fig. 9.3. This system contains four states, only two of which are affected by $u(t)$. This input only affects the states x_1 and x_2. It has no affect on x_3 and x_4. Therefore, x_3 and x_4 are uncontrollable. This means that it is impossible for $u(t)$ to change x_3 from an initial state $x_3(0)$ to final state $x_3(T)$ in a finite time interval T and the system is not completely controllable.

Let us now consider this problem more precisely and establish a criterion for determining whether a system is controllable. We limit our discussion to linear constant systems. Assume that the system is described by

$$\dot{\mathbf{x}} = \mathbf{Px} + \mathbf{Bu}, \tag{9.6}$$

$$\mathbf{c} = \mathbf{Lx}. \tag{9.7}$$

The solution of Eq. (9.6) can be expressed as Eq. (2.249):

$$\mathbf{x}(t) = \mathbf{\Phi}(t - t_0)\mathbf{x}(t_0) + \int_{t_0}^{t} \mathbf{\Phi}(t - \tau)\mathbf{Bu}(\tau)\,d\tau \qquad \text{where} \qquad t \geqslant t_0. \tag{9.8}$$

Let us assume that the desired final state of our system at $t = t_f$ is zero:

$$\mathbf{x}(t_f) = \mathbf{0}. \tag{9.9}$$

Using Eqs. (2.244) and (2.246), we can write Eq. (9.8) as

$$\mathbf{x}(t_0) = -\int_{t_0}^{t_f} \mathbf{\Phi}(t_0 - \tau)\mathbf{Bu}(\tau)\,d\tau. \tag{9.10}$$

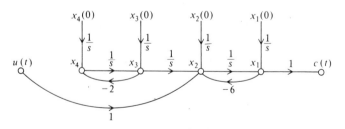

Fig. 9.3   Signal-flow diagram of a system that is not completely controllable.

The state-transition matrix can be expressed from Eq. (2.294) as

$$\mathbf{\Phi}(t) = e^{\mathbf{P}t} = \sum_{n=0}^{m-1} \alpha_n(t)\mathbf{P}^n. \tag{9.11}$$

Here, $\mathbf{x}$ is an $m \times 1$ vector, $\mathbf{P}$ is an $m \times m$ matrix, $\mathbf{u}$ is an $r \times 1$ vector, $\mathbf{B}$ is an $m \times r$ matrix, and $\alpha_n(t)$ is a scalar function of t. (This form results from application of the Cayley-Hamilton theorem.) Substituting Eq. (9.11) into Eq. (9.10), we obtain the following expression for $\mathbf{x}(t_0)$:

$$\mathbf{x}(t_0) = -\int_{t_0}^{t_f} \sum_{n=0}^{m-1} \alpha_n(t_0 - \tau)\mathbf{P}^n\mathbf{B}\mathbf{u}(\tau) \, d\tau. \tag{9.12}$$

Since the matrices $\mathbf{P}$ and $\mathbf{B}$ are not functions of τ, we can rewrite Eq. (9.12) as

$$\mathbf{x}(t_0) = -\sum_{n=0}^{m-1} \mathbf{P}^n\mathbf{B}\int_{t_0}^{t_f} \alpha_n(t_0 - \tau)\mathbf{u}(\tau) \, d\tau. \tag{9.13}$$

Equation (9.13) can be rewritten as

$$\mathbf{x}(t_0) = -[\mathbf{B} \quad \mathbf{PB} \quad \mathbf{P^2B} \quad \mathbf{P^3B} \cdots \mathbf{P}^{m-1}\mathbf{B}]\begin{bmatrix} \mathbf{A}_0 \\ \mathbf{A}_1 \\ \mathbf{A}_2 \\ \cdot \\ \cdot \\ \cdot \\ \mathbf{A}_{m-1} \end{bmatrix}, \tag{9.14}$$

where

$$\mathbf{A}_n = \int_{t_0}^{t_f} \alpha_n(t_0 - \tau)\mathbf{u}(\tau) \, d\tau. \tag{9.15}$$

If we define

$$\mathbf{D} = [\mathbf{B} \quad \mathbf{PB} \quad \mathbf{P^2B} \quad \mathbf{P^3B} \cdots \mathbf{P}^{m-1}\mathbf{B}] \tag{9.16}$$

$$\mathbf{A} = [\mathbf{A}_0 \quad \mathbf{A}_1 \quad \mathbf{A}_2 \cdots \mathbf{A}_{m-1}]^T, \tag{9.17}$$

where $\mathbf{D}$ is a $m \times mr$ matrix and $\mathbf{A}$ is a $mr \times 1$ vector, then Eq. (9.14) becomes

$$\mathbf{x}(t_0) = -\mathbf{D}\mathbf{A}. \tag{9.18}$$

For a given initial state $\mathbf{x}(t_0)$, the input $\mathbf{u}$ can be found to drive the state to $\mathbf{x}(t_f) = \mathbf{0}$ for a finite time interval $t_f - t_0$ if Eq. (9.18) has a solution. A unique solution occurs only if there is a set of m linearly independent column vectors in the matrix $\mathbf{D}$. If $\mathbf{u}$ is a scalar, then $\mathbf{D}$ is an $m \times m$ square matrix, and Eq. (9.18) represents a set of m linear independent equations which have a solution if $\mathbf{D}$ is nonsingular, or the determinant of $\mathbf{D}$ is not zero. The controllability criterion thus states that the system of Eqs. (9.6) and (9.7) is completely controllable if $\mathbf{D}$ (see Eq. 9.6) contains m linearly independent column vectors or, if $\mathbf{u}$ is a scalar, $\mathbf{D}$ is nonsingular.

In order to illustrate this mathematical controllability concept, consider a second-order system where

$$\mathbf{P} = \begin{bmatrix} -4 & 1 \\ 0 & -2 \end{bmatrix}$$

and

$$\mathbf{B} = \begin{bmatrix} 1 \\ 0 \end{bmatrix}.$$

Then, from Eq. (9.16),

$$\mathbf{D} = [\mathbf{B} \quad \mathbf{PB}] = \begin{bmatrix} 1 & -4 \\ 0 & 0 \end{bmatrix}. \tag{9.19}$$

The resulting matrix $\mathbf{D}$ is singular (its determinant is zero) and the system is therefore not completely controllable.

As a second example, consider a second-order system where

$$\mathbf{P} = \begin{bmatrix} -3 & 2 \\ 4 & 1 \end{bmatrix}$$

and

$$\mathbf{B} = \begin{bmatrix} 0 \\ 1 \end{bmatrix}.$$

Then, from Eq. (9.16)

$$\mathbf{D} = [\mathbf{B} \quad \mathbf{PB}] = \begin{bmatrix} 0 & 2 \\ 1 & 1 \end{bmatrix}.$$

The resulting matrix $\mathbf{D}$ is nonsingular and the system, therefore, is completely controllable.

9.4 OBSERVABILITY

In order to introduce the concept of observability [6, 7, 9], let us again consider the simple open-loop system illustrated in Fig. 9.2. A system is completely observable if, given the control and the output over the interval $t_0 \leqslant t \leqslant T$, one can determine the initial state $\mathbf{x}(t_0)$. Qualitatively, the system G is observable if every state variable of G affects some of the outputs in $\mathbf{c}$. It is very often desirable to determine information regarding the system states based on measurements of $\mathbf{c}$. However, if we cannot observe one or more of the states from the measurements of $\mathbf{c}$, then the system is not completely observable.

As an example of a system which is not completely observable, let us consider the signal-flow diagram illustrated in Fig. 9.4. This system contains four states, only two of which are observable. The states x_3 and x_4 are not connected to the output c in any manner. Therefore, x_3 and x_4 are not observable and the system is not completely observable.

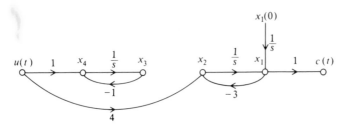

Fig. 9.4 Signal-flow diagram of a system that is not completely observable.

Let us now consider this problem more precisely and establish a criterion for determining whether a system is completely observable. Again, we limit our discussion to linear constant systems of the form

$$\dot{\mathbf{x}} = \mathbf{Px} + \mathbf{Bu}, \tag{9.20}$$

$$\mathbf{c} = \mathbf{Lx} \tag{9.21}$$

where $\mathbf{x}$ is an $m \times 1$ vector, $\mathbf{P}$ is an $m \times m$ matrix, $\mathbf{u}$ is an $r \times 1$ vector, $\mathbf{B}$ is an $m \times r$ matrix, $\mathbf{c}$ is a $p \times 1$ vector, and $\mathbf{L}$ is a $p \times r$ matrix. The solution of Eq. (9.20) is given by (see Eq. 2.249)

$$\mathbf{x}(t) = \mathbf{\Phi}(t - t_0)\mathbf{x}(t_0) + \int_{t_0}^{t} \mathbf{\Phi}(t - \tau)\mathbf{Bu}(\tau)\, d\tau. \tag{9.22}$$

It will now be shown that observability depends on the matrices $\mathbf{P}$ and $\mathbf{L}$. Substituting Eq. (9.22) into Eq. (9.21), we obtain

$$\mathbf{c}(t) = \mathbf{L\Phi}(t - t_0)\mathbf{x}(t_0) + \mathbf{L}\int_{t_0}^{t} \mathbf{\Phi}(t - \tau)\mathbf{Bu}(\tau)\, d\tau. \tag{9.23}$$

From the definition of observability we can see that the observability of $\mathbf{x}(t_0)$ depends on the term $\mathbf{L\Phi}(t - t_0)\mathbf{x}(t_0)$. Therefore, the output $\mathbf{c}(t)$ when $\mathbf{u} = \mathbf{0}$ is given by

$$\mathbf{c}(t) = \mathbf{L\Phi}(t - t_0)\mathbf{x}(t_0). \tag{9.24}$$

Substituting Eq. (9.11) into Eq. (9.24), we obtain the following expression for the output $\mathbf{c}$:

$$\mathbf{c}(t) = \sum_{n=0}^{m-1} \alpha_n(t - t_0)\mathbf{LP}^n\mathbf{x}(t_0). \tag{9.25}$$

Equation (9.25) indicates that if the output $\mathbf{c}(t)$ is known over the time interval $t_0 \leqslant t \leqslant T$, then $\mathbf{x}(t_0)$ is uniquely determined from this equation if $\mathbf{x}(t_0)$ is a linear combination of $(\mathbf{L}_j\mathbf{P}^n)^T$ for $n = 0, 1, 2, \ldots, m - 1$, and $j = 1, 2, 3, \ldots, r$. The matrix $\mathbf{L}_j$ is the $1 \times m$ matrix formed by the elements of the jth row of $\mathbf{L}$. Since $(\mathbf{L}_j\mathbf{P}^n)^T = (\mathbf{P}^T)^n\mathbf{L}_j^T$, we let $\mathbf{U}$ be the $m \times mr$ matrix defined by

$$\mathbf{U} = [\mathbf{L}^T \quad \mathbf{P}^T\mathbf{L}^T \quad (\mathbf{P}^T)^2\mathbf{L}^T \quad (\mathbf{P}^T)^3\mathbf{L}^T \cdots (\mathbf{P}^T)^{m-1}\mathbf{L}^T]. \tag{9.26}$$

The observability criterion states that the system is completely observable if there is a set of m linearly independent column vectors in $\mathbf{U}$.

In order to illustrate the observability concept mathematically, consider a second-order system where

$$\mathbf{P} = \begin{bmatrix} -4 & 0 \\ 0 & -2 \end{bmatrix}$$

and

$$\mathbf{L} = [1 \quad 0].$$

Therefore,

$$\mathbf{L}^T = \begin{bmatrix} 1 \\ 0 \end{bmatrix}$$

and

$$\mathbf{P}^T\mathbf{L}^T = \begin{bmatrix} -4 & 0 \\ 0 & -2 \end{bmatrix}\begin{bmatrix} 1 \\ 0 \end{bmatrix} = \begin{bmatrix} -4 \\ 0 \end{bmatrix}.$$

Substituting these values into Eq. (9.26), we obtain

$$\mathbf{U} = [\mathbf{L}^T \quad \mathbf{P}^T\mathbf{L}^T] = \begin{bmatrix} 1 & -4 \\ 0 & 0 \end{bmatrix}.$$

Since $\mathbf{U}$ is singular, the system is not completely observable.

As a second example, consider the second-order system where

$$\mathbf{P} = \begin{bmatrix} 4 & -2 \\ 4 & -2 \end{bmatrix}$$

and

$$\mathbf{L} = [1 \quad 1].$$

Therefore,

$$\mathbf{L}^T = \begin{bmatrix} 1 \\ 1 \end{bmatrix}$$

and

$$\mathbf{P}^T\mathbf{L}^T = \begin{bmatrix} 4 & 4 \\ -2 & -2 \end{bmatrix}\begin{bmatrix} 1 \\ 1 \end{bmatrix} = \begin{bmatrix} 8 \\ -4 \end{bmatrix}.$$

Substituting these values into Eq. (9.26), we obtain

$$\mathbf{U} = [\mathbf{L}^T \quad \mathbf{P}^T\mathbf{L}^T] = \begin{bmatrix} 1 & 8 \\ 1 & -4 \end{bmatrix}.$$

Since $\mathbf{U}$ has two independent columns, the system is completely observable.

9.5 CALCULUS OF VARIATIONS

The calculus of variations [3, 5, 10, 11] is concerned with obtaining the maxima and minima of entire functional expressions. It can be applied to a great many problems in the fields of classical mechanics, aerodynamics, optics, and control theory. We shall consider the calculus of variations only in the context of control theory. Three of its central problems are as follows:

1. The *Lagrange problem* is concerned with determining a function **u** which minimizes a certain performance criterion S. Thus, for the given set of equations

$$\dot{\mathbf{x}} = \mathbf{f}(\mathbf{x}, \mathbf{u}, t), \tag{9.27}$$

$$\mathbf{c} = \mathbf{Lx},$$

a set of initial conditions given by

$$\mathbf{x}(0) = \mathbf{p},$$

and a performance criterion given by

$$S = \int_0^T G(\mathbf{c}(t), \mathbf{u}(t), t)\, dt, \tag{9.28}$$

this problem is concerned with determining the function $\mathbf{u}(t)$ which minimizes S.

2. The *Mayer problem* is concerned with determining a function **u** which minimizes a certain performance criterion S evaluated at the endpoint and containing some variables whose final values are unspecified in advance. Thus, for the given set of equations

$$\dot{\mathbf{x}} = \mathbf{f}(\mathbf{x}, \mathbf{u}, t), \tag{9.29a}$$

$$\mathbf{c} = \mathbf{Lx}, \tag{9.29b}$$

a set of initial conditions given by

$$\mathbf{x}(0) = \mathbf{p}, \tag{9.30}$$

a set of final conditions given by

$$\mathbf{x}(T) = \mathbf{q}, \tag{9.31}$$

and a performance criterion given by

$$S = \int_0^T G(\mathbf{c}(t), \mathbf{u}(t), t)\, dt, \tag{9.32}$$

this problem is concerned with determining the function $\mathbf{u}(t)$ which minimizes S.

3. The *Bolza problem* is concerned with determining a function **u**, which contains some variables that are unspecified in advance, and minimizing a certain performance criterion at the endpoint. Thus, for the given set of equations

$$\dot{\mathbf{x}} = \mathbf{f}(\mathbf{x}, \mathbf{u}, t), \tag{9.33}$$

$$\mathbf{c} = \mathbf{Lx},$$

a set of initial conditions given by

$$\mathbf{x}(0) = \mathbf{p},$$

a set of final conditions given by

$$\mathbf{x}(T) = \mathbf{q},$$

and a performance criterion given by

$$S = F[(\mathbf{c}(t), \mathbf{u}(t), t)]_0^T + \int_0^T G(\mathbf{c}(t), \mathbf{u}(t), t)\, dt, \tag{9.34}$$

the problem is concerned with determining the function $\mathbf{u}(t)$ which minimizes S.

The Bolza problem is the most general case of variational calculus. Most optimal control problems can be formulated as one of these three fundamental problems, and it is usually possible to introduce a mathematical substitution which can transform a Lagrange or Mayer problem into a Bolza problem.

In order to solve for u^0 in a scalar problem using the calculus of variations, we must first solve the Euler-Lagrange equation. The derivation of this differential equation and demonstration of its computational difficulty can be shown by considering the Mayer scalar problem. It is assumed that $c = x$, and that Eq. (9.32) has been optimized and the optimum values x^0 and u^0 have been determined. For example, if x^0 and u^0 are substituted back into Eq. (9.32), then S^0 is optimized (let us assume that it is a minimum value).

What is the effect on S^0 if we now allow x^0 and u^0 to be perturbed by small arbitrary amounts Δx and Δu, respectively? If S^0 is indeed optimum, then ΔS must be zero if we deviate slightly from it. We limit these perturbations, however, to $\Delta x(0) = \Delta x(T) = 0$, in order to meet the constraints given by Eqs. (9.30) and (9.31). The resulting effect on S^0 in Eq. (9.32) can be determined by substituting

$$x = x^0 + \Delta x \tag{9.35}$$

$$u = u^0 + \Delta u \tag{9.36}$$

into Eq. (9.29a) as follows:

$$\frac{d}{dt}(x^0 + \Delta x) = f(x^0 + \Delta x, u^0 + \Delta u, t). \tag{9.37}$$

Using a Taylor series expansion around the optimum values, we obtain

$$\dot{x}^0 + \Delta\dot{x} = f(x^0, u^0, t) + \left(\frac{\partial f}{\partial x}\right)_0 \Delta x + \left(\frac{\partial f}{\partial u}\right)_0 \Delta u. \tag{9.38}$$

Since

$$\dot{x}^0 = f(x^0, u^0, t), \tag{9.39}$$

we have

$$\Delta\dot{x} = \left(\frac{\partial f}{\partial x}\right)_0 \Delta x + \left(\frac{\partial f}{\partial u}\right)_0 \Delta u \tag{9.40}$$

or

$$\Delta u = \frac{\Delta\dot{x} - \left(\dfrac{\partial f}{\partial x}\right)_0 \Delta x}{\left(\dfrac{\partial f}{\partial u}\right)_0}. \tag{9.41}$$

Equations (9.40) and (9.41) indicate that if Δx is arbitrarily chosen, then Δu is restricted to values which obey this equation. Now, knowing the relationship between x, Δx, u and Δu, can we say what effect Δx and Δu have on S^0? We can say that

$$\Delta S = S(x^0 + \Delta x, u^0 + \Delta u, t) - S(x^0, u^0, t). \tag{9.42}$$

Therefore,

$$\Delta S = \int_0^T G(x^0 + \Delta x, u^0 + \Delta u, t) \, dt - \int_0^T G(x^0, u^0, t) \, dt. \tag{9.43}$$

Combining terms under the integral sign, we obtain

$$\Delta S = \int_0^T [G(x^0 + \Delta x, u^0 + \Delta u, t) - G(x^0, u^0, t)] \, dt. \tag{9.44}$$

By means of the Taylor series, we can show that

$$\Delta S = \int_0^T \left[\left(\frac{\partial G}{\partial x} \right)_0 \Delta x + \left(\frac{\partial G}{\partial u} \right)_0 \Delta u \right] dt. \tag{9.45}$$

Substituting Eq. (9.41) into Eq. (9.45), we obtain the following expression:

$$\Delta S = \int_0^T \left[\left(\frac{\partial G}{\partial x} \right)_0 \Delta x + \left(\frac{\partial G}{\partial u} \right)_0 \frac{\Delta \dot{x} - (\partial f / \partial x)_0 \, \Delta x}{(\partial f / \partial u)_0} \right] dt. \tag{9.46}$$

Equation (9.46) can be simplified to

$$\Delta S = \int_0^T \left\{ \frac{(\partial G / \partial u)_0}{(\partial f / \partial u)_0} \Delta \dot{x} + \left[\left(\frac{\partial G}{\partial x} \right)_0 - \left(\frac{\partial G}{\partial u} \right)_0 \frac{(\partial f / \partial x)_0}{(\partial f / \partial u)_0} \right] \Delta x \right\} dt. \tag{9.47}$$

Integrating the first term of the integral by parts, we obtain

$$\int_0^T \frac{(\partial G / \partial u)_0}{(\partial f / \partial u)_0} \Delta \dot{x} \, dt = \left[\frac{(\partial G / \partial u)_0}{(\partial f / \partial u)_0} \Delta x \right]_0^T - \int_0^T \Delta x \frac{d}{dt} \frac{(\partial G / \partial u)_0}{(\partial f / \partial u)_0} \, dt. \tag{9.48}$$

Since $\Delta x(0) = \Delta x(T) = 0$, the first term on the right-hand side of Eq. (9.48) is zero. Therefore, returning to Eq. (9.47), and substituting Eq. (9.48) into it, we obtain the following:

$$\Delta S = \int_0^T \left[\left(\frac{\partial G}{\partial x} \right)_0 - \left(\frac{\partial G}{\partial u} \right)_0 \frac{(\partial f / \partial x)_0}{(\partial f / \partial u)_0} - \frac{d}{dt} \frac{(\partial G / \partial u)_0}{(\partial f / \partial u)_0} \right] \Delta x \, dt. \tag{9.49}$$

Equation (9.49) states the resulting effect on S for a perturbation of Δx and Δu. However, if x^0 and u^0 do indeed result in an optimum S^0, then ΔS must be zero for a perturbation of Δx. Therefore, it is necessary that, during the interval $0 < t < T$,

$$\left(\frac{\partial G}{\partial x} \right)_0 - \left(\frac{\partial G}{\partial u} \right)_0 \frac{(\partial f / \partial x)_0}{(\partial f / \partial u)_0} - \frac{d}{dt} \frac{(\partial G / \partial u)_0}{(\partial f / \partial u)_0} = 0; \tag{9.50}$$

this is the well-known Euler-Lagrange differential equation.

The results of this exercise indicate that we have to solve the Euler-Lagrange differential equation which cannot, in general, be integrated analytically. In addition, if numerical techniques are used, one finds that the Euler-Lagrange equations are unstable.

Let us illustrate the application of the calculus of variations to a simple problem that is easily solvable [11]. Consider the open-loop system illustrated in Fig. 9.5. It is desired to control the state between $x(0)$ and $x(T)$ in order that the performance criterion, which penalizes (in ISE sense) u and the deviation from the reference $x = 0$,

$$S = \int_0^T (x^2 + u^2) \, dt \tag{9.51}$$

is minimized. For this simple system, the state equation is given by

$$\dot{x} = f(x, u, t) = u. \tag{9.52}$$

In addition, to obtain the Euler-Lagrange equation, we need to determine the following:

$$\left(\frac{\partial G}{\partial x}\right)_0 = 2x, \tag{9.53}$$

$$\left(\frac{\partial G}{\partial u}\right)_0 = 2u, \tag{9.54}$$

$$\frac{\partial f}{\partial x} = 0, \tag{9.55}$$

$$\frac{\partial f}{\partial u} = 1. \tag{9.56}$$

Substituting Eqs. (9.53) through (9.56) into Eq. (9.50), we obtain the following Euler-Lagrange equation:

$$x - \dot{u} = 0. \tag{9.57}$$

For this simple system, since Eqs. (9.52) and (9.57) are linear, the Laplace transformation can be used. We have

$$sX(s) - x(0) = U(s), \tag{9.58}$$

$$X(s) - [sU(s) - u(0)] = 0. \tag{9.59}$$

Observe from Eqs. (9.58) and (9.59) that we must obey the constraints given by $x(0)$ and $u(0)$. Since we know $x(0)$ and $x(T)$ but not $u(0)$, we will proceed by using only the one initial condition we know, $x(0)$. The value of $u(0)$ will be eliminated later on by

Fig. 9.5 A simple integrating system.

attempting to meet the final condition $x(T)$. Solving for $X(s)$ from Eqs. (9.58) and (9.59), we obtain the expression

$$X(s) = \frac{sx(0) + u(0)}{s^2 - 1}. \tag{9.60}$$

Taking the inverse transform, we have

$$x(t)^0 = x(0) \cosh t + u(0) \sinh t. \tag{9.61}$$

The value of $u(0)$ must now be obtained in order that the state $x(T)$ is reached. Therefore,

$$x(T) = x(0) \cosh T + u(0) \sinh T. \tag{9.62}$$

Solving for $u(0)$, we obtain

$$u(0) = \frac{x(T) - x(0) \cosh T}{\sinh T}. \tag{9.63}$$

Substituting Eq. (9.63) into Eq. (9.61), we obtain the final expression for $x(t)^0$:

$$x(t)^0 = x(0) \cosh t + \left[\frac{x(T) - x(0) \cosh T}{\sinh T} \right] \sinh t. \tag{9.64}$$

In addition, we can determine the corresponding optimal control from Eq. (9.52):

$$u(t)^0 = \dot{x}(t)^0 = x(0) \sinh t + \left[\frac{x(T) - x(0) \cosh T}{\sinh T} \right] \cosh t. \tag{9.65}$$

In this example, notice that the Euler-Lagrange equation turned out to be linear because the state dynamic equations were linear and the performance was quadratic. In addition, the Euler-Lagrange equation resulted in an open-loop optimal control solution and is based on a priori knowledge of the final state and the control interval.

In general, design of control systems by means of the calculus of variations usually leads to the solution of a two-point boundary-value problem. Usually, one is faced with a numerical trial-and-error solution to a resulting nonlinear differential equation. The calculus of variations is not commonly used by the control system engineer for designing optimal control systems since it cannot easily handle "hard" constraints such as $|u| \leqslant u_{max}$. Its use is limited to control systems that are linear, have no constraints on x and u, and have a quadratic performance criterion.

9.6 DYNAMIC PROGRAMMING

The concept of dynamic programming [4, 8, 12, 13], as originally developed by Bellman, is based on the principle of optimality and the imbedding approach. The principle of optimality states that regardless of the initial state and the initial decision, the remaining decision must form an optimal control policy with respect to the state resulting from the first decision. The imbedding approach is one in which an optimal decision problem is imbedded in a series of smaller problems that are easier to solve.

A multistage decision process is an example of a problem that can be simplified considerably by applying the principle of dynamic programming. Utilizing the imbedding approach and the principle of optimality, the total return of an N-stage decision process is reduced to the problem of solving a sequence of N single-stage decision processes. This permits a simple, systematic solution to the problem. In practice, dynamic programming is basically an optimization procedure that proceeds backwards in time. The solution is computed first over the last stage of the process, and successive solutions are computed for the remaining stages until the entire solution is obtained.

The principle of optimality results in a partial differential equation, known as the Hamilton-Jacobi equation, whose solution results in the optimum control policy. The Hamilton-Jacobi equation can be derived from the definition of the performance criterion defined by Eq. (9.2):

$$S = \int_{t_0}^{T} G\big(\mathbf{c}(t), \mathbf{u}(t), \mathbf{r}(t), t\big) \, dt. \tag{9.66}$$

Let us assume that the optimal control input $\mathbf{u}^0(t)$ has been determined for a reference signal $\mathbf{r}(t)$ and a final time T. Therefore, the minimum value of this performance criterion S^0 is a function only of the initial state $\mathbf{c}(t_0)$ and the initial time t_0. From Eq. (9.66), we obtain S^0 as follows:

$$S^0 = S^0\big(\mathbf{c}(t_0), t_0\big) = \min_{\mathbf{u}(t)} \left[\int_{t_0}^{T} G\big(\mathbf{c}(t), \mathbf{u}(t), \mathbf{r}(t), t\big) \, dt \right]. \tag{9.67}$$

Applying the imbedding approach and assuming that the last stage of the process occurs between t' and T, Eq. (9.67) may be rewritten as

$$S^0\big(\mathbf{c}(t_0), t_0\big) = \min_{\mathbf{u}(t)} \left[\int_{t_0}^{t'} G\big(\mathbf{c}(t), \mathbf{u}(t), \mathbf{r}(t), t\big) \, dt + \int_{t'}^{T} G\big(\mathbf{c}(t), \mathbf{u}(t), \mathbf{r}(t), t\big) \, dt \right]. \tag{9.68}$$

Applying the principle of optimality over the last stage which occurs from t' to T, Eq. (9.68) can be rewritten as

$$S^0\big(\mathbf{c}(t_0), t_0\big) = \min_{\mathbf{u}(t)} \left[\int_{t_0}^{t'} G\big(\mathbf{c}(t), \mathbf{u}(t), \mathbf{r}(t), t\big) \, dt + S^0\big(\mathbf{c}(t'), t'\big) \right]. \tag{9.69}$$

The first term of Eq. (9.69) may be approximated as

$$\int_{t_0}^{t'} G\big(\mathbf{c}(t), \mathbf{u}(t), \mathbf{r}(t), t\big) dt \approx \big[G\big(\mathbf{c}(t), \mathbf{u}(t), \mathbf{r}(t), t\big) \big]_{t=t_0} \Delta t, \tag{9.70}$$

by defining

$$t' = t_0 + \Delta t. \tag{9.71}$$

Expanding the second term of Eq. (9.69) by means of a Taylor series about the point $\mathbf{c}(t_0)$, t_0 yields

$$S^0\big(\mathbf{c}(t'), t'\big) = S^0\big(\mathbf{c}(t_0), t_0\big) + \frac{\partial S^0}{\partial c_1} \Delta c_1 + \cdots + \frac{\partial S^0}{\partial c_n} \Delta c_n + \frac{\partial S^0}{\partial t_0} \Delta t. \quad (9.72)$$

Substituting Eqs. (9.70) and (9.72) into (9.69), we obtain the following relationship:

$$S^0\big(\mathbf{c}(t_0), t_0\big) = \min_{\mathbf{u}(t_0)} \bigg[G\big(\mathbf{c}(t_0), \mathbf{u}(t_0), \mathbf{r}(t_0), t_0\big) \Delta t + S^0\big(\mathbf{c}(t_0), t_0\big)$$

$$+ \frac{\partial S^0}{\partial c_1} \Delta c_1 + \cdots + \frac{\partial S^0}{\partial c_n} \Delta c_n + \frac{\partial S^0}{\partial t_0} \Delta t \bigg]. \quad (9.73)$$

Taking $S^0\big(\mathbf{c}(t_0), t_0\big)$ outside the minimization operation (since it is not a function of $\mathbf{u}(t_0)$), dividing both sides of Eq. (9.73) by Δt, and letting $\Delta t \to 0$, we obtain

$$\min_{\mathbf{u}(t_0)} \bigg[G\big(\mathbf{c}(t_0), \mathbf{u}(t_0), \mathbf{r}(t_0), t_0\big) + \frac{\partial S^0}{\partial c_1} \dot{c}_1 + \cdots + \frac{\partial S^0}{\partial c_n} \dot{c}_n + \frac{\partial S^0}{\partial t_0} \bigg] = 0. \quad (9.74)$$

Assuming that $\mathbf{u}^0(t_0)$ minimizes the bracketed term, Eq. (9.74) can be rewritten as

$$\frac{\partial S^0}{\partial t_0} + \sum_{i=1}^{n} \frac{\partial S^0}{\partial c_i} \dot{c}_i + G\big(\mathbf{c}(t_0), \mathbf{u}^0(t_0), \mathbf{r}(t_0), t_0\big) = 0. \quad (9.75)$$

Since $\dot{c}_i$ is a function of $\mathbf{c}$ and $\mathbf{u}^0$, we may write this relationship as

$$\dot{c}_i = f_i(\mathbf{c}, \mathbf{u}^0). \quad (9.76)$$

Substituting Eq. (9.76) into (9.75), we obtain the Hamilton-Jacobi equation:

$$\left(\frac{\partial S^0}{\partial t_0} + \sum_{i=1}^{n} \frac{\partial S^0}{\partial c_i} f_i(\mathbf{c}, \mathbf{u}^0) + G(\mathbf{c}, \mathbf{u}^0, \mathbf{r}, t_0) = 0. \right) \quad (9.77)$$

The Hamilton-Jacobi equation is a partial differential equation whose boundary conditions can be obtained from the definition of the performance criterion (see Eq. 9.66) and from the relationship

$$S^0\big(\mathbf{c}(t_0), t_0\big) = \int_{t_0}^{T} G(\mathbf{c}, \mathbf{u}^0, \mathbf{r}, t) \, dt. \quad (9.78)$$

If $\mathbf{c}(T)$ does not contain an impulse function at $t = T$, then the boundary condition is given by

$$\left(\lim_{t_0 \to T} S^0\big(\mathbf{c}(t_0), t_0\big) = 0. \right) \quad (9.79)$$

If $\mathbf{c}(T)$ contains an impulse function at $t = T$, then the boundary condition is given by

$$\lim_{t_0 \to T} S^0\big(\mathbf{c}(t_0), t_0\big) = [G(\mathbf{c}, \mathbf{u}^0, \mathbf{r}, t)]_{t=T}. \quad (9.80)$$

The solution to the Hamilton-Jacobi equation, with appropriate boundary conditions, results in the optimal control policy. However, since the Hamilton-Jacobi equation was derived from the imbedding approach and the principle of optimality, it represents only a necessary condition for optimality. For example, if there are constraints on the inputs or outputs to the system, these constraints must also be considered as necessary conditions for optimality.

As an example of solving an optimal control problem by means of dynamic programming, a simple regulator problem will be considered [14]. The objective in the regulator problem is to maintain the output at a fixed value. Therefore, the input $r = 0$. For this problem:

1. The controlled process is a first-order linear plant whose dynamics are given by

$$\dot{c} = Ac + Bu. \tag{9.81}$$

2. The performance criterion is given by

$$S = \int_0^\infty c^2 \, dt. \tag{9.82}$$

3. The controlled input satisfies the following relationship:

$$-1 \leqslant u \leqslant 1. \tag{9.83}$$

From Eq. (9.77), the scalar Hamilton-Jacobi equation is

$$\frac{\partial S^0}{\partial t_0} + \frac{\partial S^0}{\partial c} f(c, u^0) + G(c, u^0, r, t_0) = 0. \tag{9.84}$$

From the statement of the problem, the terms of this equation are:

$$\frac{\partial S^0}{\partial t_0} = 0, \tag{9.85}$$

$$f(c, u^0) = \dot{c} = Ac + Bu, \tag{9.86}$$

$$G(c, u^0, r, t_0) = c^2. \tag{9.87}$$

Substituting these equations into Eq. (9.84), we obtain

$$\frac{\partial S^0}{\partial c} (Ac + Bu) + c^2 = 0. \tag{9.88}$$

In order to minimize this expression with respect to u, the constraint

$$-1 \leqslant u \leqslant 1 \tag{9.89}$$

must be incorporated into the solution. Therefore,

$$\min_{u(t)} \left[\frac{\partial S^0}{\partial c} (Ac + Bu) + c^2 \right] = \min_{u(t)} \left[\frac{\partial S^0}{\partial c} Bu \right]. \tag{9.90}$$

Clearly, the optimal control must satisfy

$$u^0 = 1 \quad \text{for} \quad \frac{\partial S^0}{\partial c} B < 0,$$

$$u^0 = -1 \quad \text{for} \quad \frac{\partial S^0}{\partial c} B > 0,$$

(9.91)

or

$$u^0 = - \operatorname{sgn} \left(B \frac{\partial S^0}{\partial c} \right).$$

(9.92)

Substituting Eq. (9.92) into (9.88), we obtain

$$\frac{\partial S^0}{\partial c} \left[Ac - B \operatorname{sgn} \left(B \frac{\partial S^0}{\partial c} \right) \right] + c^2 = 0.$$

(9.93)

Simplifying this equation, we obtain

$$Ac \frac{\partial S^0}{\partial c} - \left| B \frac{\partial S^0}{\partial c} \right| + c^2 = 0,$$

(9.94)

or

$$c^2 + Ac \frac{\partial S^0}{\partial c} - \left| B \frac{\partial S^0}{\partial c} \right| = 0.$$

(9.95)

Equation (9.95) is a nonlinear partial differential equation. Except for some simple cases, the solution for S^0 requires the aid of a digital computer. The resulting solution for S^0 is then substituted into Eq. (9.92) in order to obtain the optimum u^0.

As a second example of solving an optimal control problem by means of dynamic programming, consider the following problem:

1. The controlled process is a second-order linear plant whose dynamics are given by

$$\ddot{c} + 4\dot{c} + c = u.$$

(9.96)

2. It is desired to minimize the response time so that the performance criterion is given by

$$S = \int_0^\infty dt.$$

(9.97)

3. The controlled input satisfies

$$-1 \leqslant u \leqslant 1.$$

(9.98)

By defining the states

$$x_1 = c, \qquad x_2 = \dot{c},$$

(9.99)

the state equations for this system are given by:

$$\dot{x}_1 = x_2, \qquad \dot{x}_2 = -x_1 - 4x_2 + u.$$

(9.100)

The terms of the Hamilton-Jacobi equation (see Eq. 9.75) for this problem are

$$\left.\begin{array}{c} \dfrac{\partial S^0}{\partial t_0} = 0, \\[12pt] \displaystyle\sum_{i=1}^{2} \dfrac{\partial S^0}{\partial c_i}\, \dot{c}_i = \dfrac{\partial S^0}{\partial c}\, \dot{c} + \dfrac{\partial S^0}{\partial \dot{c}}\,(-c - 4\dot{c} + u), \\[12pt] G(x,\, u^0,\, r,\, t_0) = 1. \end{array}\right\} \tag{9.101}$$

Substituting these equations into the Hamilton-Jacobi equation, Eq. (9.75), we obtain

$$\frac{\partial S^0}{\partial c}\, \dot{c} + \frac{\partial S^0}{\partial \dot{c}}\,(-c - 4\dot{c} + u) + 1 = 0. \tag{9.102}$$

In order to minimize this expression with respect to u, the constraint

$$-1 \leqslant u \leqslant 1$$

must be incorporated into the solution. Therefore,

$$\min_{u(t)} \left[\frac{\partial S^0}{\partial c}\, \dot{c} + \frac{\partial S^0}{\partial \dot{c}}\,(-c - 4\dot{c} + u) + 1 \right] = \min_{u(t)} \left[\frac{\partial S^0}{\partial \dot{c}}\,(u) \right].$$

Clearly, the optimal control satisfies

$$u^0 = -\,\text{sgn}\left(\frac{\partial S^0}{\partial \dot{c}} \right). \tag{9.103}$$

Substituting this equation into Eq. (9.102), we obtain

$$\dot{c}\left[\frac{\partial S^0}{\partial c} - 4 \frac{\partial S^0}{\partial \dot{c}} \right] - c\left[\frac{\partial S^0}{\partial \dot{c}} \right] - \left[\left| \frac{\partial S^0}{\partial \dot{c}} \right| - 1 \right] = 0. \tag{9.104}$$

Equation (9.104) is a nonlinear partial differential equation which requires a digital computer to find S^0. The resulting solution for S^0 is then substituted into Eq. (9.103) in order to obtain the optimum u^0.

9.7 PONTRYAGIN'S MAXIMUM PRINCIPLE

In 1956, the Russian mathematicians, Pontryagin, Boltyanskii, and Gamkrelidge developed the maximum principle [6, 8, 12, 15, 16]. According to Pontryagin, the maximum principle was derived originally from the calculus of variations (see Section 9.5). Pontryagin's maximum principle is very similar to the calculus of variations, and is very closely related to dynamic programming. It is possible to obtain the maximum principle from dynamic programming by a simple change of variables.

In this section, Pontryagin's maximum principle and an illustrative example of its application are presented. In the following section, it is applied to the space attitude control problem.

Let us assume that we have a process whose dynamics are given by

$$\dot{c}_i = f_i(\mathbf{c}, \mathbf{u}), \qquad i = 1, 2, \ldots, n \tag{9.105}$$

and a performance index

$$S = \int_0^T G(\mathbf{c}, \mathbf{u}, \mathbf{r}, t)\, dt \tag{9.106}$$

which is to be minimized. The maximum principle requires that the optimal control input $\mathbf{u}^0$ which minimizes S will maximize the scalar

$$H = \sum_{i=1}^n p_i f_i(\mathbf{c}, \mathbf{u}) - G(\mathbf{c}, \mathbf{u}, \mathbf{r}, t_0), \tag{9.107}$$

where the function $\dot{p}_i$ is defined as

$$\dot{p}_i = -\frac{\partial H}{\partial c_i}, \qquad i = 1, 2, \ldots, n. \tag{9.108}$$

The scalar H is called the Hamiltonian function; the vector $\mathbf{p}$ is known as the co-state. From Eqs. (9.105) and (9.107), $\dot{c}_i$ can also be expressed in terms of H and p_i, as follows:

$$\dot{c}_i = \frac{\partial H}{\partial p_i}, \qquad i = 1, 2, \ldots, n. \tag{9.109}$$

The necessary conditions of the maximum principle can be obtained from the dynamic programming equations by a simple change of variables. Let us reconsider the Hamilton-Jacobi equation (see Eq. 9.77):

$$\frac{\partial S^0}{\partial t_0} + \sum_{i=1}^n \frac{\partial S^0}{\partial c_i} f_i(\mathbf{c}, \mathbf{u}) + G(\mathbf{c}, \mathbf{u}, \mathbf{r}, t_0) = 0. \tag{9.110}$$

Let us define

$$p_i = -\frac{\partial S^0}{\partial c_i} \tag{9.111}$$

and

$$H = \frac{\partial S^0}{\partial t_0}. \tag{9.112}$$

Substituting Eqs. (9.111) and (9.112) into (9.110), we obtain the following relationship:

$$H = \sum_{i=1}^n p_i f_i(\mathbf{c}, \mathbf{u}) - G(\mathbf{c}, \mathbf{u}, \mathbf{r}, t_0), \tag{9.113}$$

which is the basic Hamiltonian function defined in Eq. (9.107). In addition, the relationship of Eq. (9.108) can be obtained from Eqs. (9.111) and (9.112) as follows:

$$\dot{p}_i = \frac{dp_i}{dt_0} = \frac{d}{dt_0}\left(-\frac{\partial S^0}{\partial c_i}\right) = -\frac{\partial^2 S^0}{\partial t_0\, \partial c_i} = -\frac{\partial H}{\partial c_i}. \tag{9.114}$$

Therefore, the basic maximum-principle relationships, as defined by Eqs. (9.107) and (9.108), have been derived from the dynamic programming relationship (the Hamilton-Jacobi equation) by a simple change of variables. However, it is important to note that this is not a proof of the maximum principle since this derivation assumed the existence of the derivatives $\partial S^0/\partial c_i$ and $\partial S^0/\partial t_0$. On the contrary, there are many cases where they do not exist.

As an example of an optimal control problem which can be solved by means of the maximum principle, let us reconsider the first-order problem in Section 9.6 which was solved by applying dynamic programming. The problem involves

1. A process whose dynamics are described by

$$\dot{c} = Ac + Bu. \tag{9.115}$$

2. A performance criterion given by

$$S = \int_0^\infty c^2\, dt. \tag{9.116}$$

3. A control input constrained by the relationship

$$-1 \leqslant u \leqslant 1. \tag{9.117}$$

From the above, the values of $f_i(\mathbf{c}, \mathbf{u})$ and $G(\mathbf{c}, \mathbf{u}, \mathbf{r}, t_0)$ in the Hamiltonian equation are given by

$$f_i(\mathbf{c}, \mathbf{u}) = Ac + Bu, \tag{9.118}$$

$$G(\mathbf{c}, \mathbf{u}, \mathbf{r}, t_0) = c^2. \tag{9.119}$$

Substituting Eqs. (9.118) and (9.119) into the Hamiltonian equation (see Eq. 9.107), we obtain the relationship

$$H = p(Ac + Bu) - c^2. \tag{9.120}$$

Applying the fundamental relationship of the maximum principle (see Eq. 9.108)

$$\dot{p}_i = -\frac{\partial H}{\partial c_i} \tag{9.121}$$

to Eq. (9.120), we obtain the first equation to be solved:

$$\dot{p} = -\frac{\partial H}{\partial c} = -pA + 2c. \tag{9.122}$$

A second equation can be obtained from the fact that the Hamiltonian is maximized when the performance criterion is minimized. It was shown in our discussion of dynamic programming (Section 9.6) that the performance criterion was minimized in this problem when (see Eq. 9.92).

$$u^0 = - \text{sgn} \left(B \frac{\partial S^0}{\partial c} \right). \tag{9.123}$$

From the definition of p_i in Eq. (9.111), we may rewrite Eq. (9.123) as

$$u^0 = \text{sgn} (Bp). \tag{9.124}$$

Substituting Eq. (9.124) into (9.115), we obtain the second necessary condition

$$\dot{c} = Ac + B \text{sgn} (Bp). \tag{9.125}$$

The solution to this optimal control problem, utilizing the maximum principle, has been reduced to the solution of the nonlinear ordinary differential equations given by Eqs. (9.122) and (9.125). In order to solve these equations, the boundary conditions must be utilized. Note that they define a two-point boundary-value problem.

It is interesting to compare the dynamic-programming solution with that obtained with the maximum principle. The dynamic-programming solution was reduced to the solution of one nonlinear partial differential equation (see Eq. 9.104). The maximum-principle solution of the same problem was reduced to the solution of two, nonlinear, ordinary differential equations (see Eqs. 9.122 and 9.125). Although both techniques result in equations which require digital computers for solution, the two, nonlinear, first-order ordinary differential equations (obtained using the maximum principle) are easier to solve than the nonlinear partial differential equation, which is a function of two variables (obtained using dynamic programming). In practice, however, the choice between these approaches will depend to a great extent on the particular problem. The application of Pontryagin's maximum principle to the synthesis of optimum attitude controllers for space vehicles is discussed in the following section.

9.8 APPLICATION OF THE MAXIMUM PRINCIPLE TO THE SPACE ATTITUDE CONTROL PROBLEM

Optimal control theory has been applied to a wide variety of important problems [8, 12, 15, 17–29]. It has been used to solve problems concerning the attitude control of space vehicles, control of traffic flow, the orbit transfer problem for interplanetary space vehicles, chemical process control problems, and problems concerned with communication systems. In this section, the use of the maximum principle for solving the attitude control problem of various space vehicles is illustrated. The objective is to synthesize the optimum strategy for controlling the space vehicle to satisfy a given performance criterion.

Attitude control of a space vehicle encompasses a very wide variety of problems. During powered flight, the attitude control system receives commands from a guidance system and controls the attitude of the vehicle. This causes the vehicle to pitch or yaw and results in changes in attitude and/or direction of the flight path. After

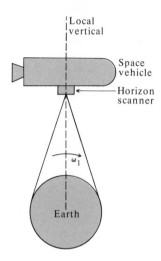

Fig. 9.6   Attitude control via horizon scanners to establish the local vertical.

the vehicle has attained the desired orbit, it is attitude-stabilized with respect to some reference such as the earth, sun, or the stars. During reentry into the earth's atmosphere, the vehicle is pitched over to the proper angle from the reference attitude by signals from a reference gyroscope. Then the firing of a retrorocket places the vehicle on a transfer orbit into the earth's atmosphere. The attitude control problem is even more complicated for manned space stations where orbit rendezvous is required for purposes of orbital refueling, crew changes and/or satellite inspection.

This section is concerned with the attitude stabilization of a manned or unmanned vehicle in orbit about the earth. It is to be stabilized perpendicular to the earth's local vertical as shown in Fig. 9.6. Consider one plane of such a space vehicle orbiting the earth which is slaved to the local vertical via horizon sensors and gyros. A physical model of the problem is illustrated in Fig. 9.7 and a block diagram of a typical

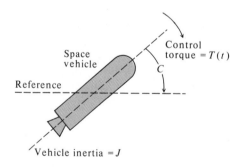

Fig. 9.7   Attitude control problem for one plane.

attitude control system is shown in Fig. 9.8. The reference input position to the horizon tracker is denoted by R and the resultant output position of the attitude control system is denoted by C. The rate of change of the local vertical with respect to the earth is denoted by ω_1 and the rate of the vehicle relative to the local vertical is denoted by ω_2. The vehicle inertial rate, which is the sum of ω_1 and ω_2, is denoted by ω_T.

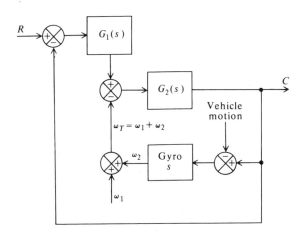

Fig. 9.8   One axis of an attitude control system.

Assuming that friction and disturbing forces are negligible, the motion of the space vehicle is given by the following simple second-order differential equation:

$$\ddot{c} = T(t)/J. \tag{9.126}$$

By defining

$$u(t) = \frac{T(t)}{J}, \qquad \text{where} \qquad |u(t)| \leqslant 1, \tag{9.127}$$

Eq. (9.126) can be rewritten as

$$\ddot{c}(t) = u(t). \tag{9.128}$$

Utilizing state-space notation, this second-order differential equation can be rewritten as two first-order differential equations. Let

$$c_1 = c, \qquad c_2 = \dot{c}; \qquad \dot{c}_1 = c_2 \qquad \dot{c}_2 = u. \tag{9.129}$$

Then

$$\dot{\mathbf{c}} = \mathbf{Pc} + \mathbf{Bu}, \tag{9.130}$$

where

$$\mathbf{c} = \begin{bmatrix} c_1 \\ c_2 \end{bmatrix}, \qquad \dot{\mathbf{c}} = \begin{bmatrix} \dot{c}_1 \\ \dot{c}_2 \end{bmatrix}, \qquad \mathbf{P} = \begin{bmatrix} 0 & 1 \\ 0 & 0 \end{bmatrix}, \qquad \mathbf{B} = \begin{bmatrix} 0 & 0 \\ 0 & 1 \end{bmatrix}, \qquad \mathbf{u} = \begin{bmatrix} 0 \\ u \end{bmatrix}. \tag{9.131}$$

The basic attitude control problem is to maintain the vehicle at a referenced attitude. The desired equilibrium state for this problem is assumed to be the stable node at the origin of the c_1c_2-plane.

In the following analysis the synthesis of the optimal attitude control system for several practical attitude control problems is considered. For example, due to a disturbance torque or the command of a new reference attitude, what is the best strategy to minimize the response time of the vehicle? Other considerations may dictate that the amount of fuel or energy be minimized. These possibilities lead to the following problems:

1. the minimum-time problem,

2. the minimum fuel-consumption problem,

3. the minimum energy problem.

In all cases, it will be assumed that u is constrained to be $1 \leqslant u \leqslant -1$.

1. The Minimum-Time Problem Although the minimum-time problem can be solved with conventional techniques [28], it is synthesized here, utilizing Pontryagin's maximum principle as an introduction to the application of optimal control concepts. In addition, its solution is useful for comparison with the other problems which are subsequently considered.

Since it is desired to minimize the response time of the space vehicle, the performance criterion for the minimum-time problem is given by

$$S = \int_0^T dt = T. \tag{9.132}$$

Here the loss function is unity: that is,

$$G(\mathbf{c}, \mathbf{u}, \mathbf{r}, t) = 1. \tag{9.133}$$

Substituting Eqs. (9.129) and (9.133) into the expression for the Hamiltonian, Eq. (9.107), the following expression is obtained:

$$H = p_1c_2 + p_2u - 1. \tag{9.134}$$

The values of p_1, p_2, and c_2 can be evaluated by applying Eqs. (9.108) and (9.109) to Eq. (9.134). The results are as follows:

$$\dot{p}_1 = -\frac{\partial H}{\partial c_1} = 0, \qquad p_1 = K_1, \tag{9.135}$$

$$\dot{p}_2 = -\frac{\partial H}{\partial c_2} = -p_1, \qquad p_2 = K_2 - p_1t = K_2 - K_1t, \tag{9.136}$$

where K_1 and K_2 represent constants of integration; and

$$\dot{c}_1 = \frac{\partial H}{\partial p_1} = c_2, \tag{9.137}$$

$$\dot{c}_2 = \frac{\partial H}{\partial p_2} = u. \tag{9.138}$$

Since the term $p_1 c_2 - 1$ in Eq. (9.134) is independent of the input u, the maximization of the Hamiltonian function is concerned only with

$$\max_{u(t)} H[p_2 u]. \tag{9.139}$$

It is seen from Eq. (9.139) that the Hamiltonian function is maximized by choosing

$$u^0 = \text{sgn} [p_2]. \tag{9.140}$$

Substituting Eq. (9.136) into Eq. (9.140), we obtain

$$u^0 = \text{sgn} [K_2 - K_1 t]. \tag{9.141}$$

The following conclusions can be drawn from this result:

1. The optimum input for minimization of the response time is piecewise constant.
2. The optimum input for minimum-time operation takes on only the values ± 1.
3. The sign of the optimum input for minimum-time operation can change its value only once.

These conclusions clearly imply that the attitude control system for the space vehicle should be bang–bang (on–off) when minimization of response time is the performance criterion. Physically this means that accelerating and then decelerating is the best that can be done in order to minimize the response time. The period of each action depends on the initial conditions of position and velocity.

The phase-plane representation and optimal switching curve can be formulated from consideration of Eqs. (9.128) and (9.129) when $u = \pm 1$. When $u = 1$, the following expressions are obtained:

$$\dot{c}_2 = 1 \tag{9.142}$$

$$c_2 = t + A_1 \tag{9.143}$$

$$\dot{c}_1 = t + A_1 \tag{9.144}$$

$$c_1 = \tfrac{1}{2} t^2 + A_1 t + A_2 \tag{9.145}$$

where A_1 and A_2 are constants of integration. By completing the square, Eq. (9.145) can be rearranged as follows:

$$c_1 = \tfrac{1}{2}(t + A_1)^2 + (A_2 - \tfrac{1}{2} A_1^2). \tag{9.146}$$

Substituting Eq. (9.144) into (9.146), we obtain the following relationship:

$$c_1 = \tfrac{1}{2}\dot{c}_1^2 + A_3, \tag{9.147}$$

where

$$A_3 = A_2 - \tfrac{1}{2}A_1^2. \tag{9.148}$$

Equation (9.147) defines the switching curve when $u = 1$. Similarly, the switching curve when $u = -1$ can be obtained as

$$c_1 = -\tfrac{1}{2}\dot{c}_1^2 + A_6, \tag{9.149}$$

where

$$A_6 = A_5 + \tfrac{1}{2}A_4^2. \tag{9.150}$$

The switching curves defined by Eqs. (9.147) and (9.149) define parabolas in the $c_1\dot{c}_1$-plane. The corresponding phase portrait and optimal switching line for the minimum time system is illustrated in Fig. 9.9. The phase trajectory travels from its initial conditions to the switching line defined by Eq. (9.147) or Eq. (9.149). At the instant the state arrives at the switching curve, the system switches its control to the opposite phase and remains at this value until the state reaches the stable node at the origin. If the initial conditions are above the curve AOB, the system is under the control of $u = -1$ until it reaches the arc BO. Then it switches over to the control $u = 1$, where it remains until the equilibrium state is reached. When the initial conditions are below AOB, the system is under the control of $u = 1$ until the state reaches arc AO. At the instant that it arrives, the system switches to the control $u = -1$ and remains at this value until the equilibrium state is reached.

Figure 9.10 shows the block diagram of the optimum control system just designed. The basic control element required has ideal relay characteristics which can easily be implemented. The control system will accelerate and then decelerate in order to minimize the response time of the attitude control system. It should be noted that the

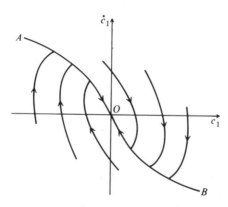

Fig. 9.9   Phase-plane representation for the minimum-time problem.

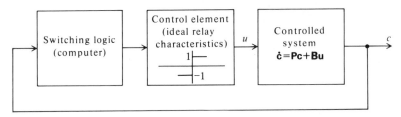

Fig. 9.10   Synthesized system to minimize response time.

exact period of each phase is dependent upon the initial conditions relative to the switching curves.

2. The Minimum Fuel-consumption Problem   In this problem it is assumed that the attitude control system is powered by a reaction jet. It is desired to minimize the fuel consumption of the space vehicle. It is also assumed that a pure minimum fuel consumption problem is inadequate from a practical viewpoint since it may result in an excessive response time. Therefore, the actual practical problem considered here is that of minimizing fuel consumption over a certain period of time.

Although response time is of secondary importance in this problem, it must be considered. It is theoretically possible to bring the state of the system to the stable node with an arbitrarily small amount of fuel if the response time were not bounded. However, the response time must be considered fixed for practical systems. Therefore, it is assumed that the total response time is limited to T seconds.

In synthesizing a meaningful optimal control system, the reaction jet system must be understood. An important parameter used to characterize the performance of a reaction jet engine is its specific impulse I_{sp}. It is defined as the ratio of the thrust F of a hypothetical engine to a propellant flow rate $\dot{\omega}$ of one pound of propellant per second:

$$I_{sp} = F/\dot{\omega}.$$

The units of I_{sp} from this relationship are pounds of thrust per pounds of propellant per second and are usually expressed in seconds. The specific impulse of an engine is indicative of how effectively each pound of propellant is utilized in producing a thrust force to the vehicle.

Since it is desired to minimize fuel consumption of the space vehicle over a period of T seconds, the performance criterion for this problem is given by

$$S = \int_0^T |u| \, dt, \tag{9.151}$$

where u is defined as the propellant rate of flow and is constrained in magnitude by Eq. (9.127). For this problem, the loss function represents a measure of the total fuel flow rate:

$$G(\mathbf{c}, \mathbf{u}, \mathbf{r}, t) = |u|. \tag{9.152}$$

Its time integral, as given by Eq. (9.151), represents a measure of the total fuel consumed in T seconds.

Substituting Eqs. (9.129) and (9.152) into the expression for the Hamiltonian function, Eq. (9.107), the following expression is obtained:

$$H = p_1 c_2 + p_2 u_2 - |u|. \tag{9.153}$$

The values of p_1, p_2, and c_2 can be evaluated by applying Eqs. (9.108) and (9.109) to Eq. (9.153). The results are the same as Eqs. (9.136) and (9.138). Since the term $p_1 c_2$ in Eq. (9.153) is independent of the input u, the maximization of the Hamiltonian function is concerned only with

$$\max_{u(t)} H[p_2 u - |u|]. \tag{9.154}$$

It is obvious from Eq. (9.154) that the Hamiltonian function is maximized by choosing u as follows:

$$\begin{aligned} u^0 &= \text{sgn}\,[p_2] \quad &&\text{for} \quad |p_2| \geqslant 1, \\ u^0 &= 0 \quad &&\text{for} \quad |p_2| < 1. \end{aligned} \tag{9.155}$$

Substituting Eq. (9.136) into Eq. (9.155), the following optimal control inputs are obtained:

$$\begin{aligned} u^0 &= \text{sgn}\,[K_2 - K_1 t] \quad &&\text{for} \quad |p_2| \geqslant 1, \\ u^0 &= 0 \quad &&\text{for} \quad |p_2| < 1. \end{aligned} \tag{9.156}$$

The following conclusions can be drawn from this result:

1. The optimum input is piecewise constant;

2. The optimum input can have values of only ± 1 and 0.

Although K_1 and K_2 are not known exactly, the fact that $p_2(t)$ is a linear function of time (see Eq. 9.136) means that u must proceed in time as a nonrepeating sequence of values having the form $(\pm 1, 0, \mp 1)$. These conclusions imply that the attitude control system should be bang–bang and incorporate a dead zone. This function can be easily implemented. Physically, this means that in order to minimize the fuel consumption over a bounded period of time, the best one can do is to accelerate, coast at a constant velocity, and then decelerate. The periods for each action depend on the initial conditions of position and velocity.

The phase-plane representation and optimal switching curves can be formulated from considerations of Eqs. (9.128) and (9.129) when $u = \pm 1$ and 0. When $u = \pm 1$, results analogous to Eqs. (9.147) and (9.149) are obtained for the minimum-time problem. When $u = 0$,

$$\dot{c}_2 = 0, \tag{9.157}$$

$$c_2 = \text{constant}, \tag{9.158}$$

and the switching curve is a horizontal line parallel to the c_1-axis. Typical phase trajectories for the minimum fuel-consumption problem when the response time is bounded are illustrated in Fig. 9.11 as curves $lmno$ and $l'm'n'o$. These results lead to the conclusion that if reaction jets are used for attitude control, then a nozzle which simply opens or closes without any intermediate settings provides the optimal system.

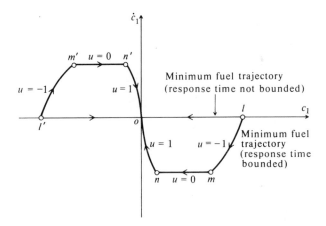

Fig. 9.11   Typical phase trajectories for the minimum fuel problem.

It is interesting to compare the phase trajectories of a pure minimum-fuel problem where the response time is not bounded and that of a minimum-fuel system where the response time is bounded. Typical phase trajectories for the pure minimum fuel optimal problem are curves lo and $l'o$ in Fig. 9.11. The corresponding time response of curves $lmno$ and lo can be obtained from an examination of the phase trajectories. It was shown in Section 8.15 that the variation of time along a phase trajectory can be obtained from (see Eq. 8.136)

$$t = \int_l^0 \frac{1}{\dot{c}_1}\, dc_1. \tag{9.159}$$

The integral represents the area under the reciprocal phase-plane trajectory. A corresponding reciprocal plot for the phase trajectories $lmno$ and lo are illustrated in Fig. 9.12. The points l and o are assumed to lie at infinity; this has a negligible effect on the area under curve $lmno$ since the area under these points approaches zero. However, the area under the rectangular curve lo is infinity. This illustrates that a pure minimum-fuel problem, where the response time is not bounded, results in an infinite response time and is not very practical.

Figure 9.13 shows the block diagram of the optimal control system. The basic control element required has the characteristics of a relay with a dead zone.

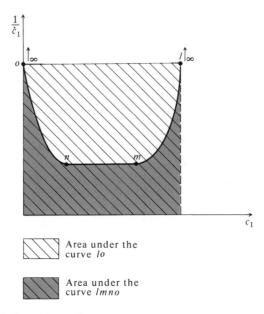

Area under the curve *lo*

Area under the curve *lmno*

Fig. 9.12   Reciprocal phase-plane plot.

3. The Minimum-energy Problem   In this problem it is assumed that the energy utilized in the attitude control system of a space vehicle is to be minimized. This is a very practical problem for space vehicles utilizing momentum-exchanging devices such as control-moment gyros, inertia wheels, fluid flywheels, and magnetic moment devices in their attitude control systems. All of these systems utilize electrical energy derived from such sources as batteries, fuel cells, and/or solar cells.

It is assumed that the input to the attitude control system is an electrical signal u. In addition, it is desired to utilize a minimum amount of electrical energy over a bounded period of time in order to accomplish a desired control. We know that the square of the voltage utilized is proportional to power, and the time integral of power is energy. Since the square of the control signal is proportional to the integral power required for control, and the energy is proportional to the integral of the square of the control signal, the minimum-energy problem can be formulated as a problem of

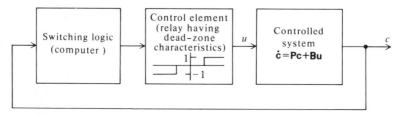

Fig. 9.13   Synthesized system to minimize fuel consumption over a specified period of time.

minimum integral square control. The performance criterion is thus given by

$$S = \int_0^T u^2 \, dt. \tag{9.160}$$

Here the loss function represents power

$$G(\mathbf{c}, \mathbf{u}, \mathbf{r}, t) = u^2 \tag{9.161}$$

and its time integral, as given by Eq. (9.160), represents a measure of energy. Substituting Eqs. (9.129) and (9.161) into the expression for the Hamiltonian function, Eq. (9.107), we obtain

$$H = p_1 c_2 + p_2 u - u^2. \tag{9.162}$$

The values of p_1, p_2, and c_2, can be evaluated by applying Eqs. (9.108) and (9.109) to Eq. (9.162). Since the term $p_1 c_2$ in Eq. (9.162) is independent of the input u, the maximization of the Hamiltonian function is only concerned with the expression

$$\max_{u(t)} H \, [p_2 u - u^2]. \tag{9.163}$$

It is obvious from Eq. (9.163) that two cases exist for the Hamiltonian function to be maximized. To find the optimal value of u^0, differentiation of Eq. (9.163) yields

$$\frac{\partial}{\partial u} [p_2 u - u^2] = 0, \tag{9.164}$$

$$u^0 = p_2/2.$$

Therefore, if $|p_2| < 2$, $u^0 = p_2/2$. When $|p_2| \geqslant 2$, then

$$u^0 = \text{sgn} \, [p_2] \tag{9.165}$$

results in a maximum value of the Hamiltonian function. Substituting Eq. (9.136) into Eqs. (9.164) and (9.165), we obtain

$$u^0 = \frac{K_2}{2} - \frac{K_1}{2} t = K_2' - K_1' t \quad \text{for} \quad |p_2| < 2, \tag{9.166}$$

$$u^0 = \text{sgn} \, [p_2] = \text{sgn} \, [K_2 - K_1 t] \quad \text{for} \quad |p_2| \geqslant 2.$$

These results indicate that the control is linear over a fixed range and saturates whenever $|p_2| \geqslant 2$. These conclusions imply that the attitude control system for the space vehicle should contain a limiter whose output is linear for small signals and saturates for large signals in order to minimize energy.

Figure 9.14 shows the block diagram of the control system. The basic control element required is a limiter which will operate linearly until saturation is reached.

The attitude control problem has been used as a convenient illustration of the application of Pontryagin's maximum principle. The results of this analysis indicated

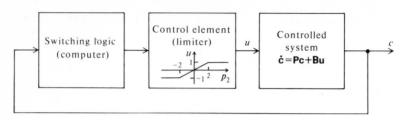

Fig. 9.14   Synthesized system to minimize energy over a specified period of time.

that the resulting structure of the control element was in the form of a relay to minimize the response time, a relay with a dead zone to minimize the fuel consumption over a bounded period of time, and a limited linear amplifier to minimize energy over a bounded period of time.

PROBLEMS

9.1   Determine whether the control system illustrated in Fig. P9.1 is controllable and observable.

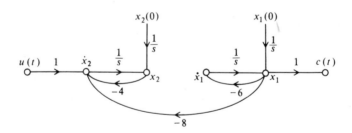

Figure P9.1

9.2   Repeat Problem 9.1 for the control system illustrated in Fig. P9.2.

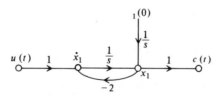

Figure P9.2

9.3 Repeat Problem 9.1 for the control system illustrated in Fig. P9.3.

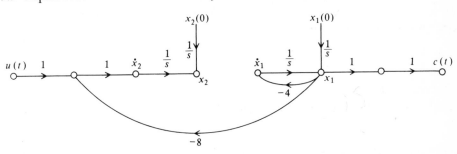

Figure P9.3

9.4 Repeat Problem 9.1 for the control system illustrated in Fig. P9.4.

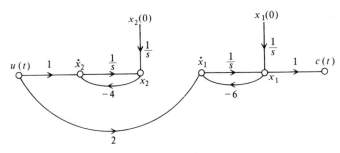

Figure P9.4

9.5 Consider a simple open-loop system that consists of an integration and an amplifier having a gain of 2. It is desired to control the state of this system between $x(0)$ and $x(T)$ in order that the performance criterion

$$S = \int_0^T (4x^2 + 2u^2)\, dt$$

is minimized. Using the calculus of variations, determine the optimal control $u^0(t)$ which can achieve this.

9.6 Repeat Problem 9.5 with the performance criterion changed to

$$S = \int_0^T (2x^2 + 4u^2)\, dt.$$

What conclusions can you draw from your results?

9.7 Using dynamic programming, determine the optimal control policy for a second-order system where

$$\ddot{c} + 4\dot{c} + c = u,$$

$$S = \int_0^\infty dt.$$

Assume the magnitude of the controlled input must be less than or equal to unity.

9.8 Repeat Problem 9.7 using the maximum principle.

9.9 Using dynamic programming, determine the optimal control policy for a third-order system where $-1 \leqslant u \leqslant 1$ and

$$\dddot{c} + 2\ddot{c} + 6\dot{c} = 4 + u,$$

$$S = \int_0^\infty dt.$$

9.10 Repeat Problem 9.9 with the dynamics given by

$$\dddot{c} + 2\ddot{c} + 6\dot{c} + 4c = u.$$

9.11 Using the maximum principle, determine the optimal control policy for a third-order system where $-1 \leqslant u \leqslant 1$ and

$$\dddot{c} + 2\ddot{c} + 6\dot{c} = 4 + u,$$

$$S = \int_0^\infty dt.$$

9.12 Repeat Problem 9.11 with the dynamics given by

$$\dddot{c} + 2\ddot{c} + 6\dot{c} + 4c = u.$$

9.13 The APOLLO 11 mission, in which Astronauts Neil Armstrong and Edwin Aldrin successfully soft-landed the Lunar Excursion Module (LEM) on the lunar surface, was a historic event. Figure P9.13(a) is a photograph of the LEM vehicle, taken by Astronaut Michael Collins from the Apollo Command Module window, after they separated. The problem of synthesizing an optimal control policy for minimal fuel thrust, during the terminal phase of a lunar soft-landing mission, is one that can best be solved using modern optimal control theory [31]. Let us first look at this problem from the basic physics involved in a lunar soft landing as shown in Fig. P9.13(b). It is assumed in this problem that the motion of the vehicle is vertical and subject to the following conditions:

a) The only forces acting on the vehicle are its own weight and the thrust which acts as a braking force.
b) The moon is flat in the vicinity of the desired landing point.
c) The propulsion system is capable of a mass-flow rate, $\dot{m}$, between zero and an upper fixed limit of α:

$$-\alpha \leqslant \dot{m}(t) \leqslant 0.$$

On these assumptions, the motion of the vehicle is governed by the relation

$$\ddot{x} = -\frac{K\dot{m}}{m} - g,$$

where

x = altitude,
m = total mass,
$\dot{m}$ = mass flow rate $\leqslant 0$,
g = acceleration of gravity at the surface of the moon,
K = velocity of exhaust gases = constant > 0.

(a)

(b)

Fig. P9.13 (a) Apollo II: Astronauts Neil Armstrong and Edwin Aldrin are inside the lunar module separated from the Apollo command module. (Official NASA Photo) (b) Forces acting on vehicle.

In addition, system performance is based on the following criterion:

$$S = \int_0^T \dot{m}(t) \, dt.$$

Utilizing the maximum principle, determine the form of the optimal control policy.

9.14 Utilizing dynamic programming, determine the optimal control policy for a control system where

$$S = \int_0^\infty (Ac^2 + u^2) \, dt,$$

assuming that the plant dynamics are given by

$$\dot{c} = Ku.$$

and that there are no constraints on the input.

9.15 A second-order space attitude control system is characterized by the following state equations:

$$\dot{x}_1 = x_2,$$
$$\dot{x}_2 = -x_2 - x_1 + u.$$

Determine the control signal u such that the system is taken from the initial state $c(t_0)$ to the equilibrium state $c(t_f) = 0$ in the shortest possible time. Assume that

$$|u(t)| \leqslant U.$$

9.16 Repeat Problem 9.15 with the state equations given by

$$\dot{x}_1 = x_2,$$
$$\dot{x}_2 = -2x_2 - 4x_1 + u.$$

9.17 Repeat Problem 9.15 with the state equations given by

$$\dot{x}_1 = x_2,$$
$$\dot{x}_2 = -4x_2 + 4x_1 + u.$$

9.18 Determine the optimal control policy for a control system whose performance criterion is given by

$$S = \int_0^T (u + c^2) \, dt,$$

and whose plant dynamics are given by

$$\dot{c} = -c + 4u.$$

Assume that the magnitude of the input $\leqslant 10$. Solve using the maximum principle.

9.19 Aquaculture, the science of farming and husbandry of fresh water and marine organisms, is a field where optimal control theory has recently been applied for optimizing its operation [32]. An optimal policy for the raising of Maine lobsters has been developed. Experimental results indicate that Maine lobsters take from five to eight years to reach maturity when raised in their natural environment, while lobsters raised in water held at 70° F can reach maturity in two years. It has been shown that lobster growth is a function of temperature and lobster weight. When the costs of heating the water and maintaining the lobsters are included, the total expense for raising lobsters can be treated as an optimal control problem. The differential equations describing growth and cost of raising lobsters are given by the following:

$$\dot{W} = K_0 W^\alpha (U - U_0)^m \qquad \text{for } U_0 \leq U \leq U_{\max},$$

$$\dot{W} = O \qquad\qquad\qquad \text{for all other temperatures,}$$

$$\dot{L} = K_1 W(U - U_0) + K_2 W + K_3,$$

where

W = weight of the lobster (state variable),

U = water temperature for the lobster,

U_0 = empirically determined zero growth temperature,

$U_{\max}$ = maximum practical temperature for growing lobsters,

L = cost of raising lobsters (loss function).

The values K_0, α, and m are constants and K_1, K_2, and K_3 represent estimates of cost.

a) Write the expressions for the Hamiltonian function and for the co-state vector for this aquaculture temperature control system.
b) Assuming that m is greater than unity, determine the optimal control temperature, U°. What kind of temperature control system results?

9.20 Almost all agricultural crops suffer damage caused by certain prey (insects) that eat or otherwise destroy the crop. Nature keeps these pests in check by subjecting them to the role of prey, with respect to other insects that act as their predators. Predators usually cause little crop damage. Man, however, has tried to control pests by utilizing insecticides that are lethal not only to the prey but to their predators as well. The unfortunate side effects of such control programs over the past are now being realized and analyzed by those concerned with the environment. Control engineers are entering the ecological domain in order to guide future pest-control programs. Any control attempt should start first with a model of the biological control system. By utilizing a biological model of the ecological environment, the effects of human control can be examined. Most prey–predator systems can be modeled by the following Lotka-Volterra equations [33]:

$$\frac{dN_1}{dt} = AN_1 - BN_1N_2,$$

$$\frac{dN_2}{dt} = -CN_2 + DN_1N_2,$$

where N_1 and N_2 are the instantaneous populations of the prey and predators, respectively, and A, B, C and D are all positive constants. This model can be simplified by nondimensionalizing all the variables and eliminating all constants except one:

$$x_1 = \frac{N_1}{N_{1r}}, \qquad N_{1r} = \frac{A}{D},$$

$$x_2 = \frac{N_2}{N_{2r}}, \qquad N_{2r} = \frac{A}{B},$$

$$\tau = \frac{t}{t_r}, \qquad t_r = \frac{1}{A},$$

$$K = \frac{C}{A}.$$

The resulting equations are as follows:

$$\frac{dx_1}{d\tau} = x_1(1 - x_2),$$

$$\frac{dx_2}{d\tau} = x_2(x_1 - K).$$

The result of human control on this biological control system is to decrease the growth rates of both prey and predators. Representing the human control effort as u, which has the physical significance of using traps and chemical spray (insecticides), then

$$\frac{dx_1}{d\tau} = x_1(1 - x_2) - ux_1,$$

$$\frac{dx_2}{d\tau} = x_2(x_1 - K) - eux_2,$$

where e is a constant proportional to the relative effectiveness of the control used on the predators as compared with that used on the prey. In order to determine an optimal control strategy, the following loss function has been proposed:

$$G = (ax_1 + u).$$

Its significance is that it assumes the two factors associated with the cost of the control program are those associated with the presence of the pests and those associated with using control—social and economic. Utilizing the Maximum Principle, determine the optimal control law for this biological system. Explain your result. Assume that $0 \leq u \leq u_{max}$.

9.21 Optimal control systems based on a quadratic performance criterion, which were discussed in Section 9.2 (see Eqs. 9.3, 9.4, and 9.5), are very frequently used in practice by the control system engineer. Consider a system with an initial displacement

$$\dot{\mathbf{x}} = \mathbf{Px}, \quad \mathbf{x}(0) = c,$$

in which it is desired to minimize the quadratic performance criterion

$$S = \int_0^\infty \mathbf{x}^T \mathbf{Q} \mathbf{x}\, dt,$$

where $\mathbf{Q}$ is a positive-definite (or positive-semidefinite) real matrix. If the eigenvalues of the companion matrix $\mathbf{P}$ have negative real parts ($\mathbf{P}$ is stable), then it can be shown that a matrix $\mathbf{A}$ exists which can be found from the following relationship that it satisfies:

$$\mathbf{P}^T\mathbf{A} + \mathbf{AP} = -\mathbf{Q}.$$

In addition, since all of the eigenvalues of **P** have real parts, then $\mathbf{x}(\infty) \to 0$ and it can also be shown that the performance criterion can be obtained in terms of the initial condition $\mathbf{x}(0)$:

$$S = \mathbf{x}^T(0)\mathbf{A}\mathbf{x}(0)$$

It is desired to adjust the damping factor, ζ, for the control system shown in Fig. P9.21 in order that the integral of the square of the error (ISE) performance criterion is minimized. Using this approach, determine the value of ζ for this second-order system in order that the following performance criterion is minimized when the control system is subjected to a unit step input:

$$S = \int_0^\infty \mathbf{e}^T\mathbf{Q}\mathbf{e}\, dt = \int_0^\infty e^2(t)\, dt,$$

where

$$\mathbf{e} = \begin{bmatrix} e_1 \\ e_2 \end{bmatrix} = \begin{bmatrix} e \\ \dot{e} \end{bmatrix}, \qquad \mathbf{Q} = \begin{bmatrix} 1 & 0 \\ 0 & 0 \end{bmatrix}.$$

Assume that the system is initially at rest. Compare your results with the value obtained using classical techniques and discussed in Section 5.6.

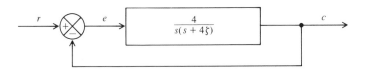

Figure P9.21

REFERENCES

1. D. McDonald, "Nonlinear techniques for improving servo performance," in *Proceedings of the National Electronics Conference*, Vol. 6, pp. 400–21 (1950).
2. C. S. Draper and V. T. Li, *Principles of optimizing control systems and an application to the internal combusion engine*, Am. Soc. Mech. Engrs., New York (September 1951).
3. G. Leitman (Ed.), *Optimization Techniques*, Academic, New York (1962).
4. L. A. Zadeh and C. A. Desoer, *Linear System Theory—The State Space Approach*, McGraw-Hill, New York (1963).
5. R. E. Kalman, *The Theory of Optimal Control and the Calculus of Variations*, RIAS Technical Report 6-13, Baltimore, Maryland (1961).
6. R. E. Kalman, "On the general theory of control systems," in *Proceedings of the First International Congress of Automatic Control*, Moscow (1960).
7. R. E. Kalman, Y. C. Ho, and K. S. Navendra, "Controllability of linear dynamical systems," *Contribution to Differential Equations* **1**, 189–213 (1961).
8. S. M. Shinners, *Techniques of System Engineering*, McGraw-Hill, New York (1967).
9. E. G. Gilbert, "Controllability and observability in multivariable control systems," *J. Control, Series A* **1**, Society for Industrial and Applied Mathematics, 128–51 (1963).

10. J. T. Tou, *Modern Control Theory*, McGraw-Hill, New York (1964).
11. O. I. Elgerd, *Control Systems Theory*, McGraw-Hill, New York (1967).
12. R. E. Bellman, *Dynamic Programming*, Princeton University Press, Princeton, N.J. (1959).
13. R. E. Bellman, *Applied Dynamic Programming*, Princeton University Press, Princeton, N.J. (1962).
14. P. H. Dosik, "Synthesis of optimal control systems," *Electro-Technology* **75**, 36–43 (1965).
15. V. G. Boltyanskii, R. V. Gamkrelidge, and L. J. Pontryagin," The theory of optimal processes, I. The maximum principle," *Izvestiya Akad. Nauk SSR., Ser. Mat.* **24**, 3 (1960).
16. L. J. Pontryagin, "Optimal control processes," *USP Mat. Nauk* **14**, 3 (1959).
17. E. L. Peterson, *Statistical Analysis and Optimization of Systems*, Wiley, New York (1961).
18. B. Friedland, "The structure of optimum control systems," *ASME Trans., Series E, J. of Basic Eng.* **84**, 1–11 (1962).
19. E. B. Lee, "Mathematical aspects of the synthesis of linear minimum response time controllers," *IRE Trans. Automatic Control* **AC-5**, 283–90 (1960).
20. M. Athanassiades and O. J. Smith, "Theory and design of high order bang-bang control systems," *IRE Trans. Automatic Control* **AC-6**, 125–34 (1961).
21. L. W. Neustadt, "Time-optimal synthesis with position and integral limits," *J. Math. Analysis Applications* **3**, 406–27 (1961).
22. I. Flugge-Lotz and H. Marback, "The optimal control of some attitude control systems for different performance criteria," in *Proceedings of the 1962 Joint Automatic Control Conference*, pp. 12-1-1 to 12-1-12.
23. B. Friedland, "The design of optimum controllers for linear processes with energy limitations," in *Proceedings of the 1962 Joint Automatic Control Conference*, pp. 12-4-1 to 12-4-12.
24. A. B. Pearson, "Synthesis of a minimum energy controller subject to an average power constraint," in *Proceedings of the 1962 Joint Automatic Control Conference*, pp. 19-4-1 to 19-4-6.
25. M. Athanassiades, "Optimal control for linear time invariant plants with time, fuel, and energy constraints," *Trans. AIEE* **81**, 321–25 (1962).
26. M. Athans, P. L. Falb, and R. I. Lacoss, "Time-, fuel-, and energy-optimal control of nonlinear norm-invariant systems," *IRE Trans. Automatic Control* **AC-8**, 196–202 (1963).
27. S. M. Shinners, "Optimal and adaptive control systems," *Electro-Technology* **74**, 63–80 (1964).
28. T. M. Stout, "Effects of friction in an optimum relay servomechanism," *Trans. AIEE* **72**, 329–335 (1953).
29. M. Athans, "The status of optimal control theory and applications for deterministic systems," *IEEE Trans. Automatic Control* **AC-11**, 580–96 (1966).
30. N. Wiener, *The Extrapolation, Interpolation, and Smoothing of Stationary Time Series*, MIT Technology Press, Cambridge, Mass. (1949).
31. J. S. Meditch, "On the problem of optimal thrust programming for a lunar soft landing," *IEEE Trans. Automatic Control* **AC-9**, 477–84 (1964).
32. L. W. Botsford, H. E. Rauch, and R. A. Shlesser, *IEEE Trans. Automatic Control* **AC-19**, 541–43 (1974).
33. T. L. Vincent, "Pest Management Programs via Optimal Control Theory," in *Proceedings of the 1972 Joint Automatic Control Conference*, pp. 658–63.

APPENDIX A
LAPLACE TRANSFORM TABLE

Laplace transform, $F(s)$	Time function, $f(t)$, $t \geqslant 0$
1	$\delta(t_0)$, unit impulse at $t = t_0$
$\dfrac{1}{s}$	$U(t)$, unit step function
$\dfrac{1}{s^2}$	t
$\dfrac{2}{s^3}$	t^2
$\dfrac{n!}{s^{n+1}}$	t^n
$\dfrac{1}{s+a}$	e^{-at}
$\dfrac{1}{(s+a)(s+b)}$	$\dfrac{e^{-at} - e^{-bt}}{b-a}$
$\dfrac{1}{(s+a)^n}$	$\dfrac{1}{(n-1)!} t^{n-1}e^{-at}$
$\dfrac{s+\alpha}{(s+a)(s+b)}$	$\dfrac{1}{(b-a)}[(\alpha - a)e^{-at} - (\alpha - b)e^{-bt}]$
$\dfrac{1}{(s+a)(s+b)(s+c)}$	$\dfrac{e^{-at}}{(b-a)(c-a)} + \dfrac{e^{-bt}}{(c-b)(a-b)} + \dfrac{e^{-ct}}{(a-c)(b-c)}$
$\dfrac{\omega}{s^2 + \omega^2}$	$\sin \omega t$
$\dfrac{s}{s^2 + \omega^2}$	$\cos \omega t$
$\dfrac{\omega}{(s+\omega)^2}$	$\omega t e^{-\omega t}$
$\dfrac{1}{(1+Ts)^n}$	$\dfrac{t^{n-1}e^{-t/T}}{T^n(n-1)!}$

Laplace transform, $F(s)$	Time function, $f(t)$, $t \geqslant 0$
$\dfrac{1}{s(1 + Ts)}$	$1 - e^{-t/T}$
$\dfrac{1}{s(1 + Ts)^2}$	$1 - \dfrac{t + T}{T} e^{-t/T}$
$\dfrac{\omega}{(s + a)^2 + \omega^2}$	$e^{-at} \sin \omega t$
$\dfrac{(s + a)}{(s + a)^2 + \omega^2}$	$e^{-at} \cos \omega t$
$\dfrac{\omega_n^2}{s(s^2 + 2\zeta\omega_n s + \omega_n^2)}$	$1 + \dfrac{e^{-\zeta\omega_n t}}{\sqrt{1 - \zeta^2}} \sin (\omega_n \sqrt{1 - \zeta^2}\, t - \alpha)$ where $\cos \alpha = -\zeta$
$\dfrac{\omega_n^2}{s^2(s^2 + 2\zeta\omega_n s + \omega_n^2)}$	$t - \dfrac{2\zeta}{\omega_n} + \dfrac{1}{\omega_n \sqrt{1 - \zeta^2}} e^{-\zeta\omega_n t} \sin (\omega_n \sqrt{1 - \zeta^2}\, t - \theta)$ where $\theta = 2 \tan^{-1} \dfrac{\sqrt{1 - \zeta^2}}{-\zeta}$
$\dfrac{s}{(1 + Ts)(s^2 + \omega_n^2)}$	$\dfrac{-1}{(1 + T^2\omega_n^2)} e^{-t/T} + \dfrac{1}{\sqrt{1 + T^2\omega_n^2}} \cos (\omega_n t - \theta)$ where $\theta = \tan^{-1} \omega_n T$
$\dfrac{s}{(s^2 + \omega_n^2)^2}$	$\dfrac{1}{2\omega_n} t \sin \omega_n t$
$\dfrac{1}{(s + b)[(s + a)^2 + \omega^2]}$	$\dfrac{e^{-bt}}{(b - a)^2 + \omega^2} + \dfrac{e^{-at} \sin (\omega t - \theta)}{\omega[(b - a)^2 + \omega^2]^{1/2}}$ where $\theta = \tan^{-1} \dfrac{\omega}{b - a}$
$\dfrac{2abs}{[s^2 + (a + b)^2][s^2 + (a - b)^2]}$	$\sin at \sin bt$
$\dfrac{1 + as + bs^2}{s^2(1 + T_1 s)(1 + T_2 s)}$	$t + (a - T_1 - T_2) + \dfrac{b - aT_1 + T_1^2}{T_1 - T_2} e^{-t/T_1}$ $- \dfrac{b - aT_2 + T_2^2}{T_1 - T_2} e^{-t/T_2}$

APPENDIX B
PROOF OF THE NYQUIST STABILITY CRITERION

The Nyquist stability criterion can be derived from Cauchy's residue theorem, which states that

$$\frac{1}{2\pi j} \int_C g(s)\, ds = \sum \text{residues of } g(s) \text{ at the poles enclosed by}$$

$$\text{the closed contour } C. \tag{B.1}$$

Let us replace $g(s)$ by $f'(s)/f(s)$ where $f(s)$ is a function of s which is single valued on and within the closed contour C and analytic on C. Observe that the singularities of $f'(s)/f(s)$ occur only at the zeros and poles of $f(s)$. The residue may be found at each singularity with multiplicity of the order of zeros and poles taken into account. The residues in the zeros of $f(s)$ are positive and the residues in the poles of $f(s)$ are negative. Therefore, if $f(s)$ is not equal to zero along C, and if there are at most a finite number of singular points that are all poles within the contour C, then

$$\frac{1}{2\pi j} \int_C \frac{f'(s)}{f(s)}\, ds = Z - P, \tag{B.2}$$

where

Z = number of zeros of $f(s)$ within C, with due regard for their multiplicity of order,
P = number of poles of $f(s)$ within C, with due regard for their multiplicity of order.
The left-hand side of Eq. (B.2) may be written as

$$\frac{1}{2\pi j} \int_C \frac{f'(s)}{f(s)}\, ds = \frac{1}{2\pi j} \int_C d[\ln f(s)]. \tag{B.3}$$

In general, $f(s)$ will have both real and imaginary parts along the contour C. Therefore, its logarithm can be rewritten as

$$\ln f(s) = \ln |f(s)| + j\underline{/f(s)}. \tag{B.4}$$

If we assume that $f(s)$ is not zero anywhere on the contour C, the integration of Eq. (B.3) results in the expression

$$\frac{1}{2\pi j} \int_C d[\ln f(s)] = \frac{1}{2\pi j} [\ln |f(s)| + j\underline{/f(s)}]_{s_1}^{s_2}, \tag{B.5}$$

where s_1 and s_2 denote the arbitrary beginning and end of the closed contour C as it is followed. Since $|f(s)|$ returns to its initial value in completing the closed curve,

$$\frac{1}{2\pi j} \int_C d[\ln f(s)] = \frac{1}{2\pi} [\underline{/f(s_2)} - \underline{/f(s_1)}].$$ (B.6)

Therefore, Eq. (B.6) can be rewritten as

$$\frac{1}{2\pi j} \int_C \frac{f'(s)}{f(s)} ds = \frac{1}{2\pi} \times \text{ net change in angle of } f(s) \text{ as } s \text{ is}$$

varied over the contour C.   (B.7)

By equating Eqs. (B.2) and (B.7), we obtain

$$Z - P = \frac{1}{2\pi} \times \text{ net change in angle of } f(s) \text{ as } s \text{ is varied}$$

over the contour C.   (B.8)

Equation (B.8) states that the excess of zeros over poles of $f(s)$ within the contour C equals $1/2\pi$ times the net change in angle (i.e., equals the number of net encirclements of the origin) of $f(s)$ as s is varied over the contour C. Let N be this number. Then Eq. (B.8) can be written as

$$Z - P = N,$$ (B.9)

where the contour C is traversed in a clockwise direction, and where an encirclement is defined as being positive if it also is in a clockwise direction. The number Z must be zero for the system to be stable.

This relationship, which is known as Cauchy's principle of the argument, is the basis of Nyquist's stability criterion. To make use of this principle in applying Nyquist's stability criterion, let us consider the feedback control system of Fig. B.1. Let

$$f(s) = 1 + G(s)H(s)$$

and examine the number of times $f(s)$ encircles the origin as s traverses the contour of Fig. B.2. (It is assumed that there are no poles on the imaginary axis, except at the origin.) Observe that the origin of $f(s)$ corresponds to $G(s)H(s) = -1$. Therefore, if $G(s)H(s)$ is plotted for the contour of Fig. (B.2), the number of times that $G(s)H(s)$ encircles the point $-1 + j0$ equals the number of zeros minus the number of poles of $1 + G(s)H(s)$ for s in the right half-plane.

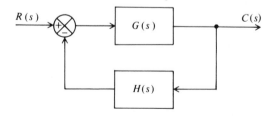

Figure B.1

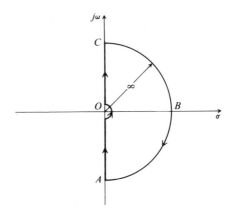

Figure B.2

Let us examine the contour of Fig. B.2 in detail. Since

$$\lim_{s \to \infty} |G(s)H(s)| = 0 \qquad (B.10)$$

for all physical systems, the points along the infinite arc ABC contract to the origin $\omega = 0$. Along the small semicircle about the origin, let

$$s = re^{j\theta} = \sigma + j\omega. \qquad (B.11)$$

Since $G(s)H(s)$ is usually a rational function with a denominator of higher power than that of the numerator, then

$$G(s)H(s) = \lim_{r \to 0} G(re^{j\theta})H(re^{j\theta}) \qquad (B.12)$$

will map into a segment of an infinite circle. Along the imaginary axis, $s = j\omega$, so that we are concerned with $G(j\omega)H(j\omega)$. Since $G(-j\omega)H(-j\omega)$ is a conjugate of $G(j\omega)H(j\omega)$, the two functions are symmetric about the real axis and $G(-j\omega)H(-j\omega)$ is a reflection of $G(j\omega)H(j\omega)$ about the real axis. Therefore, it is only necessary to plot $G(j\omega)H(j\omega)$ from $\omega = 0$ to ∞.

Therefore, N can be determined by plotting $G(j\omega)H(j\omega)$ and observing the number of encirclements of the $-1 + j0$ point; P can be determined by inspection of the $G(s)H(s)$ expression; and Z can be found by using Eq. (B.9).

ANSWERS TO SELECTED PROBLEMS

CHAPTER 1

1.1 An electrical signal proportional to the difference between the desired heading (gyroscopic setting) and original heading is amplified by the power amplifier. The amplified signal drives the motor which turns the rudder until the desired heading and actual heading of the ship are in agreement, and the corresponding electrical signal that is proportional to the difference between these two headings is zero.

1.2 The rudder's positioning can be made into a closed-loop control system by fastening a resistor in a similar manner as the resistor that is fixed to the ship's frame. An electrical feedback signal can then be obtained of the actual position, which can be appropriately compared with that of the desired, or reference, position.

1.3 If the reference temperature of the thermostat is changed, the reference input to the control system changes, and an electrical error signal results. The electric hot-water heater will then change the temperature of the water until the difference between the reference input and actual temperatures is zero.

1.4 A change in the ambient temperature surrounding the tank manifests itself as a disturbance input within the heating control system. The explanation of the system's resulting control action is similar to that discussed in the book for a disturbance occurring in an automatic speed control system (see Fig. 1.9).

1.6 The speed of an internal combustion engine can be varied by adjusting the spark setting and fuel–air mixture. By utilizing a tachometer to feed back an electrical signal proportional to the internal combustion engine's speed and comparing it with a reference voltage that is proportional to the desired speed, the spark setting and fuel–air mixture can be theoretically adjusted to control the speed of the engine.

1.7 a) The basic system would require that the elevator's actuating signal be proportional to position, velocity, and acceleration. Theoretically, this could be obtained by feeding back electrical signals proportional to position, velocity, and acceleration, and comparing these signals with reference signals representing the desired values. In practice, this can be simplified by placing integrators properly within the control system in order that only electrical signals proportional to position, and perhaps velocity, need be sensed.

b) Specifications must be placed on the velocity and acceleration capabilities in order that they be limited to safe values from the passenger's viewpoint.

1.8 Assuming that safety brakes are not utilized in the feedback control system devised, the man entering the elevator acts as a disturbance force in the feedback control system. The functioning of the control system's resulting action is similar to that discussed in Chapter 1 for a disturbance torque occurring in an automatic speed control system (see Fig. 1.9).

CHAPTER 2

2.2 a) $I(s)\left[LS + R + \dfrac{1}{Cs}\right] - Li(0^+) + \dfrac{i^{(-)}(0^+)}{Cs} = E(s),$

where $i^{(-)}(0^+) = \lim\limits_{t \to 0^+} \displaystyle\int_0^t i(t)\,dt$

b) $X(s)[Ms^2 + Bs + K] - M\left[sx(0^+) + \dfrac{dx(t)}{dt}(0^+)\right] - Bx(0^+) = \dfrac{3}{s^2}$

c) $\theta(s)[Js^2 + Bs + K] - J\left[s\theta(0^+) + \dfrac{d\theta(t)}{dt}(0^+)\right] - B\theta(0^+) = \dfrac{10\omega}{s^2 + \omega^2}$

2.3 a) $f_A(t) = -5te^{-2t} - 10e^{-2t} - 0.0315e^{-10.48t} + 10.0315e^{-1.52t}$
b) $f_B(t) = -\frac{5}{6}e^{-4t} + \frac{3}{2}e^{4t} - \frac{2}{3}e^{-t}$
c) $f_C(t) = 0.5 - 0.25e^{-7.47t} - 0.250e^{-0.53t}$

2.7 $\dfrac{C(s)}{R(s)} = \dfrac{G_1(s)G_2(s)G_3(s)G_4(s)}{[1 + G_2(s)G_3(s)H_3(s) + G_3(s)G_4(s)H_4(s) + G_1(s)G_2(s)G_3(s)H_2(s) \\ -G_1(s)G_2(s)G_3(s)G_4(s)H_1(s)]}$

2.8 $\dfrac{C(s)}{R(s)} = \dfrac{G_1(s)G_2(s)G_3(s)G_4(s)}{[1 - G_2(s)G_3(s)H_3(s) + G_3(s)G_4(s)H_4(s) + G_1(s)G_2(s)G_3(s)H_2(s) \\ -G_1(s)G_2(s)G_3(s)G_4(s)H_1(s)]}$

2.9 $\dfrac{E(s)}{R(s)} = \dfrac{1 + G_2(s)G_3(s)H_3(s) + G_3(s)G_4(s)H_4(s)}{[1 + G_2(s)G_3(s)H_3(s) + G_3(s)G_4(s)H_4(s) + G_1(s)G_2(s)G_3(s)H_2(s) \\ -G_1(s)G_2(s)G_3(s)G_4(s)H_1(s)]}$

2.10 $\dfrac{E(s)}{R(s)} = \dfrac{1 - G_2(s)G_3(s)H_3(s) + G_3(s)G_4(s)H_4(s)}{[1 - G_2(s)G_3(s)H_3(s) + G_3(s)G_4(s)H_4(s) + G_1(s)G_2(s)G_3(s)H_2(s) \\ -G_1(s)G_2(s)G_3(s)G_4(s)H_1(s)]}$

2.11 $\dfrac{C(s)}{R(s)} = \dfrac{G_1(s)G_2(s)G_3(s)G_4(s)}{[1 + G_2(s)G_3(s)H_3(s) + G_3(s)G_4(s)H_4(s) + G_1(s)G_2(s)G_3(s)H_2(s) \\ -G_1(s)G_2(s)G_3(s)G_4(s)H_1(s)]}$

2.12 $\dfrac{C(s)}{R(s)} = \dfrac{G_1(s)G_2(s)G_3(s)G_4(s)}{[1 - G_2(s)G_3(s)H_3(s) + G_3(s)G_4(s)H_4(s) + G_1(s)G_2(s)G_3(s)H_2(s) \\ -G_1(s)G_2(s)G_3(s)G_4(s)H_1(s)]}$

2.13 $\dfrac{E(s)}{R(s)} = \dfrac{1 + G_2(s)G_3(s)H_3(s) + G_3(s)G_4(s)H_4(s)}{[1 + G_2(s)G_3(s)H_3(s) + G_3(s)G_4(s)H_4(s) + G_1(s)G_2(s)G_3(s)H_2(s) \\ -G_1(s)G_2(s)G_3(s)G_4(s)H_1(s)]}$

2.14 $\dfrac{E(s)}{R(s)} = \dfrac{1 - G_2(s)G_3(s)H_3(s) + G_3(s)G_4(s)H_4(s)}{[1 - G_2(s)G_3(s)H_3(s) + G_3(s)G_4(s)H_4(s) + G_1(s)G_2(s)G_3(s)H_2(s) \\ -G_1(s)G_2(s)G_3(s)G_4(s)H_1(s)]}$

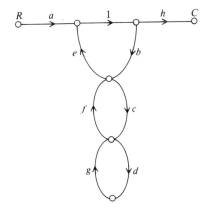

Figure A2.24

2.24 One possible solution to this problem is shown in Fig. A2.24.

2.27 a) Defining $x_1 = c$ and $x_2 = \dot{c}$, the plant dynamics become

$$\dot{x}_1 = x_2, \qquad \dot{x}_2 = -2x_2 - x_1,$$

or in vector form

$$\dot{\mathbf{x}} = \mathbf{P}\mathbf{x} + \mathbf{B}\mathbf{u}, \qquad c = \mathbf{L}\mathbf{x},$$

where

$$\mathbf{x} = \begin{bmatrix} x_1 \\ x_2 \end{bmatrix}, \qquad \dot{\mathbf{x}} = \begin{bmatrix} \dot{x}_1 \\ \dot{x}_2 \end{bmatrix}, \qquad \mathbf{P} = \begin{bmatrix} 0 & 1 \\ -1 & -2 \end{bmatrix}, \qquad \mathbf{B} = \begin{bmatrix} 0 & 0 \\ 0 & 0 \end{bmatrix}, \qquad \mathbf{u} = \begin{bmatrix} 0 \\ 0 \end{bmatrix},$$

$$\mathbf{L} = [1 \quad 0].$$

b) With $x_1 = c$ and $x_2 = \dot{c}$, the plant dynamics become

$$\dot{x}_1 = x_2, \qquad \dot{x}_2 = -2x_2 - x_1 + A,$$

or in vector form

$$\dot{\mathbf{x}} = \mathbf{P}\mathbf{x} + \mathbf{B}\mathbf{u}, \qquad c = \mathbf{L}\mathbf{x},$$

where

$$\mathbf{x} = \begin{bmatrix} x_1 \\ x_2 \end{bmatrix}, \qquad \dot{\mathbf{x}} = \begin{bmatrix} \dot{x}_1 \\ \dot{x}_2 \end{bmatrix}, \qquad \mathbf{P} = \begin{bmatrix} 0 & 1 \\ -1 & -2 \end{bmatrix}, \qquad \mathbf{B} = \begin{bmatrix} 0 & 0 \\ 0 & 1 \end{bmatrix}, \qquad \mathbf{u} = \begin{bmatrix} 0 \\ A \end{bmatrix},$$

$$\mathbf{L} = [1 \quad 0].$$

c) With $x_1 = c$, $x_2 = \dot{c}$, and $x_3 = \ddot{c}$, the plant dynamics become

$$\dot{x}_1 = x_2,$$
$$\dot{x}_2 = x_3,$$
$$\dot{x}_3 = -2x_1 - 2x_2 - 3x_3,$$

or in vector form

$$\dot{\mathbf{x}} = \mathbf{P}\mathbf{x} + \mathbf{B}\mathbf{u}, \qquad c = \mathbf{L}\mathbf{x},$$

where

$$\mathbf{x} = \begin{bmatrix} x_1 \\ x_2 \\ x_3 \end{bmatrix}, \qquad \dot{\mathbf{x}} = \begin{bmatrix} \dot{x}_1 \\ \dot{x}_2 \\ \dot{x}_3 \end{bmatrix}, \qquad \mathbf{P} = \begin{bmatrix} 0 & 1 & 0 \\ 0 & 0 & 1 \\ -2 & -2 & -3 \end{bmatrix},$$

$$\mathbf{B} = \begin{bmatrix} 0 & 0 & 0 \\ 0 & 0 & 0 \\ 0 & 0 & 0 \end{bmatrix}, \qquad \mathbf{u} = \begin{bmatrix} 0 \\ 0 \\ 0 \end{bmatrix}, \qquad \mathbf{L} = [1 \quad 0 \quad 0].$$

d) With $x_1 = c$, $x_2 = \dot{c}$, and $x_3 = \ddot{c}$, the plant dynamics become

$$\dot{x}_1 = x_2,$$
$$\dot{x}_2 = x_3,$$
$$\dot{x}_3 = -2x_1 - 2x_2 - 3x_3 + A,$$

or in vector form

$$\dot{\mathbf{x}} = \mathbf{Px} + \mathbf{Bu}, \qquad c = \mathbf{Lx},$$

where

$$\mathbf{x} = \begin{bmatrix} x_1 \\ x_2 \\ x_3 \end{bmatrix}, \qquad \dot{\mathbf{x}} = \begin{bmatrix} \dot{x}_1 \\ \dot{x}_2 \\ \dot{x}_3 \end{bmatrix}, \qquad \mathbf{P} = \begin{bmatrix} 0 & 1 & 0 \\ 0 & 0 & 1 \\ -2 & -2 & -3 \end{bmatrix},$$

$$\mathbf{B} = \begin{bmatrix} 0 & 0 & 0 \\ 0 & 0 & 0 \\ 0 & 0 & 1 \end{bmatrix}, \qquad \mathbf{u} = \begin{bmatrix} 0 \\ 0 \\ A \end{bmatrix}, \qquad \mathbf{L} = [1 \quad 0 \quad 0].$$

2.28 With $x_1 = \theta_1$, $x_2 = \dot{\theta}_1$, $x_3 = \theta_2$, $x_4 = \dot{\theta}_2$, $x_5 = \theta_3$, and $x_6 = \dot{\theta}_3$, the plant dynamics become

$$\dot{x}_1 = x_2,$$
$$\dot{x}_2 = -\omega_0 x_6 + L_1/I,$$
$$\dot{x}_3 = x_4,$$
$$\dot{x}_4 = L_2/I,$$
$$\dot{x}_5 = x_6,$$
$$\dot{x}_6 = \omega_0 x_2 + L_3/I,$$

or in vector form

$$\dot{\mathbf{x}} = \mathbf{Px} + \mathbf{Bu},$$

where

$$\mathbf{P} = \begin{bmatrix} 0 & 1 & 0 & 0 & 0 & 0 \\ 0 & 0 & 0 & 0 & 0 & -\omega_0 \\ 0 & 0 & 0 & 1 & 0 & 0 \\ 0 & 0 & 0 & 0 & 0 & 0 \\ 0 & 0 & 0 & 0 & 0 & 1 \\ 0 & \omega_0 & 0 & 0 & 0 & 0 \end{bmatrix},$$

$$\mathbf{B} = \begin{bmatrix} 0 & 0 & 0 & 0 & 0 & 0 \\ 0 & 1/I & 0 & 0 & 0 & 0 \\ 0 & 0 & 0 & 0 & 0 & 0 \\ 0 & 0 & 0 & 1/I & 0 & 0 \\ 0 & 0 & 0 & 0 & 0 & 0 \\ 0 & 0 & 0 & 0 & 0 & 1/I \end{bmatrix}, \qquad \mathbf{u} = \begin{bmatrix} 0 \\ L_1 \\ 0 \\ L_2 \\ 0 \\ L_3 \end{bmatrix}.$$

2.30 With $x_1 = v$ and $x_2 = \dot{v}$, the plant dynamics become

$$\dot{x}_1 = x_2,$$
$$\dot{x}_2 = -x_1 + u(1 - x_1^2)x_2.$$

2.32 The differential equation of the system is given by

$$\ddot{c} + 4\dot{c} + c = r(t).$$

With $x_1 = c$ and $x_2 = \dot{c}$, the plant dynamics become

$$\dot{x}_1 = x_2, \qquad \dot{x}_2 = -x_1 - 4x_2 + r(t),$$

or in vector form

$$\dot{\mathbf{x}} = \mathbf{Px} + \mathbf{B}r(t)$$

where

$$\mathbf{P} = \begin{bmatrix} 0 & 1 \\ -1 & -4 \end{bmatrix}, \quad \mathbf{B} = \begin{bmatrix} 0 \\ 1 \end{bmatrix}, \quad \mathbf{x} = \begin{bmatrix} x_1 \\ x_2 \end{bmatrix}, \quad \dot{\mathbf{x}} = \begin{bmatrix} \dot{x}_1 \\ \dot{x}_2 \end{bmatrix}.$$

2.43

$$\Phi(t) = \begin{bmatrix} 1 & 7 + 0.00305e^{-6.86t} - 7.3e^{-0.14t} & 1 + 0.0217e^{-6.86t} - 1.065e^{-0.14t} \\ 0 & -0.0209e^{-6.86t} + 1.02e^{-0.14t} & -0.149e^{-6.86t} + 0.149e^{-0.14t} \\ 0 & 0.149e^{-6.86t} - 0.149e^{-0.14t} & 1.02e^{-6.86t} - 0.0208e^{-0.14t} \end{bmatrix}.$$

CHAPTER 3

3.1 a) $f(t) = \dfrac{B_1 M_2}{K_2 + K_3}\dfrac{d^3 y(t)}{dt^3} + \left[M_2 + \dfrac{M_2 K_1}{K_2 + K_3} \right]\dfrac{d^2 y(t)}{dt^2} + B_1\dfrac{dy(t)}{dt} + K_1 y(t)$

b) $\dfrac{Y(s)}{F(s)} = \dfrac{1}{\dfrac{B_1 M_2}{K_2 + K_3}s^3 + \left[M + \dfrac{MK_1}{K_2 + K_3} \right]s^2 + B_1 s + K_1}.$

3.2 b) $\dfrac{Y(s)}{F(s)} = \dfrac{(B_1 s + K_1)(B_2 s + K_2)(B_3 s + K_3)}{[ABCD - AB(B_3 s + K_3)^2 - AD(B_2 s + K_2)^2 - CD(B_1 s + K_1)^2}$
$$+ (B_1 s + K_1)^2(B_3 s + K_3)^2]$$

where

$$A = M_1 s^2 + B_1 s + K_1,$$
$$B = M_2 s^2 + (B_1 + B_2)s + (K_1 + K_2),$$
$$C = M_3 s^2 + (B_2 + B_3)s + (K_2 + K_3),$$
$$D = M_4 s^2 + B_3 s + K_3$$

3.3 b) $\dfrac{\theta(s)}{T(s)} = \dfrac{(K_2 + B_2 s)^2 - X(s)Y(s)}{W(s)[(K_2 + B_2)^2 - X(s)Y(s)] + Y(s)(K_1 + B_1 s)^2},$

where

$$W(s) = J_1 s^2 + B_1 s + K_1,$$
$$X(s) = J_2 s^2 + (B_1 + B_2)s + (K_1 + K_2),$$
$$Y(s) = J_3 s^2 + (B_2 + B_3)s + (K_2 + K_3).$$

3.4 a) $\dfrac{d^2\theta_0(t)}{dt^2} + \dfrac{B}{J}\dfrac{d\theta_0(t)}{dt} + \dfrac{K}{J}\theta_0(t) = \dfrac{K'}{J}\dfrac{d\theta_i(t)}{dt}$

b) $\dfrac{\theta_0(s)}{\theta_i(s)} = \dfrac{(K'/J)s}{s^2 + (B/J)s + K/J}.$

3.5 a) $\dfrac{\theta_0(s)}{E_f(s)} = \dfrac{9.43}{s(1.11s + 1)(0.227s + 1)(0.028s + 1)}$

3.7 Defining the following terms

$$T_f = L_f/R_f, \qquad\qquad \gamma'' = B(R_g + R_m + R)R_f/K_T K_e,$$

$$T_a'' = (L_g + L_m)/(R_g + R_m + R), \qquad \gamma''' = BK_g R/K_T K_e,$$

$$T_0 = JK_g R/K_T K_e, \qquad\qquad T_m'' = J(R_g + R_m + R)R_f/K_T K_e,$$

a) $T_m'' T_a'' T_f \dfrac{d^4\theta_0(t)}{dt^4} + [T_m''(T_a'' + T_f) + \gamma'' T_a'' T_f]\dfrac{d^3\theta_0(t)}{dt^3}$

$$+ [T_m'' + \gamma''(T_a'' + T_f) + T_0 + L_f]\dfrac{d^2\theta_0(t)}{dt^2} + [\gamma'' + \gamma''' + R_f]\dfrac{d\theta_0(t)}{dt} = \dfrac{K_g}{K_e}e_f(t)$$

b) $\dfrac{\theta_0(s)}{E_f(s)} = \dfrac{K_g/K_e}{\{s[T_m'' T_a'' T_f s^3 + [T_m''(T_a'' + T_f) + \gamma'' T_a'' T_f]s^2}$
$${+ [T_m'' + \gamma''(T_a'' + T_f) + T_0 + L_f]s + \gamma'' + \gamma''' + R_f]\}}$$

3.8 b) $\dfrac{e_b(s)}{E_a(s)} = \dfrac{\dfrac{R_2}{R_1 + R_2}\left[-T_a T_m s^2 - (T_m' + \gamma T_a)s + \dfrac{(R_1 + R_2)}{R_2} - \gamma'\right]}{T_a T_m s^2 + (T_m'' + \gamma T_a)s + (\gamma'' + 1)}$

where

$$T_a = L_{CF}/R_{CF}, \qquad T_m = JR_{CF}/K_e K_T, \qquad \gamma = \dfrac{R_{CF}B}{K_e K_T},$$

$$T_m' = \dfrac{J[R_{CF} - (R_1/R_2)R_{AC}]}{K_e K_T}, \qquad T_m'' = \dfrac{J[R_{CF} + R_{AC}]}{K_e K_T},$$

$$\gamma' = \dfrac{B[R_{CF} - (R_1/R_2)R_{AC}]}{K_e K_T}, \qquad \gamma'' = \dfrac{B(R_{CF} + R_{AC})}{K_e K_T}.$$

3.11 $\dfrac{C(s)}{R(s)} = \dfrac{(1/V_m)(\partial Q/\partial r)}{[Vs/(K_B V_m^2)](Ms^2 + Bs + K) + (1/V_m^2)(L - \partial Q/\partial P)(Ms^2 + Bs + K) + s}.$

3.13 $\dfrac{\theta_0(s)}{E_c(s)} = \dfrac{2.29}{s(0.0102s + 1)}.$

3.14 $\dfrac{\theta_0(s)}{E_c(s)} = \dfrac{K_m'}{s(T_m's + 1)},$

where

$$K_m' = \dfrac{K_e'}{B - K_u'}, \qquad K_e' = NK_e,$$

$$T_m' = \dfrac{J_{total}}{B - K_n'}, \qquad K_n' = N^2 K_n,\ J_{total} = J_{motor}N^2 + J_2.$$

3.15 $\dfrac{\theta_0(s)}{E_c(s)} = \dfrac{0.0636}{s(0.0102s + 1)}.$

CHAPTER 4

4.1 a) $\omega_n = 47.3$ rad/sec, $\zeta = 1.028$

b) percent overshoot $= 0$, time to peak $= \infty$

c) The error as a function of time can be plotted from the following expression:

$$e(t) = 2.62e^{-37.2t} - 1.63e^{-60.2t}.$$

4.6 a) $\dfrac{C(s)}{R(s)} = \dfrac{K_A q}{s^2 + \sqrt{q}\,K_R s + qK_A}$

b) $\omega_n = 8$ rad/sec, $\zeta = 0.5$

c) maximum percent overshoot $= 16.4\%$, time to peak $= 0.452$ sec.

CHAPTER 5

5.1 a) $S_{K_1}^T = 1$

b) $S_{K_2}^T = -\dfrac{1}{10^{-3}s^2 + 10^{-3}s + 1}$

c) $S_G^T = \dfrac{s^2 + s}{s^2 + s + 1000}$.

5.2 a) $S_{G_1}^T = \dfrac{s^3 + 182s^2 + 8200s}{s^3 + 182s^2 + 8200s + 2000}$

b) $S_{G_2}^T = \dfrac{s^3 + 102s^2 + 200s}{s^3 + 182s^2 + 8200s + 2000}$

c) $S_H^T = \dfrac{-(80s^2 + 8000s)}{s^3 + 182s^2 + 8200s + 2000}$

d) $S_{K_3}^T = 1$

e) In the vicinity of $\omega = 1$: $S_{G_2}^T$, S_H^T, $S_{G_1}^T$, $S_{K_3}^T$.

5.6 a) 4, b) infinity.

5.8 a) zero b) 2.4 c) infinity.

5.11 a) On the assumption that $J_L \gg J_m N^2$,

$$G(s) = \dfrac{C(s)}{E(s)} = \dfrac{2.1}{s(0.11s + 1)}$$

b) $\omega_n = 4.37$ rad/sec; $\zeta = 1.03$

c) 0%; ∞ sec d) zero e) 0.476 f) infinity.

5.17 $K_1 = 0.112$, $\beta_1 = 0.77$, $\beta_2 = 4.36$.

CHAPTER 6

6.1 a) $\dfrac{Y(j\omega)}{X(j\omega)} = \dfrac{1}{A(j\omega)^2 + B(j\omega) + C}$.

6.3 a) stable   b) unstable with 2 poles in right half-plane   c) stable   d) stable.

6.8 $C + 0.3 - A > 0, \quad T - B + D > 0$.

6.9 a) unstable with 2 poles in right half-plane
   b) unstable with 2 poles in right half-plane
   c) unstable with 2 poles in right half-plane
   d) unstable with 2 poles in right half-plane

6.12 a) unstable with 2 poles in right half-plane
   b) unstable with 2 poles in right half-plane
   c) stable.

6.14 a) Pertinent characteristics of the Bode diagram are as follows:
   1. Crossover frequency = 6.3 rad/sec
   2. Phase margin = $-13°$
   3. Gain margin = -5.3 db.
   c) Pertinent characteristics of the Bode diagram are as follows:
   1. Crossover frequency = 2.7 rad/sec
   2. Phase margin = $-26°$
   3. Gain margin = $-\infty$.

6.15 a) phase margin = $42°$, gain margin = 14 db
   b) phase margin = $43°$, gain margin = 14 db
   c) The difference in phase margin obtained in (a) and (b) is due to the straight-line asymptotic approximation used to find it in (b)
   d) $K_p = \infty, K_v = 2, K_a = 0$
   e) 2.5 rad.

6.17 b) crossover frequency = 3.8 rad/sec, phase margin = $-10°$, system is unstable
   c) $K_p = \infty, K_v = 10, K_a = 0$
   d) 4 rad.

6.19 b) $K_a = 45$ c) System is unstable.

6.22 b) $K = 9.9$
   c) crossover frequency = 15 rad/sec, phase margin = $-32°$.

6.24 c) $\omega_p = 3.9$ rad/sec, $M_p = 7.0$ db   d) No.

6.25 c) $\omega_p = 2.5$ rad/sec, $M_p = 1.25$ db   d) No.

6.26 a) The Bode diagram indicates a crossover frequency of 23.5 rad/sec and a phase margin of $21°$
   c) $\omega_p = 23$ rad/sec, $M_p = 9$ db
   d) No.

6.27 a) $K = 1.91$ b) System is unstable.

6.28 b) $M_p = 1.8, \omega_p = 7.2$ rad/sec.

6.29 b) System is unstable.

6.30 a) Pertinent characteristics of the root locus are as follows:
1. Root locus occurs along the real axis between the origin and −1, and between −10 and −∞.
2. The asymptotes intersect the real axis at −3.67 with angles of ±60°.
3. The point of breakaway of the root locus from the real axis occurs at −0.49.
b) $K = 11$.

6.31 a) The root locus lies along the imaginary axis.
b) Pertinent characteristics of the root locus are as follows:
1. Root locus occurs along the real axis at the origin and between −1 and −10.
2. The asymptotes intersect the real axis at −4.5 with angles of ±90°.
3. The points of breakaway of the root locus from the real axis occur at 0, −2.5, and −4.
4. The root locus indicates that the system is stable.
c) Pertinent characteristics of the root locus are as follows:
1. Root locus occurs along the real axis between the origin and −1, and at −4.
2. The asymptotes intersect the real axis at −3.5 with angles of ±90°.
3. The root locus indicates that the system is stable for all values of gain.
d) Pertinent characteristics of the root locus are as follows:
1. Root locus occurs along the real axis at the origin, −0.1, and between −4 and −5.
2. The asymptotes intersect the real axis at −4.4 with angles of ±90°.
3. The points of breakaway of the root locus from the real axis occur at 0, −0.1, and −4.5.
4. The root locus indicates that the system is stable.

6.32 Pertinent characteristics of the root locus are as follows:
1. Root locus occurs along the real axis at the origin, between −0.01 and −0.1, and between −1.67 and −∞.
2. The point of breakaway of the root locus from the real axis occurs at 0 and −3.4.
3. The root locus crosses the imaginary axis at ±$j0.386$.
4. The root locus indicates that the system is stable for $0.14 < K < ∞$.

CHAPTER 7

7.1 a) The electrical network illustrated in Table 2.2, item 3, with

$$R_1 = 690 \text{ k}\Omega, \qquad R_2 = 51 \text{ k}\Omega, \qquad C_1 = 1.38 \text{ }\mu\text{F}$$

and cascaded with an amplifier whose gain is 4.6 will satisfy the specifications.
b) The electrical network illustrated in Table 2.2, item 4, with

$$R_1 = 690 \text{ k}\Omega, \qquad R_2 = 51 \text{ k}\Omega, \qquad C_1 = 1.38 \text{ }\mu\text{F}$$

and cascaded with an amplifier whose gain is 4.6 will satisfy the specifications.
c) The electrical network illustrated in Table 2.2, item 5, with

$$R_1 = 595 \text{ k}\Omega, \qquad R_2 = 25 \text{ k}\Omega, \qquad C_1 = 1.68 \text{ }\mu\text{F}, \qquad C_2 = 4 \text{ }\mu\text{F}$$

will satisfy the specifications.

7.2 a) $\zeta = 0.5$; $\omega_n = 4$ rad/sec; maximum percent overshoot = 16.4%; steady-state error = 0.25

b) $b = 0.15$

c) maximum percent overshoot = 1.5%; steady-state error = 0.4

d) Add a high-pass filter in cascade with the tachometer as illustrated in Fig. 7.10.

7.5 a) $G_c(s) = 1 + 0.382s$ b) $b = 0.382$

7.7 a) $G_c(s) = \dfrac{0.5s + 1}{0.014s + 1} \dfrac{0.05s + 1}{0.01s + 1}$; the attenuation of 1/178.6 due to the two lead net-

works is assumed to be compensated for by increasing the gain of the amplifier by 178.6.

b) $G_c(s) = \dfrac{12.5s + 1}{250s + 1}$.

7.10 A lead network whose transfer function is

$$\frac{1}{15} \frac{1 + s}{1 + 0.067s}$$

results in a phase margin of approximately 45°, assuming that its attenuation of $\frac{1}{15}$ is compensated for by boosting the gain of the amplifier by a factor of 15.

7.11 Same answer as for Problem 7.10.

7.13 Two cascaded lead networks are required to meet these specifications. Their transfer function, together with the gain found in Problem 6.22, is given by

$$G_1(s) = 9.9 \frac{1 + 0.2s}{1 + 0.01s} \frac{1 + 0.1s}{1 + 0.005s}.$$

The constant attenuation of 400 due to the characteristics of these two lead networks must be made up by an amplification of 400.

7.17 a) $G_c(s) = \dfrac{1 + 0.416s}{1 + 0.384s}$

The constant attenuation of 1.08 due to the characteristics of this lead network must be made up by an amplification of 1.08.

b) $G_c(s) = \dfrac{1 + 100s}{1 + 10,000s}$

c) For part (a), $\omega_p = 2.5$ rad/sec

For part (b), $\omega_p = 0.025$ rad/sec

d) The phase-lag network compensated system will be less susceptible to noise problems, as well as requiring less power and cost. However, the phase-lag network compensated system has a relatively very small bandwidth which may be unacceptable in some applications.

7.18 a) $G_c(s) = \dfrac{20s + 1}{1000s + 1}$

b) $G_c(s) = \dfrac{0.5s + 1}{0.001s + 1} \dfrac{0.05s + 1}{0.001s + 1}$; the attenuation due to the lead network is

assumed to be compensated for by increasing the gain of the amplifiers

c) For the phase-lag network case, $M_p = 1.2$, $\omega_p = 0.25$ rad/sec
For the phase-lead network case, $M_p = 1.2$, $\omega_p = 7$ rad/sec

d) The phase-lag compensated system with its narrower bandwidth is desirable from noise considerations, as well as requiring less power and cost, if the reduced gain at low frequencies is acceptable.

7.19 a) $b = 0.25$ b) $e_{ss} = 0.3024$

c) The transfer function of the rate feedback path is given by

$$\frac{5s}{1 + 5s} 0.25s$$

d) $e_{ss} = 0.0524$.

7.20 a) The lag network is given by

$$\frac{s + 0.0357}{s + 0.001}$$

b) The lead network is given by

$$\frac{s + 1}{s + 100}$$

7.22 a) The lag network is given by

$$\frac{s + 0.0649}{s + 0.01}$$

b) The lead network is given by

$$\frac{s + 8}{s + 48}$$

c) For part (a)

$$K_p = \infty, \quad K_v = \infty, \quad K_a = 15,$$
$$\omega_p \approx 3.58 \text{ rad/sec}, \quad M_p \approx 1.0285$$

For part (b)

$$K_p = \infty, \quad K_v = \infty, \quad K_a = 15,$$
$$\omega_p \approx 19.9 \text{ rad/sec}, \quad M_p \approx 1.0285.$$

7.24 a) The lag network is given by

$$\frac{s + 0.201}{s + 0.01}$$

b) The lead network is given by

$$\frac{s + 4}{s + 140}$$

c) For part (a)

$$K_p = \infty, \quad K_v = 15, \quad K_a = 0,$$
$$\omega_p \approx 3.18 \text{ rad/sec}, \quad M_p \approx 1.0285$$

For part (b)

$$K_p = \infty, \quad K_v = 15, \quad K_a = 0,$$
$$\omega_p \approx 65 \text{ rad/sec}, \quad M_p \approx 1.0285.$$

7.25 a) The lag network is given by

$$\frac{s + 0.1890}{s + 0.0001}$$

b) The lead network is given by

$$\frac{s + 4}{s + 16,800}$$

c) For part (a)

$$K_p = \infty, \qquad K_v = \infty, \qquad K_a = 15,$$

$$\omega_p \approx 3.92 \text{ rad/sec}, \qquad M_p \approx 1.0285$$

For part (b)

$$K_p = \infty, \qquad K_v = \infty, \qquad K_a = 15,$$

$$\omega_p \approx 7,420 \text{ rad/sec}, \qquad M_p \approx 1.0285.$$

7.31 The synthesized system is illustrated in Fig. A7.31 where

$$K = 8, \qquad h_1 = 1, \qquad h_2 = \tfrac{5}{8}, \qquad h_3 = \frac{5 - \alpha}{8}.$$

A root locus analysis indicates that a good choice of α is approximately 2.5. Therefore, h_3 becomes $\tfrac{5}{16}$.

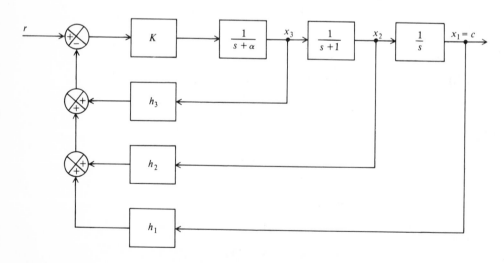

Figure A7.31

CHAPTER 8

8.1 a) Transfer function of the motor with one straight line tangent to the exact characteristics at 1000 rev/min is

$$\frac{\theta(s)}{V_c(s)} = \frac{2.36}{s(0.0083s + 1)}.$$

Transfer function of the motor with one straight line going through the two endpoints is

$$\frac{\theta(s)}{V_c(s)} = \frac{1.79}{s(0.0068s + 1)}.$$

b) Transfer function of the motor with one straight line tangent to the characteristics at 250 rev/min is

$$\frac{\theta(s)}{V_c(s)} = \frac{6.12}{s(0.0227s + 1)}.$$

Transfer function of the motor with one straight line tangent to the characteristics at 1750 rev/min is

$$\frac{\theta(s)}{V_c(s)} = \frac{2}{s(0.00334s + 1)}.$$

8.2 $N(M) = \dfrac{2K_1}{\pi}\left(-\dfrac{D}{M}\cos\sin^{-1}\dfrac{D}{M} - \sin^{-1}\dfrac{D}{M} + \dfrac{S}{M}\cos\sin^{-1}\dfrac{S}{M} + \sin^{-1}\dfrac{S}{M}\right).$

8.3 $N(M) = \dfrac{4K_1}{\pi M}\sqrt{1 - \left(\dfrac{D}{M}\right)^2}.$

8.4 $N(M) = \dfrac{2K}{\pi}\left[\sin^{-1}\dfrac{E_2}{M} - \sin^{-1}\dfrac{E_1}{M} + \left(\dfrac{2E_1}{M} + \dfrac{E_2}{M}\right)\cos\sin^{-1}\dfrac{E_2}{M} - \dfrac{E_1}{M}\cos\sin^{-1}\dfrac{E_1}{M}\right].$

8.7 A limit cycle occurs at $D/M = 0.99$ and $\omega = 0.075$ rad/sec.

8.9 System is stable and exhibits no limit cycle.

8.10 System is stable and exhibits no limit cycle.

8.11 Limit cycle occurs at $M/D = 3.15$ and $\omega = 7.68$ rad/sec.

8.13 a) Limit cycles occur at $D/M = 0.82$, $\omega = 0.39$ rad/sec; $D/M = 0.3$, $\omega = 1$ rad/sec. The former limit cycle is unstable, the latter is stable.

b) $K = 2.75$

c) $\alpha T = 0.9$ sec, $T = 0.4$ sec (assume increase in original gain by a factor of $\frac{9}{4}$)

d) $b = 0.26$.

8.16 A stable limit cycle exists at $\omega = 30$ rad/sec and $D/M = 0.62$. An unstable limit cycle exists at $\omega = 3$ rad/sec and $D/M = 0.92$.

8.17 The describing function analysis of the gain–phase diagram indicates that limit cycles do not exist for this system if the nonlinearity corresponds to coulomb friction.

8.19 a) The phase trajectory is a circle going through the point 1, 0 and whose center is the origin.

b) The system is unstable, since no viscous damping is present.

8.27 a) $\dfrac{d\dot{\theta}(t)}{d\theta(t)} = -\dfrac{B}{A} + \dfrac{D - C\theta(t)}{A\dot{\theta}(t)}$

b) $\dfrac{d\dot{\theta}(t)}{d\theta(t)} = -\dfrac{B\,|\dot{\theta}(t)|}{A} + \dfrac{D - C\theta(t)}{A\dot{\theta}(t)}$

c) $\dfrac{d\dot{\theta}(t)}{d\theta(t)} = -\dfrac{B\,|\theta(t)|}{A} + \dfrac{D - C\theta(t)}{A\dot{\theta}(t)}$

d) $\dfrac{d\dot{\theta}(t)}{d\theta(t)} = -\dfrac{B}{A} + \dfrac{D - C\theta(t)\,|\theta(t)|}{A\dot{\theta}(t)}$.

8.28 a) The isocline equation is given by

$$\frac{d\dot{\Theta}(t)}{d\theta(t)} = -2\zeta - \frac{\theta(t)}{\dot{\Theta}(t)},$$

where

$$\dot{\Theta}(t) = \frac{\dot{\theta}(t)}{\sqrt{0.4}}, \qquad \zeta = \sqrt{0.4}$$

b) The isocline equation is given by

$$\frac{d\dot{\Theta}(t)}{d\theta(t)} = -2\zeta - \frac{\theta(t)}{\dot{\Theta}(t)},$$

where

$$\dot{\Theta}(t) = \frac{\dot{\theta}(t)}{\sqrt{0.4}}, \qquad \zeta = \sqrt{0.4}$$

c) The isocline equation is given by

$$\frac{d\dot{\Theta}(t)}{d\theta(t)} = -2\zeta\,|\dot{\Theta}(t)| - \frac{\theta(t)}{\dot{\Theta}(t)},$$

where

$$\dot{\Theta}(t) = \frac{\dot{\theta}(t)}{\sqrt{0.4}}, \qquad \zeta = \sqrt{0.4}$$

d) The isocline equation is given by

$$\frac{d\dot{\Theta}(t)}{d\theta(t)} = -\zeta\,|\dot{\Theta}(t)| - \frac{\theta(t)}{\dot{\Theta}(t)},$$

where

$$\dot{\Theta}(t) = \frac{\dot{\theta}(t)}{\sqrt{0.4}}, \qquad \zeta = \sqrt{0.4} .$$

8.30 Using Lienard's construction and defining $v = \dfrac{dy}{dx}$, we find the fundamental curve is given by (for $A = 0.2$)

$$y = -2v - 0.5 \sin v.$$

8.32 Defining

$$\theta_1(t) = \theta(t) - 0.5 \qquad \text{for} \qquad \theta(t) > 0$$

and

$$\theta_1(t) = \theta(t) + 0.5 \qquad \text{for} \qquad \theta(t) < 0,$$

the isocline equation is given by

$$\frac{d\dot{\Theta}_1(t)}{d\theta_1(t)} = -0.8 - \frac{\theta_1(t)}{\dot{\Theta}_1(t)},$$

where

$$\dot{\Theta}_1(t) = \frac{\dot{\theta}_1(t)}{\omega_n(t)}$$

and

$$\omega_n = 1.$$

8.35 Popov condition satisfied if $0 < K \leqslant 5.5$.

8.39 Possible Popov sectors which result in stable systems are given by the following conditions:

$$6 < N[e(t), t] < 18.9, \qquad 8.34 < N[e(t), t] < 47.6, \qquad 33.3 < N[e(t), t] < 100.$$

CHAPTER 9

9.1 uncontrollable, unobservable.

9.2 controllable, observable.

9.7 The optimal input is given by

$$u^0 = -\text{sgn}\left[\frac{\partial S^0}{\partial \dot{c}}\right]$$

and the nonlinear partial differential equation to be solved for S^0 is given by

$$\dot{c}\left[\frac{\partial S^0}{\partial c} - 4\frac{\partial S^0}{\partial \dot{c}}\right] - c\left[\frac{\partial S^0}{\partial \dot{c}}\right] - \left[\left|\frac{\partial S^0}{\partial \dot{c}}\right| - 1\right] = 0.$$

9.8

$$u^0 = \text{sgn}\,[Ke^{2t}\cos\sqrt{3}t + \phi],$$

where K and θ are determined from the boundary conditions of the system.

9.14 The optimal control policy is given by

$$u^0 = -\frac{K}{2}\frac{\partial S^0}{\partial c}$$

and the nonlinear partial differential equation to be solved for S^0 is given by

$$\frac{-K^2}{4}\left(\frac{\partial S^0}{\partial c}\right)^2 + Ac^2 = 0.$$

9.15 $u^0(t) = U\,\text{sgn}\,[Ke^{t/2}\sin(0.866t + \theta)]$, where K and θ are determined from the boundary conditions of the system.

9.16 $u^0(t) = U\,\text{sgn}\,[Ke^t\sin(1.732t + \theta)]$, where K and θ are determined from the boundary conditions of the system.

9.17 $u^0(t) = U\,\text{sgn}\,[K_1e^t + K_2te^t]$, where K_1, K_2, and θ are determined from the boundary conditions of the system.

INDEX